*S*pringer *M*onographs in *M*athematics

For further volumes:
www.springer.com/series/3733

Moshe Jarden

Algebraic
Patching

 Springer

Moshe Jarden
Tel Aviv University
School of Mathematics
Ramat Aviv 69978
Israel
jarden@post.tau.ac.il

ISSN 1439-7382
ISBN 978-3-642-26651-5   ISBN 978-3-642-15128-6 (eBook)
DOI 10.1007/978-3-642-15128-6
Springer Heidelberg Dordrecht London New York

Mathematics Subject Classification (2010): 12E30

Cover design: deblik

Printed on acid-free paper

Springer is part of Springer Science+Business Media (www.springer.com)

לִילָדַי הָאֲהוּבִים

כְּמֶהָה, הֶמְיַת, עוּרִי וְגָאִי

Table of Contents

Introduction

The central problem of contemporary Galois theory is to describe the absolute Galois group $\mathrm{Gal}(K)$ of a given field K. A less ambitious problem, also known as the "inverse Galois problem over K" is to list all finite quotients of $\mathrm{Gal}(K)$; that is, to find which finite groups occur as Galois groups over K.

1. BACKGROUND.

1.1 The inverse problem of Galois theory over $\mathbb{Q}$. The case where K is the field $\mathbb{Q}$ of rational numbers has been the most prominent one. It is the consequence of the theory of cyclotomic extensions, developed in the 19th century, that every finite Abelian group can be realized as a Galois group over $\mathbb{Q}$. Using class field theory, Shafarevich was able in the 20th century to realize every finite solvable group over $\mathbb{Q}$. For a non-Abelian simple group, one usually tries to realize G over a finitely generated purely transcendental extension $E = \mathbb{Q}(x_1, \ldots, x_n)$ of $\mathbb{Q}$. Once this has been successfully done, it is possible to apply Hilbert's irreducibility theorem to specialize the Galois extension of E to a Galois extension of $\mathbb{Q}$ with an isomorphic Galois group. This procedure was initiated by Hilbert in [Hil1892], where he realized each one of the symmetric groups S_n and the alternating groups A_n over $\mathbb{Q}$. Further work along these lines was done by Matzat, Thompson, Völklein, and others in the fourth quarter of the 20th century. Starting from the Riemann existence theorem which gives a realization of each finite group G over $\mathbb{C}(x)$ with a detailed description of ramification, they used several criteria, most notably rigidity, to descend those realizations to Galois extensions over $\mathbb{Q}$. In this way they succeeded to realize all sporadic simple groups (except M_{23}) over $\mathbb{Q}$ and several families of simple non-Abelian finite groups. For more details, the reader is referred to [MaM99] and [Voe96].

1.2 Complex analytic methods. While a solution to the inverse Galois problem over $\mathbb{Q}$ seems to be still out of reach, interest in Galois theory has been extended to other base fields, especially to fields of local flavor. Starting again from the Riemann existence theorem, Fried and Völklein constructed for each finite group G and "ramification data" for G a complex analytic space H. Then they used a higher dimensional analog to the Riemann existence theorem due to Grauert and Remmert to prove that H is indeed an absolutely irreducible algebraic variety defined over $\mathbb{C}$. Moreover, if K is a subfield of $\mathbb{C}$ over which H is defined and $H(K) \neq \emptyset$, then G is realizable over $K(x)$ with additional information about branch points and inertia groups. If K is a **PAC** subfield of $\mathbb{C}$ (i.e. every absolutely irreducible variety defined over K has a K-rational point), then H can be chosen to be defined over K and then, by definition, $H(K) \neq \emptyset$. This shows that the inverse Galois problem over $K(x)$ has a positive solution. In the case when K is in addition Hilbertian, Fried and Völklein exploited the fact that $\mathrm{Gal}(K)$ is projective and proved that every finite split embedding problem over K is solvable. In particular, if K is countable, it follows that $\mathrm{Gal}(K)$ is isomorphic to the free profinite group $\hat{F}_\omega$ of rank $\aleph_0$, giving a satisfactory solution to the structure problem of Galois theory over K, and solving one of the open problems of Field Arithmetic [FrJ86, Problem 24.41] in characteristic 0. The reader may consult the original papers [FrV91] and [FrV92] as well as Völklein's book [Voe96] on that subject.

1.3 Formal Patching. The application of complex analytic methods restricts the results of the former subsection to fields of characteristic 0. In the general case one uses one of several methods of "patching".

The first method of this kind is called "Formal Patching". It uses Grothendieck's formal schemes and was developed by Harbater in [Hrb87] in order to prove that if R is a complete local domain which is not a field and $K = \text{Quot}(R)$, then each finite group occurs as a Galois group over $K(x)$. It follows for example that the inverse Galois problem has a positive solution over $\mathbb{Q}_p(x)$ or over $K_0((x_1, \ldots, x_n))(x)$, where K_0 is a field and $n \geq 1$. Among further applications of Formal Patching by Harbater and his collaborators let us mention the solution of the generalized Abhyankar's problem over curves in positive characteristic [Hrb94a] (see also Remark 9.2.2). Moreover, they proved that if F is a function field of one variable over a separably closed field, then $\text{Gal}(F)$ is a free profinite group of rank $\text{card}(F)$ [Hrb95].

1.4 Rigid Patching. Following an idea of Serre, Liu translated in 1990 Harbater's method to the language of rigid analytic geometry and reproved that for each complete field K under a nonarchimedean absolute value, every finite group is realizable over $K(x)$ [Liu95]. Instead of "Formal Patching" one speaks here about "Rigid Patching". An account of rigid analytic geometry can be found in [FrP04].

At the end of 1990, Pop proved that if a field K is either PAC or Henselian, then each finite group has a K-regular realization over $K(x)$. As explained in a letter from Roquette to Geyer from December 1990, this follows from the result of Harbater-Liu, because in both cases, K is existentially closed in $K((t))$. This common property of PAC fields and Henselian fields was at that time somewhat surprising, because by Frey-Prestel, a field K cannot be both PAC and Henselian except if K is separably closed [FrJ08, Cor. 11.5.5].

Pop formalizes that property in [Pop96]. He calls a field K **large** if K is existentially closed in $K((t))$. Alternatively, K is large (or **ample** as we prefer to call it) if every absolutely irreducible K-curve C with a K-rational simple point has infinitely many K-rational points [Pop96, Prop. 1.1]. In particular, $K((t))$ itself is ample, because it is Henselian.

Using rigid patching, Pop proved that every finite split embedding problem over $K((t))$ has a $K((t))$-regular solution over $K((t))(x)$ [Pop96, Lemma 1.4]. In particular, this reproves the result of Harbater and Liu. Again, if K is ample, then the same statement holds over K [Pop96, Main Theorem A].

In particular, if K is PAC, then every finite split embedding problem over $K(x)$ is solvable. If, in addition, K is Hilbertian, then every finite split embedding problem over K is solvable. Since the absolute Galois group of a PAC field is projective, each finite embedding problem over K can be reduced to a finite split embedding problem [FrJ08, Thm. 11.6.2]. It follows that every finite embedding problem over K is solvable. If further, K is countable, then by Iwasawa [FrJ08, Thm. 24.8.1], $\mathrm{Gal}(K) \cong \hat{F}_\omega$ [Pop96, Thm. 1]. This completes the proof of [FrJ86, Problem 24.41] in the general case.

If K is algebraically closed, then by Tsen Theorem (Proposition 9.4.6), $\mathrm{Gal}(K(x))$ is projective. Since K is ample, every finite split embedding problem over $K(x)$ is solvable. If in addition, K is countable, then as in the preceding paragraph, $\mathrm{Gal}(K(x)) \cong \hat{F}_\omega$. In particular, $\mathrm{Gal}(\tilde{\mathbb{F}}_p(x)) \cong \hat{F}_\omega$. Note that in the latter case, $\tilde{\mathbb{F}}_p(x)$ is obtained from $\mathbb{F}_p(x)$ by adjoining all roots of unity. Thus, the latter result appears as an analog to the still open problem of Shafarevich that asks whether $\mathrm{Gal}(\mathbb{Q}_{ab})$ is free.

Generalizing the analog of Shafarevich's conjecture to the case where K is an algebraically closed field of an arbitrary infinite cardinality m, Harbater [Hrb95] and Pop [Pop95] independently proved that if E is a function field of one variable over K, then every finite split embedding problem over E has m solutions. Adding the projectivity of $\mathrm{Gal}(E)$ a generalization of Iwasawa's theorem due to Melnikov-Chatzidakis, that implies that $\mathrm{Gal}(E) \cong \hat{F}_m$.

We note that [Pop96] was printed from a manuscript that was ready in 1993. In a subsequent manuscript [Pop93], Pop applies rigid patching once more to prove that if E is a function field of one variable over an ample field K (called "a field with a universal local-global principle" in that paper), then every finite split embedding problem over E has a "regular solution".

Harbater improved Pop's result and proved that if K is an ample field of cardinality m, E is a function field of one variable over K, and $\mathcal{E}$ is a finite split embedding problem, than the number of solutions of $\mathcal{E}$ is m [Hrb09, Thm. 3.4]. Theorem A below improves Harbater's result even further.

Another application of Rigid Patching by Pop was to reduce the general Abhyankar's conjecture to the special one over the affine line proved by Raynaud [Pop95]. A detailed account of the patching methods mentioned so far can be found in [Hrb03].

Both Formal Patching and Rigid Patching draw inspiration from "cut-and-paste" methods in topology and analysis, in which spaces are constructed on metric open sets and glued on overlaps. In the case of Formal Patching, one considers "formal opens" which are defined by rings of formal power series and patches them together in order to get a formal total space. By applying the "formal GAGA", more precisely Gorthendieck's existence theorem [Gro61, Thm. 2.1.1], one concludes that the formal total space originates actually from the "usual" algebraic geometry. In the context of Rigid Patching, one considers "affinoids" which are defined by Tate algebras and patches them together to get a total rigid analytic space. Applying the "rigid GAGA" one concludes that the total rigid analytic space originates from algebraic geometry. Both of these constructions are actually parallel to the complex analytic construction, where one concludes by using Serre's complex analytic

GAGA and the Grauert-Remmert theorem. The key technical point in proving GAGA type results is always some form of Cartan's lemma on matrix factorization [Hrb03, discussion proceeding Theorem 2.2.5].

2. ALGEBRAIC PATCHING.   Inspired by a talk of v. d. Put, Haran and Völklein realized that the formal/rigid GAGA, on which formal/rigid patching relies in an essential way, can be actually replaced by a more specialized and simpler algebraization technique. In a few words, Haran and Völklein work in [HaV96] directly with the Tate algebras defining the affinoids to be patched, and using Cartan's Lemma, show that the final patching results originates from Galois theory.

The paper [HaV96], develops all the theory needed from scratch, without any prerequisites, and solves the inverse Galois problem over $K(x)$, where K is a complete discrete valued field. It also proves that $\mathrm{Gal}(K(x)) \cong \hat{F}_\omega$ if K is an algebraically closed countable field. The method Haran and Völklein developed got the name "Algebraic Patching". That method is further developed in [HaJ98a], [HaJ98b], and [HaJ00a]. In those papers, most of the results about absolute Galois groups of fields previously achieved by Formal Patching and Rigid Patching are proved by Algebraic Patching.

The basic idea behind each of the patching methods is that every finite group G is generated by finite cyclic groups. It is not very difficult to realize each of these groups over a given rational function field $K(x)$. The question is how to construct a Galois extension F of $K(x)$ with Galois group G out of these extensions. For example, their compositum will almost never give the desired field F.

2.1 Patching data.   Algebraic patching takes an axiomatic approach to the problem of realizing finite groups, like the one used in [Pop94, Subsection (1.1)] for Rigid Patching. Starting from a field E and a finite group G, we choose finitely many cyclic subgroups G_i, $i \in I$, that generate G. For each $i \in I$ we construct a Galois extension F_i of E with Galois group G_i. Similar to the formal/rigid patching, algebraic patching proceeds in three steps: lifting, inducing, and algebraization. The first two steps are easier and of rather formal nature, whereas the the third one is more difficult and includes a GAGA type assertion.

To be more precise, algebraic patching assumes the existence of an extension field P_i of E, (which we view as an "analytic field"), $i \in I$, and a field Q containing all P_i's. The data obtained should satisfies the following conditions:

(1a) $F_i \subseteq P_i'$, where $P_i' = \bigcap_{j \neq i} P_j$, $i \in I$;

(1b) $\bigcap_{i \in I} P_i = E$; and

(1c) Let $n = |G|$. Then for every $B \in \mathrm{GL}_n(Q)$ and each $i \in I$ there exist $B_1 \in \mathrm{GL}_n(P_i)$ and $B_2 \in \mathrm{GL}_n(P_i')$ such that $B = B_1 B_2$ (Cartan's decomposition).

We call $\mathcal{E} = (E, F_i, P_i, Q; G_i, G)_{i \in I}$ a **patching data**. The lifting step takes F_i to $Q_i = P_i F_i$ and we observe that Q_i is a Galois extension of P_i

with $\mathrm{Gal}(Q_i/P_i) = G_i$. Then we consider the induced vector spaces $N_i = \mathrm{Ind}_{G_i}^G Q_i$, $i \in I$, and $N = \mathrm{Ind}_1^G Q$, and use (1c) to construct a basis for N/Q which is also a basis for N_i/Q_i for each $i \in I$. Once this is done, we prove that a certain proper E-translate F of $\bigcap_{i \in I} N_i$ into Q (called the **compound** of $\mathcal{E}$) is a Galois extension of E with $\mathrm{Gal}(F/E) \cong G$ (Lemma 1.1.7).

Next suppose E is a finite Galois extension of a field E_0 with a Galois group Γ and Γ **acts properly** on $\mathcal{E}$. This means that Γ acts on the group G, on the set I, and on the field Q in a compatible way. Thus:

(2a) The action of Γ on Q extends the action of Γ on E.

(2b) $F_i^\gamma = F_{i^\gamma}$, $P_i^\gamma = P_{i^\gamma}$, and $G_i^\gamma = G_{i^\gamma}$, for all $i \in I$ and $\gamma \in \Gamma$.

(2c) $(a^\tau)^\gamma = (a^\gamma)^{\tau^\gamma}$ for all $i \in I$, $a \in F_i$, $\tau \in G_i$, and $\gamma \in \Gamma$.

By Proposition 1.2.2, the compound F is a Galois extension of E_0 that solves the **finite split embedding problem** $\Gamma \ltimes G \to \mathrm{Gal}(E/E_0)$; that is, there is an isomorphism $\mathrm{Gal}(F/E_0) \cong \Gamma \ltimes G$ identifying the projection of $\Gamma \ltimes G$ onto Γ with the restriction map $\mathrm{Gal}(F/E_0) \to \mathrm{Gal}(E/E_0)$. Moreover, the action of Γ on F as a subgroup of $\mathrm{Gal}(F/E_0)$ coincides with the action of Γ on Q restricted to F. In this case we also have, by Lemma 1.1.7, that $P_i F = Q_i$ for each $i \in I$.

2.2 Complete fields under ultra-metric absolute values. We are able to put together a patching data for fields of the form $E = \hat{K}(x)$, where $\hat{K}$ is a complete field with respect to an ultra-metric absolute value $|\ |$. Parallel to [Pop94], we start with a finite set I. For each $i \in I$ we choose an element $c_i \in \hat{K}$ such that $c_i \neq c_j$ if $i \neq j$ and an element $r \in \hat{K}^\times$ satisfying $|r| \leq |c_i - c_j|$ for $i \neq j$. Then we set $w_i = \frac{r}{x - c_i}$ and consider the ring $R = R_I = \hat{K}\{w_i\}_{i \in I}$ of all power series $f = a_0 + \sum_{i \in I} \sum_{n=1}^\infty a_{in} w_i^n$, where $a_0, a_{in} \in \hat{K}$ and $|a_{in}|$ tends to 0 as $n \to \infty$ (Section 3.2). It turns out that R is a complete ring with respect to the norm $\|f\| = \max(|a_0|, |a_{in}|)_{i,n}$ (Lemma 3.2.1). Also, R is a principal ideal domain, hence a unique factorization domain. Moreover, for each $i \in I$, every ideal of R is generated by an element of $\hat{K}[w_i]$ (Proposition 3.2.9). We let $Q = P_I = \mathrm{Quot}(R_I)$. Similarly, we construct the fields $P_i = \mathrm{Quot}(\hat{K}\{w_j\}_{j \neq i})$ and $P_i' = \mathrm{Quot}(\hat{K}\{w_i\})$. By Corollary 3.3.2, $P_i' = \bigcap_{j \neq i} P_j$ and $\bigcap_{i \in I} P_i = E$. Thus, the "analytic" fields P_i satisfy Condition (1b) and the second part of Condition (1a). By definition, each element of R is a sum of an element of $R_{I \smallsetminus \{i\}}$ and an element of $R_{\{i\}}$. This implies that the P_i's also satisfy Condition (1c) (Corollary 3.4.4).

Given a finite group G, we choose I such that for each $i \in I$ there is a cyclic subgroup G_i whose order is a power of a prime number and $G = \langle G_i \rangle_{i \in I}$. It is classical that E has a cyclic extension F_i in $\hat{K}((x))$ with Galois group G_i [FrJ08, Section 16]. Here we construct F_i/E with control on its ramification. In particular, each prime divisor of $E/\hat{K}$ that ramifies in F_i is totally ramified (Lemma 4.2.5). Now we apply Proposition 2.4.5 saying that every power series $z \in \hat{K}[[x]]$ which is algebraic over E converges at some $c \in \hat{K}^\times$. This allows us to shift F_i into P_i'. Thus, the first half of Condition (1a) is also satisfied.

Finally, assume that $\hat{K}$ is a finite Galois extension of a field $\hat{K}_0$ with Galois group Γ that acts on G and $\hat{K}_0$ is complete with respect to the restriction of $|\ |$. Set $E_0 = \hat{K}_0(x)$. Then the proof of Proposition 4.4.2 shows how to choose the set I, the groups G_i, the fields F_i, and the fields P_i such that Γ acts properly on the patching data $\mathcal{E} = (E, F_i, P_i, Q; G_i, G)_{i \in I}$. It follows that the finite split embedding problem $\Gamma \ltimes G \to \mathrm{Gal}(\hat{K}/\hat{K}_0)$ (also called a **constant finite split embedding problem**) is solvable over E_0. Moreover, the solution field F has a $\hat{K}$-rational place φ unramified over E_0 and $\varphi(x) \in \hat{K}_0$. In particular, F is a regular extension of $\hat{K}$. Thus, F is a **regular solution** of the embedding problem.

2.3 Ample fields. If K_0 is an ample field and K is a finite Galois extension of K_0 with Galois group Γ, then K is also an ample field (Lemma 5.5.1). Let $\hat{K}_0 = K_0((t))$ and $\hat{K} = K((t))$. Then $\hat{K}/\hat{K}_0$ is a Galois extension of complete fields under the t-adic absolute value with $\Gamma = \mathrm{Gal}(\hat{K}/\hat{K}_0)$. Thus, if Γ acts on a finite group G, then $\hat{K}(x)$ has a finite extension $\hat{F}$, Galois over $\hat{K}_0(x)$ and regular over $\hat{K}$, that solves the constant finite split embedding problem $\Gamma \ltimes G \to \mathrm{Gal}(\hat{K}/\hat{K}_0)$. By definition, K_0 is existentially closed in $\hat{K}_0$. Since $\hat{K}_0/K_0$ is a regular extension, we may apply the Bertini-Noether theorem and descend from $\hat{F}$ to a field F, Galois over $K_0(x)$ and regular over K, that solves the finite split embedding problem $\Gamma \ltimes G \to \mathrm{Gal}(K/K_0)$ (Lemma 5.9.1) over $K_0(x)$. As noticed above, in the special case where K_0 is a countable Hilbertian PAC field, this reproves the isomorphism $\mathrm{Gal}(K_0) \cong \hat{F}_\omega$.

One of the equivalent conditions for a field K to be ample is that the set $V(K)$ of K-rational points of each absolutely irreducible variety V defined over K with a simple K-rational point is Zariski-dense in V (Lemma 5.3.1). It turns out that under the above conditions, $\mathrm{card}(V(K)) = \mathrm{card}(K)$ (Proposition 5.4.3). Moreover, if h is a nonconstant rational function of V, then $\mathrm{card}\{h(\mathbf{a}) \mid \mathbf{a} \in V(K)\} = \mathrm{card}(K)$ (Corollary 5.4.4).

In addition to PAC fields and Henselian fields we find that real closed fields are ample. So are the quotient fields of Henselian domains (Proposition 5.7.7). In particular, for every field K_0 and $n \geq 1$, the field of formal power series $K_0((X_1, \ldots, X_n))$ is ample. Similarly, $\mathrm{Quot}(\mathbb{Z}_p[[X_1, \ldots, X_n]])$ is ample for every prime number p and $n \geq 0$ (Remark 5.7.8). Finally, if the absolute Galois group of a field K is pro-p for some prime number p, then K is ample (Theorem 5.8.3).

2.4 Non-ample fields. Chapter 6 reveals the other side of the coin. It gives examples of nonample fields. By Corollary 5.3.3, every finite field is nonample. Elementary arguments that apply the Riemann-Roch formula show that every finitely generated transcendental extension F of a field K is nonample. Moreover, if F is a union of a directed family of function fields of one variable over K of bounded genus, then F is nonample (Theorem 6.1.8). The proof that every number field is nonample uses a deep result, namely Faltings' theorem that curves of genus at least 2 over number fields have only finitely many rational points (Proposition 6.2.5). To give examples of infinite algebraic extensions of $\mathbb{Q}$ that are nonample, we use even more advanced tools, namely the Mordell-Lang conjecture proved in characteristic 0 by Faltings and others. That theorem implies that if A is a nonzero Abelian variety defined over an ample field K of characteristic 0, then $\dim_{\mathbb{Q}}(A(K) \otimes \mathbb{Q}) = \infty$. Now we refer to an example of Kato and Rohrlich of an Abelian infinite extension K of $\mathbb{Q}$ with $\mathrm{Gal}(K/\mathbb{Q})$ finitely generated and an elliptic curve E defined over $\mathbb{Q}$ such that $E(K)$ is finitely generated. Thus, K is nonample (Example 6.5.5). Using Faltings' result again and the concept of gonality of curves, we construct for every positive integer d a linearly disjoint sequence $K_1, K_2, K_3, \ldots$ of extensions of degree d of a given number field whose compositum $K = K_1 K_2 K_3 \cdots$ is nonample (Proposition 6.8.8).

2.5 Many solutions.   While the solvability of constant finite split embedding problems for $\mathrm{Gal}(K)$ over $K(x)$ suffices to prove that $\mathrm{Gal}(K) \cong \hat{F}_\omega$ if K is a countable PAC Hilbertian field, it does not give us enough information about nonconstant embedding problems over $K(x)$ and ignores the uncountable case. Both problems are addressed in Chapter 7 over complete fields. The most effective way to create uncountably many solutions to a finite split embedding problem is to solve the embedding problem with information about the branch points. Proposition 7.3.1 considers a complete field $\hat{K}_0$ with respect to an ultrametric absolute value, a finite Galois extension E/E_0 over $E_0 = \hat{K}_0(x)$ such that $\mathrm{Gal}(E/E_0)$ acts on a finite group H (we have replaced E' appearing in Proposition 7.3.1 by E). We assume that E has a $\hat{K}$-rational place unramified over the algebraic closure $\hat{K}$ of $\hat{K}_0$ in E. Then the embedding problem $\mathrm{Gal}(E/E_0) \ltimes H \to \mathrm{Gal}(E/E_0)$ has a solution field $\hat{F}$ regular over $\hat{K}$. Moreover, if G_j, $j \in J$, are finitely many cyclic subgroups of H of prime power orders that generate H, then for each $j \in J$ the extension $\hat{F}/E_0$ has a branch point b_j with G_j as an inertia group. Moreover, if $\hat{K}_0$ is an extension of infinite transcendence degree of a field K_0, then we may choose the b_j's to be algebraically independent over K_0.

The next step is to solve a finite split embedding problem

$$\mathrm{Gal}(E/K_0(x)) \ltimes H \to \mathrm{Gal}(E/K_0(x))$$

(which we denote by $\mathcal{E}$) for an ample field K_0 in many ways. As in the case of constant split embedding problems, we go over to the field $\hat{K}_0 = K_0((t))$, let $\hat{E} = E\hat{K}_0$, and solve the finite split embedding problem $\mathrm{Gal}(\hat{E}/\hat{K}_0(x)) \ltimes H \to \mathrm{Gal}(\hat{E}/\hat{K}_0(x))$ (which we denote by $\hat{\mathcal{E}}$) as in Proposition 7.3.1 (with E replacing E'). Then we choose appropriate $u_1, \ldots, u_n \in \hat{K}_0$ and descend the embedding problem with its solution field to an embedding problem (which we denote by $\mathcal{E}_{\mathbf{u}}$) with a solution field $F_{\mathbf{u}}$ over $K_0(\mathbf{u})$, keeping the branch points and the corresponding inertia groups. The new decisive step is to reduce $F_{\mathbf{u}}$ to a solution field F of $\mathcal{E}$ with sufficient information on the reduced branch points $\bar{b}_j$. To this aim we apply good reduction to the function field of one variable $F_{\mathbf{u}}/K(\mathbf{u})$ such that the inertia group $\bar{I}_j$ over $\bar{b}_j$ contains G_j. We also use the information that b_j is transcendental over K_0 to choose the reduction such that the branch point $\bar{b}_j$ is unramified in E and in the compositum N of all solution fields of $\mathcal{E}$ obtained in a transfinite induction up to that point. This implies that $\bar{I}_j$ is contained in $\mathrm{Gal}(F/F \cap N)$. Since the G_j's were chosen to generate H (which we identify with $\mathrm{Gal}(F/E)$), the $\bar{I}_j$'s generate $\mathrm{Gal}(F/E)$. This implies that $F \cap N = E$. In this way our transfinite induction constructs a transfinite sequence $(F_\kappa)_{\kappa < \mathrm{card}(K_0)}$ of solutions to $\mathcal{E}$ that are linearly disjoint over E (see also Lemma 7.4.1).

2.6 Algebraically closed base fields. Section 9.1 surveys the classical results about fundamental groups of Riemann surfaces. Given a finite set S of prime divisors of $\mathbb{C}(x)/\mathbb{C}$, we denote the maximal extension of $\mathbb{C}(x)$ in its algebraic closure that ramifies at most over S by $\mathbb{C}(x)_S$. A consequence of the Riemann existence theorem then describes $\mathrm{Gal}(\mathbb{C}(x)_S/\mathbb{C}(x))$ by generators and relations, where each generator generates an inertia group over an element of S. We are then able to take the limit over all the sets S and deduce that $\mathrm{Gal}(\mathbb{C}(x))$ is a free profinite group of rank $\mathrm{card}(\mathbb{C})$. In particular, $\mathrm{Gal}(\mathbb{C}(x))$ is projective (Corollary 9.1.11).

That result can be carried over to a result over an arbitrary algebraically closed field C of characteristic 0. If $p = \mathrm{char}(C) > 0$, the same result holds provided one stays away from p (Proposition 9.2.1). However, it is not true any more in its general form (Proposition 9.9.4). In particular, in the notation of the preceding paragraph, $\mathrm{Gal}(C(x)_S/C(x))$ is not free. Consequently, we are not able to repeat the proof that works in characteristic 0 that $\mathrm{Gal}(C(x))$ is free in the general case.

Nevertheless the latter result is still true for each algebraically closed field. The first step is a proof that $\mathrm{Gal}(C(x))$ is projective. Since we try to be as self-contained as possible, we give a direct proof of the projectivity of $\mathrm{Gal}(C(x))$ based on simple results about homogeneous equations that we prove and basic results about Galois cohomology that we survey (Proposition 9.4.6). Combined with Proposition 8.6.3, this proves that $\mathrm{Gal}(C(x))$ is isomorphic to the free profinite group of rank $\mathrm{card}(C)$. In positive characteristic p, we generalize that result and prove that if the absolute Galois group of a field K is pro-p and F is a function field of one variable over K, then $\mathrm{Gal}(F)$ is free of rank $\mathrm{card}(K)$ (Theorem 9.4.8).

2.7 Semi-free groups. Chapter 10 develops consequences of Proposition 7.3.1 from the point of view of profinite groups. We say that a profinite group G of infinite rank m is **semi-free** if every finite split embedding problem for G with a nontrivial kernel has m independent solutions. This condition is transferred to every open subgroup of G (Lemma 10.4.1), to every closed normal subgroup N such that G/N is finitely generated (Lemma 10.4.2) or Abelian (Theorem 10.5.4), and to every closed subgroup M of G that is **contained in a diamond** (Theorem 10.5.3).

As we saw above, one of the major steps to prove that the absolute Galois group of a field is free is to show that this group is projective. If K is PAC, then this is guaranteed by an old theorem of Ax [FrJ08, Thm. 11.6.2]. Going over to a function field E of one variable over K raises the cohomological dimension by 1, in particular, $\mathrm{Gal}(E)$ is usually not projective. However, we prove a local-global principle for the Brauer group $\mathrm{Br}(E)$ of E: the restriction map $\mathrm{Br}(E) \to \prod_{\mathfrak{p}} \mathrm{Br}(E_{\mathfrak{p}})$ is injective and the image of each element of $\mathrm{Br}(E)$ in $\mathrm{Br}(E_{\mathfrak{p}})$ is 0 for all but finitely many $\mathfrak{p}$ (Lemma 11.5.4). Here $\mathfrak{p}$ ranges over all prime divisors of E/K and $E_{\mathfrak{p}}$ is the Henselian closure of E at $\mathfrak{p}$. This implies an embedding $\mathrm{Br}(F) \to \prod \mathrm{Br}(F_{\mathfrak{p}})$ for each regular extension F of K of transcendence degree 1 (Proposition 11.5.5). In particular, if $x \in F$ is transcendental over K and F contains an nth root of each monic irreducible polynomial in $K[x]$ for each n with $\mathrm{char}(K) \nmid n$, then the valuation groups of $F_{\mathfrak{p}}$ are divisible away from $\mathrm{char}(K)$. This implies that $\mathrm{Br}(F_{\mathfrak{p}}) = 0$ for each $\mathfrak{p}$ (Proposition 11.1.3). It follows that $\mathrm{Br}(F) = 0$. Since the same holds for each finite extension of F, $\mathrm{Gal}(F)$ is projective (Proposition 11.6.6).

On the other hand, we use the previous results to prove that if E is a function field of one variable over an ample field K, then $\mathrm{Gal}(E)$ is semi-free. In particular, this is the case when K is PAC. We then choose nth roots $\sqrt[n]{f}$ in a compatible way for each monic irreducible polynomial $f \in K[x]$ and every positive integer n with $\mathrm{char}(K) \nmid n$. We let $F = K(\sqrt[n]{f})_{f,n}$ and prove that F lies in a diamond over $K(x)$. This implies that $\mathrm{Gal}(F)$ is semi-free of rank $m = \mathrm{char}(K)$ and projective. It follows that $\mathrm{Gal}(F) \cong \hat{F}_m$ (Theorem 11.7.6). We call F a **special K-radical extension of $K(x)$**. In the special case where K contains all roots of unity, we get that $\mathrm{Gal}(K(x)_{\mathrm{ab}}) \cong \hat{F}_m$. This is an analog to a well known conjecture of Shafarevich that $\mathrm{Gal}(\mathbb{Q}_{\mathrm{ab}}) \cong \hat{F}_\omega$.

2.8 Hilbertian ample Krull fields. In the last chapter we consider an ample Hilbertian field K. Although every finite split embedding problem over K is solvable (Theorem 5.10.2), $\mathrm{Gal}(K)$ need not be semi-free of rank $\mathrm{card}(K)$ (Example 10.6.7). However, if K is the quotient field of a complete local Noetherian domain of height at least 2, then K is ample, Hilbertian, and $\mathrm{Gal}(K)$ is semi-free (Theorem 12.4.3). One of the main ingredients of the proof is a quantitative Chebotarev type theorem for K. We prove that K is a **Krull field**. Thus, K has a set $\mathcal{V}$ of discrete valuations such that $\{v \in \mathcal{V} \mid v(a) \neq 0\}$ is finite for each $a \in K^\times$ and for each finite Galois extension L of K there are $\mathrm{card}(K)$ elements $v \in \mathcal{V}$ that completely split in L.

The leading examples for fields K as in the preceding paragraph are $K_0((X_1, \ldots, X_n))$, for any field K_0 and $n \geq 2$, $\mathrm{Quot}(\mathbb{Z}_p[[X_1, \ldots, X_n]])$, where p is prime and $n \geq 1$, and $\mathrm{Quot}(\mathbb{Z}[[X_1, \ldots, X_n]])$, where $n \geq 1$ (Example 12.4.4). In the special case where $K = C((X_1, X_2))$ and C is algebraically closed of cardinality m, it follows that $\mathrm{Gal}(K_{\mathrm{ab}}) \cong \hat{F}_m$ (Theorem 12.4.6).

3. MAIN RESULTS.

3.1 List of main results. Each result is followed by a short reference. We expand on it in the notes at the end of the chapters where the theorems are proved.

The two main results of the book are perhaps the following theorems.

THEOREM A: *Let K be an ample field of cardinality m and E a function field of one variable over K. Then $\mathrm{Gal}(E)$ is semi-free of rank m (Theorem 11.7.1).*

See notes to Chapters 7 and 11 for the history of Theorem A.

THEOREM B: *Let K be a Hilbertian ample Krull field of cardinality m. Then $\mathrm{Gal}(K)$ is semi-free of rank m (Theorem 12.4.1).*

Pop proves this result in [Pop10], where the notion of a Krull field is introduced.

Among the consequences of Theorems A and B we mention the following:

THEOREM C: *Every Hilbertian PAC field is ω-free (Theorem 5.10.3).*

This theorem is proved by Fried-Völklein [FrV92] when $\mathrm{char}(K) = 0$ and by Pop [Pop96] in general. Haran-Jarden reprove the result using algebraic patching in [HaJ98a].

THEOREM D: *Let $\mathbb{Q}_{\mathrm{tr}}$ be the field of totally real algebraic numbers. Then $\mathrm{Gal}(\mathbb{Q}_{\mathrm{tr}}(\sqrt{-1})) \cong \hat{F}_\omega$ (Example 5.10.7).*

This example is a special case of a more general example: Let S be a finite set of primes of $\mathbb{Q}$ and let $\mathbb{Q}_S$ be the field of totally S-adic numbers. Then $\mathrm{Gal}(\mathbb{Q}_S(\mu_\infty)) \cong \hat{F}_\omega$ [Pop96]. Notice that $\mathbb{Q}_{\mathrm{tr}}(\sqrt{-1}) = \mathbb{Q}_{\mathrm{tr}}(\mu_\infty)$ is even PAC, whereas the former fields are not.

THEOREM E: *Let K be a field of characteristic p and cardinality m and let E be a function field of one variable over K. Suppose that $\mathrm{Gal}(K)$ is a pro-p group. Then $\mathrm{Gal}(E) \cong \hat{F}_m$ (Theorem 9.4.8).*

In the case where K is algebraically closed the theorem was proved independently by Harbater [Hrb95] and Pop [Pop95]. Haran-Jarden use algebraic patching in [HaJ98b] to reprove the theorem. The proof of the general case follows along the same lines.

THEOREM F: *Let K be a PAC field of cardinality m, x a variable, and F a special K-radical extension of $K(x)$. Then F is Hilbertian and $\mathrm{Gal}(F) \cong \hat{F}_m$. In particular, if K contains all roots of unity, then $\mathrm{Gal}(K(x)_{\mathrm{ab}}) \cong \hat{F}_m$ (Theorem 11.7.6).*

The result is proved by Jarden-Pop in [JaP09] and gives evidence for a conjecture of Bogomolov-Positselski.

THEOREM G: *Let R be a Noetherian domain, $\mathfrak{m}$ a prime ideal of R of height at least 2, and $\mathfrak{a}$ an ideal of R in $\mathfrak{m}$ such that R is complete in the $\mathfrak{a}$-adic topology. Then $K = \mathrm{Quot}(R)$ is Hilbertian, ample, and Krull. Moreover, $\mathrm{Gal}(K)$ is semi-free of rank $\mathrm{card}(K)$ (Theorem 12.4.3).*

The proof of Theorem G is based, among others, on Theorem B. It appears in [Pop10]. The following special cases of Theorem G are proved by Paran in [Par10] by other methods.

THEOREM H: *Let R be a complete local Noetherian domain of height at least 2 and of cardinality m. Set $K = \mathrm{Quot}(R)$. Then $\mathrm{Gal}(K)$ is semi-free of rank m (Theorem 12.4.5).*

THEOREM I: *In each of the following cases $\mathrm{Gal}(K)$ is semi-free of rank $\mathrm{card}(K)$:*
(a) $K = K_0((X_1, \ldots, X_n))$, where K_0 is an arbitrary field and $n \geq 2$;
(b) $K = \mathrm{Quot}(\mathbb{Z}_p[[X_1, \ldots, X_n]])$, where p is a prime number and $n \geq 1$;
(c) $K = \mathrm{Quot}(\mathbb{Z}[[X_1, \ldots, X_n]])$, where $n \geq 1$;
(Example 12.4.4).

Two interesting results that we prove use ingredients whose proofs are unfortunately beyond the scope of this book:

THEOREM J (Fehm-Petersen): *Let A be a nonzero Abelian variety defined over an ample field K of characteristic 0. Then, the rational rank of $A(K)$ is infinite (Theorem 6.5.2).*

THEOREM K: *Let K be a separably closed field of cardinality m. Then $\mathrm{Gal}(K((X_1, X_2))_{\mathrm{ab}}) \cong \hat{F}_m$ (Theorem 12.4.6).*

The case where K is algebraically closed is due to Harbater [Hrb09]. We have observed that Harbater's proof applies to the general case. A generalization appears in [Pop10]. ☐

3.2 Sources. The proofs of Theorems A and B are self-contained up to basic results of Field Arithmetic that we quote from [FrJ08]. A few of the applications, C – K, rely on extra information that we properly quote.

The first three chapters of the book give a quick self-contained introduction into Algebraic Patching. Over an ample field K, it leads to the result that every constant split embedding problem over $K(x)$ is solvable (Theorem 5.9.2). Using basic results of Field Arithmetic taken from [FrJ08], that part of the book culminates with the proof of Theorem C.

Theorem A generalizes Theorem 5.9.2, and its proof requires much more effort to achieve linear disjointness of the solution fields through a careful choice of the branch points with additional information about their inertia groups. Again, the proof is self-contained.

The first application of Theorem A appears in the proof of Theorem F. Here the first task is to prove that $\mathrm{Gal}(K(x)_{\mathrm{ab}})$ is projective. The proof uses some basic Galois cohomology that we survey in Section 9.3.

The proofs of Theorems G, H, and I use several results from commutative algebra. The proof of Theorem J uses, among others, the Mordel-Lang Conjecture proved by Faltings. The proof of Theorem K applies a result whose proof applies étale cohomology.

3.3 Advantages and disadvantages of Algebraic Patching. Our method of Algebraic Patching has the advantage of being quickly accessible. The cost of this convenience is its inability to deal with fundamental groups of curves over algebraically closed fields K. Indeed, every Galois extension F_i of $K(x)$

involved in patching data (see 2.1), contributes at least one extra branch point to the solution field of the given embedding problem. In addition, our patching is carried out only over rational fields $K(x)$ (geometrically, over the line) and not over algebraic function fields of one variable (geometrically, over curves). Thus, that method seems not to be suitable to handle questions like the general Abhyankar's conjecture that was reduced to the special Abhyankar's conjecture by both the Formal Patching and the Rigid Patching methods. Nevertheless, we have been able to apply several methods of descent from function fields of one variable to rational function fields and prove all of the above mentioned results about absolute Galois groups.

4. FIELD PATCHING.   David Harbater and Julia Hartmann have recently developed a new kind of patching called "field patching". In its simplest form, that method considers fields $F \subseteq F_1, F_2 \subseteq F_0$ such that $F = F_1 \cap F_2$ and for each matrix $A_0 \in \mathrm{GL}_n(F_0)$ there exist matrices $A_1 \in \mathrm{GL}_n(F_1)$ and $A_2 \in \mathrm{GL}_n(F_2)$ such that $A_0 = A_1 A_2$. Thus, these fields satisfy the second part of Condition (1a) and Conditions (1b) and (1c) in the special case where $I = \{1,2\}$, E is replaced by F, and P_1, P_2 are replaced by F_1, F_2. They prove that if for $i = 0, 1, 2$, V_i is a vector space of dimension n over F_i and $F_0 V_i = V_0$ for $i = 1, 2$, then $V = V_1 \cap V_2$ is a vector space of dimension n over F [HaH10, Prop. 2.1]. This corresponds to Lemma 1.1.7, where V_i is replaced by the algebra N_i (again for $I = \{1, 2\}$) and V is replaced by the field F' (which is a Galois extension of E of degree equal to $\dim_{P_i} N_i$).

Harbater and Hartmann verify the axioms of Field Patching over a complete discrete valuation ring T with uniformizer t and quotient field K. They consider a smooth projective curve $\hat{X}$ over T, let X be its closed fiber and F the function field of X. Then they consider proper subsets U_1, U_2 of X such that $X = U_1 \cup U_2$, and set $U_0 = U_1 \cap U_2$. For $i = 0, 1, 2$ they let R_i be the ring of rational functions of $\hat{X}$ that are regular on U_i. Now they consider the t-adic completion $\hat{R}_i$ of R_i and let $F_i = \mathrm{Quot}(\hat{R}_i)$. They prove that these objects satisfy the axioms of the preceding paragraph, hence also their conclusion [HaH10, Thm. 4.10]. In contrast to Formal Patching, the proofs rely only on elementary arguments such as the Riemann-Roch theorem for function fields of one variable.

In the special case where $\hat{X}$ is the projective line over T, $R_1 = T[x]$, $R_2 = T[x^{-1}]$, and $R_0 = T[x, x^{-1}]$, the corresponding t-adic completions are $\hat{R}_1 = T\{x\}$, $\hat{R}_2 = T\{x^{-1}\}$, and $\hat{R}_0 = T\{x, x^{-1}\}$. These rings are respectively contained in the rings $R_1' = K\{x\}$, $R_2' = K\{x^{-1}\}$, and $R_0' = K\{x, x^{-1}\}$ that appear in [HaV96] and also, in a more general form in Chapters 2 and 3. Moreover, $\mathrm{Quot}(\hat{R}_i) = \mathrm{Quot}(R_i')$ for $i = 0, 1, 2$ and $F = K(x)$.

While Algebraic Patching works only over rational function fields of one variable and aims at Galois Theory and in particular toward applications to absolute Galois groups, Field Patching works over arbitrary function fields of one variable (over $\mathrm{Quot}(T)$) and has been also applied to other areas of algebra, most notably to differential algebra, quadratic forms, and Brauer groups.

We mention here only one result: In the notation of the first paragraph of 4, let G be a connected linear algebraic group over F whose function field over F is rational and let H be an F-variety. Suppose $G(F')$ acts transitively on $H(F')$ for each extension F' of F and that $H(F_i) \neq \emptyset$ for $i = 1, 2$. Then $H(F) \neq \emptyset$ [HHK09].

For each field K let $u(K)$ be the maximal number of variables of a quadratic form over K with no nontrivial zero. As an application, [HHK09] proves that if K is a complete discrete valued field whose residue field is a C_d-field, then $u(F) \leq 2^{d+2}$ for each function field F of one variable over K. In particular, if F is a function field of one variable over $\mathbb{Q}_p$, a finite extension of $\mathrm{Quot}(\mathbb{Z}_p[[X]])$, or a finite extension of $\mathrm{Quot}(\mathbb{F}_p[[X, Y]])$, with $p \neq 2$, then $u(F) = 8$. Thus, every quadratic form in 9 variables over F has a nontrivial zero but there is a quadratic form in 8 variables over F that fails to have a nontrivial zero. See also an earlier proof of that result by Parimala and Suresh [PaS07] that use other methods and a recent generalization of the result by David Leep to function fields of several variables over $\mathbb{Q}_p$.

Acknowledgement: I would like to thank Aharon Razon for thorough reading of the whole manuscript and for many suggestions that helped me clean the book from many mistakes and typos. Elad Paran's, Lior Bary-Soroker's, Arno Fehm's, and Wulf-Dieter Geyer's many comments are also warmly appreciated. Special recognition goes to Peter Roquette for his simplification and considerable shortening of the proof of Lemma 8.2.4. Finally, my thanks go to David Harbater, Florian Pop, and Dan Haran for their constant support and essential advise.

Mevasseret Zion, Spring 2010   Moshe Jarden

Notation and Convention

$\mathbb{Z}$ = the ring of rational integers.

$\mathbb{Z}_p$ = the ring of p-adic integers.

$\mathbb{Q}$ = the field of rational numbers.

$\mathbb{R}$ = the field of real numbers.

$\mathbb{C}$ = the field of complex numbers.

$\mathbb{F}_q$ = the field with q elements.

K_s = the separable closure of a field K.

$\tilde{K}$ = the algebraic closure of a field K.

$\mathrm{Gal}(L/K)$ = the Galois group of a Galois extension L/K.

We call a polynomial $f \in K[X]$ **separable** if f has no multiple root in $\tilde{K}$.

$\mathrm{Gal}(K) = \mathrm{Gal}(K_s/K)$ = the absolute Galois group of a field K.

$\mathrm{irr}(x, K)$ = the monic irreducible polynomial of an algebraic element x over a field K.

Whenever we form the compositum EF of field extensions of a field K we tacitly assume that E and F are contained in a common field.

$\mathrm{card}(A)$ = the cardinality of a set A.

$R^\times$ = the group of invertible elements of a ring R.

$\mathrm{Quot}(R)$ = the quotient field of an integral domain R.

$A \subset B$ means "the set A is properly contained in the set B".

$a^x = x^{-1}ax$, for elements a and x of a group G.

$H^x = \{h^x \mid h \in H\}$, for a subgroup H of G.

Given subgroups A, B of a group G, we use "$A \le B$" for "A is a subgroup of B" and "$A < B$" for "A is a proper subgroup of B".

Given an Abelian (additive) group A and a positive integer n, we write A_n for the subgroup $\{a \in A \mid na = 0\}$. For a prime number p we let $A_{p^\infty} = \bigcup_{i=1}^{\infty} A_{p^i}$.

For a group B that acts on a group A from the right, we use $B \ltimes A$ to denote the semidirect product of A and B.

Bold face letters stand for n-tuples, e.g. $\mathbf{x} = (x_1, \ldots, x_n)$.

$\mathrm{ord}(x)$ is the order of an element x in a group G.

In the context of fields, ζ_n stands for a primitive root of unity of order n.

$\bigcup_{i \in I} B_i$ is the disjoint union of sets B_i, $i \in I$.

Chapter 1.
Algebraic Patching

Let E be a field, G a finite group, and $\{G_i \mid i \in I\}$ a finite set of subgroups of G with $G = \langle G_i \mid i \in I \rangle$. For each $i \in I$ we are given a Galois extension F_i of E with Galois group G_i. We suggest a general method how to 'patch' the given F_i's into a Galois extension F with Galois group G (Lemma 1.1.7). Our method requires extra fields P_i, all contained in a common field Q and satisfying certain conditions making $\mathcal{E} = (E, F_i, P_i, Q; G_i, G)_{i \in I}$ into 'patching data' (Definition 1.1.1). The auxiliary fields P_i in this data substitute, in some sense, analytic fields in rigid patching and fields of formal power series in formal patching.

If in addition to the patching data, E is a Galois extension of a field E_0 with Galois group Γ and Γ 'acts properly' (Definition 1.2.1) on the patching data $\mathcal{E}$, then we construct F above to be a Galois extension of E_0 with Galois group isomorphic to $\Gamma \ltimes G$ (Proposition 1.2.2).

1.1 Patching

Let E be a field and G a finite group, generated by finitely many subgroups G_i, $i \in I$. Suppose for each $i \in I$ we have a finite Galois field extension F_i of E with Galois group G_i. We use these extensions to construct a Galois field extension F of E (not necessarily containing F_i) with Galois group G. First we 'lift' each F_i/E to a Galois field extension Q_i/P_i, where P_i is an appropriate field extension of E such that $P_i \cap F_i = E$ and all of the Q_i's are contained in a common field Q. Then we define F to be the maximal subfield contained in $\bigcap_{i \in I} Q_i$ on which the Galois actions of $\mathrm{Gal}(Q_i/P_i)$ combine to an action of G.

The construction works if certain patching conditions on the initial data are satisfied.

Definition 1.1.1: Patching data. Let I be a finite set with $|I| \geq 2$. A **patching data**
$$\mathcal{E} = (E, F_i, P_i, Q; G_i, G)_{i \in I}$$
consists of fields $E \subseteq F_i, P_i \subseteq Q$ and finite groups $G_i \leq G$, $i \in I$, such that
(1a) F_i/E is a Galois extension with Galois group G_i, $i \in I$;
(1b) $F_i \subseteq P_i'$, where $P_i' = \bigcap_{j \neq i} P_j$, $i \in I$;
(1c) $\bigcap_{i \in I} P_i = E$; and
(1d) $G = \langle G_i \mid i \in I \rangle$.
(1e) (Cartan's decomposition) Let $n = |G|$. Then for every $B \in \mathrm{GL}_n(Q)$ and each $i \in I$ there exist $B_1 \in \mathrm{GL}_n(P_i)$ and $B_2 \in \mathrm{GL}_n(P_i')$ such that $B = B_1 B_2$. $\qquad \square$

M. Jarden, *Algebraic Patching*, Springer Monographs in Mathematics,
DOI 10.1007/978-3-642-15128-6_1, © Springer-Verlag Berlin Heidelberg 2011

We extend $\mathcal{E}$ by more fields. For each $i \in I$ let $Q_i = P_i F_i$ be the compositum of P_i and F_i in Q. Conditions (1b) and (1c) imply that $P_i \cap F_i = E$. Hence Q_i/P_i is a Galois extension with Galois group isomorphic (via the restriction of automorphisms) to $G_i = \mathrm{Gal}(F_i/E)$. We identify $\mathrm{Gal}(Q_i/P_i)$ with G_i via this isomorphism.

We need some auxiliary results from linear algebra. Let

$$(2) \qquad N = \Big\{ \sum_{\zeta \in G} a_\zeta \zeta \,\Big|\, a_\zeta \in Q \Big\}$$

be the vector space over Q with basis $(\zeta \mid \zeta \in G)$, where G is given some fixed ordering. Thus $\dim_Q N = |G|$. For each $i \in I$ we consider the following subset of N:

$$(3) \qquad N_i = \Big\{ \sum_{\zeta \in G} a_\zeta \zeta \in N \,\Big|\, a_\zeta \in Q_i,\ a_\zeta^\eta = a_{\zeta\eta} \text{ for all } \zeta \in G,\ \eta \in G_i \Big\}.$$

It is a vector space over P_i.

LEMMA 1.1.2: *Let $i \in I$. Then N has a Q-basis which is contained in N_i.*

Proof: Let $\Lambda = \{\lambda_1, \ldots, \lambda_m\}$ be a system of representatives of G/G_i and let $\eta_1, \ldots, \eta_r$ be a listing of the elements of G_i. Thus, $G = \{\lambda_k \eta_\nu \mid k = 1, \ldots, m;\ \nu = 1, \ldots, r\}$. Let z be a primitive element for Q_i/P_i. The following sequence of $|G|$ elements of N_i

$$\Big(\sum_{\nu=1}^{r} (z^{j-1})^{\eta_\nu} \lambda_k \eta_\nu \,\Big|\, j = 1, \ldots, r;\ k = 1, \ldots, m \Big)$$

(in some order) is linearly independent over Q, hence it forms a basis of N over Q.

Indeed, let $a_{jk} \in Q$ such that $\sum_{j=1}^{r} \sum_{k=1}^{m} a_{jk} \big(\sum_{\nu=1}^{r} (z^{j-1})^{\eta_\nu} \lambda_k \eta_\nu \big) = 0$. Then

$$\sum_{k=1}^{m} \sum_{\nu=1}^{r} \Big(\sum_{j=1}^{r} a_{jk}(z^{j-1})^{\eta_\nu} \Big) \lambda_k \eta_\nu = 0.$$

This gives $\sum_{j=1}^{r} a_{jk}(z^{j-1})^{\eta_\nu} = 0$ for all k, ν. Thus, for each k, $(a_{1k}, \ldots, a_{rk})$ is a solution of the homogeneous system of equations with the Vandermonde matrix $\big((z^{j-1})^{\eta_\nu}\big)$. Since this matrix is invertible, $a_{jk} = 0$ for all j, k. $\square$

LEMMA 1.1.3 (Common lemma): *N has a Q-basis in $\bigcap_{i \in I} N_i$.*

Proof: Consider a nonempty subset J of I. By induction on $|J|$ we find a Q-basis in $\bigcap_{j \in J} N_j$. For $J = I$ this gives the assertion of the lemma.

For each $i \in I$, Lemma 1.1.2 gives a Q-basis $\mathbf{v}_i$ of N in N_i, so the result follows when $|J| = 1$. Assume $|J| \geq 2$ and fix $i \in J$. By induction N has a

Q-basis $\mathbf{u}$ in $\bigcap_{j \in J \smallsetminus \{i\}} N_j$. The transition matrix $B \in \mathrm{GL}_n(Q)$ between $\mathbf{v}_i$ and $\mathbf{u}$ satisfies

(4) $$\mathbf{u} = \mathbf{v}_i B.$$

By (1e), there exist $B_1 \in \mathrm{GL}_n(P_i)$ and $B_2 \in \mathrm{GL}_n(P_i') \subseteq \bigcap_{j \in J \smallsetminus \{i\}} \mathrm{GL}_n(P_j)$. such that $B = B_1 B_2$. Then $\mathbf{u} B_2^{-1} = \mathbf{v}_i B_1$ is a Q-basis of N in $\bigcap_{j \in J} N_j$. This finishes the induction. $\qquad\square$

We introduce a special subset F of $\bigcap_{i \in I} Q_i$, call it the 'compound of the special data $\mathcal{E}$', and prove that F is a Galois extension of E with Galois group G and additional properties.

Definition 1.1.4: Compound. The **compound** of the patching data $\mathcal{E}$ is the set F of all $a \in \bigcap_{i \in I} Q_i$ for which there exists a function $f \colon G \to \bigcap_{i \in I} Q_i$ such that
(5a) $a = f(1)$ and
(5b) $f(\zeta\tau) = f(\zeta)^\tau$ for every $\zeta \in G$ and $\tau \in \bigcup_{i \in I} G_i$.

Note that for each $a \in \bigcap_{i \in I} Q_i$, the function f is uniquely determined by (5a) and (5b). Indeed, let $f' \colon G \to \bigcap_{i \in I} Q_i$ be another function such that $f'(1) = 1$ and $f(\zeta\tau) = f(\zeta)^\tau$ for all $\zeta \in G$ and $\tau \in \bigcup_{i \in I} G_i$. In particular, $f'(1) = f(1)$. By (1d), each $\sigma \in G$, $\sigma \neq 1$, can be written as $\sigma = \tau_1 \cdots \tau_m$ with $\tau_i \in \bigcup_{i \in I} G_i$, $i = 1, \ldots, m$, and $m \geq 1$. Set $\zeta = \tau_1 \cdots \tau_{m-1}$ and $\tau = \tau_m$. By induction on m we assume that $f'(\zeta) = f(\zeta)$. Then $f'(\sigma) = f'(\zeta)^\tau = f(\zeta)^\tau = f(\sigma)$.

We call f the **expansion** of a and denote it by f_a. Thus, $f_a(1) = a$ and $f_a(\zeta\tau) = f_a(\zeta)^\tau$ for all $\zeta \in G$ and $\tau \in \bigcup_{i \in I} G_i$. $\qquad\square$

We list some elementary properties of the expansions:

LEMMA 1.1.5: *Let F be the compound of $\mathcal{E}$. Then:*
(a) *Every $a \in E$ has an expansion, namely the constant function $\zeta \mapsto a$.*
(b) *Let $a, b \in F$. Then $a + b, ab \in F$; in fact, $f_{a+b} = f_a + f_b$ and $f_{ab} = f_a f_b$.*
(c) *Let $0 \neq a \in F$, then $a^{-1} \in F$. More precisely: $f_a(\zeta) \neq 0$ for all $\zeta \in G$, and $\zeta \mapsto f_a(\zeta)^{-1}$ is the expansion of a^{-1}.*
(d) *Let $a \in F$ and $\sigma \in G$. Then $f_a(\sigma) \in F$; in fact, $f_{f_a(\sigma)}(\zeta) = f_a(\sigma\zeta)$.*

Proof: Statement (a) holds, because $a^\tau = a$ for each $\tau \in \bigcup_{i \in I} G_i$. Statement (b) follows from the uniqueness of the expansions and from the observation $(f_{a+b})(1) = a + b = f_a(1) + f_b(1) = (f_a + f_b)(1)$.

Next we consider a nonzero $a \in F$ and let $\zeta \in G$. Using (1d), we write $\zeta = \tau_1 \cdots \tau_m$ with $\tau_1, \ldots, \tau_m \in \bigcup_{i \in I} G_i$ and set $\zeta' = 1$ if $m = 1$ and $\zeta' = \tau_1 \cdots \tau_{m-1}$ if $m \geq 2$. If $m = 1$, then $f_a(\zeta') = a \neq 0$, by assumption. If $m \geq 2$, then $f_a(\zeta') \neq 0$, by an induction hypothesis. In each case, $f_a(\zeta) = f_a(\zeta')^{\tau_m} \neq 0$. Since taking inverse in $\bigcap_{i \in I} Q_i$ commutes with the action of G, the map $\zeta \mapsto f_a(\zeta)^{-1}$ is the expansion of a^{-1}. This proves (c).

Finally, we check that the map $\zeta \to f_a(\sigma\zeta)$ has the value $f_a(\sigma)$ at $\zeta = 1$ and it satisfies (5b). Hence, that map is an expansion of $f_a(\sigma)$, as claimed in (d). $\square$

Definition 1.1.6: G-action on F.   For $a \in F$ and $\sigma \in G$ put

(6) $$a^\sigma = f_a(\sigma),$$

where f_a is the expansion of a. $\square$

LEMMA 1.1.7: *The compound F of the patching data $\mathcal{E}$ is a Galois field extension of E with Galois group G acting by (6). Moreover, for each $i \in I$,*
(a) *the restriction of this action to G_i coincides with the action of $G_i = \mathrm{Gal}(Q_i/P_i)$ on F as a subset of Q_i*
(b) *and $Q_i = P_i F$.*

Proof: By Lemma 1.1.5(a),(b),(c), F is a field containing E. Furthermore, (6) defines an action of G on F. Indeed, if $a \in F$ and $\sigma, \zeta \in G$, then by Lemma 1.1.5(d), $(a^\sigma)^\zeta = f_a(\sigma)^\zeta = f_{f_a^\sigma}(\zeta) = f_a(\sigma\zeta) = a^{(\sigma\zeta)}$.

Proof of (a):   Let $\tau \in G_i$ and $a \in F$. Then $f_a(\tau) = f_a(1)^\tau = a^\tau$, where τ acts as an element of $G = \mathrm{Gal}(Q_i/P_i)$. Thus, that action coincides with the action given by (6).

The rest of the proof of (a) breaks up into three parts.

PART A: $F^G = E$.   Indeed, by Lemma 1.1.5(a), elements of E have constant expansions, hence are fixed by G. Conversely, let $a \in F^G$. Then for each $i \in I$ we have $a \in Q_i^{G_i} = P_i$. Hence, by (1c), $a \in E$.

PART B: $[F : E] \geq |G|$.   We define a map $T: F \to N$ by

$$T(a) = \sum_{\zeta \in G} f_a(\zeta)\zeta.$$

By Lemma 1.1.5(a),(b), T is an E-linear map, and $\mathrm{Im}(T) = \bigcap_{i \in I} N_i$. By Lemma 1.1.3, $\mathrm{Im}(T)$ contains $|G|$ linearly independent elements over Q, hence over E. Therefore, $[F : E] = \dim_E F \geq |G|$.

PART C: F/E is Galois and $\mathrm{Gal}(F/E) = G$.   The action (6) of G on F maps G onto a subgroup $\bar{G}$ of $\mathrm{Aut}(F/E)$. By Part A, $F^{\bar{G}} = E$. Hence, by Galois theory, F/E is a Galois extension with Galois group $\bar{G}$. In particular, $[F : E] = |\bar{G}| \leq |G|$. By Part B, $[F : E] \geq |G|$, Hence $G \cong \bar{G}$. So, we may (and we will) identify $\mathrm{Gal}(F/E)$ with G.

Proof of (b):   By (a), the restriction $\mathrm{Gal}(Q_i/P_i) \to \mathrm{Gal}(F/E)$ is injective. Hence $Q_i = P_i F$. $\square$

Remark 1.1.8: The vector spaces N and N_i defined by (2) and (3) are actually induced from Q and P_i, respectively, namely $N = \operatorname{Ind}_1^G Q$ and $N_i = \operatorname{Ind}_{G_i}^G Q_i$. We may define multiplication on N componentwise:

$$\sum_{\zeta \in G} a_\zeta \zeta \sum_{\zeta \in G} b_\zeta \zeta = \sum_{\zeta \in G} a_\zeta \beta_\zeta \zeta.$$

Then N becomes a Q-Algebra and N_i becomes a P_i-algebra. By Lemma 1.1.5, the map $T\colon F \to \bigcap_{i \in I} N_i$ defined in Part B of the proof of Lemma 1.1.7, is an E-linear isomorphism of E-algebras whose inverse is the map $\sum_{\zeta \in G} a_\zeta \zeta \mapsto a_1$. Hence, by that lemma, $F' = \bigcap_{i \in I} N_i$ is a Galois extension of E with Galois group G. The following diagram describes the respective location of all fields and algebras mentioned in our construction:

(7)

□

1.2 Galois Action on Patching Data

A **finite split embedding problem** over a field E_0 is an epimorphism

(1) $$\operatorname{pr}\colon \Gamma \ltimes G \to \Gamma$$

of finite groups, where $\Gamma = \operatorname{Gal}(E/E_0)$ is the Galois group of a Galois extension E/E_0, G is a finite group on which Γ acts from the right, $\Gamma \ltimes G$ is the corresponding semidirect product, and pr is the projection on Γ. Each element of $\Gamma \ltimes G$ has a unique representation as a product $\gamma\zeta$ with $\gamma \in \Gamma$ and $\zeta \in G$. The product and the inverse operation are given in $\Gamma \ltimes G$ by the formulas $\gamma\zeta \cdot \delta\eta = \gamma\delta \cdot \zeta^\delta \eta$ and $(\gamma\zeta)^{-1} = \gamma^{-1}(\zeta^{\gamma^{-1}})^{-1}$. A **solution** of (1) is a Galois extension F of E_0 that contains E and an isomorphism $\psi\colon \operatorname{Gal}(F/E_0) \to \Gamma \ltimes G$ such that $\operatorname{pr} \circ \psi = \operatorname{res}_E$. We call F a **solution field** of (1).

Suppose the compound F of a patching data $\mathcal{E}$ (§1.1) realizes G over E. A 'proper' action of Γ on $\mathcal{E}$ will then ensure that F is even a solution field for the embedding problem (1).

Definition 1.2.1: Let E/E_0 be a finite Galois extension with Galois group Γ. Let $\mathcal{E} = (E, F_i, P_i, Q; G_i, G)_{i \in I}$ be a patching data (Definition 1.1.1). A **proper action** of Γ on $\mathcal{E}$ is a triple that consists of an action of Γ on the group G, an action of Γ on the field Q, and an action of Γ on the set I such that the following conditions hold:

(2a) The action of Γ on Q extends the action of Γ on E.

(2b) $F_i^\gamma = F_{i^\gamma}$, $P_i^\gamma = P_{i^\gamma}$, and $G_i^\gamma = G_{i^\gamma}$, for all $i \in I$ and $\gamma \in \Gamma$.

(2c) $(a^\tau)^\gamma = (a^\gamma)^{\tau^\gamma}$ for all $i \in I$, $a \in F_i$, $\tau \in G_i$, and $\gamma \in \Gamma$.

The action of Γ on G defines a semidirect product $\Gamma \ltimes G$ such that $\tau^\gamma = \gamma^{-1}\tau\gamma$ for all $\tau \in G$ and $\gamma \in \Gamma$. Let pr: $\Gamma \ltimes G \to \Gamma$ be the canonical projection. $\square$

PROPOSITION 1.2.2: *In the notation of Definition 1.2.1 suppose that* $\Gamma = \mathrm{Gal}(E/E_0)$ *acts properly on the patching data* $\mathcal{E}$ *given in Definition 1.2.1. Let F be the compound of $\mathcal{E}$. Then Γ acts on F via the restriction from its action on Q and the actions of Γ and G on F combine to an action of $\Gamma \ltimes G$ on F with fixed field E_0. This gives an identification* $\mathrm{Gal}(F/E_0) = \Gamma \ltimes G$ *such that the following diagram of short exact sequences commutes:*

$$
\begin{array}{ccccccccc}
1 & \longrightarrow & G & \longrightarrow & \Gamma \ltimes G & \overset{\mathrm{pr}}{\longrightarrow} & \Gamma & \longrightarrow & 1 \\
& & \| & & \| & & \| & & \\
1 & \longrightarrow & \mathrm{Gal}(F/E) & \longrightarrow & \mathrm{Gal}(F/E_0) & \overset{\mathrm{res}}{\longrightarrow} & \mathrm{Gal}(E/E_0) & \longrightarrow & 1
\end{array}
$$

Thus, F is a solution field of the embedding problem (1).

Proof: We break the proof of the proposition into three parts.

PART A: *The action of Γ on F.*

Let $i \in I$ and $\gamma \in \Gamma$. Then $Q_i = P_i F_i$, so by (2b), $Q_i^\gamma = Q_{i^\gamma}$. Moreover, we have identified $\mathrm{Gal}(Q_i/P_i)$ with $G_i = \mathrm{Gal}(F_i/E)$ via restriction. Hence, by (2b), for all $a \in P_i$ and $\tau \in G_i$ we have $\tau^\gamma \in G_{i^\gamma}$ and $a^\gamma \in P_{i^\gamma}$, so $(a^\tau)^\gamma = a^\gamma = (a^\gamma)^{\tau^\gamma}$. Together with (2c), this gives

$$(4) \qquad (a^\tau)^\gamma = (a^\gamma)^{\tau^\gamma} \quad \text{for all } a \in Q_i \text{ and } \tau \in G_i.$$

Consider an $a \in F$ and let f_a be the expansion of a (Definition 1.1.4). Define $f_a^\gamma \colon G \to \bigcap_{i \in I} Q_i$ by $f_a^\gamma(\zeta) = f_a(\zeta^{\gamma^{-1}})^\gamma$. Then f_a^γ is the expansion f_{a^γ} of a^γ. Indeed, $f_a^\gamma(1) = f_a(1^{\gamma^{-1}})^\gamma = a^\gamma$ and if $\zeta \in G$ and $\tau \in G_i$, then $\tau^{\gamma^{-1}} \in G_{i^{\gamma-1}}$. Hence, by (4) with $i^{\gamma^{-1}}, f_a(\zeta^{\gamma^{-1}}), \tau^{\sigma^{-1}}$, respectively, replacing i, a, τ, we have

$$f_a^\gamma(\zeta\tau) = f_a(\zeta^{\gamma^{-1}}\tau^{\gamma^{-1}})^\gamma = \left(f_a(\zeta^{\gamma^{-1}})^{\tau^{\gamma^{-1}}}\right)^\gamma$$

$$= \left(f_a(\zeta^{\gamma^{-1}})^\gamma\right)^{\tau^{\gamma^{-1}\gamma}} = \left(f_a(\zeta^{\gamma^{-1}})^\gamma\right)^\tau = f_a^\gamma(\zeta)^\tau.$$

Thus $a^\gamma \in F$. It follows that the action of Γ on Q gives an action of Γ on F.

PART B: *The action of $\Gamma \ltimes G$ on F.* Let $a \in F$ and $\gamma \in \Gamma$. We claim that

(5) $$(a^\sigma)^\gamma = (a^\gamma)^{\sigma^\gamma} \quad \text{for all } \sigma \in G,$$

where $a^\sigma = f_a(\sigma)$ (Definition 1.1.6). Indeed, write σ as a word in $\bigcup_{i \in I} G_i$. Then (5) follows from (4) by induction on the length of the word. If $\sigma = 1$, then (5) is an identity. Suppose (5) holds for some $\sigma \in G$ and let $\tau \in \bigcup_{i \in I} G_i$. Using the identification of the action of each $\tau \in G_i$ on F as an element of G_i with its action as an element of G (Lemma 1.1.7(a)) and (4) for a^σ rather that a, we have

$$(a^{\sigma\tau})^\gamma = \left((a^\sigma)^\tau\right)^\gamma = \left((a^\sigma)^\gamma\right)^{\tau^\gamma} = \left((a^\gamma)^{\sigma^\gamma}\right)^{\tau^\gamma} = (a^\gamma)^{\sigma^\gamma \tau^\gamma} = (a^\gamma)^{(\sigma\tau)^\gamma}.$$

Now we apply (5) to $a^{\gamma^{-1}}$ instead of a to find that $\left((a^{\gamma^{-1}})^\sigma\right)^\gamma = a^{\sigma^\gamma}$. It follows that the actions of Γ and G on F combine to an action of $\Gamma \ltimes G$ on F.

(6)

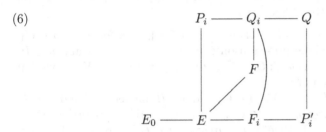

PART C: *Conclusion of the proof.* Since $F^G = E$ (Lemma 1.1.7) and $E^\Gamma = E_0$, we have $F^{\Gamma \ltimes G} = E_0$. Furthermore, $[F : E_0] = [F : E] \cdot [E : E_0] = |G| \cdot |\Gamma| = |G \times \Gamma|$. By Galois theory, $\mathrm{Gal}(F/E_0) = \Gamma \ltimes G$ and the map res: $\mathrm{Gal}(F/E_0) \to \mathrm{Gal}(E/E_0)$ coincides with the canonical map pr: $\Gamma \ltimes G \to \Gamma$. $\qquad \square$

1.3 The Compound of the Patching Data

This section offers additional useful information about the patching data $\mathcal{E} = (E, F_i, P_i, Q; G_i, G)_{i \in I}$ and the diagram (5) of §1.2.

LEMMA 1.3.1: *Let F be the compound of the patching data $\mathcal{E}$. Then:*
(a) *Suppose $1 \in I$, $G = G_1 \ltimes H$ and $H = \langle G_i \mid i \in I \smallsetminus \{1\} \rangle$. Then, $F_1 = F^H$ and the identification $\mathrm{Gal}(F/E) = G$ of Lemma 1.1.7 gives the following commutative diagram of short exact sequences:*

$$
\begin{array}{ccccccccc}
1 & \longrightarrow & H & \longrightarrow & G = G_1 \ltimes H & \xrightarrow{\ \mathrm{pr}\ } & G_1 & \longrightarrow & 1 \\
& & \| & & \| & & \| & & \\
1 & \longrightarrow & \mathrm{Gal}(F/F_1) & \longrightarrow & \mathrm{Gal}(F/E) & \xrightarrow{\ \mathrm{res}\ } & \mathrm{Gal}(F_1/E) & \longrightarrow & 1
\end{array}
$$

(b) If, in addition to the assumptions of (a), E is a finite Galois extension of a field E_0 with Galois group Γ that acts properly on $\mathcal{E}$ such that $1^\gamma = 1$ for each $\gamma \in \Gamma$, then F is a Galois extension of E_0, F_1, Q_1, and G_1 are Γ-invariant, F_1/E_0 is Galois, and we can identify groups as in the following commutative diagram:

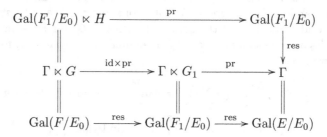

Proof of (a): The proof breaks up into several parts.

PART A: $F_1 \subseteq \bigcap_{i \in I} Q_i$. Indeed, $F_1 \subseteq F_1 P_1 = Q_1$ and $F_1 \subseteq P_i \subseteq Q_i$ for $i \neq 1$, by Condition (1b) of Section 1.1.

PART B: $F_1 \subseteq F$. Let $a \in F_1$. Then $a^\sigma \in F_1 \subseteq \bigcap_{i \in I} Q_i$ for every $\sigma \in G_1$. Every $\zeta \in G$ has a unique presentation $\zeta = \sigma\eta$, where $\sigma \in G_1$ and $\eta \in H$. Use this to define a function $f \colon G \to \bigcap_{i \in I} Q_i$ by $f(\sigma\eta) = a^\sigma$. We prove that $f = f_a$ is the expansion of a.

First note that $f(1) = a$. Fix $\sigma \in G_1$ and $\eta \in H$ and let $i \in I$ and $\tau \in G_i$. If $i = 1$, then $\sigma\eta\tau = (\sigma\tau)\eta^\tau$, $\sigma\tau \in G_1$, and $\eta^\tau \in H$. Hence $f(\sigma\eta\tau) = a^{\sigma\tau} = (a^\sigma)^\tau = f(\sigma\eta)^\tau$. If $i \neq 1$, then $\sigma\eta\tau = \sigma(\eta\tau)$, $\sigma \in G_1$, and $\eta\tau \in H$. Also, $a^\sigma \in F_1 \subseteq P_i = Q_i^{G_i}$, so $(a^\sigma)^\tau = a^\sigma$. Hence $f(\sigma\eta\tau) = a^\sigma = (a^\sigma)^\tau = f(\sigma\eta)^\tau$. Thus, in both cases $f(\sigma\eta\tau) = f(\sigma\eta)^\tau$. It follows from Definition 1.1.4 that $f = f_a$.

PART C: $a^\tau = a^{\mathrm{pr}(\tau)}$ for all $a \in F_1$ and $\tau \in G$. Since pr is a homomorphism, it suffices to prove the equality for each τ in a set of generators of G. So we may assume that $\tau \in G_i$ for some $i \in I$. If $i = 1$, then $\mathrm{pr}(\tau) = \tau$ and the assertion follows. If $i \neq 1$, then $\mathrm{pr}(\tau) = 1$, whence $a^{\mathrm{pr}(\tau)} = a$. Moreover, $a \in F_1 \subseteq P_i = Q_i^{G_i}$, so $a^\tau = a$, as claimed.

PART D: *Completion of the proof.* Part C says that res: $\mathrm{Gal}(F/E) \to \mathrm{Gal}(F_1/E)$ and pr: $G \to G_1$ coincide. Hence, the $H = \mathrm{Gal}(F/F_1)$, $F_1 = F^H$, and the diagram in (c) commutes.

Proof of (b): Note that $F_1^\gamma = F_{1^\gamma} = F_1$ and similarly $Q_1^\gamma = Q_1$ and $G_1^\gamma = G_1$ for each $\gamma \in \Gamma$. Thus, $\Gamma \ltimes G_1$ is a subgroup of $\Gamma \ltimes G = \mathrm{Gal}(F/E_0)$ that leaves F_1 invariant. The fixed field of $\Gamma \ltimes G_1$ in F_1 is E_0. Since $|\Gamma \ltimes G_1| = [F_1 : E_0]$, this implies by Galois theory that F_1/E_0 is Galois with Galois group $\Gamma \ltimes G_1$. The identification $\mathrm{Gal}(F_1/E_0) = \Gamma \ltimes G_1$ restricts further to $\mathrm{Gal}(E/E_0) = \Gamma$. This completes the commutativity of the lower part of the diagram in (b).

Since each $\gamma \in \Gamma$ fixes 1 it leaves $I \smallsetminus \{1\}$ invariant, so Γ leaves $H = \langle G_i \mid i \in I \smallsetminus \{1\} \rangle$ invariant. Thus, H can be considered as a normal subgroup

of $\Gamma \ltimes G$ with $(\Gamma \ltimes G)/H = \Gamma \ltimes G_1$. To prove the commutativity of the upper part just note that $\Gamma \ltimes G = \Gamma \ltimes (G_1 \ltimes H) = (\Gamma \ltimes G_1) \ltimes H = \mathrm{Gal}(F_1/E_0) \ltimes H$.
$\square$

Notes

We call the field $F' = \bigcap_{i \in I} N_i$ that appears in Remark 1.1.8 the **precompound** of the patching data $\mathcal{E}$ of Definition 1.1.1. The idea of a patching data as well as the notions of a 'precompound' and 'compound' of $\mathcal{E}$ appear in [HaV96, Sec. 3] (however, the precompound is denoted by F while the compound is denoted by F' in [HaV96]). It is used there in order to prove that if R is a complete local integral domain which is not a field and if $K = \mathrm{Quot}(R)$, then for every finite group G, the field $K(x)$ has a Galois extension F with Galois group G such that F is a regular extension of K and has a prime divisor of degree 1 unramified over $K(x)$ [HaV96, Thm. 4.4]. In addition, [HaV96, Cor. 4.7] states that if E is a function field of one variable over a countable algebraically closed field, then $\mathrm{Gal}(E) \cong \hat{F}_\omega$.

The action of a finite group on a patching data $\mathcal{E}$ is introduced in [HaJ98a] in order to prove that the precompounds are solution fields of finite split embedding problems [HaJ98a, Prop. 1.5]. This suffices to prove the main theorem of [HaJ98a] that every PAC Hilbertian field is ω-free. Note however, that [HaJ98a] calls a 'compound' what we call a 'precompound'.

The main advantage of the compound F on the precompound F' is that $P_i F = Q_i$ for each $i \in I$. This implies that the set of 'branch points' of F/E_0 is the union of the sets of branch points of F_i/E_0 (Proposition 7.2.3).

The presentation of the compound in Definition 1.1.4 is direct. Thus, we prove the properties of the compound, in particular the solvability of finite split embedding problems, without proving them first for the precompound (as is done in [HaV96] and [HaJ98a]). This shorter presentation is due to Dan Haran (private communication).

Lemma 1.1.3 is a workout of [HaV96, Prop. 3.4] for $|I| = 1$ and [HaJ98a, Lemma 1.2] in the general case. Lemma 1.1.7 appears as [HaV96, Lemma 3.6].

The roles of P_i and Q_i in the patching data of [HaV96], [HaJ98a], etc. have been exchanged in this book in order for the smaller fields to be named by earlier letters.

Chapter 2.
Normed Rings

Norms $\| \cdot \|$ of associative rings are generalizations of absolute values $| \cdot |$ of integral domains, where the inequality $\|xy\| \leq \|x\| \cdot \|y\|$ replaces the standard multiplication rule $|xy| = |x| \cdot |y|$. Starting from a complete normed commutative ring A, we study the ring $A\{x\}$ of all formal power series with coefficients in A converging to zero. This is again a complete normed ring (Lemma 2.2.1). We prove an analog of the Weierstrass division theorem (Lemma 2.2.4) and the Weierstrass preparation theorem for $A\{x\}$ (Corollary 2.2.5). If A is a field K and the norm is an absolute value, then $K\{x\}$ is a principal ideal domain, hence a factorial ring (Proposition 2.3.1). Moreover, $\mathrm{Quot}(K\{x\})$ is a Hilbertian field (Theorem 2.3.3). It follows that $\mathrm{Quot}(K\{x\})$ is not a Henselian field (Corollary 2.3.4). In particular, $\mathrm{Quot}(K\{x\})$ is not separably closed in $K((x))$. In contrast, the field $K((x))_0$ of all formal power series over K that converge at some element of K is algebraically closed in $K((x))$ (Proposition 2.4.5).

2.1 Normed Rings

In Section 4.4 we construct patching data over fields $K(x)$, where K is a complete ultrametric valued field. The 'analytic' fields P_i will be the quotient fields of certain rings of convergent power series in several variables over K. At a certain point in a proof by induction we consider a ring of convergent power series in one variable over a complete ultrametric valued ring. So, we start by recalling the definition and properties of the latter rings.

Let A be a commutative ring with 1. An **ultrametric absolute value** of A is a function $| \ |: A \to \mathbb{R}$ satisfying the following conditions:

(1a) $|a| \geq 0$, and $|a| = 0$ if and only if $a = 0$.
(1b) There exists $a \in A$ such that $0 < |a| < 1$.
(1c) $|ab| = |a| \cdot |b|$.
(1d) $|a + b| \leq \max(|a|, |b|)$.

By (1a) and (1c), A is an integral domain. By (1c), the absolute value of A extends to an absolute value on the quotient field of A (by $|\frac{a}{b}| = \frac{|a|}{|b|}$). It follows also that $|1| = 1$, $|-a| = |a|$, and

(1d') if $|a| < |b|$, then $|a + b| = |b|$.

Denote the ordered additive group of the real numbers by $\mathbb{R}^+$. The function $v \colon \mathrm{Quot}(A) \to \mathbb{R}^+ \cup \{\infty\}$ defined by $v(a) = -\log|a|$ satisfies the following conditions:

(2a) $v(a) = \infty$ if and only if $a = 0$.
(2b) There exists $a \in \mathrm{Quot}(A)$ such that $0 < v(a) < \infty$.
(2c) $v(ab) = v(a) + v(b)$.

M. Jarden, *Algebraic Patching*, Springer Monographs in Mathematics,
DOI 10.1007/978-3-642-15128-6_2, © Springer-Verlag Berlin Heidelberg 2011

(2d) $v(a + b) \geq \min\{v(a), v(b)\}$ (and $v(a + b) = v(b)$ if $v(b) < v(a)$).

In other words, v is a **real valuation** of $\mathrm{Quot}(A)$. Conversely, every real valuation $v\colon \mathrm{Quot}(A) \to \mathbb{R}^+ \cup \{\infty\}$ gives rise to a nontrivial ultrametric absolute value $|\cdot|$ of $\mathrm{Quot}(A)$: $|a| = \varepsilon^{v(a)}$, where ε is a fixed real number between 0 and 1.

An attempt to extend an absolute value from A to a larger ring A' may result in relaxing Condition (1c), replacing the equality by an inequality. This leads to the more general notion of a 'norm'.

Definition 2.1.1: Normed rings. Let R be an associative ring with 1. A **norm** on R is a function $\| \ \|\colon R \to \mathbb{R}$ that satisfies the following conditions for all $a, b \in R$:

(3a) $\|a\| \geq 0$, and $\|a\| = 0$ if and only if $a = 0$; further $\|1\| = \| -1\| = 1$.

(3b) There is an $x \in R$ with $0 < \|x\| < 1$.

(3c) $\|ab\| \leq \|a\| \cdot \|b\|$.

(3d) $\|a + b\| \leq \max(\|a\|, \|b\|)$.

The norm $\| \ \|$ naturally defines a topology on R whose basis is the collection of all sets $U(a_0, r) = \{a \in R \mid \|a - a_0\| < r\}$ with $a_0 \in R$ and $r > 0$. Both addition and multiplication are continuous under that topology. Thus, R is a **topological ring**. $\qquad\square$

Definition 2.1.2: Complete rings. Let R be a normed ring. A sequence $a_1, a_2, a_3, \ldots$ of elements of R is **Cauchy** if for each $\varepsilon > 0$ there exists m_0 such that $\|a_n - a_m\| < \varepsilon$ for all $m, n \geq m_0$. We say that R is **complete** if every Cauchy sequence converges. $\qquad\square$

Lemma 2.1.3: Let R be a normed ring and let $a, b \in R$. Then:

(a) $\| -a\| = \|a\|$.

(b) If $\|a\| < \|b\|$, then $\|a + b\| = \|b\|$.

(c) A sequence $a_1, a_2, a_3, \ldots$ of elements of R is Cauchy if for each $\varepsilon > 0$ there exists m_0 such that $\|a_{m+1} - a_m\| < \varepsilon$ for all $m \geq m_0$.

(d) The map $x \to \|x\|$ from R to $\mathbb{R}$ is continuous.

(e) If R is complete, then a series $\sum_{n=0}^{\infty} a_n$ of elements of R converges if and only if $a_n \to 0$.

(f) If R is complete and $\|a\| < 1$, then $1 - a \in R^{\times}$. Moreover, $(1-a)^{-1} = 1 + b$ with $\|b\| < 1$.

Proof of (a): Observe that $\| -a\| \leq \| -1\| \cdot \|a\| \leq \|a\|$. Replacing a by $-a$, we get $\|a\| \leq \| -a\|$, hence the claimed equality.

Proof of (b): Assume $\|a + b\| < \|b\|$. Then, by (a), $\|b\| = \|(-a) + (a+b)\| \leq \max(\| -a\|, \|a+b\|) < \|b\|$, which is a contradiction.

Proof of (c): With m_0 as above let $n > m \geq m_0$. Then

$$\|a_n - a_m\| \leq \max(\|a_n - a_{n-1}\|, \ldots, \|a_{m+1} - a_m\|) < \varepsilon.$$

Proof of (d): By (3d), $\|x\| = \|(x - y) + y\| \leq \max(\|x - y\|, \|y\|) \leq \|x - y\| + \|y\|$. Hence, $\|x\| - \|y\| \leq \|x - y\|$. Symmetrically, $\|y\| - \|x\| \leq \|y - x\| =$

$\|x - y\|$. Therefore, $|\|x\| - \|y\|| \leq \|x - y\|$. Consequently, the map $x \mapsto \|x\|$ is continuous.

Proof of (e): Let $s_n = \sum_{i=0}^{n} a_i$. Then $s_{n+1} - s_n = a_{n+1}$. Thus, by (c), $s_1, s_2, s_3, \ldots$ is a Cauchy sequence if and only if $a_n \to 0$. Hence, the series $\sum_{n=0}^{\infty} a_n$ converges if and only if $a_n \to 0$.

Proof of (f): The sequence a^n tends to 0. Hence, by (e), $\sum_{n=0}^{\infty} a^n$ converges. The identities $(1 - a) \sum_{i=0}^{n} a^i = 1 - a^{n+1}$ and $\sum_{i=0}^{n} a^i (1 - a) = 1 - a^{n+1}$ imply that $\sum_{n=0}^{\infty} a^n$ is both the right and the left inverse of $1 - a$. Moreover, $\sum_{n=0}^{\infty} a^n = 1 + b$ with $b = \sum_{n=1}^{\infty} a^n$ and $\|b\| \leq \max_{n \geq 1} \|a\|^n < 1$. □

Example 2.1.4:

(a) Every field K with an ultrametric absolute value is a normed ring. For example, for each prime number p, $\mathbb{Q}$ has a p-adic absolute value $|\cdot|_p$ which is defined by $|x|_p = p^{-m}$ if $x = \frac{a}{b} p^m$ with $a, b, m \in \mathbb{Z}$ and $p \nmid a, b$.

(b) The ring $\mathbb{Z}_p$ of p-adic integers and the field $\mathbb{Q}_p$ of p-adic numbers are complete with respect to the p-adic absolute value.

(c) Let K_0 be a field and let $0 < \varepsilon < 1$. The ring $K_0[[t]]$ (resp. field $K_0((t))$) of formal power series $\sum_{i=0}^{\infty} a_i t^i$ (resp. $\sum_{i=m}^{\infty} a_i t^i$ with $m \in \mathbb{Z}$) with coefficients in K_0 is complete with respect to the absolute value $|\sum_{i=m}^{\infty} a_i t^i| = \varepsilon^{\min(i \mid a_i \neq 0)}$.

(d) Let $\|\cdot\|$ be a norm of a commutative ring A. For each positive integer n we extend the norm to the associative (and usually not commutative) ring $M_n(A)$ of all $n \times n$ matrices with entries in A by

$$\|(a_{ij})_{1 \leq i,j \leq n}\| = \max(\|a_{ij}\|_{1 \leq i,j \leq n}).$$

If $b = (b_{jk})_{1 \leq j,k \leq n}$ is another matrix and $c = ab$, then $c_{ik} = \sum_{j=1}^{n} a_{ij} b_{jk}$ and $\|c_{ik}\| \leq \max(\|a_{ij}\| \cdot \|b_{jk}\|) \leq \|a\| \cdot \|b\|$. Hence, $\|c\| \leq \|a\| \|b\|$. This verifies Condition (3c). The verification of (3a), (3b), and (3d) is straightforward. Note that when $n \geq 2$, even if the initial norm of A is an absolute value, the extended norm satisfies only the weak condition (3c) and not the stronger condition (1c), so it is not an absolute value.

If A is complete, then so is $M_n(A)$. Indeed, let $a_i = (a_{i,rs})_{1 \leq r,s \leq n}$ be a Cauchy sequence in $M_n(A)$. Since $\|a_{i,rs} - a_{j,rs}\| \leq \|a_i - a_j\|$, each of the sequences $a_{1,rs}, a_{2,rs}, a_{3,rs}, \ldots$ is Cauchy, hence converges to an element b_{rs} of A. Set $b = (b_{rs})_{1 \leq r,s \leq n}$. Then $a_i \to b$. Consequently, $M_n(A)$ is complete.

(e) Let $\mathfrak{a}$ be a proper ideal of a Noetherian domain A. By a theorem of Krull, $\bigcap_{n=0}^{\infty} \mathfrak{a}^n = 0$ [AtM69, p. 110, Cor. 10.18]. We define an $\mathfrak{a}$-adic norm on A by choosing an ε between 0 and 1 and setting $\|a\| = \varepsilon^{\max(n \mid a \in \mathfrak{a}^n)}$. If $\|a\| = \varepsilon^m$ and $\|b\| = \varepsilon^n$, and say $m \leq n$, then $\mathfrak{a}^n \subseteq \mathfrak{a}^m$, so $a + b \in \mathfrak{a}^m$, hence $\|a + b\| \leq \varepsilon^m = \max(\|a\|, \|b\|)$. Also, $ab \in \mathfrak{a}^{m+n}$, so $\|ab\| \leq \|a\| \cdot \|b\|$. □

Like absolute valued rings, every normed ring has a completion:

LEMMA 2.1.5: *Every normed ring $(R, \| \ \|)$ can be embedded into a complete normed ring $(\hat{R}, \| \ \|)$ such that R is dense in $\hat{R}$ and the following universal condition holds:*

(4) *Each continuous homomorphism f of R into a complete ring S uniquely extends to a continuous homomorphism $\hat{f} \colon \hat{R} \to S$.*

The normed ring $(\hat{R}, \| \ \|)$ is called the **completion** *of $(R, \| \ \|)$.*

Proof: We consider the set A of all Cauchy sequences $\mathbf{a} = (a_n)_{n=1}^{\infty}$ with $a_n \in R$. For each $\mathbf{a} \in A$, the values $\|a_n\|$ of its components are bounded. Hence, A is closed under componentwise addition and multiplication and contains all constant sequences. Thus, A is a ring. Let $\mathfrak{n}$ be the ideal of all sequences that converge to 0. We set $\hat{R} = A/\mathfrak{n}$ and identify each $x \in R$ with the coset $(x)_{n=1}^{\infty} + \mathfrak{n}$.

If $\mathbf{a} \in A \smallsetminus \mathfrak{n}$, then $\|a_n\|$ eventually becomes constant. Indeed, there exists $\beta > 0$ such that $\|a_n\| \geq \beta$ for all sufficiently large n. Choose n_0 such that $\|a_n - a_m\| < \beta$ for all $n, m \geq n_0$. Then, $\|a_n - a_{n_0}\| < \beta \leq \|a_{n_0}\|$, so $\|a_n\| = \|(a_n - a_{n_0}) + a_{n_0}\| = \|a_{n_0}\|$. We define $\|\mathbf{a}\|$ to be the eventual absolute value of a_n and note that $\|\mathbf{a}\| \neq 0$. If $\mathbf{b} \in \mathfrak{n}$, we set $\|\mathbf{b}\| = 0$ and observe that $\|\mathbf{a} + \mathbf{b}\| = \|\mathbf{a}\|$. It follows that $\|\mathbf{a} + \mathfrak{n}\| = \|\mathbf{a}\|$ is a well defined function on $\hat{R}$ which extends the norm of R.

One checks that $\| \ \|$ is a norm on $\hat{R}$ and that R is dense in $\hat{R}$. Indeed, if $\mathbf{a} = (a_n)_{n=1}^{\infty} \in A$, then $a_n + \mathfrak{n} \to \mathbf{a} + \mathfrak{n}$. To prove that $\hat{R}$ is complete under $\| \ \|$ we consider a Cauchy sequence $(\mathbf{a}_k)_{k=1}^{\infty}$ of elements of $\hat{R}$. For each k we choose an element $b_k \in R$ such that $\|b_k - \mathbf{a}_k\| < \frac{1}{k}$. Then $(b_k)_{k=1}^{\infty}$ is a Cauchy sequence of R and the sequence $(\mathbf{a}_k)_{k=1}^{\infty}$ converges to the element $(b_k)_{k=1}^{\infty} + \mathfrak{n}$ of $\hat{R}$.

Finally, let S be a complete normed ring and $f \colon R \to S$ a continuous homomorphism. Then, for each $\mathbf{a} = (a_n)_{n=1}^{\infty} \in A$, the sequence $(f(a_n))_{n=1}^{\infty}$ of S is Cauchy, hence it converges to an element s. Define $\hat{f}(\mathbf{a} + \mathfrak{n}) = s$ and check that $\hat{f}$ has the desired properties. $\qquad\qquad\square$

Example 2.1.6: Let A be a commutative ring. We consider the ring $R = A[x_1, \ldots, x_n]$ of polynomials over A in the variables $x_1, \ldots, x_n$ and the ideal $\mathfrak{a}$ of R generated by $x_1, \ldots, x_n$. The completion of R with respect to $\mathfrak{a}$ is the ring $\hat{R} = A[[x_1, \ldots, x_n]]$ of all formal power series $f(x_1, \ldots, x_n) = \sum_{i=0}^{\infty} f_i(x_1, \ldots, x_n)$, where $f_i \in A[x_1, \ldots, x_n]$ is a homogeneous polynomial of degree i. Moreover, $\hat{R} = A[[x_1, \ldots, x_{n-1}]][[x_n]]$ and $\hat{R}$ is complete with respect to the ideal $\hat{\mathfrak{a}}$ generated by $x_1, \ldots, x_n$ [Lan93, Chap. IV, Sec. 9]. If R is a Noetherian integral domain, then so is $\hat{R}$ [Lan93, p. 210, Cor. 9.6]. If $A = K$ is a field, then $\hat{R}$ is a unique factorization domain [Mat94, Thm. 20.3].

If A is an integral domain, then the function $v \colon \hat{R} \to \mathbb{Z} \cup \{\infty\}$ defined for f as in the preceding paragraph by $v(f) = \min_{i \geq 0}(f_i \neq 0)$ satisfies Condition (2), so it extends to a discrete valuation of $\hat{K} = \mathrm{Quot}(\hat{R})$. However, by Weissauer, $\hat{K}$ is Hilbertian if $n \geq 2$. [FrJ08, Example 15.5.2]. Hence, $\hat{K}$

is Henselian with respect to no valuation [FrJ08, Lemma 15.5.4]. Since v is discrete, $\hat{K}$ is not complete with respect to v. $\qquad\square$

2.2 Rings of Convergent Power Series

Let A be a complete normed commutative ring and x a variable. Consider the following subset of $A[[x]]$:

$$A\{x\} = \{\sum_{n=0}^{\infty} a_n x^n \mid a_n \in A, \lim_{n\to\infty} \|a_n\| = 0\}.$$

For each $f = \sum_{n=0}^{\infty} a_n x^n \in A\{x\}$ we define $\|f\| = \max(\|a_n\|)_{n=0,1,2,\dots}$. This definition makes sense because $a_n \to 0$, hence $\|a_n\|$ is bounded.

We prove the Weierstrass division and the Weierstrass preparation theorems for $A\{x\}$ in analogy to the corresponding theorems for the ring of formal power series in one variable over a local ring.

LEMMA 2.2.1:
(a) $A\{x\}$ is a subring of $A[[x]]$ containing A.
(b) The function $\| \ \|\colon A\{x\} \to \mathbb{R}$ is a norm.
(c) The ring $A\{x\}$ is complete under that norm.
(d) Let B be a complete normed ring extension of A. Then each $b \in B$ with $\|b\| \le 1$ defines an **evaluation homomorphism** $A\{x\} \to B$ given by

$$f = \sum_{n=0}^{\infty} a_n x^n \mapsto f(b) = \sum_{n=0}^{\infty} a_n b^n.$$

Proof of (a): We prove only that $A\{x\}$ is closed under multiplication. To that end let $f = \sum_{i=0}^{\infty} a_i x^i$ and $g = \sum_{j=0}^{\infty} b_j x^j$ be elements of $A\{x\}$. Consider $\varepsilon > 0$ and let n_0 be a positive number such that $\|a_i\| < \varepsilon$ if $i \ge \frac{n_0}{2}$ and $\|b_j\| < \varepsilon$ if $j \ge \frac{n_0}{2}$. Now let $n \ge n_0$ and $i + j = n$. Then $i \ge \frac{n_0}{2}$ or $j \ge \frac{n_0}{2}$. It follows that $\|\sum_{i+j=n} a_i b_j\| \le \max(\|a_i\| \cdot \|b_j\|)_{i+j=n} \le \varepsilon \cdot \max(\|f\|, \|g\|)$. Thus, $fg = \sum_{n=0}^{\infty} \sum_{i+j=n} a_i b_j x^n$ belongs to $A\{x\}$, as claimed.

Proof of (b): Standard checking.

Proof of (c): Let $f_i = \sum_{n=0}^{\infty} a_{in} x^n$, $i = 1, 2, 3, \dots$, be a Cauchy sequence in $A\{x\}$. For each $\varepsilon > 0$ there exists i_0 such that $\|a_{in} - a_{jn}\| \le \|f_i - f_j\| < \varepsilon$ for all $i, j \ge i_0$ and for all n. Thus, for each n, the sequence $a_{1n}, a_{2n}, a_{3n}, \dots$ is Cauchy, hence converges to an element $a_n \in A$. If we let j tend to infinity in the latter inequality, we get that $\|a_{in} - a_n\| < \varepsilon$ for all $i \ge i_0$ and all n. Set $f = \sum_{n=0}^{\infty} a_n x^n$. Then $a_n \to 0$ and $\|f_i - f\| = \max(\|a_{in} - a_n\|)_{n=0,1,2,\dots} < \varepsilon$ if $i \ge i_0$. Consequently, the f_i's converge in $A\{x\}$.

Proof of (d): Note that $\|a_n b^n\| \le \|a_n\| \to 0$, so $\sum_{n=0}^{\infty} a_n b^n$ is an element of B. $\qquad\square$

Definition 2.2.2: Let $f = \sum_{n=0}^{\infty} a_n x^n$ be a nonzero element of $A\{x\}$. We define the **pseudo degree** of f to be the integer $d = \max\{n \geq 0 \,|\, \|a_n\| = \|f\|\}$ and set pseudo.deg$(f) = d$. The element a_d is the **pseudo leading coefficient** of f. Thus, $\|a_d\| = \|f\|$ and $\|a_n\| < \|f\|$ for each $n > d$. If $f \in A[x]$ is a polynomial, then pseudo.deg$(f) \leq \deg(f)$. If a_d is invertible in A and satisfies $\|ca_d\| = \|c\| \cdot \|a_d\|$ for all $c \in A$, we call f **regular**. In particular, if A is a field and $\|\ \|$ is an ultrametric absolute value, then each $0 \neq f \in A\{x\}$ is regular. The next lemma implies that in this case $\|\ \|$ is an absolute value of $A\{x\}$. □

LEMMA 2.2.3 (Gauss' Lemma): *Let $f, g \in A\{x\}$. Suppose f is regular of pseudo degree d and $f, g \neq 0$. Then $\|fg\| = \|f\| \cdot \|g\|$ and pseudo.deg$(fg) =$ pseudo.deg$(f) +$ pseudo.deg(g).*

Proof: Let $f = \sum_{i=0}^{\infty} a_i x^i$ and $g = \sum_{j=0}^{\infty} b_j x^j$. Let a_d (resp. b_e) be the pseudo leading coefficient of f (resp. g). Then $fg = \sum_{n=0}^{\infty} c_n x^n$ with $c_n = \sum_{i+j=n} a_i b_j$.

If $i + j = d + e$ and $(i, j) \neq (d, e)$, then either $i > d$ or $j > e$. In each case, $\|a_i b_j\| \leq \|a_i\| \|b_j\| < \|f\| \cdot \|g\|$. By our assumption on a_d, we have $\|a_d b_e\| = \|a_d\| \cdot \|b_e\| = \|f\| \cdot \|g\|$. By Lemma 2.1.3(b), this implies $\|c_{d+e}\| = \|f\| \cdot \|g\|$.

If $i+j > d+e$, then either $i > d$ and $\|a_i\| < \|f\|$ or $j > e$ and $\|b_j\| < \|g\|$. In each case $\|a_i b_j\| \leq \|a_i\| \cdot \|b_j\| < \|f\| \cdot \|g\|$. Hence, $\|c_n\| < \|c_{d+e}\|$ for each $n > d + e$. Therefore, c_{d+e} is the pseudo leading coefficient of fg, and the lemma is proved. □

PROPOSITION 2.2.4 (Weierstrass division theorem): *Let $f \in A\{x\}$ and let $g \in A\{x\}$ be regular of pseudo degree d. Then there are unique $q \in A\{x\}$ and $r \in A[x]$ such that $f = qg + r$ and $\deg(r) < d$. Moreover,*

$$(1) \qquad \|qg\| = \|q\| \cdot \|g\| \leq \|f\| \qquad and \qquad \|r\| \leq \|f\|$$

Proof: We break the proof into several parts.

PART A: *Proof of (1).* First we assume that there exist $q \in A\{x\}$ and $r \in A[x]$ such that $f = qg + r$ with $\deg(r) < d$. If $q = 0$, then (1) is clear. Otherwise, $q \neq 0$ and we let $e =$ pseudo.deg(q). By Lemma 2.2.3, $\|qg\| = \|q\| \cdot \|g\|$ and pseudo.deg$(qg) = e+d > \deg(r)$. Hence, the coefficient c_{d+e} of x^{d+e} in qg is also the coefficient of x^{d+e} in f. It follows that $\|qg\| = \|c_{d+e}\| \leq \|f\|$. Consequently, $\|r\| = \|f - qg\| \leq \|f\|$.

PART B: *Uniqueness.* Suppose $f = qg + r = q'g + r'$, where $q, q' \in A\{x\}$ and $r, r' \in A[x]$ are of degrees less than d. Then $0 = (q - q')g + (r - r')$. By Part A, applied to 0 rather than to f, $\|q - q'\| \cdot \|g\| = \|r - r'\| = 0$. Hence, $q = q'$ and $r = r'$.

PART C: *Existence if g is a polynomial of degree d.* Write $f = \sum_{n=0}^{\infty} b_n x^n$ with $b_n \in A$ converging to 0. For each $m \geq 0$ let $f_m = \sum_{n=0}^{m} b_n x^n \in$

$A[x]$. Then the $f_1, f_2, f_3, \ldots$ converge to f, in particular they form a Cauchy sequence. Since g is regular of pseudo degree d, its leading coefficient is invertible. Euclid's algorithm for polynomials over A produces $q_m, r_m \in A[x]$ with $f_m = q_m g + r_m$ and $\deg(r_m) < \deg(g)$. Thus, for all k, m we have $f_m - f_k = (q_m - q_k)g + (r_m - r_k)$. By Part A, $\|q_m - q_k\| \cdot \|g\|, \|r_m - r_k\| \leq \|f_m - f_k\|$. Thus, $\{q_m\}_{m=0}^{\infty}$ and $\{r_m\}_{m=0}^{\infty}$ are Cauchy sequences in $A\{x\}$. Since $A\{x\}$ is complete (Lemma 2.2.1), the q_m's converge to some $q \in A\{x\}$. Since A is complete, the r_m's converge to an $r \in A[x]$ of degree less than d. It follows that $f = qg + r$

PART D: *Existence for arbitrary g.* Let $g = \sum_{n=0}^{\infty} a_n x^n$ and set $g_0 = \sum_{n=0}^{d} a_n x^n \in A[x]$. Then $\|g - g_0\| < \|g\|$. By Part C, there are $q_0 \in A\{x\}$ and $r_0 \in A[x]$ such that $f = q_0 g_0 + r_0$ and $\deg(r_0) < d$. By Part A, $\|q_0\| \leq \frac{\|f\|}{\|g\|}$ and $\|r_0\| \leq \|f\|$. Thus, $f = q_0 g + r_0 + f_1$, where $f_1 = -q_0(g - g_0)$, and $\|f_1\| \leq \frac{\|g - g_0\|}{\|g\|} \cdot \|f\|$.

Set $f_0 = f$. By induction we get, for each $k \geq 0$, elements $f_k, q_k \in A\{x\}$ and $r_k \in A[x]$ such that $\deg(r_k) < d$ and

$$f_k = q_k g + r_k + f_{k+1}, \quad \|q_k\| \leq \frac{\|f_k\|}{\|g\|}, \quad \|r_k\| \leq \|f_k\|, \quad \text{and}$$

$$\|f_{k+1}\| \leq \frac{\|g - g_0\|}{\|g\|} \|f_k\|.$$

It follows that $\|f_k\| \leq \left(\frac{\|g - g_0\|}{\|g\|}\right)^k \|f\|$, so $\|f_k\| \to 0$. Hence, also $\|q_k\|, \|r_k\| \to 0$. Therefore, $q = \sum_{k=0}^{\infty} q_k \in A\{x\}$ and $r = \sum_{k=0}^{\infty} r_k \in A[x]$. By construction, $f = \sum_{n=0}^{k} q_n g + \sum_{n=0}^{k} r_n + f_{k+1}$ for each k. Taking k to infinity, we get $f = qg + r$ and $\deg(r) < d$. $\qquad\square$

COROLLARY 2.2.5 (Weierstrass preparation theorem): *Let $f \in A\{x\}$ be regular of pseudo degree d. Then $f = qg$, where q is a unit of $A\{x\}$ and $g \in A[x]$ is a monic polynomial of degree d with $\|g\| = 1$. Moreover, q and g are uniquely determined by these conditions.*

Proof: By Proposition 2.2.4 there are $q' \in A\{x\}$ and $r' \in A[x]$ of degree $< d$ such that $x^d = q'f + r'$ and $\|r'\| \leq \|x^d\| = 1$. Set $g = x^d - r'$. Then g is monic of degree d, $g = q'f$, and $\|g\| = 1$. It remains to show that $q' \in A\{x\}^{\times}$.

Note that g is regular of pseudo degree d. By Proposition 2.2.4, there are $q \in A\{x\}$ and $r \in A[x]$ such that $f = qg + r$ and $\deg(r) < d$. Thus, $f = qq'f + r$. Since $f = 1 \cdot f + 0$, the uniqueness part of Proposition 2.2.4 implies that $qq' = 1$. Hence, $q' \in A\{x\}^{\times}$.

Finally suppose $f = q_1 g_1$, where $q \in A\{x\}^{\times}$ and $g_1 \in A[x]$ is monic of degree d with $\|g_1\| = 1$. Then $g_1 = (q_1^{-1} q_2)g$ and $g_1 = 1 \cdot g + (g_1 - g)$, where $g_1 = g$ is a polynomial of degree at most $d - 1$. By the uniqueness part of Proposition 2.2.4, $q_1^{-1} q_2 = 1$, so $q_1 = q_2$ and $g_1 = g$. $\qquad\square$

COROLLARY 2.2.6: Let $f = \sum_{n=0}^{\infty} a_n x^n$ be a regular element of $A\{x\}$ such that $\|a_0 b\| = \|a_0\| \cdot \|b\|$ for each $b \in A$. Then $f \in A\{x\}^{\times}$ if and only if pseudo.deg$(f) = 0$ and $a_0 \in A^{\times}$.

Proof: If there exists $g \in \sum_{n=0}^{\infty} b_n x^n$ in $A\{x\}$ such that $fg = 1$, then pseudo.deg(f) + pseudo.deg$(g) = 0$ (Lemma 2.2.3 applied to 1 rather than to f), so pseudo.deg$(f) = 0$. In addition, $a_0 b_0 = 1$, so $a_0 \in A^{\times}$.

Conversely, suppose pseudo.deg$(f) = 0$ and $a_0 \in A^{\times}$. Then f is regular. Hence, by Corollary 2.2.5, $f = q \cdot 1$ where $q \in A\{x\}^{\times}$.

Alternatively, $a_0^{-1} f = 1 - h$, where $h = -\sum_{n=1}^{\infty} a_0^{-1} a_n x^n$. By our assumption on a_0, we have $\|a_0^{-1}\| \cdot \|a_0\| = \|a_0^{-1} a_0\| = 1$, so $\|a_0^{-1}\| = \|a_0\|^{-1}$. Since pseudo.deg$(f) = 0$, we have $\|a_0\| < \|a_n\|$, so $\|a_0^{-1} a_n\| \leq \|a_0\|^{-1} \|a_n\| < 1$ for each $n \geq 1$. It follows that $\|h\| = \max(\|a_0^{-1} a_n\|)_{n=1,2,3,\dots} < 1$. By Lemma 2.1.3(f), $a_0^{-1} f \in A\{x\}^{\times}$, so $f \in A\{x\}^{\times}$.   □

2.3 Properties of the Ring $K\{x\}$

We turn our attention in this section to the case where the ring A of the previous sections is a complete field K under an ultrametric absolute value $|\ |$ and $O = \{a \in K \mid |a| \leq 1\}$ its **valuation ring**. We fix K and O for the whole section and prove that $K\{x\}$ is a principal ideal domain and that $F = \mathrm{Quot}(K\{x\})$ is a Hilbertian field.

Note that in our case $|ab| = |a| \cdot |b|$ for all $a, b \in K$ and each nonzero element of K is invertible. Hence, each nonzero $f \in K\{x\}$ is regular. It follows from Lemma 2.2.3 that the norm of $K\{x\}$ is multiplicative, hence it is an absolute value which we denote by $|\ |$ rather than by $\|\ \|$.

PROPOSITION 2.3.1:
(a) $K\{x\}$ is a principal ideal domain. Moreover, each ideal in $K\{x\}$ is generated by an element of $O[x]$.
(b) $K\{x\}$ a unique factorization domain.
(c) A nonzero element $f \in K\{x\}$ is invertible if and only if pseudo.deg$(f) = 0$.
(d) pseudo.deg(fg) = pseudo.deg(f) + pseudo.deg(g) for all $f, g \in K\{x\}$ with $f, g \neq 0$.
(e) Every prime element f of $K\{x\}$ can be written as $f = ug$, where u is invertible in $K\{x\}$ and g is an irreducible element of $K[x]$.
(f) If a $g \in K[x]$ is monic of degree d, irreducible in $K[x]$, and $|g| = 1$, then g is irreducible in $K\{x\}$.
(g) There are irreducible polynomials in $K[x]$ that are not irreducible in $K\{x\}$.
(h) There are reducible polynomials in $K[x]$ that are irreducible in $K\{x\}$.

Proof of (a): By the Weierstrass preparation theorem (Corollary 2.2.5) (applied to K rather than to A) each nonzero ideal $\mathfrak{a}$ of $K\{x\}$ is generated by the ideal $\mathfrak{a} \cap K[x]$ of $K[x]$. Since $K[x]$ is a principal ideal domain, $\mathfrak{a} \cap K[x] = f K[x]$

for some nonzero $f \in K[x]$. Consequently, $\mathfrak{a} = fK\{x\}$ is a principal ideal. Moreover, dividing f by one of its coefficients with highest absolute value, we may assume that $f \in O[x]$.

Proof of (b): Since every principal ideal domain has a unique factorization, (b) is a consequence of (a).

Proof of (c): Apply Corollary 2.2.6.

Proof of (d): Apply Lemma 2.2.3.

Proof of (e): By (a), $f = u_1 f_1$ with $u_1 \in K\{x\}^\times$ and $f_1 \in K[x]$. Write $f_1 = g_1 \cdots g_n$ with irreducible polynomials $g_1, \ldots, g_n \in K[x]$. Then $f = u_1 g_1 \cdots g_n$. Since f is irreducible in $K\{x\}$, one of the g_i's, say g_n is irreducible in $K\{x\}$ and all the others, that is $g_1, \ldots, g_{n-1}$, are invertible in $K\{x\}$. Set $u = u_1 g_1 \cdots g_{n-1}$ and $g = g_n$. Then $f = ug$ is the desired presentation.

Proof of (f): The irreducibility of g in $K[x]$ implies that $d > 0$. Our assumptions imply that pseudo.deg$(g) = d$. Hence, by Corollary 2.2.6, $g \notin K\{x\}^\times$.

Now assume $g = g_1 g_2$, where $g_1, g_2 \in K\{x\}$ are nonunits. By Corollary 2.2.5, we may assume that $g_1 \in K[x]$ is monic, say of degree d_1, and $|g_1| = 1$. Thus pseudo.deg$(g_1) = d_1$. By Euclid's algorithm, there are $q, r \in K[x]$ such that $g = qg_1 + r$ and deg$(r) < d_1$. Applying the additional presentation $g = g_2 g_1 + 0$ and the uniqueness part of Proposition 2.2.4, we get that $g_2 = q \in K[x]$. Thus, either $g_1 \in K[x]^\times \subseteq K\{x\}^\times$ or $g_1 \in K[x]^\times \subseteq K\{x\}^\times$. In both cases we get a contradiction.

Proof of (g): Let a be an element of K with $|a| < 1$. Then $ax - 1$ is irreducible in $K[x]$. On the other hand, pseudo.deg$(ax - 1) = 0$, so, by (c), $ax - 1 \in K\{x\}^\times$. In particular, $ax - 1$ is not irreducible in $K\{x\}$.

Proof of (h): We choose a as in the proof of (f) and consider the reducible polynomial $f(x) = (ax-1)(x-1)$. By the proof of (f), $ax-1 \in K\{x\}^\times$. Next we note that pseudo.deg$(x - 1) = 1$, so by (d) and (c), $x - 1$ is irreducible in $K\{x\}$. Consequently, $f(x)$ is irreducible in $K\{x\}$. $\square$

Let $E = K(x)$ be the field of rational functions over K in the variable x. Then $K[x] \subseteq K\{x\}$ and the restriction of $|\ |$ to $K[x]$ is an absolute value. By the multiplicativity of $|\ |$, it extends to an absolute value of E. Let $\hat{E}$ be the completion of E with respect to $|\ |$ [CaF67, p. 47]. For each $\sum_{n=0}^\infty a_n x^n \in K\{x\}$ we have, by definition, $a_n \to 0$, hence $\sum_{n=0}^\infty a_n x^n = \lim_{n \to \infty} \sum_{i=0}^n a_i x^i$. Thus, $K[x]$ is dense in $K\{x\}$. Since $K\{x\}$ is complete (Lemma 2.2.1(c)), this implies that $K\{x\}$ is the closure of $K[x]$ in $\hat{E}$.

Remark 2.3.2:
    (a) $|x| = 1$.
    (b) Let $\bar{K} \subseteq \bar{E}$ be the residue fields of $K \subseteq E$ with respect to $|\ |$. Denote the image in $\bar{E}$ of an element $u \in K(x)$ with $|u| \leq 1$ by $\bar{u}$. Then $\bar{x}$ is transcendental over $\bar{K}$. Indeed, let h be a monic polynomial over $\bar{K}$. Choose

a monic polynomial p with coefficients in the valuation ring of K such that $\bar{p} = h$. Since $|p(x)| = 1$, we have $h(\bar{x}) = \bar{p}(\bar{x}) \neq 0$. It follows that $\bar{K}(\bar{x})$ is the field of rational functions over $\bar{K}$ in the variable $\bar{x}$ and $\bar{K}(\bar{x}) \subseteq \bar{E}$. Moreover, $\bar{K}(\bar{x}) = \bar{E}$. Indeed, let $u = \frac{f(x)}{g(x)}$ with $f = \sum_{i=0}^{m} a_i x^i$, $g = \sum_{j=0}^{n} b_j x^j \neq 0$, and $a_i, b_j \in K$ such that $|u| \leq 1$. Then $\max_i |a_i| \leq \max_j |b_j|$. Choose $c \in K$ with $|c| = \max_j |b_j|$. Then replace a_i with $c^{-1}a_i$ and b_j with $c^{-1}b_j$, if necessary, to assume that $|a_i|, |b_j| \leq 1$ for all i, j and there exists k with $|b_k| = 1$. Under these assumptions, $\bar{u} = \frac{\bar{f}(\bar{x})}{\bar{g}(\bar{x})} \in \bar{K}(\bar{x})$, as claimed.

(c) If $|\cdot|'$ is an absolute value of E which coincides with $|\ |$ on K and the residue x' of x with respect to $|\ |'$ is transcendental over $\bar{K}$, then $|\ |'$ coincides with $|\ |$.

Indeed, let $p(x) = \sum_{i=0}^{n} a_i x^i$ be a nonzero polynomial in $K[x]$. Choose a $c \in K^\times$ with $|c| = \max_i |a_i|$. Then $(c^{-1}p(x))' = \sum_{i=0}^{n}(c^{-1}a_i)'(x')^i \neq 0$ (the prime indicates the residue with respect to $|\ |'$), hence $|c^{-1}p(x)|' = 1$, so $|p(x)|' = |c| = |p(x)|$.

(d) It follows from (c) that if γ is an automorphism of E that leaves K invariant, preserves the absolute value of K, and $\overline{x^\gamma}$ is transcendental over $\bar{K}$, then γ preserves the absolute value of E.

In particular, γ is $|\ |$-continuous. Moreover, if $(x_1, x_2, x_3, \ldots)$ is a $|\ |$-Cauchy sequence in E, then so is $(x_1^\gamma, x_2^\gamma, x_3^\gamma, \ldots)$. Hence γ extends uniquely to a continuous automorphism of the $|\ |$-completion $\hat{E}$ of E.

(e) Now suppose K is a finite Galois extension of a complete field K_0 with respect to $|\ |$ and set $E_0 = K_0(x)$. Let $\gamma \in \mathrm{Gal}(K/K_0)$ and extend γ in the unique possible way to an element $\gamma \in \mathrm{Gal}(E/E_0)$. Then γ preserves $|\ |$ on K. Indeed, $|z|' = |z^\gamma|$ is an absolute valued of K. Since K_0 is complete with respect to $|\ |$, K_0 is Henselian, so $|\ |'$ is equivalent to $|\ |$. Thus, there exists $\varepsilon > 0$ with $|z^\gamma| = |z|^\varepsilon$ for each $z \in K$. In particular, $|z| = |z|^\varepsilon$ for each $z \in K_0$, so $\varepsilon = 1$, as claimed. In addition $x^\gamma = x$. By (d), γ preserves $|\ |$ also on E.

(f) Under the assumptions of (e) we let $\hat{E}_0$ and $\hat{E}$ be the $|\ |$-completions of E and E_0, respectively. Then $\hat{E}_0 E$ is a finite separable extension of $\hat{E}_0$ in $\hat{E}$. As such $\hat{E}_0 E$ is complete [CaF67, p. 57, Cor. 2] and contains E, so $\hat{E}_0 E = \hat{E}$. Thus, $\hat{E}/\hat{E}_0$ is a finite Galois extension.

By (d) and (e) each $\gamma \in \mathrm{Gal}(E/E_0)$ extends uniquely to a continuous automorphism γ of $\hat{E}$. Every $x \in \hat{E}_0$ is the limit of a sequence $(x_1, x_2, x_3, \ldots)$ of elements of E_0. Since $x_i^\gamma = x_i$ for each i, we have $x^\gamma = x$. It follows that $\mathrm{res} \colon \mathrm{Gal}(\hat{E}/\hat{E}_0) \to \mathrm{Gal}(E/E_0)$ is an isomorphism.

(g) Finally suppose $y = \frac{ax+b}{cx+d}$ with $a, b, c, d \in K$ such that $|a|, |b|, |c|, |d| \leq 1$ and $\bar{a}\bar{d} - \bar{b}\bar{c} \neq 0$. Then $\bar{a}\bar{x} + \bar{b}$ and $\bar{c}\bar{x} + \bar{d}$ are nonzero elements of $\bar{K}(\bar{x})$, so $\bar{y} = \frac{\bar{a}\bar{x}+\bar{b}}{\bar{c}\bar{x}+\bar{d}} \in \bar{K}(\bar{x})$. Moreover, $\bar{K}(\bar{x}) = \bar{K}(\bar{y})$, hence $\bar{y}$ is transcendental over $\bar{K}$. We conclude from (c) that the map $x \mapsto y$ extends to a K-automorphism of $K(x)$ that preserves the absolute value. It therefore extends to an isomorphism $\sum a_n x^n \to \sum a_n y^n$ of $K\{x\}$ onto $K\{y\}$. $\qquad\square$

In the following theorem we refer to an equivalence class of a valuation of a field F as a **prime** of F. For each prime $\mathfrak{p}$ we choose a valuation $v_\mathfrak{p}$ representing the prime and let $O_\mathfrak{p}$ be the corresponding valuation ring.

We say that an ultrametric absolute value $|\;|$ of a field K is **discrete**, if the group of all values $|a|$ with $a \in K^\times$ is isomorphic to $\mathbb{Z}$.

THEOREM 2.3.3: *Let K be a complete field with respect to a nontrivial ultrametric absolute value $|\;|$. Then $F = \mathrm{Quot}(K\{x\})$ is a Hilbertian field.*

Proof: Let $O = \{a \in K \mid |a| \le 1\}$ be the valuation ring of K with respect to $|\;|$ and let $D = O\{x\} = \{f \in K\{x\} \mid |f| \le 1\}$. Each $f \in K\{x\}$ can be written as af_1 with $a \in K$, $f_1 \in D$, and $|f_1| = 1$. Hence, $\mathrm{Quot}(D) = F$.

We construct a set S of prime divisors of F that satisfies the following conditions:

(1a) For each $\mathfrak{p} \in S$, $v_\mathfrak{p}$ is a real valuation (i.e. $v_\mathfrak{p}(F) \subseteq \mathbb{R}$).

(1b) The valuation ring $O_\mathfrak{p}$ of $v_\mathfrak{p}$ is the local ring of D at the prime ideal
$$\mathfrak{m}_\mathfrak{p} = \{f \in D \mid v_\mathfrak{p}(f) > 0\}.$$

(1c) $D = \bigcap_{\mathfrak{p} \in S} O_\mathfrak{p}$.

(1d) For each $f \in F^\times$ the set $\{\mathfrak{p} \in S \mid v_\mathfrak{p}(f) \ne 0\}$ is finite.

(1e) The Krull dimension of D is at least 2.

Then D is a **generalized Krull domain of dimension exceeding** 1. A theorem of Weissauer [FrJ05, Thm. 15.4.6] will then imply that F is Hilbertian.

THE CONSTRUCTION OF S: The absolute value $|\;|$ of $K\{x\}$ extends to an absolute value of F. The latter determines a prime $\mathfrak{M}$ of F with a real valuation $v_\mathfrak{M}$ (Section 2.1). Each $u \in F$ with $|u| \le 1$ can be written as $u = a\frac{f_1}{g_1}$ with $a \in O$ and $f_1, g_1 \in D$, $|f_1| = |g_1| = 1$. Hence, $O_\mathfrak{M} = D_\mathfrak{m}$, where $\mathfrak{m} = \{f \in D \mid |f| < 1\}$.

By Proposition 2.3.1, each nonzero prime ideal of $K\{x\}$ is generated by a prime element $p \in K\{x\}$. Divide p by its pseudo leading coefficient, if necessary, to assume that $|p| = 1$. Then let v_p be the discrete valuation of F determined by p and let $\mathfrak{p}_p$ be its equivalence class. We prove that p is a prime element of D. This will prove that pD is a prime ideal of D and its local ring will coincide with the valuation ring of v_p.

Indeed, let f, g be nonzero elements of D such that p divides fg in D. Write $f = af_1$, $g = bg_1$ with nonzero $a, b \in O$, $f_1, g_1 \in D$, $|f_1| = |g_1| = 1$. Then p divides $f_1 g_1$ in $K\{x\}$ and therefore it divides, say, f_1 in $K\{x\}$. Thus, there exists $q \in K\{x\}$ with $pq = f_1$. But then $|q| = 1$, so $q \in D$. Consequently, p divides f in D, as desired.

Let P be the set of all prime elements p as in the paragraph before the preceding one. Then $S = \{\mathfrak{p}_p \mid p \in P\} \cup \{\mathfrak{M}\}$ satisfies (1a) and (1b).

By Proposition 2.3.1(b), $K\{x\}$ is a unique factorization domain, hence $K\{x\} = \bigcap_{\mathfrak{p} \in P} O_p$, hence $\bigcap_{\mathfrak{p} \in S} O_\mathfrak{p} = \{f \in K\{x\} \mid |f| \le 1\} = D$. This settles (1c).

Next observe that for each $f \in F^\times$ there are only finitely many $p \in P$ such that $v_p(f) \neq 0$, so (1d) holds.

Finally note that if $f = \sum_{n=0}^\infty a_n x^n$ is in D, then $|a_n| \leq 1$ for all n and $|a_n| < 1$ for all large n. Hence $D/\mathfrak{m} \cong \bar{K}[\bar{x}]$, where $\bar{K}$ and $\bar{x}$ are as in Remark 2.3.2(b). Since $\bar{x}$ is transcendental over $\bar{K}$, $\mathfrak{m}$ is a nonzero prime ideal and $\mathfrak{m} + Ox$ is a prime ideal of D that properly contains $\mathfrak{m}$. This proves (1e) and concludes the proof of the theorem. $\square$

COROLLARY 2.3.4: $\mathrm{Quot}(K\{x\})$ is not a Henselian field.

Proof: Since $K\{x\}$ is Hilbertian (Theorem 2.3.3), $K\{x\}$ can not be Henselian [FrJ08, Lemma 15.5.4]. $\square$

2.4 Convergent Power Series

Let K be a complete field with respect to an ultrametric absolute value $|\ |$. We say that a formal power series $f = \sum_{n=m}^\infty a_n x^n$ in $K((x))$ **converges** at an element $c \in K$, if $f(c) = \sum_{n=m}^\infty a_n c^n$ converges, i.e. $a_n c^n \to 0$. In this case f converges at each $b \in K$ with $|b| \leq |c|$. For example, each $f \in K\{x\}$ converges at 1. We say that f **converges** if f converges at some $c \in K^\times$.

We denote the set of all convergent power series in $K((x))$ by $K((x))_0$ and prove that $K((x))_0$ is a field that contains $K\{x\}$ and is algebraically closed in $K((x))$.

LEMMA 2.4.1: A power series $f = \sum_{n=m}^\infty a_n x^n$ in $K((x))$ converges if and only if there exists a positive real number γ such that $|a_n| \leq \gamma^n$ for each $n \geq 0$.

Proof: First suppose f converges at $c \in K^\times$. Then $a_n c^n \to 0$, so there exists $n_0 \geq 1$ such that $|a_n c^n| \leq 1$ for each $n \geq n_0$. Choose

$$\gamma = \max\{|c|^{-1}, |a_k|^{1/k} \mid k = 0, \ldots, n_0 - 1\}.$$

Then $|a_n| \leq \gamma^n$ for each $n \geq 0$.

Conversely, suppose $\gamma > 0$ and $|a_n| \leq \gamma^n$ for all $n \geq 0$. Increase γ, if necessary, to assume that $\gamma > 1$. Then choose $c \in K^\times$ such that $|c| \leq \gamma^{-1.5}$ and observe that $|a_n c^n| \leq \gamma^{-0.5n}$ for each $n \geq 0$. Therefore, $a_n c^n \to 0$, hence f converges at c. $\square$

LEMMA 2.4.2: $K((x))_0$ is a field that contains $\mathrm{Quot}(K\{x\})$, hence also $K(x)$.

Proof: The only difficulty is to prove that if $f = 1 + \sum_{n=1}^\infty a_n x^n$ converges, then also $f^{-1} = 1 + \sum_{n=1}^\infty a_n' x^n$ converges.

Indeed, for $n \geq 1$, a_n' satisfies the recursive relation $a_n' = -a_n - \sum_{i=1}^{n-1} a_i a_{n-i}'$. By Lemma 2.4.1, there exists $\gamma > 1$ such that $|a_i| \leq \gamma^i$ for each $i \geq 1$. Set $a_0' = 1$. Suppose, by induction, that $|a_j'| \leq \gamma^j$ for $j = 1, \ldots, n-1$. Then $|a_n'| \leq \max_i(|a_i| \cdot |a_{n-i}'|) \leq \gamma^n$. Hence, f^{-1} converges. $\square$

Let v be the valuation of $K((x))$ defined by

$$v\left(\sum_{n=m}^{\infty} a_n x^n\right) = m \qquad \text{for } a_m, a_{m+1}, a_{m+2}, \ldots \in K \text{ with } a_m \neq 0.$$

It is discrete, complete, its valuation ring is $K[[x]]$, and $v(x) = 1$. The residue of an element $f = \sum_{n=0}^{\infty} a_n x^n$ of $K[[x]]$ at v is a_0, and we denote it by $\bar{f}$. We also consider the valuation ring $O = K[[x]] \cap K((x))_0$ of $K((x))_0$ and denote the restriction of v to $K((x))_0$ also by v. Since $K((x))_0$ contains $K(x)$, it is v-dense in $K((x))$. Finally, we also denote the unique extension of v to the algebraic closure of $K((x))$ by v.

Remark 2.4.3: $K((x))_0$ is not complete. Indeed, choose $a \in K$ such that $|a| > 1$. Then there exists no $\gamma > 0$ such that $|a^{n^2}| \leq \gamma^n$ for all $n \geq 1$. By Lemma 2.4.1, the power series $f = \sum_{n=0}^{\infty} a^{n^2} x^n$ does not belong to $K((x))_0$. Therefore, the valued field $(K((x))_0, v)$ is not complete.   □

LEMMA 2.4.4: *The field $K((x))_0$ is separably algebraically closed in $K((x))$.*

Proof: Let $y = \sum_{n=m}^{\infty} a_n x^n$, with $a_n \in K$, be an element of $K((x))$ which is separably algebraic of degree d over $K((x))_0$. We have to prove that $y \in K((x))_0$.

PART A: *A shift of y.* Assume that $d > 1$ and let $y_1, \ldots, y_d$, with $y = y_1$, be the (distinct) conjugates of y over $K((x))_0$. In particular $r = \max(v(y - y_i) \mid i = 2, \ldots, d)$ is an integer. Choose $s \geq r + 1$ and let

$$y_i' = \frac{1}{x^s}\left(y_i - \sum_{n=m}^{s} a_n x^n\right), \qquad i = 1, \ldots, d.$$

Then $y_1', \ldots, y_d'$ are the distinct conjugates of y_1' over $K((x))_0$. Also, $v(y_1') \geq 1$ and $y_i' = \frac{1}{x^s}(y_i - y) + y_1'$, so $v(y_i') \leq -1$, $i = 2, \ldots, d$. If y_1' belongs to $K((x))_0$, then so does y, and conversely. Therefore, we replace y_i by y_i', if necessary, to assume that

(1) $v(y) \geq 1$ and $v(y_i) \leq -1$, $i = 2, \ldots, d$.

In particular $y = \sum_{n=0}^{\infty} a_n x^n$ with $a_0 = 0$. The elements $y_1, \ldots, y_d$ are the roots of an irreducible separable polynomial

$$h(Y) = p_d Y^d + p_{d-1} Y^{d-1} + \cdots + p_1 Y + p_0$$

with coefficients $p_i \in O$. Let $e = \min(v(p_0), \ldots, v(p_d))$. Divide the p_i, if necessary, by x^e, to assume that $v(p_i) \geq 0$ for each i between 0 and d and that $v(p_j) = 0$ for at least one j between 0 and d.

PART B: *We prove that* $v(p_0), v(p_d) > 0$, $v(p_k) > v(p_1)$ *if* $2 \leq k \leq d-1$ *and* $v(p_1) = 0$. Indeed, since $v(y) > 0$ and $h(y) = 0$, we have $v(p_0) > 0$. Since $v(y_2) < 0$ and $h(y_2) = 0$, we have $v(p_d) > 0$. Next observe that

$$\frac{p_1}{p_d} = \pm y_2 \cdots y_d \pm \sum_{i=2}^{d} \frac{y_1 \cdots y_d}{y_i}.$$

If $2 \leq i \leq d$, then $v(y_i) < v(y_1)$, so $v(y_2 \cdots y_d) < v(\frac{y_1}{y_i}) + v(y_2 \cdots y_d) = v(\frac{y_1 \cdots y_d}{y_i})$. Hence,

(2) $$v\left(\frac{p_1}{p_d}\right) = v(y_2 \cdots y_d).$$

For k between 1 and $d - 2$ we have

(3) $$\frac{p_{d-k}}{p_d} = \pm \sum_{\sigma} \prod_{i=1}^{k} y_{\sigma(i)},$$

where σ ranges over all monotonically increasing maps from $\{1, \ldots, k\}$ to $\{1, \ldots, d\}$. If $\sigma(1) \neq 1$, then $\{y_{\sigma(1)}, \ldots, y_{\sigma(k)}\}$ is properly contained in $\{y_2, \ldots, y_d\}$. Hence, $v(\prod_{i=1}^{k} y_{\sigma(i)}) > v(y_2 \cdots y_d)$. If $\sigma(1) = 1$, then

$$v\left(\prod_{i=1}^{k} y_{\sigma(i)}\right) > v\left(\prod_{i=2}^{k} y_{\sigma(i)}\right) > v(y_2 \cdots y_d).$$

Hence, by (2) and (3), $v(\frac{p_{d-k}}{p_d}) > v(\frac{p_1}{p_d})$, so $v(p_{d-k}) > v(p_1)$. Since $v(p_j) = 0$ for some j between 0 and d, since $v(p_i) \geq 0$ for every i between 0 and d, and since $v(p_0), v(p_d) > 0$, we conclude that $v(p_1) = 0$ and $v(p_i) > 0$ for all $i \neq 1$. Therefore,

(4) $$p_k = \sum_{n=0}^{\infty} b_{kn} x^n, \qquad k = 0, \ldots, d$$

with $b_{kn} \in K$ such that $b_{1,0} \neq 0$ and $b_{k,0} = 0$ for each $k \neq 1$. In particular, $|b_{1,0}| \neq 0$ but unfortunately, $|b_{1,0}|$ may be smaller than 1.

PART C: *Making* $|b_{1,0}|$ *large.* We choose $c \in K$ such that $|c^{d-1}b_{1,0}| \geq 1$ and let $z = cy$. Then z is a zero of the polynomial $g(Z) = p_d Z^d + cp_{d-1} Z^{d-1} + \cdots + c^{d-1} p_1 Z + c^d p_0$ with coefficients in O. Relation (4) remains valid except that the zero term of the coefficient of Z in g becomes $c^{d-1} b_{1,0}$. By the choice of c, its absolute value is at least 1. So, without loss, we may assume that

(5) $$|b_{1,0}| \geq 1.$$

PART D: *An estimate for* $|a_n|$. By Lemma 2.4.1, there exists $\gamma > 0$ such that $|b_{kn}| \leq \gamma^n$ for all $0 \leq k \leq d$ and $n \geq 1$. By induction we prove that $|a_n| \leq \gamma^n$ for each $n \geq 0$. This will prove that $y \in O$ and will conclude the proof of the lemma.

Indeed, $|a_0| = 0 < 1 = \gamma^0$. Now assume that $|a_m| \leq \gamma^m$ for each $0 \leq m \leq n-1$. For each k between 0 and d we have that $p_k y^k = \sum_{n=0}^{\infty} c_{kn} x^n$, where

$$c_{kn} = \sum_{\sigma \in S_{kn}} b_{k,\sigma(0)} \prod_{j=1}^{k} a_{\sigma(j)},$$

and

$$S_{kn} = \{\sigma \colon \{0,\ldots,k\} \to \{0,\ldots,n\} \mid \sum_{j=0}^{k} \sigma(j) = n\}.$$

It follows that

(6) $c_{0n} = b_{0n}$ and $c_{1n} = b_{1,0} a_n + b_{11} a_{n-1} + \cdots + b_{1,n-1} a_1$.

For $k \geq 2$ we have $b_{k,0} = 0$. Hence, if a term $b_{k,\sigma(0)} \prod_{j=1}^{k} a_{\sigma(j)}$ in c_{kn} contains a_n, then $\sigma(0) = 0$, so $b_{k,\sigma(0)} = 0$. Thus,

(7) $c_{kn} = $ sum of products of the form $b_{k,\sigma(0)} \prod_{j=1}^{k} a_{\sigma(j)}$,

$$\text{with } \sigma(j) < n, \ j = 1, \ldots, k.$$

From the relation $\sum_{k=0}^{d} p_k y^k = h(y) = 0$ we conclude that $\sum_{k=0}^{d} c_{kn} = 0$ for all n. Hence, by (6),

$$b_{1,0} a_n = -b_{0n} - b_{11} a_{n-1} - \cdots - b_{1,n-1} a_1 - c_{2n} - \cdots - c_{dn}.$$

Therefore, by (7),

(8) $b_{1,0} a_n = $ sum of products of the form $- b_{k,\sigma(0)} \prod_{j=1}^{k} a_{\sigma(j)}$,

$$\text{with } \sigma \in S_{kn}, \ 0 \leq k \leq d, \text{ and } \sigma(j) < n, \ j = 1, \ldots, k.$$

Note that $b_{k,0} = 0$ for each $k \neq 1$ (by (4)), while $b_{1,0}$ does not occur on the right hand side of (8). Hence, for a summand in the right hand side of (8) indexed by σ we have

$$|b_{k,\sigma(0)} \prod_{j=1}^{k} a_{\sigma(j)}| \leq \gamma^{\sum_{j=0}^{k} \sigma(j)} = \gamma^n.$$

We conclude from $|b_{1,0}| \geq 1$ that $|a_n| \leq \gamma^n$, as contended. $\square$

PROPOSITION 2.4.5: *The field $K((x))_0$ is algebraically closed in $K((x))$. Thus, each $f \in K((x))$ which is algebraic over $K(x)$ converges at some $c \in K^\times$. Moreover, there exists a positive integer m such that f converges at each $b \in K^\times$ with $|b| \leq \frac{1}{m}$.*

Proof: In view of Lemma 2.4.4, we have to prove the proposition only for char$(K) > 0$. Let $f = \sum_{n=m}^{\infty} a_n x^n \in K((x))$ be algebraic over $K((x))_0$. Then $K((x))_0(f)$ is a purely inseparable extension of a separable algebraic extension of $K((x))_0$. By Lemma 2.4.4, the latter coincides with $K((x))_0$. Hence, $K((x))_0(f)$ is a purely inseparable extension of $K((x))_0$.

Thus, there exists a power q of char(K) such that $\sum_{n=m}^{\infty} a_n^q x^{nq} = f^q \in K((x))_0$. By Lemma 2.4.1, there exists $\gamma > 0$ such that $|a_n^q| \leq \gamma^{nq}$ for all $n \geq 1$. It follows that $|a_n| \leq \gamma^n$ for all $n \geq 1$. By Lemma 2.4.1, $f \in K((x))_0$, so there exists $c \in K^\times$ such that f converges at c. If $\frac{1}{m} \leq |c|$, then f converges at each $b \in K^\times$ with $|b| \leq \frac{1}{m}$. $\qquad\square$

COROLLARY 2.4.6: *The valued field $(K((x))_0, v)$ is Henselian.*

Proof: Consider the valuation ring $O = K[[x]] \cap K((x))_0$ of $K((x))_0$ at v. Let $f \in O[X]$ be a monic polynomial and $a \in O$ such that $v(f(a)) > 0$ and $v(f'(a)) \neq 0$. Since $(K((x)), v)$ is Henselian, there exists $z \in K[[x]]$ such that $f(z) = 0$ and $v(z - a) > 0$. By Proposition 2.4.5, $z \in K((x))_0$, hence $z \in O$. It follows that $(K((x))_0, v)$ is Henselian. $\qquad\square$

2.5 The Regularity of $K((x))/K((x))_0$

Let K be a complete field with respect to an ultrametric absolute value $|\ |$. We extend $|\ |$ in the unique possible way to $\tilde{K}$. We also consider the discrete valuation v of $K(x)/K$ defined by $v(a) = 0$ for each $a \in K^\times$ and $v(x) = 1$. Then $K((x))$ is the completion of $K(x)$ at v. Let $K((x))_0$ be the subfield of $K((x))$ of all convergent power series.

Proposition 2.4.5 states that $K((x))_0$ is algebraically closed in $K((x))$. In this section we prove that $K((x))$ is even a regular extension of $K((x))_0$. To do this, we only have to assume that $p = \text{char}(K) > 0$ and prove that $K((x))/K((x))_0$ is a separable extension. In other words, we have to prove that $K((x))$ is linearly disjoint from $K((x))_0^{1/p}$ over $K((x))_0$. We do that in several steps.

LEMMA 2.5.1: *The fields $K((x))$ and $K((x^{1/p}))_0$ are linearly disjoint over $K((x))_0$.*

Proof: First note that $1, x^{1/p}, \ldots, x^{p-1/p}$ is a basis for $K(x^{1/p})$ over $K(x)$. Then $1, x^{1/p}, \ldots, x^{p-1/p}$ have distinct v-values modulo $\mathbb{Z} = v(K((x)))$, so they are linearly independent over $K((x))$.

Next we observe that $1, x^{1/p}, \ldots, x^{p-1/p}$ also generate $K((x^{1/p}))$ over $K((x))$. Indeed, each $f \in K((x^{1/p}))$ may be multiplied by an appropriate power of x to be presented as

$$(1) \qquad f = \sum_{n=0}^{\infty} a_n x^{n/p},$$

with $a_0, a_1, a_2, \ldots \in K$. We write each n as $n = kp + l$ with integers $k \geq 0$ and $0 \leq l \leq p - 1$ and rewrite f as

$$(2) \qquad f = \sum_{l=0}^{p-1} \Big(\sum_{k=0}^{\infty} a_{kp+l} x^k \Big) x^{l/p}.$$

If $f \in K((x^{1/p}))_0$, then there exists $b \in K^{\times}$ such that $\sum_{n=0}^{\infty} a_n b^{n/p}$ converges in K, hence $a_n b^{n/p} \to 0$ as $n \to \infty$, so $a_{kp+l} b^k b^{l/p} \to 0$ as $k \to \infty$ for each l. Therefore, for each l, we have $a_{kp+l} b^k \to 0$ as $k \to \infty$, hence $\sum_{k=0}^{\infty} a_{kp+l} x^k$ converges, so belongs to $K((x))_0$.

It follows that $1, x^{1/p}, \ldots, x^{p-1/p}$ form a basis for $K((x^{1/p}))_0 / K((x))_0$ as well as for $K((x^{1/p})) / K((x))$. Consequently, $K((x))$ is linearly disjoint from $K((x^{1/p}))_0$ over $K((x))_0$. $\qquad\square$

We set $K[[x]]_0 = K[[x]] \cap K((x))_0$.

LEMMA 2.5.2: Let $u_1, \ldots, u_m \in \tilde{K}[[x]]_0$ and $f_1, \ldots, f_m \in K[[x]]$. Set $u_{i0} = u_i(0)$ for $i = 1, \ldots, m$ and

$$(3) \qquad f = \sum_{i=1}^{m} f_i u_i.$$

Suppose $u_{10}, \ldots, u_{m0}$ are linearly independent over K, $f \in \tilde{K}[[x]]_0$, and $f(0) = 0$. Then $f_1, \ldots, f_m \in K[[x]]_0$.

Proof: We break up the proof into several parts.

PART A: Comparison of norms. We consider the K-vector space $V = \sum_{i=1}^{m} K u_{i0}$ and define a function $\mu \colon V \to \mathbb{R}$ by

$$(4) \qquad \mu\Big(\sum_{i=1}^{m} a_i u_{i0} \Big) = \max(|a_1|, \ldots, |a_m|).$$

It satisfies the following rules:
(5a) $\mu(v) > 0$ for each nonzero $v \in V$.
(5b) $\mu(v + v') \leq \max(\mu(v), \mu(v'))$ for all $v, v' \in V$.
(5c) $\mu(av) = |a| \mu(v)$ for all $a \in K$ and $v \in V$.

Thus, v is a **norm** of V. On the other hand, $|\ |$ extends to an absolute value of $\tilde{K}$ and its restriction to V is another norm of V. Since K is complete under $|\ |$, there exists a positive real number s such that
(6) $\mu(v) \leq s|v|$ for all $v \in V$
[CaF67, p. 52, Lemma].

PART B: *Power series.* For each i we write $u_i = u_{i0} + u'_i$ where $u'_i \in \tilde{K}[[x]]_0$ and $u'_i(0) = 0$. Then

$$(7a) \qquad f = \sum_{n=1}^{\infty} a_n x^n \quad \text{with } a_1, a_2, \ldots \in \tilde{K},$$

$$(7b) \qquad u'_i = \sum_{n=1}^{\infty} b_{in} x^n \quad \text{with } b_{i1}, b_{i2}, \ldots \in \tilde{K}, \text{ and}$$

$$(7c) \qquad f_i = \sum_{n=0}^{\infty} a_{in} x^n \quad \text{with } a_{i0}, a_{i1}, a_{i2}, \ldots \in K.$$

If a power series converges at a certain element of $\tilde{K}^\times$, it converges at each element with a smaller absolute value. Since to each element of $\tilde{K}^\times$ there exists an element of $K^\times$ with a smaller absolute value, there exists $d \in K^\times$ such that $\sum_{n=1}^{\infty} a_n d^n$ and $\sum_{n=1}^{\infty} b_{in} d^n$, $i = 1, \ldots, m$, converge. In particular, the numbers $|a_n d^n|$ and $|b_{in} d^n|$ are bounded. It follows from the identities $|a_n c^n| = |a_n d^n| \cdot \left|\frac{c}{d}\right|^n$ and $|b_{in} c^n| = |b_{in} d^n| \cdot \left|\frac{c}{d}\right|^n$ that there exists $c \in K^\times$ such that

$$(8) \qquad \max_{n \geq 1} |a_n c^n| \leq s^{-1} \quad \text{and} \quad \max_{n \geq 1} |b_{in} c^n| \leq s^{-1}$$

for $i = 1, \ldots, m$.

PART C: *Claim:* $|a_{in} c^n| \leq 1$ for $i = 1, \ldots, m$ and $n = 0, 1, 2, \ldots$. To prove the claim we substitute the presentations (7) of f, u'_i, f_i in the relation (3) and get:

$$(9) \qquad \sum_{n=1}^{\infty} a_n x^n = \sum_{n=0}^{\infty} \sum_{j=1}^{m} a_{jn} u_{j0} x^n + \sum_{n=1}^{\infty} \sum_{j=1}^{m} \sum_{k=0}^{n-1} a_{jk} b_{j,n-k} x^n.$$

In particular, for $n = 0$ we get $0 = \sum_{j=1}^{m} a_{j0} u_{j0}$. Since $u_{10}, \ldots, u_{m0}$ are linearly independent over K and $a_{10}, \ldots, a_{m0} \in K$, we get $a_{10} = \cdots = a_{m0} = 0$, so our claim holds in this case.

Proceeding by induction, we assume $|a_{ik} c^k| \leq 1$ for $i = 1, \ldots, m$ and $k = 0, \ldots, n-1$. By (5) and (6),

$$|a_{in}| \leq \max(|a_{1n}|, \ldots, |a_{mn}|) = \mu(\sum_{j=1}^{m} a_{jn} u_{j0}) \leq s |\sum_{j=1}^{m} a_{jn} u_{j0}|,$$

hence

$$(10) \qquad |a_{in} c^n| \leq s |\sum_{j=1}^{m} a_{jn} u_{j0} c^n|.$$

Next we compare the coefficients of x^n on both sides of (9),

$$a_n = \sum_{j=1}^{m} a_{jn} u_{j0} + \sum_{j=1}^{m} \sum_{k=0}^{n-1} a_{jk} b_{j,n-k},$$

change sides and multiply the resulting equation by c^n:

$$\sum_{j=1}^{m} a_{jn} u_{j0} c^n = a_n c^n - \sum_{j=1}^{m} \sum_{k=0}^{n-1} a_{jk} c^k \cdot b_{j,n-k} c^{n-k}.$$

By the induction hypothesis and by (8),

$$(11) \quad \Big| \sum_{j=1}^{m} a_{jn} u_{j0} c^n \Big| \leq \max \Big(|a_n c^n|, \max_{1 \leq j \leq m} \max_{0 \leq k \leq n-1} |a_{jk} c^k| \cdot |b_{j,n-k} c^{n-k}| \Big)$$

$$\leq \max(s^{-1}, 1 \cdot s^{-1}) = s^{-1}$$

It follows from (10) and (11) that $|a_{in} c^n| \leq 1$. This concludes the proof of the claim.

PART D: *End of the proof.* We choose $a \in K^\times$ such that $|a| < |c|$. Then $|a_{in} a^n| = |a_{in} c^n \left(\frac{a}{c}\right)^n| \leq |\frac{a}{c}|^n$. Since the right hand side tends to 0 as $n \to \infty$, so does the left hand side. We conclude that f_i converges at a. $\qquad \square$

LEMMA 2.5.3: *The fields $K((x))$ and $K^{1/p}((x))_0$ are linearly disjoint over $K((x))_0$.*

Proof: We have to prove that every finite extension F' of $K((x))_0$ in $K^{1/p}((x))_0$ is linearly disjoint from $K((x))$ over $K((x))_0$.

If $F' = K((x))_0$, there is nothing to prove, so we assume F' is a proper extension of $K((x))$. Each element $f' \in F'$ has the form $f' = \sum_{i=k}^{\infty} b_i x^i$ with $b_i \in K^{1/p}$ and $\sum_{i=k}^{\infty} b_i c^i$ converges for some $c \in (K^{1/p})^\times$. Thus, $(f')^p = \sum_{i=k}^{\infty} b_i^p x^{ip} \in K((x))$ and $\sum_{i=k}^{\infty} b_i^p (c^p)^i$ converges, so $(f')^p \in K((x))_0$. We may therefore write $F' = F(f)$, where F is a finite extension of $K((x))_0$ in F' and $[F' : F] = p$.

By induction on the degree, F is linearly disjoint from $K((x))$ over $K((x))_0$. Let $m = [F : K((x))_0]$.

Moreover, $K((x))$ is the completion of $K(x)$, so also of $K((x))_0$. Hence, $\hat{F} = K((x))F$ is the completion of F under v. By the linear disjointness, $[\hat{F} : K((x))] = m$.

The residue field of $K((x))$ and of $K((x))_0$ is K and the residue field of $\hat{F}$ is equal to the residue field $\bar{F}$ of F. Both $K((x))$ and $K^{1/p}((x))$ have the same valuation group under v, namely $\mathbb{Z}$. Therefore, also $v(\hat{F}^\times) = \mathbb{Z}$, so $e(\hat{F}/K((x))) = 1$. Since $K((x))$ is complete and discrete, $[\hat{F} : K((x))] = e(\hat{F} : K((x)))[\bar{F} : K] = [\bar{F} : K]$ [CaF65, p. 19, Prop. 3].

Now we choose a basis $u_{10}, \ldots, u_{m0}$ for $\bar{F}/K$ and lift each u_{i0} to an element u_i of $F \cap \tilde{K}[[x]]_0$. Then, $u_1, \ldots, u_m$ are linearly independent over $K((x))_0$ and over $K((x))$, hence they form a basis for $F/K((x))_0$ and for $\hat{F}/K((x))$.

As before, $\widehat{F'} = K((x))F'$ is the completion of F'. Again, both F' and $\widehat{F'}$ have the same residue field $\overline{F'}$ and $[\widehat{F'} : \hat{F}] = [\overline{F'} : \bar{F}]$. Note that $\overline{F'} \subseteq K^{1/p}$ and $[\overline{F'} : \bar{F}] \leq [F' : F] = p$. Therefore, $\overline{F'} = \bar{F}$ or $[\overline{F'} : \bar{F}] = p$.

In the first case $f \in \hat{F}$, so by the paragraph before the preceding one, there exist $f_1, \ldots, f_m \in K((x))$ such that $f = \sum_{i=1}^m f_i u_i$. Multiplying both sides by a large power of x, we may assume that $f_1, \ldots, f_m \in K[[x]]$ and $f(0) = 0$. By Lemma 2.5.2, $f_1, \ldots, f_m \in K((x))_0$, hence $f \in F$. This contradiction to the choice of f implies that $[\overline{F'} : \bar{F}] = p$. Hence, $[K((x))F' : K((x))F] = [\widehat{F'} : \hat{F}] = p = [F' : F]$. This implies that $\hat{F}$ and F' are linearly disjoint over F. By the tower property of linear disjointness, $K((x))$ and F' are linearly disjoint over $K((x))_0$, as claimed. □

PROPOSITION 2.5.4: *Let K be a complete field under an ultrametric absolute value $| \; |$ and denote the field of all convergent power series in x with coefficients in K by $K((x))_0$. Then $K((x))$ is a regular extension of $K((x))_0$.*

Proof: In view of Proposition 2.4.5, it suffices to assume that $p = \mathrm{char}(K) > 0$ and to prove that $K((x))$ is linearly disjoint from $K((x))_0^{1/p}$ over $K((x))_0$.

Indeed, by Lemma 2.5.3, $K((x))$ is linearly disjoint from $K^{1/p}((x))_0$ over $K((x))_0$. Next observe that $K^{1/p}$ is also complete under $| \; |$. Hence, by Lemma 2.5.1, applied to $K^{1/p}$ rather than to K, $K^{1/p}((x))$ is linearly disjoint from $K^{1/p}((x^{1/p}))_0$ over $K^{1/p}((x))_0$.

$$
\begin{array}{ccccc}
K((x)) & \!\!\!\text{---}\!\!\! & K^{1/p}((x)) & \!\!\!\text{---}\!\!\! & K((x))^{1/p}\!\!= K^{1/p}((x^{1/p})) \\
| & & | & & | \\
K((x))_0 & \!\!\!\text{---}\!\!\! & K^{1/p}((x))_0 & \!\!\!\text{---}\!\!\! & K((x))_0^{1/p}\!\!= K^{1/p}((x^{1/p}))_0
\end{array}
$$

Finally we observe that $K((x))_0^{1/p} = K^{1/p}((x^{1/p}))_0$ to conclude that $K((x))$ is linearly disjoint from $K((x))_0^{1/p}$ over $K((x))_0$. □

Notes

The rings of convergent power series in one variable introduced in Section 2.2 are the rings of holomorphic functions on the closed unit disk that appear in [FrP04, Example 2.2]. Weierstrass Divison Theorem (Proposition 2.2.4) appears in [FrP, Thm. 3.1.1]. Our presentation follows the unpublished manuscript [Har05].

Proposition 2.4.5 appears as [Art67, p. 48, Thm. 14]. The proof given by Artin uses the method of Newton polynomials.

The property of $K\{x\}$ of being a principle ideal domain appears in [FrP, Thm. 2.2.9].

The proof that $K((x))/K((x))_0$ is a separable extension (Proposition 2.5.4) is due to Kuhlmann and Roquette [KuR96].

Chapter 3.
Several Variables

Starting from a complete valued field $(K, |\ |)$, we choose an element $r \in K^\times$, a finite set I, and for each $i \in I$ an element $c_i \in K$ such that $|r| \leq |c_i - c_j|$ if $i \neq j$. Then we set $w_i = \frac{r}{x-c_i}$, with an indeterminate x, and consider the ring $R = K\{w_i \mid i \in I\}$ of all series

$$f = a_0 + \sum_{i \in I} \sum_{n=1}^{\infty} a_{in} w_i^n,$$

with $a_0, a_{in} \in K$ such that for each i the element a_{in} tends to 0 as $n \to \infty$. The ring R is complete under the norm defined by $\|f\| = \max_{i,n}(|a_0|, |a_{in}|)$ (Lemma 3.2.1). We prove that R is a principal ideal domain (Proposition 3.2.9) and denote its quotient field by P. More generally for each subset J of I, we denote the quotient field of $K\{w_i \mid i \in J\}$ by P_J. We deduce (Proposition 3.3.1) that $P_J \cap P_{J'} = P_{J \cap J'}$ if $J, J' \subseteq I$ have a nonempty intersection and $P_J \cap P_{J'} = K(x)$ if $J \cap J' = \emptyset$. Thus, setting $P_i = P_{I \smallsetminus \{i\}}$ for $i \in I$, we conclude that $\bigcap_{i \in I} P_i = K(x)$. The fields $E = K(x)$ and P_i are the first objects of patching data (Definition 1.1.1) that we start to assemble.

3.1 A Normed Subring of $K(x)$

Let $E = K(x)$ be the field of rational functions in the variable x over a field K. Let I be a finite set and r an element of $K^\times$. For each $i \in I$ let c_i be an element of K. Suppose $c_i \neq c_j$ if $i \neq j$. For each $i \in I$ let $w_i = \frac{r}{x-c_i} \in K(x)$. We consider the subring $R_0 = K[w_i \mid i \in I]$ of $K(x)$, prove that each of its elements is a linear combination of the powers w_i^n with coefficients in K, and define a norm on R_0.

LEMMA 3.1.1:

(a) For all $i \neq j$ in I and for each nonnegative integer m

(1)
$$w_i w_j^m = \frac{r^m}{(c_i - c_j)^m} w_i - \sum_{k=1}^{m} \frac{r^{m+1-k}}{(c_i - c_j)^{m+1-k}} w_j^k.$$

(b) Given nonnegative integers m_i, $i \in I$, not all zero, there exist $a_{ik} \in K$ such that

(2)
$$\prod_{i \in I} w_i^{m_i} = \sum_{i \in I} \sum_{k=1}^{m_i} a_{ik} w_i^k.$$

M. Jarden, *Algebraic Patching*, Springer Monographs in Mathematics, DOI 10.1007/978-3-642-15128-6_3, © Springer-Verlag Berlin Heidelberg 2011

(c) *Every* $f \in K[w_i \mid i \in I]$ *can be uniquely written as*

$$(3) \qquad f = a_0 + \sum_{i \in I} \sum_{n=1}^{\infty} a_{in} w_i^n$$

where $a_0, a_{in} \in K$ *and almost all of them are zero.*

(d) *Let* $i \neq j$ *be elements of* I. *Then* $\frac{w_i}{w_j} = 1 + \frac{c_i - c_j}{r} w_i \in K[w_i]$ *is invertible in* $K[w_i, w_j]$.

Proof of (a) and (b):   Starting from the identity

$$(4) \qquad w_i w_j = \frac{r}{c_i - c_j} w_i - \frac{r}{c_i - c_j} w_j$$

one proves (1) by induction on m. Then one proceeds by induction on $|I|$ and $\max_{i \in I} m_i$ to prove (2).

Proof of (c):   The existence of the presentation (3) follows from (b). To prove the uniqueness we assume that $f = 0$ in (3) but $a_{jk} \neq 0$ for some $j \in I$ and $k \in \mathbb{N}$. Then, $\sum_{n=1}^{\infty} a_{jn} w_j^n = -a_0 - \sum_{i \neq j} \sum_{n=1}^{\infty} a_{in} w_i^n$. The left hand side has a pole at c_j while the right hand side has not. This is a contradiction.

Proof of (d):   Multiplying $\frac{r}{w_j} - \frac{r}{w_i} = c_i - c_j$ by $\frac{w_i}{r}$ we get that

$$\frac{w_i}{w_j} = 1 + \frac{c_i - c_j}{r} w_i$$

is in $K[w_i]$. Similarly, $\frac{w_j}{w_i} \in K[w_j]$. Hence $\frac{w_i}{w_j}$ is invertible in $K[w_i, w_j]$.   □

Now we make an assumption for the rest of this chapter:

Assumption 3.1.2:   The field K is complete with respect to a nontrivial ultrametric absolute value $|\ |$ and

$$(5) \qquad |r| \leq |c_i - c_j| \quad \text{for all } i \neq j.$$ □

Geometrically, Condition (5) means that the open disks $\{a \in K \mid |a - c_i| < r\}$, $i \in I$, of K are disjoint.

Let $E = K(x)$ be the field of rational functions over K in the variable x. We define a function $\|\ \|$ on $R_0 = K[w_i \mid i \in I]$ using the unique presentation (3):

$$\left\| a_0 + \sum_{i \in I} \sum_{n \geq 1} a_{in} w_i^n \right\| = \max_{i,n} \{ |a_0|, |a_{in}| \}.$$

Then $\|f\| \geq 0$ for each $f \in R_0$, $\|f\| = 0$ if and only if $f = 0$ (Lemma 3.1.1(c)), and $\|f + g\| \leq \max(\|f\|, \|g\|)$ for all $f, g \in R_0$. Moreover, $\|w_i\| = 1$

for each $i \in I$ but $\|w_i w_j\| = \frac{|r|}{|c_i - c_j|}$ (by (4)) is less than 1 if $|r| < |c_i - c_j|$. Thus, $\| \ \|$ is in general not an absolute value. However, by (1) and (5)

$$\|w_i w_j^m\| \leq \max_{1 \leq k \leq m} \left(\left| \frac{r}{c_i - c_j} \right|^m, \left| \frac{r}{c_i - c_j} \right|^{m+1-k} \right) \leq 1.$$

By induction, $\|w_i^k w_j^m\| \leq 1$ for each k, so $\|fg\| \leq \|f\| \cdot \|g\|$ for all $f, g \in R_0$. Moreover, if $a \in K$ and $f \in R_0$, then $\|af\| = \|a\| \|f\|$. Therefore, $\| \ \|$ is a norm on R_0 in the sense of Definition 2.1.1.

3.2 Mitagg-Leffler Series

We keep the notation of Section 3.1 and Assumption 3.1.2 and proceed to define rings of convergent power series of several variables over K. In the language of rigid geometry, these are the rings of holomorphic functions on the complements of finitely many open discs of the projective line $\mathbb{P}^1(K)$.

Let $R = K\{w_i \mid i \in I\}$ be the completion of $R_0 = K[w_i \mid i \in I]$ with respect to $\| \ \|$ (Lemma 2.1.5). Our first result gives a Mitagg-Leffler decomposition of each $f \in R$. It generalizes Lemma 3.1.1(c):

LEMMA 3.2.1: *Each element f of R has a unique presentation as a* **Mitagg-Leffler series**

(1) $$f = a_0 + \sum_{i \in I} \sum_{n=1}^{\infty} a_{in} w_i^n,$$

where $a_0, a_{in} \in K$, and $|a_{in}| \to 0$ as $n \to \infty$. Moreover,

$$\|f\| = \max_{i,n}\{|a_0|, |a_{in}|\}.$$

Proof: Each f as in (1) is the limit of the sequence $(f_d)_{d \geq 1}$ of its partial sums $f_d = a_0 + \sum_{i \in I} \sum_{n=1}^{d} a_{in} w_i^n \in R_0$, so $f \in R$. Since $\|f_d\| = \max_{i,n}(|a_0|, |a_{in}|)$ for each sufficiently large d, we have $\|f\| = \max_{i,n}(|a_0|, |a_{in}|)$. If $f = 0$ in (1), then $0 = \max_{i,n}(|a_0|, |a_{in}|)$, so $a_0 = a_{in} = 0$ for all i and n. It follows that the presentation (1) is unique.

On the other hand, let $g \in R$. Then there exists a sequence of elements $g_k = a_{k,0} + \sum_{i \in I} \sum_{n=1}^{\infty} a_{k,in} w_i^n$, $k = 1, 2, 3, \ldots$, in R_0, that converges to g. In particular, for each pair (k, i) we have $a_{k,in} = 0$ if n is sufficiently large. Also, the sequence $(g_k)_{k=1}^{\infty}$ is Cauchy. Hence, each of the sequences $\{a_{k,0} \mid k = 1, 2, 3, \ldots\}$ and $\{a_{k,in} \mid k = 1, 2, 3, \ldots\}$ is Cauchy. Since K is complete, $a_{k,0} \to a_0$ and $a_{k,in} \to a_{in}$ for some $a_0, a_{in} \in K$. Fix $i \in I$ and let $\varepsilon > 0$ be a real number. There is an m such that for all $k \geq m$ and all n we have $|a_{k,in} - a_{m,in}| \leq \|g_k - g_m\| \leq \varepsilon$. If n is sufficiently large, then $a_{m,in} = 0$, and hence $|a_{k,in}| \leq \varepsilon$. Therefore, $|a_{in}| \leq \varepsilon$. It follows that $|a_{in}| \to 0$. Define f by (1). Then $f \in R$ and $g_k \to f$ in R. Consequently, $g = f$. $\square$

If $I = \emptyset$, then $R = R_0 = K$.

We call the partial sum $\sum_{n=1}^{\infty} a_{in} w_i^n$ in (1) the *i*-**component** of f.

Remark 3.2.2: Let $i \in I$. Then $K\{w_i\} = \{\sum_{n=0}^{\infty} a_n w_i^n \mid a_n \to 0\}$ is a subring of R, the completion of $K[w_i]$ with respect to the norm. Consider the ring $K\{x\}$ of converging power series over K. By Lemma 2.2.1(d), there is a homomorphism $K\{x\} \to K\{w_i\}$ given by $\sum_{n=0}^{\infty} a_n x^n \mapsto \sum_{n=0}^{\infty} a_n w_i^n$. By Lemma 3.2.1, this is an isomorphism of normed rings. $\square$

LEMMA 3.2.3: *Let $i, j \in I$ be distinct, let $p \in K[w_i] \subseteq R$ be a polynomial of degree $\leq d$ in w_i, and let $f \in K\{w_j\} \subseteq R$. Then $pf \in K\{w_i, w_j\}$ and the i-component of pf is a polynomial of degree $\leq d$ in w_i.*

Proof: Presenting p as the sum of its monomials we may assume that p is a power of w_i, say, $p = w_i^d$.

The assertion is obvious, if $d = 0$.

Let $d \geq 1$ and assume, by induction, that $w_i^{d-1} f = p' + f'$, where $p' \in K[w_i]$ is of degree $\leq d - 1$ and $f' \in K\{w_j\}$. Then $w_i^d f = w_i p' + w_i f'$. Here $w_i p' \in K[w_i]$ is of degree $\leq d$ and the i-component of $w_i f'$ is, by (1) of Section 3.1, a polynomial of degree ≤ 1. Thus, the i-component of $w_i^d f$ is of degree $\leq d$. $\square$

Remark 3.2.4: Let $(L, | \ |)$ be a complete valued field extending $(K, | \ |)$. Each $c \in L$ with $|c - c_i| \geq |r|$, for all $i \in I$, defines a continuous **evaluation homomorphism** $R \to L$ given by $f = a_0 + \sum_{i \in I} \sum_n a_{in} w_i^n \mapsto f(c) = a_0 + \sum_{i \in I} \sum_n a_{in} (\frac{r}{c - c_i})^n$. Indeed, $x \mapsto c$ defines a K-homomorphism $\varphi \colon K[x] \to L$. Let P be its kernel. Then φ extends to the localization $K[x]_P$. Since $\varphi(x - c_i) = c - c_i \neq 0$, we have $w_i \in K[x]_P$, for each $i \in I$. Thus, φ restricts to a homomorphism $R_0 \to L$, given by the above formula. Since $\left|\frac{r}{c - c_i}\right| \leq 1$ for each i, we have $|f(c)| \leq \|f\|$ for each $f \in R_0$. Hence, φ uniquely extends to a continuous homomorphism $\varphi \colon R \to L$. $\square$

LEMMA 3.2.5 (Degree shifting): *Let $f \in R$ be given by (1). Fix $i \neq j$ in I. Let $\sum_{n=1}^{\infty} a'_{in} w_i^n$ be the i-component of $\frac{w_j}{w_i} f \in R$. Then*

$$(2) \qquad a'_{in} = - \sum_{\nu = n+1}^{\infty} \frac{a_{i\nu} r^{\nu - n}}{(c_j - c_i)^{\nu - n}}$$

$$= \frac{-r}{c_j - c_i} \sum_{\nu = n+1}^{\infty} a_{i\nu} \left(\frac{r}{c_j - c_i}\right)^{\nu - (n+1)}, \qquad n = 1, 2, 3, \ldots.$$

Furthermore, let $m \geq 1$ be an integer, and let $\sum_{n=1}^{\infty} b_{in} w_i^n$ be the i-component of $(\frac{w_j}{w_i})^m f$. Let $\varepsilon \geq 0$ be a real number and let d be a positive integer.

(a) *If $|a_{in}| \leq \varepsilon$ for each $n \geq d+1$, then $|b_{in}| \leq |\frac{r}{c_j - c_i}|^m \varepsilon$ for each $n \geq d+1-m$.*

(b) *Suppose $d > m$. If $|a_{in}| < \varepsilon$ for each $n \geq d + 1$ and $|a_{id}| = \varepsilon$, then $|b_{in}| < |\frac{r}{c_j - c_i}|^m \varepsilon$ for each $n \geq d + 1 - m$ and $|b_{i, d-m}| = |\frac{r}{c_j - c_i}|^m \varepsilon$.*

(c) $\sum_{n=1}^{\infty} a_{in} w_i^n$ is a polynomial in w_i if and only if $\sum_{n=1}^{\infty} b_{in} w_i^n$ is.

Proof: By Lemma 3.1.1(d), $\frac{w_j}{w_i} \in R^\times$, so $(\frac{w_j}{w_i})^m f \in R$ for each m and the above statements make sense.

PROOF OF (2): We may assume that $a_0 = a_{i1} = 0$ and $a_{k\nu} = 0$ for each $k \neq i$ and each ν. Indeed, $\frac{w_j}{w_i} = 1 + (c_j - c_i)\frac{w_j}{r} \in K\{w_j\}$. Hence, $\frac{w_j}{w_i} \cdot w_k^\nu \in K\{w_l \mid l \neq i\}$. Furthermore, $\frac{w_j}{w_i} \cdot w_i = w_j \in K\{w_l \mid l \neq i\}$. Hence, by (1), a_0, a_{i1}, and the $a_{k\nu}$ do not contribute to the i-component of $\frac{w_j}{w_i} f$.

Thus, $f = \sum_{\nu=2}^{\infty} a_{i\nu} w_i^\nu$. Hence, by (1) of Section 3.1,

$$\frac{w_j}{w_i} f = \sum_{\nu=2}^{\infty} a_{i\nu} w_j w_i^{\nu-1} = \sum_{\nu=2}^{\infty} a_{i\nu} \left[\frac{r^{\nu-1}}{(c_j - c_i)^{\nu-1}} w_j - \sum_{n=1}^{\nu-1} \frac{r^{\nu-n}}{(c_j - c_i)^{\nu-n}} w_i^n \right]$$

$$= \sum_{\nu=2}^{\infty} \frac{a_{i\nu} r^{\nu-1}}{(c_j - c_i)^{\nu-1}} w_j - \sum_{n=1}^{\infty} \sum_{\nu=n+1}^{\infty} \frac{a_{i\nu} r^{\nu-n}}{(c_j - c_i)^{\nu-n}} w_i^n$$

from which (2) follows.

PROOF OF (a) AND (b): By induction on m it suffices to assume that $m = 1$. In this case we have to prove: (a) If $|a_{in}| \leq \varepsilon$ for each $n \geq d+1$, then $|a'_{in}| \leq |\frac{r}{c_j - c_i}| \varepsilon$ for each $n \geq d$; (b) assuming $d \geq 2$, if $|a_{in}| < \varepsilon$ for each $n \geq d+1$ and $|a_{id}| = \varepsilon$, then $|a'_{in}| < |\frac{r}{c_j - c_i}| \varepsilon$ for each $n \geq d$ and $|a'_{i,d-1}| = |\frac{r}{c_j - c_i}| \varepsilon$. By Condition (5) of Section 3.1, $|\frac{r}{c_i - c_j}| \leq 1$. Hence, (a) follows from (2) with $n = d, d+1, d+2, \ldots$ and (b) follows from (2) with $n = d-1, d, d+1, \ldots$.

PROOF OF (c): Again, it suffices to prove that $\sum_{n=1}^{\infty} a_{in} w_i^n$ is a polynomial if and only if $\sum_{n=1}^{\infty} a'_{in} w_i^n$ is a polynomial.

If $\sum_{n=1}^{\infty} a_{in} w_i^n$ is a polynomial, then $a_{i\nu} = 0$ for all large ν. It follows from (2) that $a'_{i,n} = 0$ for all large n. Hence, $\sum_{n=1}^{\infty} a'_{in} w_i^n$ is a polynomial.

If $\sum_{n=1}^{\infty} a_{in} w_i^n$ is not a polynomial, then for each d_0 there exists $d > d_0$ such that $a_{id} \neq 0$. Since $|a_{in}| \to 0$ as $n \to \infty$, there are only finitely many $n \geq d$ with $|a_{in}| \geq |a_{id}|$. Replacing d with the largest of those n's, if necessary, we may assume that $|a_{in}| < |a_{id}|$ for each $n \geq d+1$. By (b), $a'_{i,d-1} \neq 0$. Consequently, $\sum_{n=1}^{\infty} a'_{in} w_i^n$ is not a polynomial. $\square$

We apply degree shifting to generalize Weierstrass preparation theorem (Corollary 2.2.5) to Mitagg-Leffler series.

LEMMA 3.2.6: Suppose $I \neq \emptyset$ and let $0 \neq f \in R$. Then there is an $l \in I$ such that $f = pu$ with $p \in K[w_l]$ and $u \in R^\times$.

Proof: If $I = \emptyset$, then $f \in K^\times = R^\times$. We therefore suppose that $|I| \geq 1$ and continue by induction on $|I|$.

Write f in the form (1). Then, there is a coefficient with absolute value $\|f\|$. Thus we are either in Case I or Case II below:

CASE I: $|a_0| = \|f\| > |a_{in}|$ for all i and n.   Multiply f by a_0^{-1} to assume that $a_0 = 1$. Then $\|1 - f\| < 1$. By Lemma 2.1.3(f), $f \in R^\times$, and $l = i$ satisfies the claim of the lemma.

CASE II: *There exist i and $d \geq 1$ such that $|a_{id}| = \|f\|$.*   Increase d, if necessary, to assume that $|a_{in}| < |a_{id}| = \|f\|$ for all $n > d$.

Let $A = K\{w_k \mid k \neq i\}$. This is a complete subring of R. We introduce a new variable z, and consider the ring $A\{z\}$ of convergent power series in z over A (Lemma 2.2.1(c)). Since $a_{id} \in K^\times \subseteq A^\times$, the element

$$\hat{f} = \Big(a_0 + \sum_{k \neq i} \sum_{n=1}^{\infty} a_{kn} w_k^n\Big) + \sum_{n=1}^{\infty} a_{in} z^n$$

of $A\{z\}$ is regular of pseudo degree d. By Corollary 2.2.5, we have $\hat{f} = \hat{p}\hat{u}$, where $\hat{u}$ is a unit of $A\{z\}$ and $\hat{p}$ is a monic polynomial of degree d in $A[z]$.

By definition, $\|w_i\| = 1$. By Lemma 2.2.1(d), the evaluation homomorphism $\theta \colon A\{z\} \to R$ defined by $\sum c_n z^n \mapsto \sum c_n w_i^n$, with $c_n \in A$, maps $\hat{f}$ onto f, $\hat{u}$ onto a unit of R, and $\hat{p}$ onto a polynomial p of degree d in $A[w_i]$. Replacing f by p and using Lemma 3.1.1, we may assume that $f \in A[w_i] = A + K[w_i]$ is a polynomial of degree d in w_i, that is,

$$f = \Big(a_0 + \sum_{k \neq i} \sum_{n=1}^{\infty} a_{kn} w_k^n\Big) + \sum_{n=1}^{d} a_{in} w_i^n.$$

If $I = \{i\}$, then $A[w_i] = K[w_i]$, and we are done. If $|I| \geq 2$, we choose a $j \in I$ distinct from i. By Lemma 3.1.1(d), $\frac{w_j}{w_i} = 1 + \frac{c_j - c_i}{r} w_j$ is invertible in R_0, hence in R. Since $\frac{w_j}{w_i} \in A$, we have $\frac{w_j}{w_i}\big(\sum_{k \neq i} \sum_{n=1}^{\infty} a_{kn} w_k^n\big) \in A$. In addition, by Lemma 3.1.1,

$$\frac{w_j}{w_i} \sum_{n=1}^{d} a_{in} w_i^n = \sum_{n=1}^{d} a_{in} w_i^{n-1} w_j$$

is a polynomial in $A[w_i]$ of degree $\leq d - 1$. Using induction on d, we may assume that $f \in A$. Finally, we apply the induction hypothesis (on $|I|$) to conclude the proof. $\square$

LEMMA 3.2.7: *Let $j \in I$. Then each $f \in R$ can be written as $f = pu$ with $p \in K[w_j]$, $\|p\| = 1$, and $u \in R^\times$.*

Proof: Lemma 3.2.6 gives a decomposition $f = p_1 u_1$ with $u_1 \in R^\times$ and $p_1 \in K[w_i]$ for some $i \in I$. If $i = j$, we are done. If $i \neq j$, we may assume that $f \in K[w_i]$. Thus, $f = \sum_{n=0}^{d} a_n w_i^n$ with $a_d \neq 0$. By Lemma 3.1.1(d), $\frac{w_i}{w_j}$ is invertible in R_0, hence in R. Multiplying f by $\big(\frac{w_j}{w_i}\big)^d$ gives

$$\Big(\frac{w_j}{w_i}\Big)^d f = \sum_{n=0}^{d} a_n \Big(\frac{w_j}{w_i}\Big)^{d-n} w_j^n = \sum_{n=0}^{d} a_n \Big(1 + \frac{c_j - c_i}{r} w_j\Big)^{d-n} w_j^n \in K[w_j].$$

Thus, $f = pu$ with $p \in K[w_j]$ and $u \in R^\times$. Finally, we may divide p by a coefficient with the highest absolute value to get that $\|p\| = 1$.   □

COROLLARY 3.2.8: Let $0 \neq g \in R$. Then $R_0 + gR = R$.

Proof: Since $R = \sum_{i \in I} K\{w_i\}$ and $R_0 = K[w_i \mid i \in I] = \sum_{i \in I} K[w_i]$ (Lemma 3.1.1), it suffices to prove for each $i \in I$ and for every $f \in K\{w_i\}$ that there is $r \in K[w_i]$ such that $f - r \in gR$. By Lemma 3.2.7, we may assume that $g \in K[w_i]$. By Remark 3.2.2, there is a K-isomorphism $K\{z\} \to K\{w_i\}$ that maps $K[z]$ onto $K[w_i]$. Therefore the assertion follows from the Weierstrass Division Theorem (Proposition 2.2.4) for the ring $K\{z\}$.   □

The next result generalizes Proposition 2.3.1 to Mitagg-Leffler series.

PROPOSITION 3.2.9: The ring $R = K\{w_i \mid i \in I\}$ is a principal ideal domain, hence a unique factorization domain. Moreover, for each $i \in I$, each ideal $\mathfrak{a}$ of R is generated by an element $p \in K[w_i]$ such that $\mathfrak{a} \cap K[w_i] = pK[w_i]$.

Proof: Let $f_1, f_2 \in R$ with $f_1 f_2 = 0$. Choose an $i \in I$. By Lemma 3.2.7, $f_1 = p_1 u_1$ and $f_2 = p_2 u_2$ with $p_1, p_2 \in K[w_i]$ and $u_1, u_2 \in R^\times$. Then $p_1 p_2 = f_1 f_2 (u_1 u_2)^{-1} = 0$, and hence either $p_1 = 0$ or $p_2 = 0$. Therefore, either $f_1 = 0$ or $f_2 = 0$. Consequently, R is an integral domain.

By Lemma 3.2.7, each ideal $\mathfrak{a}$ of R is generated by the ideal $\mathfrak{a} \cap K[w_i]$ of $K[w_i]$. Since $K[w_i]$ is a principal ideal domain, $\mathfrak{a} \cap K[w_i] = pK[w_i]$ for some $p \in K[w_i]$. Consequently, $\mathfrak{a} = pR$ is a principal ideal.   □

Remark 3.2.10: Lower bound.   Haran proves in [Har05, Prop. 3.11] that for each $f \in R$ there exists an $\varepsilon > 0$ such that for every $g \in R$ we have $\varepsilon \|f\| \cdot \|g\| \leq \|fg\|$. He uses this bound rather than the multiplicativity of the absolute value in order to decompose matrices in $\mathrm{GL}_n(P)$ as is done in Corollary 3.4.4.   □

3.3 Fields of Mitagg-Leffler Series

In the notation of Sections 3.1 and 3.2 we consider for each nonempty subset J of I the integral domain $R_J = K\{w_i \mid i \in J\}$ (Proposition 3.2.9) and let $P_J = \mathrm{Quot}(R_J)$. For $J = \emptyset$, we set $P_J = K(x)$. All of these fields are contained in the field $Q = P_I$. The fields $P_i = P_{I \smallsetminus \{i\}}$, $i \in I$, will be our 'analytic' fields in patching data over $E = K(x)$ that we start to assemble. As in Definition 1.1.1, the fields $P'_i = \bigcap_{j \neq i} P_j$ will be useful auxiliary fields.

PROPOSITION 3.3.1: Let J and J' be subsets of I. If $J \cap J' \neq \emptyset$, then $P_J \cap P_{J'} = P_{J \cap J'}$.

Proof: If either $J = \emptyset$ or $J' = \emptyset$, then $P_J \cap P_{J'} = K(x)$, by definition. We therefore assume that $J, J' \neq \emptyset$. Let $j \in J$. Then $K[w_j] \subseteq R_J$, hence $K(x) = K(w_j) \subseteq P_J$. Similarly $K(x) \subseteq P_{J'}$. Hence $K(x) \subseteq P_J \cap P_{J'}$. If $J \cap J' \neq \emptyset$, then, by the unique representation for the elements of R appearing in (1) of Lemma 3.2.1, we have $R_{J \cap J'} = R_J \cap R_{J'}$, so $P_{J \cap J'} \subseteq P_J \cap P_{J'}$.

For the converse inclusion, let $0 \neq f \in P_J \cap P_{J'}$. Fix $j \in J$ and $j' \in J'$; if $J \cap J' \neq \emptyset$, take $j, j' \in J \cap J'$. Write f as f_1/g_1 with $f_1, g_1 \in R_J$. By Lemma 3.2.7, $g_1 = p_1 u_1$, where $0 \neq p_1 \in K[w_j]$ and $u_1 \in R_J^\times$. Replace f_1 by $f_1 u_1^{-1}$ to assume that $g_1 \in K[w_j]$. Similarly $f = f_2/g_2$ with $f_2 \in R_{J'}$ and $g_2 \in K[w_{j'}]$.

If $J \cap J' \neq \emptyset$, then $g_1, g_2 \in R_J \cap R_{J'} = R_{J \cap J'}$. Thus $g_2 f_1 = g_1 f_2 \in R_J \cap R_{J'} = R_{J \cap J'} \subseteq P_{J \cap J'}$, and hence $f = \frac{f_1 g_2}{g_1 g_2} \in P_{J \cap J'}$.

Now suppose $J \cap J' = \emptyset$. Let $g_1 = \sum_{n=0}^{d_1} b_n w_j^n$ with $b_n \in K$. Put $h_1 = (\frac{w_{j'}}{w_j})^{d_1} g_1$. Since $\frac{w_{j'}}{w_j} \in K[w_{j'}]$ (Lemma 3.1.1(d)), we have $h_1 = \sum_{n=0}^{d_1} b_n (\frac{w_{j'}}{w_j})^{d_1 - n} w_{j'}^n \in K[w_{j'}]$. Similarly there is an integer $d_2 \geq 0$ such that $h_2 = (\frac{w_j}{w_{j'}})^{d_2} g_2 \in K[w_j]$. Let $d = d_1 + d_2$. Then, for each $k \in J$

$$(1) \qquad f_1 h_2 \cdot \left(\frac{w_{j'}}{w_k} \right)^d = f_2 h_1 \cdot \left(\frac{w_j}{w_k} \right)^d.$$

Note that $f_1 h_2 \in R_J$ while $f_2 h_1 \in R_{J'}$. In particular, the k-component of $f_2 h_1$ is zero. By Lemma 3.2.5(c), the k-component of $f_2 h_1 \cdot \left(\frac{w_j}{w_k} \right)^d$ is a polynomial in w_k. By (1), the k-component of $f_1 h_2 \cdot \left(\frac{w_{j'}}{w_k} \right)^d$ is a polynomial in w_k. Hence, again by Lemma 3.2.5(c), the k-component of $f_1 h_2$ is a polynomial in w_k.

We conclude that $f_1 h_2 \in K[w_k \mid k \in J]$, so $f = \frac{f_1 h_2}{g_1 h_2} \in K(x)$. $\qquad \square$

COROLLARY 3.3.2: For each $i \in I$ let $P_i' = P_{\{i\}}$. Then, $P_i' = \bigcap_{j \neq i} P_j$ and $\bigcap_{j \in I} P_j = K(x)$.

Proof: We apply Proposition 3.3.1 several times:

$$\bigcap_{j \neq i} P_j = \bigcap_{j \neq i} P_{I \smallsetminus \{j\}} = P_{\bigcap_{j \neq i} I \smallsetminus \{j\}} = P_{\{i\}} = P_i'.$$

For the second equality we choose an $i \in I$. Then

$$\bigcap_{j \in I} P_j = P_{I \smallsetminus \{i\}} \cap \bigcap_{j \neq i} P_{I \smallsetminus \{j\}} = P_{I \smallsetminus \{i\}} \cap P_{\{i\}} = K(x),$$

as claimed. $\qquad \square$

Remark 3.3.3: Proper inclusion. Let J be a nonempty proper subset of I. Then R_J is a proper subset of $R_I \cap P_J$.

Indeed, choose an $i \in I \smallsetminus J$. By definition, $w_i \in R_I$. In addition, $w_i \in K(x) \subseteq P_J$. Thus, $w_i \in R_I \cap P_J$. However, $w_i \notin R_J$. Otherwise $w_i = a_0 + \sum_{k \in J} \sum_{n=1}^{\infty} a_{kn} w_k^n$ with $a_0, a_{kn} \in K$ and $a_{kn} \to 0$ as $n \to 0$, contradicting the uniqueness in Lemma 3.2.1. $\qquad \square$

3.4 Factorization of Matrices over Complete Rings

We show in this section how to decompose a matrix over a complete ring into a product of matrices over certain complete subrings. This will establish the decomposition condition in the definition of the patching data (Definition 1.1.1) in our setup.

LEMMA 3.4.1: *Let* $(M, \| \ \|)$ *be a complete normed ring and let* $0 < \varepsilon < 1$. *Consider elements* $a_1, a_2, a_3, \ldots \in M$ *such that* $\|a_i\| \leq \varepsilon$ *for each* i *and* $\|a_i\| \to 0$. *Let*

$$p_i = (1 - a_1) \cdots (1 - a_i), \qquad i = 1, 2, 3, \ldots \quad .$$

Then the sequence $(p_i)_{i=1}^{\infty}$ *converges to an element of* $M^{\times}$.

Proof: Let $p_0 = 1$. Then $\|p_i\| \leq \|1 - a_1\| \cdots \|1 - a_i\| \leq 1$. Also, $p_i = p_{i-1}(1 - a_i)$. Hence,

$$\|p_i - p_{i-1}\| \leq \|p_{i-1}\| \cdot \|a_i\| \leq \|a_i\| \to 0.$$

Thus, $(p_i)_{i=1}^{\infty}$ is a Cauchy sequence, so converges to some $p \in M$. Furthermore,

$$\|p_k - 1\| = \| \sum_{i=1}^{k} (p_i - p_{i-1}) \| \leq \max \|a_i\| \leq \varepsilon.$$

Consequently, $\|p - 1\| < 1$. By Lemma 2.1.3(f), $p \in M^{\times}$. $\qquad\qquad\square$

LEMMA 3.4.2 (Cartan's Lemma): *Let* $(M, \| \ \|)$ *be a complete normed ring. Let* M_1 *and* M_2 *be complete subrings of* M. *Suppose*
(1) *for each* $a \in M$ *there are* $a^+ \in M_1$ *and* $a^- \in M_2$ *with* $\|a^+\|, \|a^-\| \leq \|a\|$
    such that $a = a^+ + a^-$.
Then for each $b \in M$ *with* $\|b - 1\| < 1$ *there exist* $b_1 \in M_1^{\times}$ *and* $b_2 \in M_2^{\times}$
such that $b = b_1 b_2$.

Proof: Let $a_1 = b - 1$ and $\varepsilon = \|a_1\|$. Then $0 \leq \varepsilon < 1$. The condition

(2) $$1 + a_{j+1} = (1 - a_j^+)(1 + a_j)(1 - a_j^-),$$

with a_j^+, a_j^- associated to a_j by (1), recursively defines a sequence $(a_j)_{j=1}^{\infty}$ in M. Use the relation $a_j = a_j^+ + a_j^-$ to rewrite (2):

(3) $$a_{j+1} = a_j^+ a_j^- - a_j^+ a_j - a_j a_j^- + a_j^+ a_j a_j^-.$$

Inductively assume that $\|a_j\| \leq \varepsilon^{2^{j-1}}$. Since $\|a_j^+\|, \|a_j^-\| \leq \|a_j\|$, (3) implies that $\|a_{j+1}\| \leq \max(\|a_j\|^2, \|a_j\|^3) = \|a_j\|^2 \leq \varepsilon^{2^j}$. Therefore, $a_j \to 0$, $a_j^- \to 0$, and $a_j^+ \to 0$. Further, by (2),

(4) $$1 + a_{j+1} = (1 - a_j^+) \cdots (1 - a_1^+) \, b \, (1 - a_1^-) \cdots (1 - a_j^-).$$

By Lemma 3.4.1, the partial products $(1 - a_1^-) \cdots (1 - a_j^-)$ converge to some $b_2' \in M_2^\times$. Similarly, the partial products $(1 - a_j^+) \cdots (1 - a_1^+)$ converge to some $b_1' \in M_1^\times$. Passing to the limit in (4) we get $1 = b_1' b b_2'$. Therefore, $b = (b_1')^{-1}(b_2')^{-1}$, as desired. $\qquad\square$

LEMMA 3.4.3: *Let A be a complete integral domain with respect to an absolute value $|\ |$, A_1, A_2 complete subrings of A, and A_0 a dense subring of A. Set $E_i = \mathrm{Quot}(A_i)$ for $i = 0, 1, 2$ and $E = \mathrm{Quot}(A)$. Suppose these objects satisfy the following conditions:*

(5a) *For each $a \in A$ there are $a^+ \in A_1$ and $a^- \in A_2$ with $|a^+|, |a^-| \le |a|$ such that $a = a^+ + a^-$.*

(5b) *$A = A_0 + gA$ for each nonzero $g \in A_0$.*

(5c) *For every $f \in A$ there are $p \in A_0$ and $u \in A^\times$ such that $f = pu$.*

(5d) *$E_0 \subseteq E_2$.*

Then, for each $b \in \mathrm{GL}_n(E)$ there are $b_1 \in \mathrm{GL}_n(E_1)$ and $b_2 \in \mathrm{GL}_n(E_2)$ such that $b = b_1 b_2$.

Proof: As in Example 2.1.4(d), we define the norm of a matrix $a = (a_{ij}) \in M_n(A)$ by $\|a\| = \max_{ij} |a_{ij}|$ and note that $M_n(A)$ is a complete normed ring, $M_n(A_1,), M_n(A_2)$ are complete normed subrings of $M_n(A)$, and $M_n(A_0)$ is a dense subring of $M_n(A)$. Moreover, by (5a), for each $a \in M_n(A)$ there are $a^+ \in M_n(A_1)$ and $a^- \in M_n(A_2)$ with $\|a^+\|, \|a^-\| \le \|a\|$ such that $a = a^+ + a^-$.

By Condition (5c) each element of E is of the form $\frac{1}{h}f$, where $f \in A$ and $h \in A_0$, $h \ne 0$. Hence, there is $h \in A_0$ such that $hb \in M_n(A)$ and $h \ne 0$. If $hb = b_1 b_2'$, where $b_1 \in \mathrm{GL}_n(E_1)$ and $b_2' \in \mathrm{GL}_n(E_2)$, then $b = b_1 b_2$ with $b_2 = \frac{1}{h}b_2' \in GL_n(E_2)$. Thus, we may assume that $b \in M_n(A)$.

Let $d \in A$ be the determinant of b. By Condition (5c) there are $g \in A_0$ and $u \in A^\times$ such that $d = gu$. Let $b'' \in M_n(A)$ be the adjoint matrix of b, so that $bb'' = d1$. Let $b' = u^{-1}b''$. Then $b' \in M_n(A)$ and $bb' = g1$.

We set

$$V = \{a' \in M_n(A) \mid ba' \in gM_n(A)\} \qquad \text{and} \qquad V_0 = V \cap M_n(A_0).$$

Then V is an additive subgroup of $M_n(A)$ and $gM_n(A) \le V$. By (5b), $M_n(A) = M_n(A_0) + gM_n(A)$. Hence $V = V_0 + gM_n(A)$. Since $M_n(A_0)$ is dense in $M_n(A)$, and therefore $gM_n(A_0)$ is dense in $gM_n(A)$, it follows that $V_0 = V_0 + gM_n(A_0)$ is dense in $V = V_0 + gM_n(A)$. Since $b' \in V$, there is $a_0 \in V_0$ such that $\|b' - a_0\| < \frac{|g|}{\|b\|}$. In particular, $a_0 \in M_n(A_0)$ and $ba_0 \in gM_n(A)$.

Put $a = \frac{1}{g}a_0 \in M_n(E_0)$. Then $ba \in M_n(A)$ and $\|1 - ba\| = \|\frac{1}{g}b(b' - a_0)\| \le \frac{1}{|g|}\|b\| \cdot \|b' - a_0\| < 1$. It follows from Lemma 2.1.3(f) that $ba \in \mathrm{GL}_n(A)$. In particular $\det(a) \ne 0$ and therefore $a \in \mathrm{GL}_n(E_0) \le \mathrm{GL}(E_2)$. By Lemma 3.4.2, there are $b_1 \in \mathrm{GL}_n(A_1)$ and $b_2' \in \mathrm{GL}_n(A_2) \le \mathrm{GL}_n(E_2)$

such that $ba = b_1 b_2'$. Thus $b = b_1 b_2$, where $b_1 \in \mathrm{GL}_n(A_1) \leq \mathrm{GL}_n(E_1)$ and $b_2 = b_2' a^{-1} \in \mathrm{GL}_n(E_2)$. $\qquad \square$

We apply Corollary 3.4.3 to the rings and fields of 3.3.

COROLLARY 3.4.4: *Let $B \in \mathrm{GL}_n(Q)$.*
(a) *For each partition $I = J \cup J'$ into nonempty sets J and J' there exist $B_1 \in \mathrm{GL}_n(P_J)$ and $B_2 \in \mathrm{GL}_n(P_{J'})$ such that $B = B_1 B_2$.*
(b) *For each $i \in I$ there exist $B_1 \in \mathrm{GL}_n(P_i)$ and $B_2 \in \mathrm{GL}_n(P_i')$ such that $B = B_1 B_2$.*

Proof: We may assume without loss that both J and J' are nonempty and apply Lemma 3.4.3 to the rings $R, R_J, R_{J'}, R_0$ rather than A, A_1, A_2, A_0, where $R_0 = K[w_i \mid i \in I]$.

By definition, R, R_J, and $R_{J'}$ are complete rings (Second paragraph of Section 3.2). Given $f \in R$, say, $f = a_0 + \sum_{i \in I} \sum_{k=1}^{\infty} a_{ik} w_i^k$ (Lemma 3.2.1), we let $f_1 = a_0 + \sum_{i \in J} \sum_{k=1}^{\infty} a_{ik} w_i^k$ and $f_2 = \sum_{i \in J'} \sum_{k=1}^{\infty} a_{ik} w_i^k$. Then $|f_i| \leq |f|$, $i = 1, 2$ and $f = f_1 + f_2$. This proves condition (5a) in our context.

By definition, R is the completion of R_0, so R_0 is dense in R and $K(x) = \mathrm{Quot}(R_0)$ is contained in both $P_J = \mathrm{Quot}(R_j)$ and $P_{J'} = \mathrm{Quot}(R_{J'})$. Conditions (5b) and (5c) are Corollary 3.2.8 and Lemma 3.2.7, respectively. Our Corollary is therefore a special case of Lemma 3.4.3. $\qquad \square$

We apply Corollary 3.3.2 and Corollary 3.4.4 to put together patching data whose analytic fields are the fields P_i introduced above.

PROPOSITION 3.4.5: *Let K be a complete field with respect to an ultrametric absolute value $|\ |$. Let x be an indeterminate, G a finite group, r and element of $K^\times$, and I a finite set with $|I| \geq 2$. For each $i \in I$ let G_i be a subgroup of G, F_i a finite Galois extension of $E = K(x)$ with $\mathrm{Gal}(F_i/K) \cong G_i$, and $c_i \in K^\times$ such that $|r| \leq |c_i - c_j|$ if $i \neq j$. Set $w_i = \frac{r}{x - c_i}$, $P_i = \mathrm{Quot}(K\{w_j \mid j \in I \setminus \{i\}\})$, $P_i' = \mathrm{Quot}(K\{w_i\})$, and $Q = \mathrm{Quot}(K\{w_i \mid i \in I\})$. Suppose $G = \langle G_i \mid i \in I \rangle$ and $F_i \subseteq P_i'$ for each $i \in I$. Then $\mathcal{E} = (E, F_i, P_i, Q, G_i, G)_{i \in I}$ is patching data.*

Proof: Our assumptions imply conditions (1a) and (1d) of Definition 1.1.1. By Corollary 3.3.2, $P_i' = P_{\{i\}} = \bigcap_{j \neq i} P_{I \setminus \{j\}} = \bigcap_{j \neq i} P_j$ and $\bigcap_{i \in I} P_i = E$. Thus, Conditions (1b) and (1c) of Definition 1.1.1 hold. Finally, Condition (1e) of Definition 1.1.1 holds by Corollary 3.4.4. It follows that $\mathcal{E}$ is patching data. $\qquad \square$

Notes

The whole chapter is a rewrite of Sections 2–4 of [Har05], which for themselves are a revision of Sections 2–4 of [HaJ98a]. Starting from a complete valued field $(K, |\ |)$, [HaJ98a] chooses $c_i \in K$, $i \in I$, such that $|c_i| \leq |c_i - c_j| = 1$ for all distinct $i, j \in I$. Then [HaJ98a] extends $|\ |$ to an absolute value of $K\{w_i \mid i \in I\}$, with $w_i = \frac{1}{x - c_i}$. Here we multiply w_i by an element $r \in K^\times$

and replace the above condition on the c_i's by $|r_i| \leq |c_i - c_j|$ for all $i \neq j$. The absolute value of K extends in this case only to a norm of $K\{w_i \mid i \in I\}$. Yet, the main results of [HaJ98a] at this stage (Corollary 3.3.2 and Corollary 3.4.4)) are still attained.

Remark 3.3.3 is due to Elad Paran.

Cartan's lemma has many forms and applications. See [Hrb03] for an extensive discussion. The author learned about the lemma from [FrP81, (III.6.3)]. Our version for fields seems to appear for the first time in [HaJ98a].

There is an overlap between Chapter 3 and [FrP04, Sections 2.1 and 2.2] in the case where K is algebraically closed.

Chapter 4.
Constant Split Embedding
Problems over Complete Fields

Let K_0 be a complete field under a discrete ultrametric absolute value and
x an indeterminate. We prove that each finite split embedding problem
over K_0 has a **rational solution**. Thus, given a finite Galois extension K
of K_0 with Galois group Γ that acts on a finite group G, there is a finite
Galois extension F of $K_0(x)$ which contains $K(x)$ with $\mathrm{Gal}(F/K(x)) \cong G$
and $\mathrm{Gal}(F/K_0(x)) \cong \Gamma \ltimes G$ such that res: $\mathrm{Gal}(F/K_0(x)) \to \mathrm{Gal}(K/K_0)$
corresponds to the projection $\Gamma \ltimes G \to \Gamma$. Moreover, F has a K-rational
place unramified over $K(x)$ whose decomposition group over $K_0(x)$ is Γ.

To construct F we choose finitely many cyclic subgroups C_i, $i \in I$,
of G which generate G. For each $i \in I$ we construct a Galois extension
$F_i = K(x, z_i)$ of $K(x)$ with Galois group C_i in $K((x))$. Then we consider the
ring $R = K\{w_i \mid i \in I\}$ as in Section 3.2, where $w_i = \frac{r}{x-c_i}$, $r \in K_0$, $c_i \in K$,
and $|r| \le |c_i - c_j|$ for all $i \ne j$, and shift F_i into the field $P_i' = \mathrm{Quot}(K\{w_i\})$
(Lemma 4.3.5). Choosing the c_i's in an appropriate way (Claim A of the proof
of Proposition 4.4.2), we establish patching data $\mathcal{E}$ with a proper action of Γ
and apply Proposition 1.2.2 to solve the given embedding problem.

4.1 Tame Realization of Cyclic Groups over $K(x)$

Given a field K and an indeterminate x, we construct for each finite cyclic
group A a Galois extension F of $K(x)$ with Galois group A and with good
control on the ramification (Lemma 4.2.5). In particular, each prime divisor
of $K(x)/K$ that ramifies in F is totally ramified, so F is a regular extension
of K.

In this section we handle the tame case where $\mathrm{char}(K)$ does not divide
the order of A. In the next section we treat the wild case where the order of
A is a power of $\mathrm{char}(K)$.

Remark 4.1.1: Branch points. Let K be a field, $E = K(x)$, and F a finite
separable extension of E. A **prime divisor** $\mathfrak{P}$ of F/K is an equivalence
class of valuations of F that are trivial on K. Let $v_{\mathfrak{P}}$ be a representative
of $\mathfrak{P}$ and let $\bar{F}_{\mathfrak{P}}$ be the residue field of F at $\mathfrak{P}$. The restriction of $v_{\mathfrak{P}}$ to
E represents a prime divisor $\mathfrak{p}$ of E/K which is said to **lie under** $\mathfrak{P}$. Let
$e_{\mathfrak{P}/\mathfrak{p}} = (v_{\mathfrak{P}}(F^\times) : v_{\mathfrak{p}}(E^\times))$ be the ramification index and $\bar{F}_{\mathfrak{P}}/\bar{E}_{\mathfrak{p}}$ the residue
extension of $\mathfrak{P}/\mathfrak{p}$. We say that $\mathfrak{P}$ **ramifies over** $\mathfrak{p}$ (or **over** E) if $e_{\mathfrak{P}/\mathfrak{p}} > 1$
or $\bar{F}_{\mathfrak{P}}/\bar{E}_{\mathfrak{p}}$ is an inseparable extension. Next, $\mathfrak{p}$ is said to **ramify** if F has a
prime divisor $\mathfrak{P}$ which ramifies over $\mathfrak{p}$. If $e_{\mathfrak{P}/\mathfrak{p}} = [F : E] > 1$, we say that $\mathfrak{p}$
totally ramifies in F. Moreover, the formula $\sum e_i f_i = [F : E]$ that holds
for the ramification indices and residue degrees [Deu73, p. 97, Thm.] implies

M. Jarden, *Algebraic Patching*, Springer Monographs in Mathematics,
DOI 10.1007/978-3-642-15128-6_4, © Springer-Verlag Berlin Heidelberg 2011

that $\mathfrak{p}$ has a unique extension $\mathfrak{P}$ to F and the residue degree is 1. We denote the set of all prime divisors of E/K that ramify in F by $\mathrm{Ram}(F/E)$.

Each $\mathfrak{p} \in \mathrm{Ram}(F/E)$ corresponds either to a monic irreducible polynomial $p_{\mathfrak{p}} \in K[x]$ or to ∞. In the former case $v_{\mathfrak{p}}(y) = m$ if $y = p_{\mathfrak{p}}^m \frac{g}{h}$, where $m \in \mathbb{Z}$ and $g, h \in K[x]$ are relatively prime to $p_{\mathfrak{p}}$. In the latter case $v_{\mathfrak{p}}\left(\frac{g}{h}\right) = \deg(h) - \deg(g)$, where $g, h \in K[x]$ and $g, h \neq 0$. A **branch point** of F/E **with respect to** x is either a zero of $p_{\mathfrak{p}}$ in $\tilde{K}$ for some finite $\mathfrak{p} \in \mathrm{Ram}(F/E)$ or ∞, if $\infty \in \mathrm{Ram}(F/E)$. We denote the set of all branch points of F/E with respect to x by $\mathrm{Branch}(F/E, x)$ or by $\mathrm{Branch}(F/E)$ if x is obvious from the context. We call a branch point of F/E **separable** if it belongs to $K_s \cup \{\infty\}$. If $F = E(y)$ and y is separable over E and integral over $K[x]$, then $\mathrm{discr}(\mathrm{irr}(y, K(x)))$ is a polynomial $h \in K[x]$ and each finite branch point of F/E is a zero of h [Lan70, p. 62], hence $\mathrm{Branch}(F/E)$ is a finite subset of $\tilde{K} \cup \{\infty\}$.

If F' is a finite extension of E that contains F, then the multiplicativity of the ramification index and the inseparable degree of the residue field extension imply that $\mathrm{Ram}(F/E) \subseteq \mathrm{Ram}(F'/E)$, hence $\mathrm{Branch}(F/E) \subseteq \mathrm{Branch}(F'/E)$. If F_1 and F_2 are finite separable extensions of E and a prime divisor $\mathfrak{p}$ of E/K is unramified in both F_1 and F_2, then $\mathfrak{p}$ is unramified in $F_1 F_2$ [FrJ08, Lemma 2.3.6]. Thus, $\mathrm{Ram}(F_1 F_2/E) = \mathrm{Ram}(F_1/E) \cup \mathrm{Ram}(F_2/E)$ and $\mathrm{Branch}(F_1 F_2/E) = \mathrm{Branch}(F_1/E) \cup \mathrm{Branch}(F_2/E)$. In particular, let L be a finite separable extension of K. Then $L(x)/K(x)$ is unramified [Deu73, p. 113], so $\mathrm{Branch}(F/E) = \mathrm{Branch}(FL/L(x))$.   □

Definition 4.1.2: Let F/K be a field extension. A K**-place** of F is a place $\varphi \colon F \to \tilde{K} \cup \{\infty\}$ such that $\varphi(a) = a$ for each $a \in K$. We say that φ is a K**-rational place** if φ is a K-place of F and K is its residue field. In this case F/K is a regular extension [FrJ08, Lemma 2.6.9].   □

Notation 4.1.3: Each $a \in \tilde{K} \cup \{\infty\}$ defines a K-place $\varphi_{x,a} \colon K(x) \to \tilde{K} \cup \{\infty\}$ by $\varphi_{x,a}(x) = a$. We denote the corresponding normalized valuation by $v_{x,a}$. Thus, $v_{x,a}(\mathrm{irr}(a, K)) = 1$ if $a \in \tilde{K}$ and $v_{x,a}(x) = -1$ if $a = \infty$. We denote the corresponding prime divisor of $K(x)/K$ by $\mathfrak{p}_{x,a}$. Note that if $a' \in \tilde{K} \cup \{\infty\}$, then $\mathfrak{p}_{x,a} = \mathfrak{p}_{x,a'}$ and $v_{x,a} = v_{x,a'}$ if and only if a and a' are conjugate over K. If we wish to emphasize the dependence of $\mathfrak{p}_{x,a}$ on K, we write $\mathfrak{p}_{K,x,a}$ rather than $\mathfrak{p}_{x,a}$. Similar adjustments apply to $v_{x,a}$ and $\varphi_{x,a}$. Note also that for $a = 0$ and $a = \infty$, we get that $\mathfrak{p}_{x,0} = \mathrm{div}_0(x)$ and $\mathfrak{p}_{x,\infty} = \mathrm{div}_\infty(x)$ are the divisor of zeros and the divisor of poles of x in $K(x)$, respectively.   □

Remark 4.1.4: Every K-automorphism θ of $E = K(x)$ is given by $\theta(x) = \frac{ax+b}{cx+d}$, where $\left(\begin{smallmatrix} a & b \\ c & d \end{smallmatrix}\right) \in \mathrm{GL}_2(K)$. It induces

(1a) a permutation θ' of $\tilde{K} \cup \{\infty\}$ by $\theta'(\alpha) = \frac{a\alpha+b}{c\alpha+d}$; and

(1b) a permutation θ^* of the set of prime divisors of E/K by mapping the equivalence class of the place φ onto the equivalence class of $\varphi \circ \theta$.

In particular, $\theta(x)$ is another generator of E/K. By definition, $\varphi_{x,\alpha} \circ \theta =$

$\varphi_{x,\theta'(\alpha)}$ and $\varphi_{\theta(x),\theta'(\alpha)} = \varphi_{x,\alpha}$, so

(2) $$\theta^*(\mathfrak{p}_{x,\alpha}) = \mathfrak{p}_{x,\theta'(\alpha)} \quad \text{and} \quad \mathfrak{p}_{\theta(x),\theta'(\alpha)} = \mathfrak{p}_{x,\alpha}.$$

Furthermore, let F/E be a finite extension and extend θ to an isomorphism $\theta \colon F \to F'$ of fields. Then F' is a finite extension of E and we have

(3)
$$\theta'\big(\mathrm{Branch}(F'/E,x)\big) = \mathrm{Branch}(F/E,x),$$
$$\theta'\big(\mathrm{Branch}(F/E,x)\big) = \mathrm{Branch}(F/E,\theta(x)).$$

Indeed, let $\alpha \in \tilde{K} \cup \{\infty\}$. Then $\varphi_{x,\alpha} \circ \theta \colon E \to \tilde{K} \cup \{\infty\}$ represents $\theta^*(\mathfrak{p}_{x,\alpha}) = \mathfrak{p}_{x,\theta'(\alpha)}$. If $\psi' \colon F' \to \tilde{K} \cup \{\infty\}$ extends $\varphi_{x,\alpha}$, then $\psi' \circ \theta \colon F \to \tilde{K} \cup \{\infty\}$ extends $\varphi_{x,\alpha} \circ \theta$ and the ramification indices remain unchanged. Thus $\alpha \in \mathrm{Branch}(F'/E,x)$ if and only if $\mathfrak{p}_{x,\alpha}$ is ramified in F'/E if and only if $\mathfrak{p}_{x,\theta'(\alpha)}$ is ramified in F/E if and only if $\theta'(\alpha) \in \mathrm{Branch}(F/E,x)$. This proves the first equality of (3).

Furthermore, it follows from $\varphi_{\theta(x),\theta'(\alpha)} = \varphi_{x,\alpha}$ that $\alpha \in \mathrm{Branch}(F/E,x)$ if and only if $\mathfrak{p}_{x,\alpha}$ is ramified in F/E if and only if $\mathfrak{p}_{\theta(x),\theta'(\alpha)}$ is ramified in F/E if and only if $\theta'(\alpha) \in \mathrm{Branch}(F/E,\theta(x))$. This proves the second equality of (3). $\square$

We denote by ζ_n a root of unity of order n. If k is an integer, then ζ_n^k has order n if and only if $\gcd(k,n) = 1$. Thus, whenever one root of unity of order n belongs to a field E, all roots of unity belong to E. Hence, a condition $\zeta_n \in E$ does not depend on the choice of ζ_n.

LEMMA 4.1.5: *Let E be a field, $a \in E$, v a normalized discrete valuation of E, $n \in \mathbb{N}$, and $z \in E_s$. Suppose $\mathrm{char}(E) \nmid n$, $\zeta_n \in E$, $\gcd(n,v(a)) = 1$, and $z^n = a$. Then $F = E(z)$ is a cyclic extension of E of degree n and v is totally ramified in F. If v' is another normalized discrete valuation of E and $n|v'(a)$, then v' is unramified in F.*

Proof: By Kummer theory [Lan93, p. 288, Thm. 6.1], F is a cyclic extension of E of degree at most n. Let w be a valuation of F lying over v. Then $nw(z) = e_{w/v}v(a)$. Hence, $n|e_{w/v}$. On the other hand, $e_{w/v} \le [F:E] \le n$, so $e_{w/v} = [F:E] = n$. Consequently, v is totally ramified in F.

Finally, one can find the statement about v' in [FrJ08, Example 2.3.8]. $\square$

LEMMA 4.1.6: *Let K be a field, x an indeterminate, n a positive integer with $\mathrm{char}(K) \nmid n$, and $a,b \in K^\times$ with $b \ne a$. Set $L = K(\zeta_n)$ and $G = \mathrm{Gal}(L/K)$. Then $K(x)$ has a cyclic extension F of degree n in $K((x))$ such that each $\mathfrak{p} \in \mathrm{Ram}(F/K(x))$ is totally ramified. Moreover,*

(4) $$\mathrm{Branch}(F/K(x)) = \begin{cases} \{a,b\} & \text{if } \zeta_n \in K \\ \{a\zeta_n^\sigma \mid \sigma \in G\} & \text{if } \zeta_n \notin K \end{cases}$$

Proof: First we consider the case where $\zeta_n \in K$ and let $u = \frac{1-a^{-1}x}{1-b^{-1}x}$. Then $u \in K[[x]]$ and $u \equiv 1 \mod xK[[x]]$. Since $\mathrm{char}(K) \nmid n$, Hensel's lemma gives a $z \in K[[x]]$ with $z^n = u$. Note that $v_{x,a}(u) = 1$ and $v_{x,b}(u) = -1$. Hence, by Lemma 4.1.5, $F = K(z)$ is a cyclic extension of $K(x)$ of degree n and both $v_{x,a}$ and $v_{x,b}$ are totally ramified in F. Moreover, if v is another valuation of $K(x)/K$, then $v(u) = 0$, so v is unramified in F (Lemma 4.1.5). It follows that $\mathrm{Branch}(F/K(x)) = \{a, b\}$. This completes the proof in this case.

From now on suppose $\zeta_n \notin K$. We construct a Kummer extension F' of $L(x)$ of degree n in $L((x))$, extend the action of G on $L(x)$ to an action of G on $L((x))$, and prove that the fixed field F of G in F' has the desired properties.

PART A: *Construction of F'.* Let χ be the map of G into the set of all integers between 1 and $n-1$ that are relatively prime to n defined by $\zeta_n^{\chi(\sigma)} = \zeta_n^\sigma$. This map satisfies

(5) $$\chi(\sigma\tau) \equiv \chi(\sigma)\chi(\tau) \mod n$$

for all $\sigma, \tau \in G$. By [FrJ08, Example 3.5.1], $K((x))$ is a regular extension of K and $L((x)) = K((x))(\zeta_n)$. Thus, we may identify G with $\mathrm{Gal}(L((x))/K((x)))$.

We consider the element

$$u = \prod_{\sigma \in G} \left(\frac{1 - a^{-1}\zeta_n^{-\sigma}x}{1 - b^{-1}x} \right)^{\chi(\sigma^{-1})}$$

of $L(x)$ and observe that $\frac{1-a^{-1}\zeta_n^{-1}x}{1-b^{-1}x} \in L[[x]]$. Since $\mathrm{char}(K) \nmid n$, Hensel's lemma [FrJ08, Proposition 3.5.2] gives a $y \in L[[x]]$ with $y^n = \frac{1-a^{-1}\zeta_n^{-1}x}{1-b^{-1}x}$. Then $z = \prod_{\sigma \in G}(y^\sigma)^{\chi(\sigma^{-1})} \in L[[x]]$ and

$$z^n = \prod_{\sigma \in G} ((y^n)^\sigma)^{\chi(\sigma^{-1})} = \prod_{\sigma \in G} \left(\frac{1 - a^{-1}\zeta_n^{-\sigma}x}{1 - b^{-1}x} \right)^{\chi(\sigma^{-1})} = u.$$

For each $\sigma \in G$ we have $v_{L,x,a\zeta_n^\sigma}(u) = \chi(\sigma^{-1})$. Hence, by Lemma 4.1.5, $F' = L(x, z)$ is a cyclic extension of $L(x)$ of degree n and each of the valuations $v_{L,x,a\zeta_n^\sigma}$ is totally ramified in F'. Since $\zeta_n \notin K$, there exists $\tau \in G$ with $\chi(\tau) \neq 1$. Hence, by (5),

$$\sum_{\sigma \in G} \chi(\sigma^{-1}) \equiv \sum_{\sigma \in G} \chi(\tau\sigma^{-1}) \equiv \chi(\tau) \sum_{\sigma \in G} \chi(\sigma^{-1}) \mod n,$$

so $\sum_{\sigma \in G} \chi(\sigma^{-1}) \equiv 0 \mod n$. Therefore, $v_{L,x,b}(u) = \sum_{\sigma \in G} -\chi(\sigma^{-1}) \equiv 0 \mod n$. By Lemma 4.1.5, $v_{L,x,b}$ is unramified in F'. If v is another valuation of $L(x)/L$, then $v(u) = 0$, so v is unramified in F'. It follows that $\mathrm{Branch}(F'/L(x)) = \{a_n\zeta_n^\sigma \mid \sigma \in G\}$.

The Galois group $\mathrm{Gal}(F'/L(x))$ is generated by an element ω satisfying $z^\omega = \zeta_n^{-1}z$.

PART B: *Construction of F.* By (5) there exist for all $\tau, \rho \in G$ a nonnegative integer $k(\tau, \rho)$ and a rational function $f_\tau \in L(x)$ such that

$$(6)= \prod_{\sigma \in G} (y^{\sigma\tau})^{\chi(\sigma^{-1})} = \prod_{\rho \in G} (y^\rho)^{\chi(\tau\rho^{-1})}$$

$$= \prod_{\rho \in G} (y^\rho)^{\chi(\tau)\chi(\rho^{-1})+k(\tau,\rho)n} = z^{\chi(\tau)} \prod_{\rho \in G} \left(\frac{1 - a^{-1}\zeta_n^{-\rho}x}{1 - b^{-1}x} \right)^{k(\tau,\rho)} = z^{\chi(\tau)} f_\tau(x).$$

It follows that G leaves F' invariant. Let F be the fixed field of G in F'. Then $F \subseteq K((x))$.

$$
\begin{array}{ccc}
K((x)) & \text{---} & L((x)) \\
| & & | \\
F & \text{------} & F' = L(x, z) \\
| & & | \\
K(x) & \text{------} & L(x) \\
| & & | \\
K & \text{------} & L = K(\zeta_n)
\end{array}
$$

Let H be the subgroup of $\mathrm{Aut}(F'/K(x))$ generated by $\mathrm{Gal}(F'/L(x))$ and G. Then the fixed field of H is $K(x)$, so $F'/K(x)$ is a Galois extension with $\mathrm{Gal}(F'/K(x)) = G \cdot \mathrm{Gal}(F'/L(x))$. Moreover, given $\tau \in G$, we set $m = \chi(\tau)$. By (6), $z^\tau = z^m f_\tau(x)$, hence $z^{\omega\tau} = (\zeta_n^{-1}z)^\tau = \zeta_n^{-m} z^m f_\tau(x) = (z^m)^\omega f_\tau(x) = (z^m f_\tau(x))^\omega = z^{\tau\omega}$. Thus, $\tau\omega = \omega\tau$, so G commutes with $\mathrm{Gal}(F'/L(x))$. Moreover, the map res: $\mathrm{Gal}(F'/F) \to \mathrm{Gal}(L(x)/K(x))$ is an isomorphism, so F is linearly disjoint from $L(x)$ over $K(x)$. Therefore, $F/K(x)$ is a Galois extension with $\mathrm{Gal}(F/K(x)) \cong \mathrm{Gal}(F'/L(x)) \cong \mathbb{Z}/n\mathbb{Z}$.

By Remark 4.1.1, $\mathrm{Branch}(F/K(x)) = \mathrm{Branch}(F'/L(x))$, so

$$\mathrm{Branch}(F/K(x)) = \{a\zeta_n^\sigma \mid \sigma \in G\}.$$

Note that the elements $a\zeta_n^\sigma$ with $\sigma \in G$ are conjugate over K. Hence, all of the valuations $v_{L,x,a\zeta_n^\sigma}$ lie over the same valuation $v = v_{K,x,a\zeta_n}$ of $K(x)/K$. Moreover, since v is unramified in $L(x)$, v totally ramifies in F, by the multiplicity of the ramification index. $\square$

4.2 Realization of Cyclic Groups of Order p^k

Again, let K be a field and x an indeterminate. Our next task is to construct cyclic extensions of $K(x)$ of degree p^n with $p = \mathrm{char}(K)$ and with information about the branch points.

LEMMA 4.2.1: *Let A be principal ideal domain, $E = \mathrm{Quot}(A)$, F a finite separable extension of E, and B the integral closure of A in F. Suppose B is unramified over A. Then there exists $b \in B$ such that $\mathrm{trace}_{F/E}(b) = 1$.*

Proof: We write trace for $\mathrm{trace}_{F/E}$ and notice that the set $B' = \{x \in F \mid \mathrm{trace}(xB) \subseteq A\}$ is a fractional ideal of B that contains B [Lan70, p. 58, Cor.]. Hence, $D = (B')^{-1}$ is an ideal of B (called the **Different** of B over A). Moreover, a prime ideal P of B is ramified over A if and only if P divides D [Lan70, p. 62, Prop. 8]. Since, by assumption, no P is ramified over A, we have $D = B$, so $B' = B$.

Next observe that since trace: $F \to E$ is an E-linear map, $\mathrm{trace}(B)$ is an ideal of A, hence there exists $a \in A$ with $\mathrm{trace}(B) = aA$. Since F/E is separable, the map trace: $F \to E$ is nonzero [Lan93, p. 286, Thm. 5.2], so $a \neq 0$. Therefore, $\mathrm{trace}(a^{-1}B) = A$, hence $a^{-1} \in B' = B$. It follows that $a^{-1}B = B$, so $\mathrm{trace}(B) = A$. Consequently, there exists $b \in B$ with $\mathrm{trace}(b) = 1$. $\square$

LEMMA 4.2.2: *Suppose $p = \mathrm{char}(K) > 0$. Let F be a cyclic extension of $K(x)$ of degree p^n in $K((x))$, $n \geq 1$. Suppose the integral closure O of $K[x]$ in F is unramified over $K[x]$. Then $K(x)$ has a cyclic extension F' of degree p^{n+1} in $K((x))$ which contains F such that the integral closure O' of $K[x]$ in F' is unramified over $K[x]$.*

Proof: We define F' to be $F(z)$, where z is a zero of $Z^p - Z - a$ with a suitable element $a \in O$. The three parts of the proof produce a, and then show that F' has the desired properties.

PART A: *Construction of a and z.*   We apply Lemma 4.2.1 to choose $b \in O$ with $\mathrm{trace}_{F/K(x)}(b) = 1$. Then we set $c = b - b^p$ and notice that

$$\mathrm{trace}_{F/K(x)}(c) = \mathrm{trace}_{F/K(x)}(b) - (\mathrm{trace}_{F/K(x)}(b))^p = 0.$$

Let σ be a generator of $\mathrm{Gal}(F/K(x))$. Set $q = p^n$ and

$$a_1 = \sum_{i=1}^{q-1} \sum_{j=0}^{i-1} b^{\sigma^i} c^{\sigma^j}.$$

Then $a_1 \in O$ and

$$a_1^\sigma = \sum_{i=1}^{q-1} \sum_{j=0}^{i-1} b^{\sigma^{i+1}} c^{\sigma^{j+1}} = \sum_{i=2}^{q} \sum_{j=1}^{i-1} b^{\sigma^i} c^{\sigma^j}.$$

Hence,

$$a_1 - a_1^\sigma = b^\sigma c + b^{\sigma^2} c + \cdots + b^{\sigma^{q-1}} c - b^{\sigma^q} \sum_{j=1}^{q-1} c^{\sigma^j}$$

$$= \sum_{i=0}^{q-1} b^{\sigma^i} c - b \sum_{j=0}^{q-1} c^{\sigma^j} = \mathrm{trace}_{F/K(x)}(b)c - b \cdot \mathrm{trace}_{F/K(x)}(c) = c.$$

Now note that O is integral over $K[x]$ and is contained in $K((x))$, so $O \subseteq K[[x]]$ (because $K[[x]]$ is integrally closed), in particular $a_1 \in K[[x]]$. Let v be the K-valuation of $K((x))$ with $v(x) = 1$. Since $K[x]$ is v-dense in $K[[x]]$, there is an $a_0 \in K[x]$ with $v(a_1 - a_0) > 0$. Set $a = a_1 - a_0$. Then $a \in O$, $v(a) > 0$, and

$$(1) \qquad\qquad a^\sigma - a = b^p - b.$$

Thus, the polynomial $f(Z) = Z^p - Z - a$ satisfies $v(f(0)) = v(-a) > 0$ and $v(f'(0)) = v(-1) = 0$. By Hensel's lemma for $K((x))$, there exists $z \in K[[x]]$ such that $z^p - z - a = 0$.

PART B: *Irreducibility of $Z^p - Z - a$.* Assume $Z^p - Z - a$ is reducible over F. Then $z \in F$ [Lan93, p. 290, Thm. 6.4(b)]. By (1),

$$(2) \quad (z^\sigma - z)^p - (z^\sigma - z) - (b^p - b) = (z^\sigma - z)^p - (z^\sigma - z) - (a^\sigma - a)$$
$$= (z^p - z - a)^\sigma - (z^p - z - a) = 0.$$

Since b is a root of $Z^p - Z - (b^p - b)$, there is an integer i with $z^\sigma - z = b + i$ [Lan93, p. 290, Thm. 6.4(b)]. Apply $\mathrm{trace}_{F/K(x)}$ to both sides to get 0 on the left and 1 on the right. This contradiction proves that $Z^p - Z - a$ is irreducible.

It follows that $f = \mathrm{irr}(z, F) \in O[Z]$, $f(z) = 0$, and $f'(z) = -1$, in particular $\mathrm{discr}(f)$ is a unit of O [FrJ08, p. 109]. Hence, $O' = O[z]$ is the integral closure of O in $F' = F(z)$ and O'/O is a ring cover in the sense of [FrJ08, Def. 6.1.3]. In particular, no prime ideal of O is ramified in O' [FrJ08, Lemma 6.1.8(b)]. It follows from our assumption on O that O' is unramified over $K[x]$.

PART C: *Extension of σ to σ' that maps z to $z + b$.* Equality (1) implies that $z + b$ is a zero of $Z^p - Z - a^\sigma$. Thus, by Part B, σ extends to an automorphism σ' of F' with $z^{\sigma'} = z + b$. We need only to prove that the order of σ' is p^{n+1}. Induction shows $z^{(\sigma')^j} = z + b + b^\sigma + \cdots + b^{\sigma^{j-1}}$ for each $j \geq 1$. In particular,

$$z^{(\sigma')^q} = z + \mathrm{trace}_{F/K(x)}(b) = z + 1.$$

Hence, $z^{(\sigma')^{iq}} = z + i$ for $i = 1, \ldots, p$, so the order of σ' is p^{n+1}. Consequently, F' is a cyclic extension of $K(x)$ of order p^{n+1}. □

Remark 4.2.3: Inertia group. Let $(F, w)/(E, v)$ be a finite Galois extension of discrete valued fields. The corresponding **inertia group** is the subgroup

$$I_{w/v} = \{\sigma \in \mathrm{Gal}(F/E) \mid w(x^\sigma - x) > 0 \text{ for each } x \in O_w\}$$

of $\mathrm{Gal}(F/E)$. Its order satisfies $|I_{w/v}| = e_{w/v}[\bar{F}_w : \bar{E}_v]_i$, where the second factor on the right is the inseparability index of the residue field extension

[Ser79, p. 22, Prop. 21(a)]. Thus, w/v is ramified if and only if $I_{w/v}$ is non-trivial. If w/v is totally ramified, then $e_{w/v} = [F : E]$, so $I_{w/v} = \mathrm{Gal}(F/E)$ and $\bar{E}_v = \bar{F}_w$ (because, in general, $e_{w/v}[\bar{F}_w : \bar{E}_v] \leq [F : E]$). If (F', w') is a finite extension of (F, w) and F'/E is Galois, then the restriction map res: $\mathrm{Gal}(F'/E) \to \mathrm{Gal}(F/E)$ maps $I_{w'/v}$ onto $I_{w/v}$ [Ser79, p. 22, Prop. 22(b)]. Taking inverse limit, the latter assertion generlizes also to arbitrary Galois extensions F'/E.

Note that there may be several valuations of F lying over v and they are conjugate to one another. Thus, we refer to $I_{w/v}$ also as **an inertia group of v in $\mathrm{Gal}(F/E)$**.

Now consider an extension E_1 of E in F in which v is unramified. Then v is unramified in each E-conjugate of E_1. Hence, by [FrJ08, Cor. 2.3.7(c)], v is unramified in the compositum $\hat{E}$ of all conjugates of E_1 over E. It follows by the first paragraph of this remark that the restriction of $I_{w/v}$ to $\hat{E}$ is trivial. In particular, E_1 is contained in the fixed field of $I_{w/v}$ in F.

Let θ be an isomorphism of F onto a field F' and set $E' = E^\theta$. Then F'/E' is a finite Galois extension and $\sigma \mapsto \sigma^\theta$ is an isomorphism of $\mathrm{Gal}(F/E)$ onto $\mathrm{Gal}(F'/E')$. Let w^θ and v^θ be the corresponding valuations of F' and E' defined by $w^\theta(x') = w((x')^{\theta^{-1}})$ and $v^\theta(x') = v((x')^{\theta^{-1}})$. A direct check reveals that $I^\theta_{w/v} = I_{w^\sigma/v^\sigma}$.

If, in addition, $E = K(x)$ is a rational function field and $b \in \tilde{K} \cup \{\infty\}$, then we call each inertia group of $v_{x,b}$ in $\mathrm{Gal}(F/E)$ also an **inertia group of b in $\mathrm{Gal}(F/E)$** (with respect to x). $\square$

Lemma 4.2.4: *Let K be a field of positive characteristic p, x an indeterminate, n a positive integer, and $a \in K^\times$. Then $K(x)$ has a cyclic extension F in $K((x))$ of degree p^n such that $\mathrm{Branch}(F/K(x)) = \{a\}$ and $\mathfrak{p}_{x,a}$ is totally ramified in F.*

Proof: We reduce the general case to the case where K is an algebraic extension of a finite field.

PART A: *We assume that K is perfect.* Let $u = \frac{x}{x-a}$ and observe that $K(x) = K(u)$. Note that $x - a$ is a unit of $K[[x]]$. Hence, $K((x)) = K((u))$ and $K[[x]] = K[[u]]$.

In the case where $n = 1$ we consider the polynomial $f(Y) = Y^p - Y - u$. Then $v_{x,0}(f(0)) = v_{x,0}(u) = 1$ and $v_{x,0}(f'(0)) = v_{x,0}(-1) = 0$. Hence, by Hensel's lemma, there exists $y \in K((x))$ with $y^p - y = u$.

Next note that $v_{x,a} = v_{u,\infty}$ and $\mathfrak{p}_{x,a} = \mathfrak{p}_{u,\infty}$. If $y \in K(x)$, then $v_{x,a}(y) < 0$, so $pv_{x,a}(y) = v_{x,a}(y^p - y) = v_{x,a}(u) = -1$, which is a contradiction. Hence, by Artin-Schreier, $f(Y)$ is irreducible and $K(y)/K(x)$ is a cyclic extension of degree p [Lan93, p. 290, Thm. 6.4(b)]. Moreover, $\mathfrak{p}_{x,a}$ totally ramifies in $K(y)$ [FrJ08, Example 2.3.9]. If v is another valuation of $K(x)/K$, then $v(u) \geq 0$, so v is unramified in $F_1 = K(y)$. It follows that $\mathrm{Ram}(F_1/K(x)) = \{\mathfrak{p}_{x,a}\}$. Thus, if we denote the integral closure of $K[u]$ in F_1 by O_1, we find that O_1 is unramified over $K[u]$.

Inductively suppose we have constructed for $n \geq 1$ a cyclic extension F of $K(x)$ in $K((x))$ of degree p^n containing F_1 such that $\mathfrak{p}_{x,a}$ is the only prime divisor of $K(x)/K$ that ramifies in F. By Lemma 4.2.2, $K(u) = K(x)$ has a cyclic extension F' of degree p^{n+1} in $K((u)) = K((x))$. Moreover, the integral closure of $K[u]$ in F' is unramified over $K[u]$. Since $\mathfrak{p}_{x,a}$ ramifies in F_1, it also ramifies in F'. Thus, $\mathrm{Branch}(F'/K(x)) = \{a\}$.

In order to prove that $\mathfrak{p}_{x,a}$ is totally ramified in F' note that our construction gives a unique prime divisor $\mathfrak{q}$ of F_1/K lying over $\mathfrak{p} = \mathfrak{p}_{x,a}$. Let $\mathfrak{q}'$ be a prime divisor of F' lying over $\mathfrak{q}$. By Remark 4.2.3, the restriction map $\mathrm{res}\colon \mathrm{Gal}(F'/K(x)) \to \mathrm{Gal}(F_1/K(x))$ maps the inertia group $I_{\mathfrak{q}'/\mathfrak{p}}$ onto the inertia group $I_{\mathfrak{q}/\mathfrak{p}}$. Since $\mathrm{Gal}(F'/K(x)) \cong \mathbb{Z}/p^{n+1}\mathbb{Z}$ and $\mathrm{Gal}(F_1/K(x)) \cong \mathbb{Z}/p\mathbb{Z}$, the restriction map is a **Frattini cover** [FrJ08, Def. 22.5.1]. This implies that $I_{\mathfrak{q}'/\mathfrak{p}} = \mathrm{Gal}(F'/K(x))$. Since K is perfect, $e_{\mathfrak{q}'/\mathfrak{p}} = |I_{\mathfrak{q}'/\mathfrak{p}}| = [F' : K(x)]$ (Remark 4.2.3). This means that $\mathfrak{p}$ is totally ramified in F'.

PART B: *The general case.* Denote the algebraic closure of $\mathbb{F}_p$ in K by K_0. Then K_0 is a perfect field, so K/K_0 is a regular extension. Part A gives a cyclic extension F_0 of $K_0(x)$ in $K_0((x))$ of degree p^n such that $\mathfrak{p}_0 = \mathfrak{p}_{K_0,x,1}$ is totally ramified in F_0 and $\mathrm{Branch}(F_0/K(x)) = \{1\}$. Since F_0 and K are linearly disjoint over K_0, $F = F_0K$ is a cyclic extension of $K(x)$ of degree p^n. Moreover, $F \subseteq K((x))$.

Finally we choose a Möbius transformation θ of $K(x)/K$ with $\theta(0) = 0$ and $\theta(a) = 1$. Then we extend θ to an isomorphism of F onto a field F'. Then F' is a cyclic extension of $K(x)$ of degree p^n in $K((x))$, the prime divisor $\mathfrak{p}_{K,x,a}$ of $K(x)/K$ is totally ramified in F', and $\mathrm{Branch}(F'/K(x)) = \{a\}$, as desired. $\qquad\square$

We combine Lemma 4.1.6 and Lemma 4.2.4 into the following result.

LEMMA 4.2.5: *Let K be a field of characteristic p, x an indeterminate, n a positive integer, and $a, b \in K^\times$ with $b \neq a$. Suppose either $p \nmid n$ or n is a power of p. Then $K(x)$ has a finite cyclic extension F in $K((x))$ of order n such that*

$$\mathrm{Branch}(F/K(x)) = \begin{cases} \{a, b\} & \text{if } p \nmid n \text{ and } \zeta_n \in K \\ \{a\zeta_n^\sigma \mid \sigma \in \mathrm{Gal}(K(\zeta_n)/K)\} & \text{if } p \nmid n \text{ and } \zeta_n \notin K \\ \{a\} & \text{if } p = p > 0 \text{ and } n = p^k. \end{cases}$$

Moreover, each $\mathfrak{p} \in \mathrm{Ram}(F/K(x))$ is totally ramified in F.

Remark 4.2.6: *Finite abelian groups.* For each finite Abelian group A the field $K(x)$ has a Galois extension F such that F/K is regular and $\mathrm{Gal}(F/K(x)) \cong A$ [FrJ08, Prop. 16.3.5]. When K is infinite, a field crossing argument then shows how to choose F in $K((x))$ [HaV96, Lemma 4.5].

Alternatively, let $A = \prod_{i=1}^r A_i$ be the direct product of cyclic groups A_i, where $|A_i| = p_i^{k_i}$ and p_i is a prime number. By Lemma 4.2.5, we may construct for each i a cyclic extension F_i of degree $|A_i|$ of $K(x)$ in $K((x))$ such that the sets $\text{Branch}(F_i/K(x))$ are disjoint. Then for each $1 \le k \le r-1$,

$$\text{Branch}(F_1 \cdots F_k/K(x)) \cap \text{Branch}(F_{k+1}/K(x)) = \emptyset.$$

Therefore, $E = F_1 \cdots F_k \cap F_{k+1}$ is an unramified extension of $K(x)$. By Riemann-Hurwitz, $E = K(x)$ [FrJ08, Remark 3.6.2(d)]. It follows that $\text{Gal}(F_1 \cdots F_r/K(x)) \cong A$. $\qquad \square$

4.3 Rational Realization of Cyclic Groups over Complete Fields

When the basic field K is complete under an ultrametric absolute value, we are able to construct the cyclic extensions of $K(x)$ in the fields of converging power series. This is necessary for patching of cyclic groups into arbitrary finite groups.

Remark 4.3.1: Contraction. For $c \in K^\times$ let μ_c be the K-automorphism of $K((x))$ that maps $f = \sum_{n=m}^\infty a_n x^n$ onto $\mu_c(f) = \sum_{n=m}^\infty a_n c^n x^n$. In particular, $\mu_c(h(x)) = h(cx)$ for each $h \in K[x]$, so μ_c leaves $K[x]$ and $K(x)$ invariant. If K is complete under an ultrametric absolute value $|\ |$ and f converges at c, then $a_n c^n \to 0$, hence $\mu_c(f) \in \text{Quot}(K\{x\})$ and $\mu_c(f)$ converges at 1. Moreover, if $f \in K[[x]]$, then $\mu_c(f) \in K\{x\}$. Thus, μ_c "contracts the radius of convergence" of the functions from $|c|$ to 1. $\qquad \square$

Notation 4.3.2: Let K be a field, $|\ |$ an a ultrametric absolute value, and π_0 an element of $K^\times$ with $|\pi_0| < 1$. For each positive integer n with $\text{char}(K) \nmid n$ let ζ_n be a primitive root of unity of order n. Then let

$$\Pi(K, \pi_0) = \{\pi_0^{-i}\zeta_n \mid i = 0, 1, 2, \ldots; \ n \in \mathbb{N}, \ \text{char}(K) \nmid n\}$$

and note that $\Pi(K, \pi_0)$ is countable. $\qquad \square$

LEMMA 4.3.3: *Let K be an infinite field, x an indeterminate, $E = K(x)$, F a Galois extension of E in $K((x))$ of a finite degree n, and $b_1, \ldots, b_n$ distinct elements of K. Let S be the integral closure of $K[x]$ in F. Then S has n distinct prime ideals $\mathfrak{m}_1, \ldots, \mathfrak{m}_n$ lying over $K[x]x$ and F/E has a primitive element $y \in S$ such that $y \equiv b_i \bmod \mathfrak{m}_i$, $i = 1, \ldots, n$. Moreover, the polynomials $g = \text{irr}(y, E) \in K[x, Y]$ and $h = \text{discr}(\text{irr}(y, E)) \in K[x]$ satisfy $h'(0) \ne 0$ and $\frac{\partial g}{\partial Y}(0, b_i) \ne 0$ for $i = 1, \ldots, n$.*

Proof: Since $F \subseteq K((x))$ and since F/E is Galois, S has n distinct prime ideals $\mathfrak{m}_1, \ldots, \mathfrak{m}_n$ that lie over $K[x]x$ and $S/\mathfrak{m}_i \cong K$, $i = 1, \ldots, n$. Moreover, for each i there exists $\sigma_i \in \text{Gal}(F/E)$ such that $\sigma_i \mathfrak{m}_i = \mathfrak{m}_1$ and $\sigma_1 = \text{id}_F$. By the Chinese remainder theorem, there exists $y \in S$ such that $y \equiv b_i \bmod \mathfrak{m}_i$, $i = 1, \ldots, n$. Hence, $y_i = \sigma_i y$ satisfies $y_i \equiv b_i \bmod \mathfrak{m}_1$, $i = 1, \ldots, n$. In particular, $y_1, \ldots, y_n$ are n distinct roots of $g = \text{irr}(y, E)$.

It follows that $F = E(y)$. Since F/E is Galois, $y_1, \ldots, y_n \in S$. It follows that $h = \mathrm{discr}(\mathrm{irr}(y, E)) = \pm \prod_{i \neq j}(y_i - y_j) = \sum_{i=0}^{m} c_i x^i$ is in $K[x]$ and $c_0 = h(0) = \pm \prod_{i \neq j}(b_i - b_j) \neq 0$. Finally, $h(x) = \pm \prod_{i=1}^{n} \frac{\partial g}{\partial Y}(x, y_i)$ and $h(0) = \pm \prod_{i=1}^{n} \frac{\partial g}{\partial Y}(0, b_i)$. Consequently, $\frac{\partial g}{\partial Y}(0, b_i) \neq 0$ for $i = 1, \ldots, n$. $\qquad \square$

LEMMA 4.3.4: *Let K_0 be a complete field under an ultrametric absolute value $| \ |$, π_0 an element of $K_0^{\times}$ with $|\pi_0| < 1$, and C a finite cyclic group of a prime power order. Set $E_0 = K_0(x)$. Then for each $a \in K_0^{\times}$ there exist $\pi \in \Pi(K_0, \pi_0)$ and a Galois extension $F_0 = E_0(z)$ of E_0 in $K_0((x))$ with Galois group C such that all conjugates of z over E_0 are in $K_0\{x\}$, $\mathrm{discr}(\mathrm{irr}(z, E_0)) \in K_0\{x\}^{\times}$, and $a\pi \in \mathrm{Branch}(F_0/E_0)$. Moreover, $\mathfrak{p}_{x, a\pi}$ is totally ramified in F_0.*

Proof: Let $n = |C|$ and $a \in K_0^{\times}$. By Lemma 4.2.5 (with $a\zeta_n$ replacing a if $\zeta_n \in K$), there exist a Galois extension E_0' of E_0 in $K_0((x))$ with $\mathrm{Gal}(E_0'/E_0) \cong C$ and $a\zeta_n \in \mathrm{Branch}(E_0'/E_0)$. Moreover, $\mathfrak{p}_{x, a\zeta_n}$ is totally ramified in E_0'.

Let S_0 be the integral closure of $K_0[x]$ in E_0'. By Lemma 4.3.3 the extension E_0'/E_0 has a primitive element $y \in S_0$ such that

$$h(x) = \mathrm{discr}(\mathrm{irr}(y, E_0)) = \sum_{i=0}^{l} c_i x^i \in K_0[x]$$

and $c_0 \neq 0$. Denote the E_0-conjugates of y by $y_1, \ldots, y_n$. All of them lie in S_0.

Next note that S_0, as the integral closure of $K_0[x]$ in E_0', is contained in every valuation ring of E_0' containing $K_0[x]$, so $S_0 \subseteq K_0[[x]]$. By Proposition 2.4.5, every y_i converges at $c = \pi_0^m$ if m is sufficiently large. Then $z_i = \mu_c(y_i) \in K_0\{x\}$, $i = 1, \ldots, n$ (Remark 4.3.1). Setting $F_0 = \mu_c(E_0')$ and $z = \mu_c(y)$, we find that F_0 is a Galois extension of E_0 with $\mathrm{Gal}(F_0/E_0) \cong C$, $F_0 = E_0(z)$, and $z_1, \ldots, z_n$ are the E_0-conjugates of z. Moreover, $h(cx) = \mu_c(h(x)) = \mathrm{discr}(\mathrm{irr}(z, E_0))$, and $h(cx) = c_0 + c_1 cx + \cdots + c_l c^l x^l$. If m is sufficiently large, then $|c_k c^k| < |c_0|$, $k = 1, \ldots, l$. Thus, pseudo.deg$(h(cx)) = 0$, so $h(cx) \in K_0\{x\}^{\times}$ (Proposition 2.3.1(c)), as desired. By Remark 4.1.4, $a\zeta_n c^{-1} = (\mu_c')^{-1}(a\zeta_n) \in \mathrm{Branch}(F_0/E_0)$. Hence, $\pi = \zeta_n \pi_0^{-m} \in \Pi(K, \pi_0)$, $a\pi = a\zeta_n c^{-1} \in \mathrm{Branch}(F_0/E_0)$, and $\mathfrak{p}_{x, a\pi}$ is totally ramified in F_0 (Remark 4.2.3). $\qquad \square$

As in Chapter 3, let r be an element of $K^{\times}$, I a finite set, and c_i distinct elements of K for $i \in I$ such that $|r| \leq |c_i - c_j|$ for $i \neq j$. Recall that $w_i = \frac{r}{x - c_i}$, $P_i' = \mathrm{Quot}(K\{w_i\})$, and $R = K\{w_i \mid i \in I\}$.

LEMMA 4.3.5: *Let K_0 be a complete field under an ultrametric absolute value $| \ |$, π_0 an element of $K_0^{\times}$ with $|\pi_0| < 1$, K a finite Galois extension of K_0, and C a finite cyclic group of a prime power order. Set $E_0 = K_0(x)$ and $E = K(x)$. Then for each $i \in I$ and each $a \in K_0^{\times}$ there exists $\pi \in \Pi(\pi_0, K_0)$*

and a Galois extension $F = E(z)$ of E in $K((w_i))$ with Galois group C such that $z^\tau \in P_i' \cap R$ for all $\tau \in \text{Gal}(F/E)$, $\text{discr}(\text{irr}(z, E)) \in R^\times$, $b = \frac{\pi a c_i + r}{\pi a} \in \text{Branch}(F/E)$, and $\mathfrak{p}_{K,x,b}$ is totally ramified in F.

Proof: Since $w_i = \frac{r}{x-c_i}$, we have $K_0(w_i) = E_0$. By Lemma 2.2.1(d), the map $x \mapsto w_i$ extends to a K_0-homomorphism $\alpha_i : K_0\{x\} \to K_0\{w_i\}$. By Lemma 3.2.1, α_i is an isomorphism. Now let $a \in K_0^\times$. Then, by Lemma 4.3.4, there exists $\pi \in \Pi(K_0, \pi_0)$ and E_0 has a finite Galois extension $F_0 = E_0(z_0)$ in $K_0((x))$ with Galois group C such that all conjugates of z_0 over E_0 are in $K_0\{x\}$, $\text{discr}(\text{irr}(z_0, E_0)) \in K_0\{x\}^\times$, and $a\pi \in \text{Branch}(F_0/E_0)$. Moreover, $\mathfrak{p}_{K_0,x,a\pi}$ is totally ramified in F_0. It follows that $F_0' = \alpha_i(F_0)$ is a Galois extension of E_0 with Galois group C and with a primitive element $z = \alpha_i(z_0)$ such that $z^\tau \in K_0\{w_i\}$ for all $\tau \in \text{Gal}(F_0'/E_0)$, $\text{discr}(\text{irr}(z, E_0)) \in K_0\{w_i\}^\times$, and $b = (\alpha_i')^{-1}(a\pi) = \frac{\pi a c_i + r}{\pi a} \in \text{Branch}(F_0'/E_0)$ (Remark 4.1.4). Since $K_0\{w_i\} \subseteq P_i' \cap R$, all of the E_0-conjugates of z belong to $P_i' \cap R$ and $\text{discr}(\text{irr}(z, E_0)) \in R^\times$. Moreover, $\mathfrak{p}_{K_0,x,b}$ is totally ramified in F_0' (Remark 4.2.3).

By construction, $F_0' \subseteq K_0\{w_i\} \subseteq K_0((w_i))$. Since $K_0((w_i))$ is a regular extension of K_0 [FrJ08, Example 3.5.1] and K/K_0 is a finite Galois extension, $F = F_0'K$ is a Galois extension of E with Galois group C, $b \in \text{Branch}(F/E)$, and $\mathfrak{p}_{K,x,b}$ is totally ramified in F. $\square$

LEMMA 4.3.6: Let K be a complete field under an ultrametric absolute value $|\ |$. Then, for each $b \in K^\times$, each of the sets $D_b = \{a \in K \mid |a| \le |b|\}$, $D_b^0 = \{a \in K \mid |a| < |b|\}$, and $U_b = \{a \in K \mid |a| = |b|\}$ has cardinality $\text{card}(K)$.

Proof: The map $a \mapsto a^{-1}$ maps D_b bijectively onto the set $D_b' = \{a \in K \cup \{\infty\} \mid |a| \ge |b|\}$. Since K is complete, K is infinite. It follows from $K \cup \{\infty\} = D_b \cup D_b'$ that $\text{card}(D_b) = \text{card}(K)$. Next we choose $c \in K^\times$ with $|c| < |b|$. Then $D_c \subseteq D_b^0$. Since $\text{card}(D_c) = \text{card}(K)$, we have $\text{card}(D_b^0) = \text{card}(K)$. Finally, the map $a \mapsto a + b$ maps D_b^0 injectively into U_b. Therefore, $\text{card}(U_b) = \text{card}(K)$. $\square$

LEMMA 4.3.7: Let K be a complete field under an ultrametric absolute value $|\ |$, x an indeterminate, $E = K(x)$, and F a finite Galois extension of E. Suppose F/K has a prime divisor $\mathfrak{P}$ of degree 1 unramified over E.
(a) There is a K-automorphism of E that extends to a K-embedding of F into $K((x))$.
(b) There is a K-automorphism of E that extends to an isomorphism θ of F onto a field $F' = E(z)$ such that all conjugates of z over E are in $K\{x\}$ and $\text{discr}(\text{irr}(z, E)) \in K\{x\}^\times$.
(c) Suppose $F \subseteq K((x))$. Let B be a subset of $K \cup \{\infty\}$ with $\text{card}(B) < \text{card}(K)$. Then we can choose θ such that in addition to the requirements of (b), it satisfies $\text{Branch}(F'/E) \cap B \subseteq \{\infty\}$.

Proof of (a): Let $\mathfrak{p}$ be the prime of E/K below $\mathfrak{P}$. Let $\hat{F}$ be the completion of F at $\mathfrak{P}$, and let $\hat{E} \subseteq \hat{F}$ be the completion of E at $\mathfrak{p}$. Then $[\hat{F} : \hat{E}] =$

$e_{\mathfrak{P}/\mathfrak{p}} f_{\mathfrak{P}/\mathfrak{p}} = 1$, so $\hat{F} = \hat{E}$. Moreover, there is a K-automorphism θ_0 of E that maps $\mathfrak{p}$ to $\mathfrak{p}_{x,0}$. The completion of E at $\mathfrak{p}_{x,0}$ is $K((x))$. Hence, θ_0 extends to an automorphism θ' of $\hat{F} = \hat{E}$ onto $K((x))$.

Proof of (b): By (a), we may assume that $F \subseteq K((x))$. Let $n = [F : E]$ and let S be the integral closure of $K[x]$ in F. We choose a primitive element y for F/E in S such that $h = \operatorname{discr}(\operatorname{irr}(y, E)) = c_0 + c_1 x + \cdots + c_m x^m \in K[x]$ and $h(0) \neq 0$ (Lemma 4.3.3). Then we denote the E-conjugates of y by $y_1, \ldots, y_n$. All of them belong to S.

Next we note that S is contained in $K[[x]]$. By Proposition 2.4.5, there exists $c \in K^\times$ at which all y_i converge. By Remark 4.3.1, μ_c is an automorphism of $K((x))$, so $\theta = \mu_c|_F$ is an isomorphism of F onto a subfield $F' = E(z)$ of $K((x))$, where $z = \mu_c(y)$, and $z_i = \mu_c(y_i) \in K\{x\}$, $i = 1, \ldots, n$ are all the E-conjugates of z. Moreover, $h(cx) = \mu_c(h(x)) = \operatorname{discr}(\operatorname{irr}(z, E))$, where $h(cx) = c_0 + cc_1 x + \cdots + cc_m x^m$. Replacing c with an element of smaller absolute value, we may assume that $|cc_j| < |c_0|$, $j = 1, \ldots, m$. Thus, pseudo.deg$(h(cx)) = 0$ and therefore $h(cx) \in K\{x\}^\times$ (Proposition 2.3.1(c)), as desired.

Proof of (c): Denote the set of all zeros of h by A. Since A is a finite set, $\operatorname{card}(AB^{-1}) < \operatorname{card}(K)$. By Lemma 4.3.6, the set of all $c \in K^\times$ with $|c| < \min_{1 \leq j \leq m} |c_j^{-1} c_0|$ has cardinality $\operatorname{card}(K)$. In the proof of (b) we may therefore choose c outside AB^{-1}. Then Branch$(F'/E) \cap B \subseteq \{\infty\}$.

Indeed, let $b \in \operatorname{Branch}(F'/E) \cap B$, $b \neq \infty$. Then b is a zero of discr$((\operatorname{irr}(z, E))$ (Remark 4.1.1), that is $h(cb) = 0$, so $a = cb \in A$. Since $h(0) \neq 0$, we have $b \neq 0$. Hence, $c = ab^{-1} \in AB^{-1}$. This contradiction to the choice of c proves our assertion. $\qquad\square$

4.4 Solution of Embedding Problems

Let K/K_0 be a finite Galois extension of fields with Galois group Γ acting on a finite group G. Consider an indeterminate x and set $E_0 = K_0(x)$ and $E = K(x)$. Then E/E_0 is a Galois extension and we identify $\operatorname{Gal}(E/E_0)$ with $\Gamma = \operatorname{Gal}(K/K_0)$ via restriction. We refer to

$$\text{(1)} \qquad\qquad \operatorname{pr}: \Gamma \ltimes G \to \Gamma$$

as a **constant finite split embedding problem over** E_0. We prove that if K_0 is complete under an ultrametric absolute value, then (1) has a solution field (Section 1.2) equipped with a K-rational place whose decomposition group is Γ.

LEMMA 4.4.1: *Let F be a solution field of (1). Let $F_0 = F^\Gamma$ be the fixed field of Γ in F. Let $\varphi: F \to \widetilde{K_0} \cup \{\infty\}$ be a K-place with $\varphi(x) \in K_0 \cup \{\infty\}$. Suppose φ is unramified over E_0, and let D_φ and $\bar{F}_\varphi$ be its decomposition group (over E_0) and residue field, respectively. Then $\bar{F}_\varphi \supseteq K$ and the following assertions are equivalent:*

(a) $\bar{F}_\varphi = K$ and $\Gamma = D_\varphi$;
(b) $\Gamma \supseteq D_\varphi$;
(c) $\bar{F}_{0,\varphi} = K_0$;
(d) $\bar{F}_\varphi = K$ and $\varphi(f^\gamma) = \varphi(f)^\gamma$ for each $\gamma \in \Gamma$ and $f \in F$ with $\varphi(f) \neq \infty$.

Proof: Since $K \subseteq F$, we have $K = \varphi(K) \subseteq \bar{F}_\varphi$. Since the inertia group of φ in $\mathrm{Gal}(F/E_0)$ is trivial, we have an isomorphism $\theta \colon D_\varphi \to \mathrm{Gal}(\bar{F}_\varphi/K_0)$ given by

$$(2) \qquad \varphi(f^\gamma) = \varphi(f)^{\theta(\gamma)}, \qquad \gamma \in D_\varphi, \ f \in F, \ \varphi(f) \neq \infty.$$

Hence, $|D_\varphi| = [\bar{F}_\varphi : K_0] \geq [K : K_0] = |\Gamma|$. This gives (a) $\Longleftrightarrow$ (b).

Since φ is unramified over E_0, the decomposition field F^{D_φ} is the largest intermediate field of F/E_0 with residue field K_0. Thus, if $\bar{F}_{0,\varphi} = K_0$, then $F^\Gamma = F_0 \subseteq F^{D_\varphi}$, so $D_\varphi \subseteq \Gamma$. Conversely, if $D_\varphi \subseteq \Gamma$, then $F_0 = F^\Gamma \subseteq F^{D_\varphi}$, hence $\bar{F}_{0,\varphi} = K_0$. Consequently, (b) is equivalent to (c).

Assertion (d) implies (c) by (2). If $\bar{F}_\varphi = K$, then $f^\gamma = \varphi(f^\gamma) = \varphi(f)^{\theta(\gamma)} = f^{\theta(\gamma)}$ for each $f \in K$ (by (2)), so that $\theta(\gamma) = \gamma$ for all $\gamma \in D_\varphi$. Consequently, (a) $\Longrightarrow$ (d). $\qquad\square$

PROPOSITION 4.4.2: Let K_0 be a complete field with respect to an ultra-metric absolute value $|\ |$. Let K/K_0 be a finite Galois extension with Galois group Γ acting on a finite group G from the right. Then E has a Galois extension F such that
(3a) F/E_0 is Galois;
(3b) there is an isomorphism $\psi \colon \mathrm{Gal}(F/E_0) \to \Gamma \ltimes G$ such that $\mathrm{pr} \circ \psi = \mathrm{res}_E$; and
(3c) F has a K-rational place φ (so F/K is regular) unramified over E_0 such that $\varphi(x) \in K_0$, $\bar{F}_\varphi = K$, and $D_\varphi = \Gamma$.

Proof: Our strategy is to attach patching data $\mathcal{E}$ to the embedding problem, to define a proper action of Γ on $\mathcal{E}$. Then we apply Proposition 1.2.2 to conclude that the compound F of $\mathcal{E}$ gives a solution to the embedding problem.

We fix a finite set I on which Γ acts from the right and a system of generators $\{\tau_i \mid i \in I\}$ of G such that for each $i \in I$
(4a) $\{\gamma \in \Gamma \mid i^\gamma = i\} = \{1\}$;
(4b) the order of the group $G_i = \langle \tau_i \rangle$ is a power of a prime number;
(4c) $\tau_i^\gamma = \tau_{i^\gamma}$, for every $\gamma \in \Gamma$; and
(4d) $|I| \geq 2$.
(E.g. assuming $G \neq 1$, let G_0 be the set of all elements of G whose order is a power of a prime number and note that Γ leaves G_0 invariant. Let $I = G_0 \times \Gamma$ and for each $(\sigma, \gamma) \in I$ and $\gamma' \in \Gamma$ let $(\sigma, \gamma)^{\gamma'} = (\sigma, \gamma\gamma')$ and $G_{(\sigma,\gamma)} = \langle \sigma^\gamma \rangle$.)

Then $G_i^\gamma = G_{i^\gamma}$ for each $\gamma \in \Gamma$ and $G = \langle G_i \mid i \in I \rangle$. Choose a system of representatives J for the Γ-orbits of I. Then every $i \in I$ can be uniquely written as $i = j^\gamma$ with $j \in J$ and $\gamma \in \Gamma$.

CLAIM A: *There exists a subset $\{c_i \mid i \in I\} \subseteq K$ such that $c_i^\gamma = c_{i^\gamma}$ and $c_i \neq c_j$ for all distinct $i, j \in I$ and $\gamma \in \Gamma$.*

Indeed, it suffices to find $\{c_j \mid j \in J\} \subseteq K$ (and then define c_i, for $i = j^\gamma \in I$, as c_j^γ) such that $c_j^\delta \neq c_j^\varepsilon$ for all $j \in J$ and all distinct $\delta, \varepsilon \in \Gamma$, and $c_j^\delta \neq c_k$ for all distinct $j, k \in J$ and all $\delta \in \Gamma$.

The first condition says that c_j is a primitive element for K/K_0; the second condition means that distinct c_j and c_k are not conjugate over K_0. Thus it suffices to show that there are infinitely many primitive elements for K/K_0. But if $c \in K^\times$ is primitive, then so is $c + a$, for each $a \in K_0$. Since K_0 is complete, hence infinite, the claim follows.

CONSTRUCTION B: *Patching data.*

We choose $r \in K_0^\times$ such that $|r| \leq |c_i - c_j|$ for all distinct $i, j \in I$. For each $i \in I$ we set $w_i = \frac{r}{x - c_i} \in K(x)$. As in Section 3.2, consider the ring $R = K\{w_i \mid i \in I\}$ and let $Q = \operatorname{Quot}(R)$. For each $i \in I$ let

$$P_i = P_{I \setminus \{i\}} = \operatorname{Quot}(K\{w_j \mid j \neq i\}) \quad \text{and} \quad P_i' = P_{\{i\}} = \operatorname{Quot}(K\{w_i\})$$

(we use the notation of Section 3.3).

Let $\gamma \in \Gamma$. By our definition, $w_i^\gamma = \frac{r}{x - c_i^\gamma} = w_{i^\gamma}$, $i \in I$. Hence, γ leaves $R_0 = K[w_i \mid i \in I]$ invariant. Since $| \ |$ is complete on K_0, it has a unique extension to K, so $|a^\gamma| = |a|$ for each $a \in K$. Moreover, for each $f = a_0 + \sum_{i \in I} \sum_{n=1}^\infty a_{in} w_i^n \in R_0$, we have

$$(5) \qquad\qquad f^\gamma = a_0^\gamma + \sum_{i \in I} \sum_{n=1}^\infty a_{in}^\gamma (w_i^\gamma)^n$$

and

$$\|f^\gamma\| = \|a_0^\gamma + \sum_{i \in I} \sum_{n=1}^\infty a_{in}^\gamma (w_i^\gamma)^n\| = \|a_0^\gamma + \sum_{i \in I} \sum_{n=1}^\infty a_{in}^\gamma w_{i^\gamma}^n\|$$
$$= \max(|a_0^\gamma|, |a_{in}^\gamma|)_{i,n} = \max(|a_0|, |a_{in}|)_{i,n} = \|f\|.$$

By Lemma 2.1.5, γ uniquely extends to a continuous automorphism of the completion R of R_0, by formula (5) for $f \in R$. Hence, Γ lifts to a group of continuous automorphisms of R. Therefore, Γ extends to a group of automorphisms of $P = \operatorname{Quot}(R)$. In addition, $P_i^\gamma = P_{i^\gamma}$ and $(P_i')^\gamma = P_{i^\gamma}'$.

For each $j \in J$, Lemma 4.3.5 gives a cyclic extension $F_j = E(z_j)$ of E with Galois group $G_j = \langle \tau_j \rangle$ such that $z_j^\tau \in P_j' \cap R$ for each $\tau \in G_j$ and $\operatorname{discr}(\operatorname{irr}(z_j, E)) \in R^\times$.

For an arbitrary $i \in I$ there exist unique $j \in J$ and $\gamma \in \Gamma$ such that $i = j^\gamma$ (by (4a)). Since γ acts on P and leaves E invariant, $F_i = F_j^\gamma = E(z_i)$ with $z_i = z_j^\gamma$ is a Galois extension of E in P.

The isomorphism $\gamma \colon F_j \to F_i$ gives an isomorphism

$$\operatorname{Gal}(F_j/E) \cong \operatorname{Gal}(F_i/E)$$

that maps each $\tau \in \mathrm{Gal}(F_j/E)$ onto $\gamma^{-1} \circ \tau \circ \gamma \in \mathrm{Gal}(F_i/E)$ (notice that the elements of the Galois groups act from the right). In particular, it maps τ_j onto $\gamma^{-1} \circ \tau_j \circ \gamma$. We can therefore identify G_i with $\mathrm{Gal}(F_i/E)$ such that τ_i coincides with $\gamma^{-1} \circ \tau_j \circ \gamma$. This means that $(a^\tau)^\gamma = (a^\gamma)^{\tau^\gamma}$ for all $a \in F_j$ and $\tau \in G_j$. Therefore, $z_i^\tau \in P_i' \cap R$ for each $\tau \in G_i$ and $\mathrm{discr}(\mathrm{irr}(z_i, E)) \in R^\times$. In particular, $F_i \subseteq P_i'$ for each $i \in I$. It follows from Proposition 3.4.5 that $\mathcal{E} = (E, F_i, P_i, Q; G_i, G)_{i \in I}$ is patching data. By construction, Γ acts properly on $\mathcal{E}$ (Definition 1.2.1).

Let $Q_i = P_i F_i$. diagram (3) of Section 1.3. By Propositions 1.1.7 and 1.2.2, the compound F of $\mathcal{E}$ satisfies (3a) and (3b). Now we verify (3c).

CLAIM C: F/K has a set of prime divisors of degree 1 unramified over E_0 with cardinality $\mathrm{card}(K_0)$.

Each $b \in K_0$ with

$$(6) \qquad |b| > \max_{i \in I}(|r|, |c_i|)$$

satisfies $\left|\frac{r}{b-c_i}\right| < 1$ for each $i \in I$, hence induces the evaluation homomorphism $x \mapsto b$ from R to K that maps w_i onto $\frac{r}{b-c_i}$ (Remark 3.2.4). Since R is a principal ideal domain (Proposition 3.2.9), this homomorphism extends to a K-rational place $\varphi_b \colon P \to K \cup \{\infty\}$. Thus, $\varphi_b|_F$ is a K-rational place of F with $\varphi_b(x) = b \in K_0$, so it corresponds to a prime divisor of F/K of degree 1. By Lemma 4.3.6, the cardinality of the set $\{b \in K_0 \mid |b| > |r|, |c_i|, \ i \in I\}$ is equal to $\mathrm{card}(K_0)$. Since $\varphi_b(x) = b$, distinct b give distinct $\varphi_b|_F$. All but finitely many of the corresponding prime divisors are unramified over E_0.

CLAIM D: $D_{\varphi_b} = \Gamma$. We fix $b \in K_0$ satisfying (6) such that $\varphi = \varphi_b|_F$ is unramified over E_0 and verify that φ satisfies Condition (d) of Lemma 4.4.1. By that Lemma, D_φ will coincide with Γ and the proof will be complete.

By the proof of Claim C, $\bar{F}_\varphi = K$. It remains to prove that

(7) $\varphi_b(f^\gamma) = \varphi_b(f)^\gamma$ for each $\gamma \in \Gamma$

and every $f \in F$ with $\varphi_b(f) \neq \infty$. Of course, it suffices to prove (7) for every $f \in P$ with $\varphi_b(f) \neq \infty$. Since R is a principal ideal domain (Proposition 3.2.9), the valuation ring of φ_b in Q is the local ring of R at a certain prime ideal, so it suffices to prove (7) for each $f \in R$.

Each $f \in R$ has the form $f = a_0 + \sum_{i \in I} \sum_{n=1}^{\infty} a_{in} w_i^n$, with $a_0, a_{in} \in K$ and $a_{in} \to 0$ as $n \to \infty$ (Lemma 3.2.1). By Construction B, each $\gamma \in \Gamma$ acts continuously on R, hence $f^\gamma = a_0^\gamma + \sum_{i \in I} \sum_{n=1}^{\infty} a_{in}^\gamma w_{i\gamma}^n$. Applying φ_b we get $\varphi_b(f) = a_0 + \sum_{i \in I} \sum_{n=1}^{\infty} a_{in} \left(\frac{r}{b-c_i}\right)^n$ and $\varphi_b(f^\gamma) = a_0^\gamma + \sum_{i \in I} \sum_{n=1}^{\infty} a_{in}^\gamma \left(\frac{r}{b-c_{i\gamma}}\right)^n$. Since also the action of γ on K is continuous, $\varphi_b(f)^\gamma = a_0^\gamma + \sum_{i \in I} \sum_{n=1}^{\infty} a_{in}^\gamma \left(\frac{r}{b-c_i^\gamma}\right)^n$. Finally, by Claim A, $c_{i\gamma} = c_i^\gamma$ for each $i \in I$. Consequently, $\varphi_b(f^\gamma) = \varphi_b(f)^\gamma$, as claimed. $\qquad \square$

Notes

Lemma 4.1.6 adds a description of the branch points to [FrJ08, Lemma 16.3.1] with total ramification. Lemma 4.2.2 improves [FrJ08, Lemma 16.3.2] in that it constructs F' inside $K((x))$ with exactly one branch point which is totally ramified.

Lemma 4.2.1 goes back at least to Dedekind. It can be found in Hecke's book [Hec23, Satz 101] and in [Noe27]. Lemma 4.2.4 is due to Geyer building on classical results of Albert, Witt, and H. L. Schmidt.

Lemma 4.3.5 is a variant of [HaV96, Lemma 4.2] and of [HaJ98a, Prop. 5.1]. Finally, Proposition 4.4.2 strengthens [HaJ98a, Prop. 5.2].

The history of Proposition 4.4.2 may be traced to Harbater's realization of every finite group over $K(x)$, where K is the quotient field of an arbitrary complete local domain which is not a field [Hrb87, Thm. 2.3]. Thus, $K(x)$ has a Galois extension F with $\mathrm{Gal}(F/K(x)) \cong G$. Moreover, one can deduce from the proof of Harbater's theorem that F/K is regular. This means that G has a **K-regular realization over** $K(x)$. As mentioned in the introduction, Harbater used formal patching to prove his result.

Following ideas of Serre, Liu used rigid analytic geometry to reprove Harbater's result over each field complete with respect to a nontrivial discrete valuation. The regularity of F/K is mentioned in Liu's theorem explicitly. Liu distributed a preprint containing the proof in 1990. The paper itself was published only in 1995 [Liu95].

The importance of the K-regular realization of G over $K(x)$ lies in the simple observation that it implies an L-regular realization of G over $L(x)$ for each field extension L of K. Conversely, if G has an L-regular realization over $L(x)$ and K is existentially closed in L (Definition 5.2.4), then G has a K-regular realization over $K(x)$. Combining these observations, Pop deduced from the result of Serre-Liu (or from that of Harbater) that every finite group G has a K-regular realization over $K(x)$ for each field K which is either PAC or Henselian. This result is contained in a letter from Roquette to Geyer from December 1990.

Inspired by the method of Serre-Liu, Pop used rigid geometry to prove for any field K_{00} that every constant finite split embedding problem over $K_0 = K_{00}((t))(x)$ has a regular solution. Thus, if K is a finite Galois extension of K_0 and $\mathrm{Gal}(K/K_0)$ acts on a finite group G, than the embedding problem $\mathrm{Gal}(K(x)/K_0(x)) \ltimes G \to \mathrm{Gal}(K(x)/K_0(x))$ has a solution field F which is regular over K [Pop96, Lemma 1.4]. In particular, taking $K = K_0$, the group G has a K-regular realization over $K(x)$. This generalizes Harbater's result in the case where $R = K_{00}[[t]]$. As we shall see later, Pop's result actually implies that of Harbater in each case.

Proposition 5.2 of [HaJ98a] mentioned above applies algebraic patching to reprove Pop's result in the case where K_0 is a complete field with respect to a non-trivial discrete ultrametric absolute value with infinite residue field and K/K_0 is a finite Galois unramified extension. The solution field obtained in that proposition has in addition a K-rational place.

The extra conditions on K/K_0 mentioned in the preceding paragraph were needed in the proof of [HaJ98a, Prop. 5.2] in order to choose the c_i's in K such that $|c_i - c_j| = 1$ for all distinct $i, j \in I$. The latter condition was needed in order to extend the absolute value on K to an absolute value of $R = K[w_i \mid i \in I]$, where the w_i in [HaJ98a, Prop. 5.2] were defined to be $\frac{1}{x-c_i}$. The approach we take in this book eliminates the extra conditions by defining the w_i's as $\frac{r}{x-c_i}$ with c_i distinct and $r \in K_0^\times$ such that $|r| \leq |c_i - c_j|$ for all distinct $i, j \in I$. By doing so, R is not a complete absolute valued ring any more but only a complete normed ring. Nevertheless, the machinery still works, leading to a successful proof of Proposition 4.4.2. The shift to normed rings in this framework is due to [Har05], which for itself is borrowed from [PrP04, Sec. 2.2].

Note that eventually also [HaJ98a] gets rid of the extra conditions of [HaJ98a, Prop. 5.2]. It replaces the field extension K/K_0 by $K((t))/K_0((t))$ with the t-adic absolute value where the extra conditions are satisfied. The solution over $K_0((t))(x)$ of the constant embedding problem coming from K/K_0 is then specialized to a solution over $K_0(x)$ using the "ampleness" of K_0. This step, which is important in its own sake, is explained in the next chapter.

Chapter 5.
Ample Fields

One of the major problems of Field Arithmetic was whether the absolute Galois group of every countable PAC Hilbertian field K is free of countable rank. By Iwasawa, that means that every finite embedding problem of $\mathrm{Gal}(K)$ is solvable. The PAC property of K implies that $\mathrm{Gal}(K)$ is projective, so it suffices to solve finite split embedding problems over K. Since K is Hilbertian, it suffices to solve finite split constant embedding problems over $K(x)$, where x is transcendental over K. Since K is PAC, it is existentially closed in the field of formal power series $\hat{K} = K((t))$. By Bertini-Noether, it suffices to solve each finite split constant embedding problem over $\hat{K}(x)$. Thus, the initial problem of proving that $\mathrm{Gal}(K) \cong \hat{F}_\omega$ is reduced to a problem that Proposition 4.4.2 settles.

The property of being existentially closed in $K((t))$ that each PAC field K has is shared by all Henselian fields. We call a field K which is existentially closed in $K((t))$ **ample**. In that case, the arguments of the preceding paragraph prove that each finite split constant embedding problem over $K(x)$ is solvable (Theorem 5.9.2).

It turns out that ample fields can be characterized in diophantine terms: A field K is ample if and only if every absolutely irreducible curve over K with a simple K-rational point has infinitely many K-rational points (Lemma 5.3.1). Surprisingly enough, each field K such that $\mathrm{Gal}(K)$ is a pro-p group for a single prime number p has the latter property and is therefore ample (Theorem 5.8.3). On the other hand, the theorems of Faltings and Grauert-Manin imply that number fields and function fields of several variables are not ample (Proposition 6.2.5).

5.1 Varieties

Several types of fields K which appear in Field Arithmetic like PAC fields, PRC fields, PpC fields, and Henselian fields have a common feature: K is existentially closed in $K((t))$. This property has a remarkable consequence: Every finite split constant embedding problem over $K(x)$ has a regular solution (Theorem 5.9.2). In Section 5.3, we prove several conditions on K to be equivalent to the above mentioned one. This section and the next one are devoted to various preparations for the proof of the equivalence of those conditions. Here we recall basic notions of algebraic geometry, primarily that of an absolutely irreducible variety, and prove a few results to be used later. The proofs are not self-contained, at crucial points we refer to other sources in the literature.

By a K-**variety** we mean a separated reduced irreducible scheme V of finite type over K. For every field extension F of K, we denote the F-rational

M. Jarden, *Algebraic Patching*, Springer Monographs in Mathematics,
DOI 10.1007/978-3-642-15128-6_5, © Springer-Verlag Berlin Heidelberg 2011

points of V by $V(F)$. If $\mathbf{x} \in V(F)$ and trans.deg$(K(\mathbf{x})/K) = \dim(V)$, then $\mathbf{x}$ is a **generic point** of V (and V is **generated** by $\mathbf{x}$). In this case $K(\mathbf{x})$ is isomorphic to the **function field** of V and each point of $V(\tilde{K})$ is a K-specialization of $\mathbf{x}$. We say that V is **absolutely irreducible** if $V \times_K \tilde{K}$ is a $\tilde{K}$-variety. This is the case, if and only if $K(\mathbf{x})$ is a regular extension of K for each generic point $\mathbf{x}$ of V [FrJ08, Cor. 10.2.2(a)]. Finally, in order to be compatible with the convention of [FrJ08], we also write "let V be an absolutely irreducible variety **defined over** K" instead of "let V be an absolutely irreducible K-variety".

Every K-variety is a union of finitely many affine open sub-K-varieties. It is absolutely irreducible if and only if each of its affine open sub-K-varieties is absolutely irreducible.

Each affine K-variety V can be represented by a closed embedding into $\mathbb{A}_K^n$ for some positive integer n as $\operatorname{Spec}(K[\mathbf{X}]/\mathfrak{p})$, where $\mathbf{X} = (X_1, \ldots, X_n)$ and $\mathfrak{p}$ is a prime ideal of $K[\mathbf{X}]$. Then V is absolutely irreducible if and only if $\mathfrak{p}\tilde{K}[\mathbf{X}]$ is a prime ideal of $\tilde{K}[\mathbf{X}]$.

Let $V = \operatorname{Spec}(K[\mathbf{X}]/\langle f_1, \ldots, f_m \rangle)$ be an affine K-variety, where $\langle f_1, \ldots, f_m \rangle$ is a prime ideal of $K[\mathbf{X}]$. Let L be a field extension of K. We view each point $\mathbf{a}$ of $V(L)$ as an n-tuple $(a_1, \ldots, a_n)$ of elements of L. Then $K(\mathbf{a}) = K(a_1, \ldots, a_n)$ is the field generated by $a_1, \ldots, a_n$ over K. We call $\mathbf{a}$ **simple** (also known as **nonsingular**) if it satisfies the Jacobian criterion:

$$\operatorname{rank}\left(\frac{\partial f_i}{\partial X_j}(\mathbf{a})\right) = n - \dim(V).$$

For an arbitrary K-variety V and a point $\mathbf{a} \in V(L)$ we say that $\mathbf{a}$ is **simple** if $\mathbf{a}$ has an affine open neighborhood on which it is simple. We denote the set of all simple points of $V(L)$ by $V_{\mathrm{simp}}(L)$.

LEMMA 5.1.1: *Let V be a K-variety. If V has a simple K-rational point, then V is absolutely irreducible and defined over K.*

Proof: Assume without loss that V is affine and let $\mathbf{x}$ be a generic point of V. Then the specialization $\mathbf{x} \to \mathbf{a}$ extends to a K-rational place of $K(\mathbf{x})$ [JaRo80, Cor. A2], so $K(\mathbf{x})/K$ is regular [FrJ08, Lemma 2.6.9]. Hence V is absolutely irreducible and defined over K. $\qquad\square$

Let F be a field extension of K. We view each $\mathbf{a} \in \mathbb{P}_K^n(F)$ as an equivalence class of $(n+1)$-tuples $(a_0, \ldots, a_n) \in F^{n+1}$ with $a_j \neq 0$ for at least one j, modulo factors $c \in F^\times$, and denote that class by $(a_0 : \cdots : a_n)$. Then, $K(\mathbf{a}) = K\left(\frac{a_0}{a_j}, \ldots, \frac{a_n}{a_j}\right)$.

LEMMA 5.1.2: *Let K be an infinite field and C an absolutely irreducible K-curve which has a K-rational simple point $\mathbf{p}$. Then there is a birational correspondence θ between C and an affine plane K-curve defined by an absolutely irreducible equation $f(X, Y) = 0$ over K such that $\theta(\mathbf{p}) = (0, 0)$, $f(0, 0) = 0$, and $\frac{\partial f}{\partial Y}(0, 0) \neq 0$.*

Proof: Assume without loss that C is a projective K-curve in $\mathbb{P}^n_K$ and that $n \geq 2$. If $n = 2$, apply a linear automorphism of $\mathbb{P}^2_K$ over K to assume that $\mathbf{p} = (1\!:\!0\!:\!0)$. Then take f as the polynomial that defines an affine open neighborhood of $\mathbf{p}$ in C, so that $\mathbf{p} = (0,0)$. Exchange the coordinates X and Y, if necessary, to obtain the condition $\frac{\partial f}{\partial Y}(0,0) \neq 0$.

Assume therefore that $n \geq 3$. Then, there is a nonempty Zariski-open subset U of $\mathbb{P}^n_K$ such that for each point $\mathbf{o} \in U(\tilde{K})$, the projection $\pi\colon \mathbb{P}^n_K \to \mathbb{P}^{n-1}_K$ from $\mathbf{o}$ maps C onto an absolutely irreducible K-curve C' such that $\pi|_C$ is a birational map and $\pi(\mathbf{p})$ is simple on C'. One may take U to be the complement in $\mathbb{P}^n_K$ of the union of all lines going through $\mathbf{p}$ and another point of $C(\tilde{K})$ and the tangent to C at $\mathbf{p}$ (See [GeJ89, Lemma 9.4] or [Mum88, p. 262, Thm. 1].) Since K is infinite, we may choose $\mathbf{o} \in U(K)$. Then $\pi|_C$ and C' are defined over K, $\pi(\mathbf{p}) \in C'(K)$, and $\pi(\mathbf{p})$ is simple. Now apply induction on n. $\qquad\square$

LEMMA 5.1.3: *Let K be an infinite field, $V \subseteq \mathbb{P}^n_K$ an absolutely irreducible K-variety of positive dimension, and P a finite subset of $V(K_s)$. Then, there exists an absolutely irreducible curve $C \subseteq V$ defined over K that passes through each of the points of P. Moreover, if a point $\mathbf{p}$ of P is simple on V, then $\mathbf{p}$ is also simple on C.*

Proof: Let $r = \dim(V)$. If $r = 1$, there is nothing to prove. Suppose $r \geq 2$ and the lemma holds for $r - 1$.

Assume without loss that P is closed under the action of $\mathrm{Gal}(K)$. The method of blowing up gives a hypersurface H over K in $\mathbb{P}^n_K$ such that $W = H \cap V$ is an absolutely irreducible variety of dimension $r - 1$ over K, $P \subset W(K_s)$, and each simple point of P on V is also simple on W [JaR98, Lemma 10.1]. Now apply the induction hypothesis on W. $\qquad\square$

Let F/K be a finitely generated extension of fields of transcendence degree 1. A **model** of F/K is a K-curve C whose function field is F. If $\mathbf{x} = (x_1, \ldots, x_n)$ is a generic point over K of an affine open subcurve of C, then $F = K(\mathbf{x})$. If $\varphi\colon F \to \tilde{K} \cup \{\infty\}$ is a K-place of F (in particular $\varphi(a) = a$ for each $a \in K$) and $\mathbf{p} = \varphi(\mathbf{x})$ is finite, then $\mathbf{p}$ is a point of $C(\tilde{K})$ called the **center** of φ at C.

LEMMA 5.1.4: *Let F/K be a finitely generated extension of transcendence degree 1.*

(a) *Let $\varphi\colon F \to K \cup \{\infty\}$ be a K-rational place. Denote the valuation ring of φ by O. Then F/K is regular and has an affine model C such that the center of φ at C is a simple K-rational point $\mathbf{a}$ of C whose local ring coincides with O.*

(b) *Conversely, suppose C is a model of F/K. Then, for each point $\mathbf{p} \in C_{\mathrm{simp}}(K)$ there exists a unique K-rational place $\varphi\colon F \to K \cup \{\infty\}$ whose center at C is $\mathbf{p}$.*

Proof of (a): By [FrJ08, Lemma 2.6.9(b)], F/K is a regular extension. Let t be a separating transcendence element for F/K. Replace t by t^{-1}, if necessary, to assume that $c = \varphi(t) \in K$. Let $R = K[t]$ and let $\mathfrak{p} = (t - c)R$. Then the local ring $R_\mathfrak{p}$ is a valuation ring. The integral closure S of R in F is a finitely generated R-module [Lan58, p. 120]. In particular $S = K[x_1, \ldots, x_n]$ is finitely generated over K as a ring and φ is finite at $x_1, \ldots, x_n$. Since $K(\mathbf{x})/K$ as a subextension of F/K is regular, the curve $C = \mathrm{Spec}(S)$ generated by $\mathbf{x}$ over K is absolutely irreducible.

Let $S_\mathfrak{p}$ be the local ring of S with respect to the multiplicative set $R \smallsetminus \mathfrak{p}$. Then $S_\mathfrak{p}$ is the integral closure of $R_\mathfrak{p}$ in F. In particular $S \subseteq S_\mathfrak{p} \subseteq O$. Hence, $\mathbf{a} = \varphi(\mathbf{x}) \in C(K)$. Moreover, let $\mathfrak{M}$ be the maximal ideal of O, let $\mathfrak{m} = S_\mathfrak{p} \cap \mathfrak{M}$, and let $\mathfrak{n} = S \cap \mathfrak{M}$. By [Lan58, p. 18, Thm. 4], $O_{C,\mathbf{a}} = S_\mathfrak{n} = (S_\mathfrak{p})_\mathfrak{m} = O$. Finally, as a discrete valuation ring, O is a regular ring. Consequently, $\mathbf{a} \in C_{\mathrm{simp}}(K)$ [Lan58, p. 204, Thm. 2].

Proof of (b): The local ring $O = O_{C,\mathfrak{p}}$ is regular [Lan58, p. 201]. Since $\dim(O) = 1$, this means that the maximal ideal of O is principal. By [AtM69, p. 94, Prop. 9.2], O is a discrete valuation ring. The corresponding place is the desired one. $\qquad\qquad\qquad\qquad\qquad\qquad\qquad\qquad\qquad\qquad\qquad\qquad\qquad\square$

5.2 Existentially Closed Extensions

Let K be a discrete Henselian field and $\hat{K}$ its completion. We prove that if $\hat{K}/K$ is a separable extension, than K is existentially closed in $\hat{K}$. In particular, K is algebraically closed in $\hat{K}$. The separability assumption is satisfied if K is 'defectless'.

A valued field (K, v) is **defectless** if each finite extension L of K satisfies

$$(1) \qquad\qquad\qquad [L : K] = \sum_{w|v} e_{w/v} f_{w/v},$$

where w ranges over all of the valuations of L that extend v, $e_{w/v}$ is the ramification index, and $f_{w/v}$ is the relative residue degree of w/v. If (K, v) is Henselian, then v has a unique extension w to L. In that case we write $e_{L/K}$ (resp. $f_{L/K}$) rather than $e_{w/v}$ (resp. $f_{w/v}$). Then condition (1) simplifies to

$$(2) \qquad\qquad\qquad [L : K] = e_{L/K} f_{L/K}.$$

For example, each complete discrete valued field (K, v) is defectless [Rbn64, p. 236].

LEMMA 5.2.1: *Let (K, v) be a defectless discrete Henselian valued field and let $(\hat{K}, \hat{v})$ be its completion. Then $\hat{K}/K$ is a regular extension.*

Proof: We have to prove that each finite extension L of K is linearly disjoint from $\hat{K}$ over K.

Indeed, since $\hat{K}/K$ is an immediate extension, $e_{\hat{K}/K} = 1$ and $f_{\hat{K}/K} = 1$. Thus, $e_{\hat{K}L/\hat{K}} = e_{\hat{K}L/K} = e_{\hat{K}L/L} e_{L/K} \geq e_{L/K}$. Similarly, we have $f_{\hat{K}L/\hat{K}} \geq f_{L/K}$ for the residue degrees. Hence, by (2)

$$[\hat{K}L : \hat{K}] \leq [L : K] = e_{L/K} f_{L/K} \leq e_{\hat{K}L/\hat{K}} f_{\hat{K}L/\hat{K}} \leq [\hat{K}L : \hat{K}].$$

Thus, $[\hat{K}L : \hat{K}] = [L : K]$. Consequently, L is linearly disjoint from $\hat{K}$ over K. □

When we speak about a **function field of one variable over a field** K, we mean a finitely generated regular extension F of K of transcendence degree 1.

Remark 5.2.2: More examples of defectless valuations. Consider a discrete valued field (K, v). Let O be its valuation ring, L a finite extension of K, and O' the integral closure of K in L. If O' is a finitely generated O-module, then (1) holds [Ser79, p. 14, Prop. 10]. This is in particular the case if L/K is separable [Ser79, p. 13]. Hence, if $\mathrm{char}(K) = 0$, then K is defectless with respect to v.

If K is a function field of one variable over a field K_0, and v is a valuation of K trivial on K_0, then there exists a finitely generated ring R over K_0 and a prime ideal $\mathfrak{p}$ of R such that $R_\mathfrak{p}$ is the valuation ring of v (Proof of Lemma 5.1.4(a)). Since the integral closure of R in L is finitely generated as an R-module [Lan58, p. 120], also the integral closure of $R_\mathfrak{p}$ in L is finitely generated as an $R_\mathfrak{p}$-module. It follows that (K, v) is defectless. □

LEMMA 5.2.3: *Let (K, v) be a discrete Henselian valued field and let $(\hat{K}, \hat{v})$ be the completion of (K, v). Then (K, v) is defectless in each of the following cases:*
(a) $\mathrm{char}(K) = 0$.
(b) *(K, v) is the Henselization of a valued field (K_1, v_1), where K_1 is a function field of one variable over a field K_0 and v_1 is a valuation of K_1 which is trivial on K_0.*
Hence, by Lemma 5.2.1, in each of these cases, $\hat{K}/K$ is a regular extension.

Proof: By Remark 5.2.2, it suffices to consider only Case (b). Let L be a finite extension of K. Since (2) holds if L/K is separable, it suffices to prove (2) only when L/K is a purely inseparable extension of degree q. In that case there exist a finite extension K_2 of K_1 in K and a finite purely inseparable extension L_2 of K_2 of degree q such that $K \cap L_2 = K_2$ and $KL_2 = L$. Since K_2 is a function field of one variable over a finite extension of K_0 and the restriction of v to that extension is trivial, K_2 is defectless (Remark 5.2.2). Also, $v_2 = v|_{K_2}$ has a unique extension w_2 to L_2. Hence, $e_{w_2/v_2} f_{w_2/v_2} = q$.

Now denote the unique extension of v to L by w. Then $w|_{L_2} = w_2$. Since (K, v) is also the Henselization of (K_2, v_2), we have $e_{L/K} = e_{L/K_2} \geq e_{w_2/v_2}$ and $f_{L/K} \geq f_{w_2/v_2}$. Hence,

$$q = [L : K] \geq e_{L/K} f_{L/K} \geq e_{w_2/v_2} f_{w_2/v_2} = q,$$

so (2) holds, as desired. □

Definition 5.2.4: Let $\hat{K}/K$ be a field extension. We denote the first order language of rings with constants for the elements of K by $\mathcal{L}(\mathrm{ring}, K)$ [FrJ08,

Example 7.3.1]. We say that K is **existentially closed** in $\hat{K}$ if every existential sentence θ in $\mathcal{L}(\mathrm{ring}, K)$ that holds in $\hat{K}$ also holds in K. Since are models are fields, θ is equivalent to a sentence of the form

$$(3) \qquad (\exists X_1)\cdots(\exists X_n)\Big[\bigvee_{i=1}^{r}\bigwedge_{j=1}^{r(i)} f_{ij}(\mathbf{X}) = 0 \wedge g_i(\mathbf{X}) \neq 0\Big]$$

where $\mathbf{X} = (X_1, \ldots, X_n)$ and $f_{ij}, g_i \in K[\mathbf{X}]$. Further, (3) is equivalent to the sentence

$$(4) \quad (\exists X_1)\cdots(\exists X_n)(\exists Y_1)\cdots(\exists Y_r)\Big[\bigvee_{i=1}^{n}\bigwedge_{j=1}^{r} f_{ij}(\mathbf{X}) = 0 \wedge g_i(\mathbf{X})Y_i - 1 = 0\Big].$$

The equalities in the brackets of (4) define a Zariski-closed subset of $\mathbb{A}_K^{n+r}$. Replacing $n + r$ by n, we conclude that K is existentially closed in $\hat{K}$ if and only if for each Zariski-closed subset A of $\mathbb{A}_K^n$, $A(\hat{K}) \neq \emptyset$ implies $A(K) \neq \emptyset$. Equivalently, each $\mathbf{x} \in \hat{K}^n$ has a K-**rational specialization**. This is a point $\mathbf{a} \in K^n$ such that $f(\mathbf{a}) = 0$ for each $f \in K[\mathbf{X}]$ that satisfies $f(\mathbf{x}) = 0$. $\qquad\square$

Remark 5.2.5: Model theoretic characterization of existential closedness. A field K is existentially closed in an extension $\hat{K}$ if and only if $\hat{K}$ can be K-embedded in a field K^* that is an **elementary extension** of K [FrJ08, Lemma 27.1.4]. The latter condition means that for each formula $\varphi(X_1, \ldots, X_n)$ of the language of rings and for all $a_1, \ldots, a_n \in K$ the truth of $\varphi(a_1, \ldots, a_n)$ in K is equivalent to its truth in K^*. It is possible to choose K^* as an ultrapower of K [BeS74, p. 187, Lemma 3.9]. $\qquad\square$

LEMMA 5.2.6: *Let $\hat{K}/K$ be a field extension such that K is existentially closed in $\hat{K}$. Then $\hat{K}/K$ is a regular extension.*

Proof: It suffices (and also necessary) to prove that K is algebraically closed in $\hat{K}$ and that $\hat{K}/K$ is separable [FrJ08, Lemma 2.6.4].

PART A: K is algebraically closed in $\hat{K}$. Indeed, consider $x \in \hat{K} \cap \tilde{K}$ and let $f = \mathrm{irr}(x, K)$. Then $f(x) = 0$, so there exists $x_0 \in K$ such that $f(x_0) = 0$. It follows that $\deg(f) = 1$, hence $x \in K$.

PART B: $\hat{K}/K$ is a separable extension. We have to prove this claim only if $\mathrm{char}(K) > 0$. Indeed, let q be a power of $\mathrm{char}(K)$. Suppose $a_1, \ldots, a_n$ are elements of $K^{1/q}$ that are linearly dependent over $\hat{K}$. Thus, there exist $x_1, \ldots, x_n \in \hat{K}$ with $\sum_{i=1}^{n} a_i x_i = 0$, and, say, $x_1 \neq 0$. Then $\sum_{i=1}^{n} a_i^q x_i^q = 0$, $a_i^q \in K$, $i = 1, \ldots, n$. Since K is existentially closed in $\hat{K}$, there exist $y_1, \ldots, y_n \in K$ such that $\sum_{i=1}^{n} a_i^q y_i^q = 0$ and $y_1 \neq 0$. Hence, $\sum_{i=1}^{n} a_i y_i = 0$ and therefore $a_1, \ldots, a_n$ are linearly dependent over K. $\qquad\square$

LEMMA 5.2.7: *Let (K, v) be a Henselian valued field and let $(\hat{K}, \hat{v})$ be its completion. Suppose $\hat{K}/K$ is a separable extension. Then K is existentially closed in $\hat{K}$. In particular, if $\mathrm{char}(K) = 0$ or K is a Henselization of a function field K_1 of one variable over a field K_0 at $v|_{K_1}$ and v is trivial on K_0, then K is existentially closed in $\hat{K}$.*

Proof: Let $x_1, \ldots, x_n \in \hat{K}$. Choose a separating transcendence base $u_1, \ldots, u_r$ for $K(\mathbf{x})/K$ and let z be a primitive element for the finite separable extension $K(\mathbf{x})/K(\mathbf{u})$ which is integral over $K[\mathbf{u}]$. Replacing some of the u_i's by their inverses, if necessary, we may assume that $\hat{v}(u_i) \geq 0$, $i = 1, \ldots, r$, hence also $\hat{v}(z) \geq 0$. Let $f \in K[U_1, \ldots, U_r, Z]$ be an irreducible polynomial such that $f(\mathbf{u}, z) = 0$ and $f'(\mathbf{u}, z) \neq 0$ (the prime stands for derivative with respect to Z). In addition, let $h_i \in K[\mathbf{U}, Z]$ and $0 \neq h_0 \in K[\mathbf{U}]$ be such that $x_i = \frac{h_i(\mathbf{u}, z)}{h_0(\mathbf{u})}$, $i = 1, \ldots, n$.

Since (K, v) is dense in $(\hat{K}, \hat{v})$, we may approximate $u_1, \ldots, u_r, z$ by elements of K to any desired degree. Since K is Henselian, there exist $b_1, \ldots, b_r, c \in K$ such that $f(\mathbf{b}, c) = 0$ and $h_0(\mathbf{b}) \neq 0$. It follows that $(\mathbf{b}, c)$ is a K-rational specialization of $(\mathbf{u}, z)$.

Now let $a_i = \frac{h_i(\mathbf{b}, c)}{h_0(\mathbf{b})}$, $i = 1, \ldots, n$. Then $\mathbf{a}$ is a K-rational specialization of $\mathbf{x}$. By Definition 5.2.4, K is existentially closed in $\hat{K}$.

Finally, if $\mathrm{char}(K) = 0$ or K is a Henselization of a function field of one variable K_1 over a field K_0 at $v|_{K_1}$ and v is trivial on K_0, then $\hat{K}/K$ is separable (Lemma 5.2.3). Consequently, K is existentially closed in $\hat{K}$. $\square$

We apply Lemma 5.2.7 to the field of convergent power series.

PROPOSITION 5.2.8: *Let K be a complete field under an ultrametric absolute value. Denote the field of all convergent power series in x with coefficients in K by $K((x))_0$. Then $K((x))_0$ is existentially closed in $K((x))$.*

Proof: By Corollary 2.4.6, the field $K((x))_0$ is Henselian at the restriction of the x-adic valuation of $K((x))$. By Proposition 2.5.4, $K((x))/K((x))_0$ is a regular extension. Hence, by Lemma 5.2.7, $K((x))_0$ is existentially closed in $K((x))$. $\square$

5.3 Ample Fields

We give several equivalent definitions of ample fields and derive some of their properties.

The *t*-**adic valuation** of the field of rational functions $K(t)$ is the discrete valuation v with $v(t) = 1$ which is trivial on K.

LEMMA 5.3.1: *The following conditions on a field K are equivalent:*
(a) *For each absolutely irreducible polynomial $f \in K[X, Y]$, the existence of a point $(a, b) \in K^2$ such that $f(a, b) = 0$ and $\frac{\partial f}{\partial Y}(a, b) \neq 0$ implies the existence of infinitely many such points.*

(b) *Every absolutely irreducible K-curve C with a simple K-rational point has infinitely many K-rational points.*

(c) *If an absolutely irreducible K-variety V has a simple K-rational point, then $V(K)$ is Zariski-dense in V.*

(d) *Every function field of one variable over K that has a K-rational place has infinitely many K-rational places.*

(e) *K is existentially closed in each Henselian closure $K(t)^h$ of $K(t)$ with respect to the t-adic valuation.*

(f) *K is existentially closed in $K((t))$.*

Definition 5.3.2: Whenever a field K satisfies one of the (equivalent) conditions of Lemma 5.3.1, we say that K is an **ample field**. □

Proof of (a) $\Longrightarrow$ (b): Let C be an absolutely irreducible K-curve with a simple K-rational point $\mathbf{p}$. Lemma 5.1.2 gives a plane curve Γ defined by an absolutely irreducible equation $f(X, Y) = 0$ with coefficients in K, a point $(a, b) \in K^2$ such that $f(a, b) = 0$ and $\frac{\partial f}{\partial Y}(a, b) \neq 0$, and an isomorphism θ between a Zariski-open neighborhood C_0 of $\mathbf{p}$ in C and a Zariski-open neighborhood Γ_0 of (a, b) in Γ. By assumption, $\Gamma_0(K)$ is infinite. Hence, so is $C_0(K)$.

Proof of (b) $\Longrightarrow$ (c): Let $\mathbf{p}$ be a simple K-rational point of an absolutely irreducible K-variety V and let U be a nonempty Zariski-open subset of V. We assume without loss that $\dim(V) > 0$. By Section 5.1, V is defined over K. Lemma 5.1.3 gives an absolutely irreducible K-curve C on V such that $C \cap U \neq \emptyset$ and $\mathbf{p} \in C_{\mathrm{simp}}(K)$. By (b), $C_{\mathrm{simp}}(K)$ is infinite. Since the complement of $C \cap U$ in C is finite, $(C_{\mathrm{simp}} \cap U)(K) \neq \emptyset$. Hence, $U(K) \neq \emptyset$, so $V(K)$ is Zariski-dense in V.

Proof of (c) $\Longrightarrow$ (d): Let F be a function field of one variable over K with a K-rational place φ. Then F/K is regular and has an affine model C with a simple K-rational point $\mathbf{p}$ (Lemma 5.1.4(a)). By (c), C has infinitely many simple K-rational points $\mathbf{q}$. For each $\mathbf{q}$ Lemma 5.1.4(b) gives a K-rational place ψ of F whose center at C is $\mathbf{q}$. Thus, F has infinitely many K-rational places.

Proof of (d) $\Longrightarrow$ (e): We may assume that $K((t))$ is the completion of $K(t)^h$ with respect to the t-adic valuation. Let $\mathbf{x} = (x_1, \ldots, x_n)$ be a point with coordinates in $K(t)^h$ and set $F = K(\mathbf{x})$. If $\mathbf{x} \in \tilde{K}^n$, then $\mathbf{x} \in K^n$ (because K is algebraically closed in $K((t))$), so $\mathbf{x}$ is a K-rational specialization of itself. Otherwise, trans.deg$(F/K) = 1$. Let φ_0 be the restriction to F of the place associated with the t-adic valuation of $K((t))$. Then φ_0 is K-rational. By (d), F/K has infinitely many K-rational places φ. All but finitely many of them are finite at $\mathbf{x}$, so the specialization $\mathbf{a} = \varphi(\mathbf{x})$ belong to K^n. We conclude that K is existentially closed in $K(t)^h$.

Proof of (e) $\Longrightarrow$ (f): By (e), K is existentially closed in $K(t)^h$. By Lemma 5.2.7, $K(t)^h$ is existentially closed in $K((t))$. Hence, K is existentially closed in $K((t))$.

Proof of (f) $\Longrightarrow$ (a): Let $f \in K[X,Y]$ be an absolutely irreducible polynomial. Suppose there exists a point $(a,b) \in K^2$ such that $f(a,b) = 0$ and $\frac{\partial f}{\partial Y}(a,b) \neq 0$. Denote the set of all these points by A. We have to prove that A is an infinite set.

Indeed, denote the t-adic valuation of $K((t))$ by v. Let $(a_i, b_i) \in A$, $i = 1, \ldots, n$. Choose $a'_{n+1} \in K((t)) \smallsetminus \{a_1, \ldots, a_n\}$ such that $v(a'_{n+1} - a)$ is a large positive integer so that $v(\frac{\partial f}{\partial Y}(a'_{n+1}, b)) = v(\frac{\partial f}{\partial Y}(a,b)) < \infty$ and $v(f(a'_{n+1}, b)) = v(f(a'_{n+1}, b) - f(a,b)) > 2v(\frac{\partial f}{\partial Y}(a'_{n+1}, b))$. Since $K((t))$ is Henselian, there exists $b'_{n+1} \in K((t))$ such that $f(a'_{n+1}, b'_{n+1}) = 0$ and $v(b'_{n+1} - b)$ is a large positive integer. Hence, $\frac{\partial f}{\partial Y}(a'_{n+1}, b'_{n+1}) \neq 0$. Since K is existentially closed in $K((t))$, there exist $a_{n+1}, b_{n+1} \in K$ such that $f(a_{n+1}, b_{n+1}) = 0$, $\frac{\partial f}{\partial Y}(a_{n+1}, b_{n+1}) \neq 0$, and $a_{n+1} \neq a_1, \ldots, a_n$. Hence, (a_{n+1}, b_{n+1}) is a new point of A. We conclude by induction that A is infinite. $\square$

COROLLARY 5.3.3: *Every ample field is infinite.*

Proof: Let K be an ample field and consider the affine line $\mathbb{A}_K^1$. It is a smooth curve with K-rational points (e.g. 0). By Lemma 5.3.1(a), $K = \mathbb{A}_K^1(K)$ is infinite. $\square$

Section 5.4 strengthens (b) and (c) of Lemma 5.3.1 and generalizes Corollary 5.3.3 considerably.

5.4 Many Points

We prove that if K is an ample field and V is an absolutely irreducible K-variety of positive dimension with a simple K-rational point, then not only $V(K)$ is Zariski-dense in V but $\mathrm{card}(V(K)) = \mathrm{card}(K)$ (Proposition 5.4.3(b)). Moreover, the points of $V(K)$ generate K over every subfield of K (Proposition 5.4.3(a)) over which V is defined.

LEMMA 5.4.1: *Let K be an infinite field, V a vector space over K, n a positive integer, and I a set. For each $i \in I$ let W_i be a proper subspace of V such that $V = \bigcup_{i \in I} W_i$ and $\dim(W_i) < n$ for each $i \in I$. Then $\mathrm{card}(K) \leq \mathrm{card}(I)$.*

Proof: The case $\dim(V) \leq 1$ cannot occur, so we first suppose $d = \dim(V)$ is finite but at least 2. Let $v_1, \ldots, v_d$ be a basis of V. If we let c range on all elements of K, we get $\mathrm{card}(K)$ distinct subspaces $Kv_1 + \cdots + Kv_{d-2} + K(v_{d-1} + cv_d)$ of dimension $d-1$. If $\mathrm{card}(K) > \mathrm{card}(I)$, then V has a subspace W of dimension $d-1$ such that $W \neq W_i$ for each $i \in I$. Therefore, $W \cap W_i$ is a proper affine subspace of W and $W = \bigcup_{i \in I} W \cap W_i$. An induction hypothesis on d gives $\mathrm{card}(K) \leq \mathrm{card}(I)$. This contradicts our previous assumption.

Now we consider the case where $\dim(V) = \infty$ and choose a subspace V' of dimension n. Then $\dim(V' \cap W_i) \leq \dim(W_i) < n = \dim(V')$ and

$V' = \bigcup_{i \in I} V' \cap W_i$. It follows from the previous case that $\mathrm{card}(K) \leq \mathrm{card}(I)$.
$\square$

LEMMA 5.4.2 (Fehm): *Let K be an ample field, $f \in K[X, Y]$ an absolutely irreducible polynomial, and $y_0 \in K$ such that $f(0, y_0) = 0$, $\frac{\partial f}{\partial Y}(0, y_0) \neq 0$. We denote the absolutely irreducible K-curve defined by the equation $f(X, Y) = 0$ by C and let $\pi \colon C \to \mathbb{A}_K^1$ be the projection on the first coordinate. Then $\mathrm{card}(\pi(C(K)) \smallsetminus K_0) \geq \max(\aleph_0, \mathrm{card}(K_0))$ for each proper subfield K_0 of K.*

Proof: Since there are infinitely many pairs $(a, b) \in K^2$ with $f(a, b) = 0$ and $f(a, Y) \neq 0$ (Lemma 5.3.1(c)) and for each of them $\pi^{-1}(a)$ is finite, $\pi(C(K))$ is infinite. Thus, our inequality holds if K_0 is finite. Therefore, it suffices to prove that $\mathrm{card}(\pi(C(K)) \smallsetminus K_0) \geq \mathrm{card}(K_0)$ under the assumeption that K_0 is infinite.

Consider an indeterminate t. For each $c \in K$ let $u_1 = ct$ and $u_2 = t$. Applying Hensel's lemma to the field $K((t))$ and to the polynomials $f(u_1, Y), f(u_2, Y) \in K[[t]][Y]$, we find elements $v_1, v_2 \in K((t))$ such that $f(u_1, v_1) = 0$ and $f(u_2, v_2) = 0$. By definition, $u_2 \neq 0$ and $c = \frac{u_1}{u_2}$. Since K is existentially closed in $K((t))$ (Definition 5.3.2), there exist $a_1, b_1, a_2, b_2 \in K$ such that $f(a_1, b_1) = 0$, $f(a_2, b_2) = 0$, $a_2 \neq 0$, and $c = \frac{a_1}{a_2}$. It follows that $K = \{ \frac{x_1}{x_2} \mid \exists y_1, y_2 \colon (x_1, y_1), (x_2, y_2) \in C(K), \ x_2 \neq 0 \}$.

Let $A = \{ \mathbf{p} \in C(K) \mid \pi(\mathbf{p}) \in K_0 \}$, $A' = A \smallsetminus \pi^{-1}(0)$, and $B = \{ \mathbf{p} \in C(K) \mid \pi(\mathbf{p}) \notin K_0 \}$. Then $A \cup B = C(K)$, $A' \cup B = C(K) \smallsetminus \pi^{-1}(0)$, and $\pi(B) = \pi(C(K)) \smallsetminus K_0$. Moreover,

$$K = \left\{ \frac{\pi(\mathbf{p})}{\pi(\mathbf{q})} \;\middle|\; \mathbf{p} \in C(K), \ \mathbf{q} \in C(K) \smallsetminus \pi^{-1}(0) \right\}$$

$$= \left\{ \frac{\pi(\mathbf{p})}{\pi(\mathbf{q})} \;\middle|\; \mathbf{p} \in A, \ \mathbf{q} \in A' \right\} \cup \left\{ \frac{\pi(\mathbf{p})}{\pi(\mathbf{q})} \;\middle|\; \mathbf{p} \in A, \ \mathbf{q} \in B \right\}$$

$$\cup \left\{ \frac{\pi(\mathbf{p})}{\pi(\mathbf{q})} \;\middle|\; \mathbf{p} \in B, \ \mathbf{q} \in A' \right\} \cup \left\{ \frac{\pi(\mathbf{p})}{\pi(\mathbf{q})} \;\middle|\; \mathbf{p} \in B, \ \mathbf{q} \in B \right\}$$

$$\subseteq K_0 \cup \bigcup_{\mathbf{q} \in B} K_0 \frac{1}{\pi(\mathbf{q})} \cup \bigcup_{\mathbf{p} \in B} K_0 \pi(\mathbf{p}) \cup \bigcup_{\mathbf{p}, \mathbf{q} \in B} K_0 \cdot \frac{\pi(\mathbf{p})}{\pi(\mathbf{q})}.$$

The right hand side is a union of $1 + 2\,\mathrm{card}(B) + \mathrm{card}(B)^2$ K_0-subspaces of K of dimensions 1 and 0. Since K_0 is a proper subfield of K, the dimension of K as a vector space over K_0 is at least 2, so each of the above affine K_0-subspaces of K is proper. Hence, by Lemma 5.4.1, $\mathrm{card}(K_0) \leq 1 + 2\,\mathrm{card}(B) + \mathrm{card}(B)^2$. Since K_0 is infinite, so is B, hence $\mathrm{card}(K_0) \leq \mathrm{card}(B)$. Since the fibers of π are finite and $\pi(B) = \pi(C(K)) \smallsetminus K_0$, we have $\mathrm{card}(K_0) \leq \mathrm{card}(B) = \mathrm{card}(\pi(C(K)) \smallsetminus K_0)$, as claimed. $\square$

PROPOSITION 5.4.3: *Let K be an ample field, K_0 a subfield of K, and V an absolutely irreducible variety of positive dimension defined over K_0 with a K-rational simple point. Then:*

(a) $K = K_0(V(K))$ (Fehm),
(b) $\text{card}(V(K)) = \text{card}(K)$ (Pop).

Proof of (a): Replacing K_0 by $K_0(\mathbf{p})$, if necessary, we may assume that $\mathbf{p}$ is K_0-rational. Lemma 5.1.3 gives an absolutely irreducible K_0-curve C on V such that $\mathbf{p}$ is simple on C. It suffices to prove that $K_0(C(K)) = K$.

Assume that $K_2 = K_0(C(K))$ is a proper subfield of K. Lemma 5.1.2 gives an absolutely irreducible polynomial $f \in K_0[X, Y]$ with $f(0,0) = 0$ and $\frac{\partial f}{\partial Y}(0,0) \neq 0$, and a birational map φ of C onto the plane curve C' defined by $f(X, Y) = 0$ over K_0. In particular, there are open subsets C_0 of C and C_0' of C' such that φ maps $C_0(K)$ bijectively onto $C_0'(K)$ and $C_0(K_0)$ bijectively onto $C_0'(K_0)$. Hence, $K_1 = K_0(C_0'(K)) = K_0(C_0(K)) \subseteq K_0(C(K)) = K_2$ is a proper subfield of K. By Lemma 5.4.2, $\text{card}(\pi(C'(K)) \smallsetminus K_1) \geq \aleph_0$. Since $\pi(C_0'(K))$ and $\pi(C'(K))$ differ only by finitely many elements, also $\text{card}(\pi(C_0'(K)) \smallsetminus K_1) \geq \aleph_0$. This contradicts the fact that $\pi(C_0'(K)) \subseteq K_0(C_0'(K)) = K_1$ and completes the proof of (a).

Proof of (b): First we note that $\text{card}(V(K)) \leq \text{card}(K)$. By Corollary 5.3.3, $\text{card}(V(K)) \geq \aleph_0$. Hence, if $\text{card}(K) = \aleph_0$, then $\text{card}(V(K)) = \text{card}(K)$.

Next suppose $\text{card}(K) > \aleph_0$. Let K_0 be the prime field of K. Then

$$\text{card}(K_0) \leq \aleph_0 < \text{card}(K).$$

Hence, if $\text{card}(V(K)) < \text{card}(K)$, then $\text{card}(K_0(V(K))) < \text{card}(K)$. This contradiction to (a) proves that $\text{card}(V(K)) = \text{card}(K)$. $\square$

COROLLARY 5.4.4: *Let K be an ample field, V an absolutely irreducible K-variety of positive dimension with a simple K-rational point, and h a nonconstant rational function of V defined over K. Then, $\text{card}\{h(\mathbf{a}) \mid \mathbf{a} \in V(K)\} = \text{card}(K)$.*

Proof: The assumption on h gives distinct points $\mathbf{q}_1, \mathbf{q}_2 \in V(K_s)$ with $h(\mathbf{q}_1) \neq h(\mathbf{q}_2)$. There is also a point $\mathbf{p} \in V_{\text{simp}}(K)$. By Lemma 5.1.3, there exists an absolutely irreducible curve C on V defined over K that passes through $\mathbf{p}, \mathbf{q}_1, \mathbf{q}_2$ and $\mathbf{p} \in C_{\text{simp}}(K)$. In particular, the restriction of h to C is nonconstant. We may therefore replace V by C, if necessary, to assume that V is a curve. Removing the finitely many points of V at which h is undefined, we may assume that h is defined at each point of V. Since h is nonconstant on V, each of the fibers of $h \colon V(K) \to K$ is finite. By Lemma 5.4.3(b), $\text{card}(V(K)) = \text{card}(K)$. Consequently, $\text{card}\{h(\mathbf{a}, \mathbf{b}) \mid (\mathbf{a}, \mathbf{b}) \in V(K)\} = \text{card}(K)$. $\square$

COROLLARY 5.4.5 (Fehm): *Let K be an ample field, L a Galois extension of K, and V an absolutely irreducible variety of positive dimension defined over K with a simple K-rational point. Then $\text{Gal}(L/K)$ acts faithfully on the set $V(L)$.*

Proof: Assume $\sigma \in \text{Gal}(L/K)$ fixes each point of $V(L)$ and $\sigma \neq 1$. Then $V(L)$ is contained in the fixed field L_0 of σ in L and L_0 is properly contained in L. This contradicts Proposition 5.4.3(a). $\square$

5.5 Algebraic Extensions

Like the PAC and Henselian properties of fields, being ample is preserved under algebraic field extensions.

LEMMA 5.5.1: *Let K be a field, L an algebraic extension of K, $\hat{K}$ a field extension of K, and $\hat{L} = L\hat{K}$.*
(a) *If K is existentially closed in $\hat{K}$, then L is existentially closed in $\hat{L}$.*
(b) *If K is ample, then L is ample.*

Proof of (a): Let A be a Zariski-closed subset of $\mathbb{A}^n_L$ which has an $\hat{L}$-rational point $\mathbf{x} = (x_1, \ldots, x_n)$. We have to prove that $A(L) \neq \emptyset$ (Definition 5.2.4). Since A is defined by finitely many polynomials, we may assume that $[L : K] < \infty$.

We choose polynomials $f_1, \ldots, f_m \in L[X_1, \ldots, X_n]$ that define A in $\mathbb{A}^n_L$ and a basis $w_1, \ldots, w_d$ for L/K. By Lemma 5.2.6, $\hat{K}$ is a regular extension of K, so L is linearly disjoint from $\hat{K}$ over K, hence $w_1, \ldots, w_d$ is also a basis for $\hat{L}/\hat{K}$. We choose variables Y_{ij} such that $X_i = \sum_{j=1}^d Y_{ij}w_j$, $i = 1, \ldots, n$, $j = 1, \ldots, d$. Then, $f_r(\mathbf{X}) = \sum_{j=1}^d f_{rj}(\mathbf{Y})w_j$, $r = 1, \ldots, m$, where $f_{rj}(\mathbf{Y})$ are polynomials with coefficients in K. Also, $x_i = \sum_{j=1}^d y_{ij}w_j$, $i = 1, \ldots, m$, with $y_{ij} \in \hat{K}$. It follows that $\sum_{j=1}^d f_{rj}(\mathbf{y})w_j = f_r(\mathbf{x}) = 0$, $r = 1, \ldots, m$. Since $f_{rj}(\mathbf{y}) \in \hat{K}$, we have $f_{rj}(\mathbf{y}) = 0$, $r = 1, \ldots, m$, $j = 1, \ldots, d$. Since K is existentially closed in $\hat{K}$, there exist $b_{ij} \in K$ such that $f_{rj}(\mathbf{b}) = 0$, $r = 1, \ldots, m$, $j = 1, \ldots, d$. Let $a_i = \sum_{j=1}^d b_{ij}w_j$, $i = 1, \ldots, n$. Then $\mathbf{a} \in L^n$ and $f_r(\mathbf{a}) = \sum_{j=1}^d f_{rj}(\mathbf{b})w_j = 0$, $r = 1, \ldots, m$. Consequently, $\mathbf{a} \in A(L)$, as desired.

Alternatively, we may use Remark 5.2.5 to K-embed $\hat{K}$ into an ultrapower $K^I/\mathcal{D}$. Then $\hat{L}$ is L-embeddable in $L^I/\mathcal{D}$, so L is existentially closed in $\hat{L}$.

Proof of (b): We may assume that L is a finite extension of K. In this case $L((t)) = L \cdot K((t))$. By Definition 5.3.2, K is existentially closed in $K((t))$. By (a), L is existentially closed in $L((t))$. Consequently, L is ample.   □

PROPOSITION 5.5.2: *Let $K \subseteq L$ be fields such that K is existentially closed in L and L is ample. Then K is also ample.*

Proof: Let $f \in K[X, Y]$ be an absolutely irreducible polynomial and $(a, b) \in K^2$ a pair such that $f(a, b) = 0$ and $\frac{\partial f}{\partial Y}(a, b) \neq 0$. Since L is ample, f has infinitely many L-rational zeros. Since K is existentially closed in L, for each positive integer n, the polynomial f has at least n K-rational zeros. It follows from Lemma 5.3.1(a) that K is ample.   □

We ask about a converse of Lemma 5.5.1.

PROBLEM 5.5.3: *Let L/K be a finite separable extension such that L is ample. Is K ample?*

Remark 5.5.4: Descending purely inseparable extensions.

(a) If L/K is a purely inseparable extension of positive characteristic p such that $L \subseteq K^{1/p^n}$ and L is ample, then so is K^{1/p^n} (Lemma 5.5.1). Since $K \cong K^{1/p^n}$, it follows that K is ample.

(b) A theorem of Hrushovski gives for each prime number p a countable non-PAC field E of characteristic p such that E_{ins} is PAC [FrJ08, Thm. 11.7.8]. In order to prove that E is non-PAC the proof chooses algebraically independent elements t_1, t_2, t_3, t_4, t_5 over $\mathbb{F}_p$, constructs an absolutely irreducible curve C defined over $K = \mathbb{F}_p(t_1, t_2, t_3, t_4, t_5)$, and constructs E as a regular extension of K such that $C(E)$ is a finite set and E_{ins} is PAC. If $p \neq 2$, then C is the hyperelliptic curve defined by the equation $Y^2 = \prod_{i=1}^{5}(X - t_i)$. In particular, $(t_1, 0)$ is a simple E-rational point of C. Therefore, E is nonample.

If $p = 2$, then C is defined by the equation $Y^2 + Y = \frac{1}{\prod(X-t_i)}$. This curve has two K-rational points at infinity. Unfortunately, each of them is singular. However, if we choose (x, y) with y transcendental over E and $y^2 + y = \frac{1}{\prod_{i=1}^{5}(x-t_i)}$, we find that (y, x) is a generic point of C over K, $[K(y,x) : K(y)] = 5$ and the K-rational place $\varphi_{y,\infty}$ of $K(y)$ (Notation 4.1.3) extends to five distinct places φ_i of $K(y,x)$ with $\varphi_i(x) = t_i$, $i = 1, \ldots, 5$. They give rise to five distinct prime divisors $\mathfrak{p}_1, \ldots, \mathfrak{p}_5$ of $K(y,x)/K$ lying over the prime divisor $\mathfrak{p}_{y,\infty}$ of $K(y)/K$. It follows that φ_i is K-rational. Since E is a regular extension of K and $E, K(y,x)$ are algebraically independent over K, they are linearly disjoint over K [FrJ08, Lemma 2.6.7]. Hence, $E(y,x)$ is a regular extension of E and each φ_i extends to an E-rational place of $E(y,x)$. If E were ample, then $E(y,x)$ would have infinitely many E-rational places, (Lemma 5.3.1) so $C(E)$ would be infinite, in contrast to the construction of C.

Alternatively, we can make the substitution $Z = Y(X - t_1)$ to birationally transfer C to a curve C' defined over K by the equation $f(X, Z) = 0$, where

$$f(X, Z) = \left(Z^2 + Z(X - t_1)\right) \prod_{i=2}^{5}(X - t_i) - (X - t_1).$$

Then we observe that $f(t_1, 0) = 0$ and $\frac{\partial f}{\partial X}(t_1, 0) = -1$, so $(t_1, 0)$ is a simple K-rational (hence, also E-rational) point of C'. Since $C(E)$ and $C'(E)$ differ only by finitely many points, $C'(E)$ is finite. It follows again that E is nonample.

Of course, E_{ins} as a PAC field is ample. □

5.6 Examples of Ample Fields

Various, seemingly unrelated, types of fields turn out to be ample.

Example 5.6.1: PAC fields. A field K is **PAC** if every nonvoid absolutely irreducible K-variety V has a K-rational point. By Rabinovich trick, $V(K)$

is then Zariski-dense in V [FrJ08, Prop. 11.1.1]. Hence, by Lemma 5.3.1(c), K is ample. In particular, every separably closed field is PAC [Lan58, p. 76, prop. 10] and therefore ample.

By Weil, every infinite algebraic extension of a finite field is PAC [FrJ08, Cor. 11.2.4]. For an arbitrary field K and elements $\sigma_1, \ldots, \sigma_e$ of the absolute Galois group $\mathrm{Gal}(K)$ of K, we denote the fixed field of $\sigma_1, \ldots, \sigma_e$ in K_s by $K_s(\boldsymbol{\sigma})$. We also denote the maximal Galois extension of K in $K_s(\boldsymbol{\sigma})$ by $K_s[\boldsymbol{\sigma}]$. If K is countable and Hilbertian, then $K_s[\boldsymbol{\sigma}]$ is PAC for almost all $\boldsymbol{\sigma} \in \mathrm{Gal}(K)^e$ [FrJ08, Thm. 18.10.2]. Here **almost all** is used in the sense of the Haar measure on $\mathrm{Gal}(K)^e$ with respect to Krull topology of $\mathrm{Gal}(K)$ [FrJ08, Chap. 18]. Let K_{symm} be the compositum of all Galois extensions of K with Galois group isomorphic to some S_n. Then K_{symm} is PAC [FrV92, p. 475]. However, it is unknown if the compositum of all Galois extensions of K with Galois group isomorphic to A_n for some positive integer n is PAC [FrV92, p. 476]. Finally, we note that every algebraic extension of a PAC field is PAC [FrJ08, Cor. 11.2.5]. $\square$

Example 5.6.2: Henselian fields. Let (K, v) be a Henselian valued field. As in the proof of "(f) $\Longrightarrow$ (a)" of Lemma 5.3.1 we consider an absolutely irreducible polynomial $f \in K[X, Y]$ and a point $(a, b) \in K^2$ such that $f(a, b) = 0$ and $\frac{\partial f}{\partial Y}(a, b) \neq 0$. Replacing a, b by $\pi a, \pi b$ and $f(X, Y)$ by $\pi^{d+1} f(\pi^{-1}X, \pi^{-1}Y)$, where $d = \deg(f)$ and π is an element of K with $v(\pi)$ sufficiently large, we may assume that $f \in O_v[X, Y]$, where O_v is the valuation ring of (K, v), and $a, b \in O_v$. Let $(a_i, b_i) \in K^2$ with $f(a_i, b_i) = 0$, $i = 1, \ldots, n$. We choose $\pi \in O_v$ with $v(\pi)$ sufficiently large such that $a' = a + \pi$ is different from $a_1, \ldots, a_n$, $v(f(a', b)) > 0$ and $v(f(a', b)) > 2v\left(\frac{\partial f}{\partial Y}(a', b)\right)$. Since (K, v) is Henselian, there is a $b' \in K$ such that $f(a', b') = 0$. By the choice of a', we have $(a', b') \neq (a_i, b_i)$ for $i = 1, \ldots, n$. It follows from Lemma 5.3.1(a) that K is ample.

In particular, if (K, v) is a complete discrete valued field, then (K, v) is Henselian [CaF67, p. 83], hence K is ample. This applies in particular to each p-adic field $\mathbb{Q}_p$ and each field $K_0((t))$ of formal power series over an arbitrary field K_0.

The following observation is a straightforward consequence of the definition of a Henselian field: If a field K is algebraically closed in a Henselian field (L, v) and $v|_K$ is nontrivial, then $(K, v|_K)$ is Henselian. In particular, the algebraic part $\mathbb{Q}_{p,\mathrm{abs}}$ of the p-adic field $\mathbb{Q}_p$ is Henselian but not complete. Similarly, the algebraic closure of $K_0(t)$ in $K_0((t))$ is Henselian but not complete. Another example of a Henselian field that is not complete is the field $K_0((x))_0$ of all convergent power series $\sum_{n=0}^{\infty} a_n x^n$ with coefficients a_n in a complete field K_0 under an ultrametric absolute value (Corollary 2.4.6 and Remark 2.4.3). By the first paragraph of the example, each of these fields is ample. $\square$

Example 5.6.3: Real closed Fields. Let K be a real closed field and let $f, a, b, a_1, \ldots, a_n$ be as in Example 5.6.2. Then $f(a, Y)$ obtains in the neighborhood of b both positive and negative values. Choosing a' close enough to a, $f(a', Y)$ obtains in the neighborhood of b positive and negative values and $a' \neq a_i$ for $i = 1, \ldots, n$. Hence, there exists b' close to b such that $f(a', b') = 0$ [Lan93, p. 453, Thm. 2.5]. Again, by Lemma 5.3.1(a), K is ample. $\square$

Example 5.6.4: PKC Fields. Let K be a field and let $\mathcal{K}$ be a family of field extensions of K. We say that K is P$\mathcal{K}$C if every nonempty absolutely irreducible K-variety V with a simple $\bar{K}$-rational point for each $\bar{K} \in \mathcal{K}$ has a K-rational point [Jar91, Sec. 7] (Thus K satisfies a **local-global principle** with respect to $\mathcal{K}$.) If, in addition, each of the fields $\bar{K} \in \mathcal{K}$ is ample, then K is also ample.

The proof of the latter statement applies Lemma 5.3.1(c). Let V be an absolutely irreducible K-variety, $\mathbf{p} \in V_{\mathrm{simp}}(K)$, and U a nonempty Zariski-open subset of V. Consider $\bar{K} \in \mathcal{K}$ and let $\bar{V} = V \times_K \bar{K}$. Then, $\mathbf{p} \in V_{\mathrm{simp}}(\bar{K}) = \bar{V}_{\mathrm{simp}}(\bar{K})$. Since V_{simp} is Zariski-open in V [Lan58, p. 199, Prop. 5], so is $U \cap V_{\mathrm{simp}}$. Hence, $(U \cap V_{\mathrm{simp}}) \times_K \bar{K}$ is Zariski-open in $\bar{V}$. By assumption, $\bar{K}$ is ample, so $(U \cap V_{\mathrm{simp}})(\bar{K}) \neq \emptyset$. Applying the local-global principle to the absolutely irreducible K-variety $U \cap V_{\mathrm{simp}}$, we conclude that $(U \cap V_{\mathrm{simp}})(K) \neq \emptyset$. Consequently, K is ample.

We note for latter use that if E is a finite separable extension of K and K is P$\mathcal{K}$C, then E is P$\mathcal{K}(E)$C, where $\mathcal{K}(E) = \{\bar{K}E \mid \bar{K} \in \mathcal{K}\}$ [Jar91, Lemma 7.2]. $\square$

Example 5.6.5: Local primes. Let K be a field. A **local prime** $\mathfrak{p}$ of K is an equivalence class of absolute values of K such that the completion $\hat{K}_{\mathfrak{p}}$ of K at $\mathfrak{p}$ is a separable extension of K. Moreover, we demand that if the absolute values in $\mathfrak{p}$ are metric, then $\hat{K}_{\mathfrak{p}}$ is either $\mathbb{R}$ or $\mathbb{C}$. If the absolute values in $\mathfrak{p}$ are ultrametric, then $\hat{K}_{\mathfrak{p}}$ is a finite extension of $\mathbb{F}_p((t))$ or a finite extension of $\mathbb{Q}_p$ for some prime p.

Now consider a finite set S of local primes of K. For each $\mathfrak{p} \in S$ choose an embedding of $\tilde{K}$ in the algebraic closure of $\hat{K}_{\mathfrak{p}}$ and let $K_{\mathfrak{p}} = \tilde{K} \cap \hat{K}_{\mathfrak{p}}$. If $\mathfrak{p}$ is metric, then $K_{\mathfrak{p}}$ is either $\tilde{K}$ or a real closure of K. If $\mathfrak{p}$ is ultrametric, then $K_{\mathfrak{p}}$ is a Henselian closure of K at $\mathfrak{p}$. The **field of totally S-adic numbers** is:

$$K_{\mathrm{tot}, S} = \bigcap_{\mathfrak{p} \in S} \bigcap_{\sigma \in \mathrm{Gal}(K)} K_{\mathfrak{p}}^{\sigma}.$$

Pop [Pop96, p. 25, Thm. 6] proves that $K_{\mathrm{tot}, S}$ is **PSC**, that is $K_{\mathrm{tot}, S}$ is P$\mathcal{K}$C with

$$\mathcal{K} = \{K_{\mathfrak{p}}^{\sigma} \mid \mathfrak{p} \in S, \ \sigma \in \mathrm{Gal}(K)\}$$

By Examples 5.6.2, 5.6.3, and 5.6.4, $K_{\mathrm{tot}, S}$ is an ample field. $\square$

Example 5.6.6: Almost all σ. Now suppose that K is a countable Hilbertian field and let S be a finite set of local primes of K. Then for almost all $\sigma \in \mathrm{Gal}(K)^e$, the field $K_{\mathrm{tot}, S}[\sigma] = K_s[\sigma] \cap K_{\mathrm{tot}, S}$ is PSC [GeJ02, Thm. A]. Again, by Examples 5.6.2, 5.6.3, and 5.6.4, $K_{\mathrm{tot}, S}$ is an ample field.

By Lemma 5.5.1, for almost all $\boldsymbol{\sigma} \in \mathrm{Gal}(K)^e$, each algebraic extension of $K_{\mathrm{tot},S}[\boldsymbol{\sigma}]$ is ample. In particular, $K_s[\boldsymbol{\sigma}]$, $K_{\mathrm{tot},S}(\boldsymbol{\sigma}) = K_{\mathrm{tot},S} \cap K_s(\boldsymbol{\sigma})$, $K_s(\boldsymbol{\sigma})$, and $K_{\mathrm{tot},S}$ are ample. Thus, the present example generalizes Examples 5.6.1 and 5.6.5.   □

Example 5.6.7: Fields with pro-p absolute Galois groups. A completely different type of ample fields has a Galois theoretic flavor. Each field K whose absolute Galois group is a pro-p group for a prime number p is ample. This is proved in Section 5.8.   □

5.7 Henselian Pairs

The basic definition of Henselian fields can be generalized to the category of commutative rings, giving rise to "Henselian pairs" $(A, \mathfrak{a})$. A simple argument shows that if in those pairs A is an integral domain, then $\mathrm{Quot}(A)$ is an ample field. This gives a new large class of ample fields. The argument is based on the following version of the definition of an ample field.

LEMMA 5.7.1: *A necessary and sufficient condition for a field K to be ample is that every absolutely irreducible polynomial*

$$(1) \qquad\qquad f(X,Y) = Y + \sum_{i+j \geq 2} c_{ij} X^i Y^j$$

with $c_{ij} \in K$ has infinitely many K-rational zeros.

Proof: Note that $f(0,0) = 0$ and $\frac{\partial f}{\partial Y}(0,0) = 1$. Hence, if K is ample, f has infinitely many K-rational zeros.

Conversely, suppose every absolutely irreducible polynomial f as in (1) has infinitely many K-rational zeros. In order to prove that K is ample, it suffices to prove that each absolutely irreducible polynomial $g \in K[X,Y]$ with a simple K-rational zero has infinitely many K-rational zeros (Lemma 5.3.1(a)). Thus, we may assume that there exist $a, b \in K$ with $g(a,b) = 0$ and $\frac{\partial g}{\partial Y}(a,b) \neq 0$. Replacing X, Y by $X - a, Y - b$, if necessary, we may assume that $(a,b) = (0,0)$. Then $g(X,Y) = cX + dY + \sum_{i+j \geq 2} a_{ij} X^i Y^j$ with $c, d, a_{ij} \in K$ and $d \neq 0$. Now we define a new variable Z by $Z = cX + dY$. Then $h(X, Z) = g(X,Y) = Z + \sum_{i+j \geq 2} b_{ij} X^i Z^j$ for some $b_{ij} \in K$. By assumption $h(X, Z)$ has infinitely many K-rational zeros. Therefore, $g(X,Y)$ also has infinitely many K-rational zeros, as contended.   □

Each of the rings appearing in this section will be commutative with 1. A **ring-ideal pair** is a pair $(A, \mathfrak{a})$ consisting of a ring A and a nonzero ideal $\mathfrak{a}$ of A.

Definition 5.7.2: We say that a ring-ideal pair $(A, \mathfrak{a})$ is **weakly Henselian** if for every polynomial $f \in A[X]$ satisfying

$$f(0) \equiv 0 \ \mathrm{mod} \ \mathfrak{a} \quad \text{and} \quad f'(0) \equiv 1 \ \mathrm{mod} \ \mathfrak{a}$$

there exists $x \in A$ such that $f(x) = 0$.   □

PROPOSITION 5.7.3 (Pop): *If A is an integral domain and $(A, \mathfrak{a})$ is a weakly Henselian pair, then $K = \mathrm{Quot}(A)$ is an ample field.*

Proof: For a polynomial f as in (1) we choose an $e \in \mathfrak{a}$ such that $e \neq 0$ and $a_{ij} = ec_{ij} \in \mathfrak{a}$ for all i, j. Then we consider the polynomial

$$g(X, Y) = \frac{f(eX, eY)}{e} = \frac{1}{e}\left(eY + \sum_{i+j\geq 2} c_{ij}e^{i+j}X^iY^j\right)$$

$$= Y + \sum_{i+j\geq 2} a_{ij}e^{i+j-2}X^iY^j.$$

For each $a \in A$ we observe that

$$h(Y) = g(a, Y) = Y + \sum_{i+j\geq 2} a_{ij}e^{i+j-2}a^iY^j$$

$$= b_0 + (1 + b_1)Y + \sum_{k\geq 2} b_kY^k$$

with $b_k \in \mathfrak{a}$ and $h'(Y) = (1+b_1)+\sum_{k\geq 2} kb_kY^{k-1}$. Then $h(0) = b_0 \equiv 0 \mod \mathfrak{a}$ and $h'(0) = 1 + b_1 \equiv 1 \mod \mathfrak{a}$. The assumption on $(A, \mathfrak{a})$ gives $b \in A$ such that $h(b) = 0$, that is $g(a, b) = 0$, so $f(ea, eb) = 0$. Thus, $f(X, Y)$ has infinitely many zeros in K. Consequently, K is ample. $\square$

The basic examples of weakly Henselian pairs are complete pairs. Actually complete domain-ideals pairs $(A, \mathfrak{a})$ are even "Henselian pairs" in the following sense:

Definition 5.7.4: A ring-ideal pair $(A, \mathfrak{a})$ is said to be a **Henselian pair** if for every polynomial $f \in A[X]$ satisfying

(2) $\qquad\qquad f(0) \equiv 0 \mod \mathfrak{a} \quad \text{and} \quad f'(0)$ is a unit mod $\mathfrak{a}$

there exists $x \in \mathfrak{a}$ such that $f(x) = 0$. $\square$

Example 5.7.5: Completions. By definition, every Henselian pair $(A, \mathfrak{a})$ is a weak Henselian pair. We show below that if A is complete with respect to $\mathfrak{a}$, then $(A, \mathfrak{a})$ is also a Henselian pair. To this end recall that a sequence $(a_1, a_2, a_3, \ldots)$ of elements of a commutative ring A **converges** (with respect to an ideal $\mathfrak{a}$) to an element $a \in A$ if for each positive integer r there exists n_0 such that $a_n - a \in \mathfrak{a}^r$ for each $n \geq n_0$. We say that $(a_1, a_2, a_3, \ldots)$ is a **Cauchy sequence** (with respect to $\mathfrak{a}$) if for each positive integer r there exists n_0 such that $a_n - a_m \in \mathfrak{a}^r$ for all $m, n \geq n_0$. Finally, the pair $(A, \mathfrak{a})$ is **complete** if every Cauchy sequence converges to a unique element of A. In particular, $\bigcap_{i=1}^n \mathfrak{a}^n = 0$. Also, if a sequence $a_1, a_2, a_3, \ldots$ of A converges to 0, then the partial sums of the infinite series $\sum_{n=1}^\infty a_n$ form a Cauchy sequence, so $\sum_{n=1}^\infty a_n$ converges in A.

For an arbitrary ring-ideal pair $(A, \mathfrak{a})$ we consider the ring $\hat{A} = \varprojlim A/\mathfrak{a}^n$ and its ideal $\hat{\mathfrak{a}} = \varprojlim \mathfrak{a}/\mathfrak{a}^n$. We assume that $\bigcap_{n=1}^{\infty} \mathfrak{a}^n = 0$ (This condition is satisfied for example if A is a Noetherian domain [AtM69, Cor. 10.18].) Then A may be embedded into $\hat{A}$ by mapping each $a \in A$ onto the constant sequence $(a, a, a, \ldots)$. Then one observes that every Cauchy sequence $(a_1, a_2, a_3, \ldots)$ of elements of A converges to an element of $\hat{A}$. Approximating each Cauchy sequence of $\hat{A}$ by a Cauchy sequence of A, we find that every Cauchy sequence of $\hat{A}$ with respect to $\hat{\mathfrak{a}}$ converges. Thus, $(\hat{A}, \hat{\mathfrak{a}})$ is complete. Moreover, if $(B, \mathfrak{b})$ is a complete pair and $\alpha \colon A \to B$ is a homomorphism such that $\alpha(\mathfrak{a}) \subseteq \mathfrak{b}$, then α can be uniquely extended to a homomorphism $\hat{\alpha} \colon \hat{A} \to B$ such that $\alpha(\hat{\mathfrak{a}}) \subseteq \mathfrak{b}$. This means that $(\hat{A}, \hat{\mathfrak{a}})$ is the **completion** of $(A, \mathfrak{a})$.

For example, let $R = A[X_1, \ldots, X_n]$ be the ring of polynomials over a ring A and let $\mathfrak{a} = \sum_{i=1}^{n} RX_i$. Then $\hat{R} = \varprojlim R/\mathfrak{a}^n$ is naturally isomorphic to the ring $A[[X_1, \ldots, X_n]]$ of formal power series over A. If $f = (f_1 + \mathfrak{a}, f_2 + \mathfrak{a}^2, f_3 + \mathfrak{a}^3, \ldots)$ is an element of $\hat{R}$, then $f_k \equiv g_0 + \cdots + g_{k-1} \bmod \mathfrak{a}^k$, where g_i is a homogeneous polynomial of degree i independent of k. The above mentioned isomorphism maps f onto $\sum_{k=0}^{\infty} g_k$. It maps $\hat{\mathfrak{a}} = \varprojlim \mathfrak{a}/\mathfrak{a}^n$ onto the ideal I of $A[[X_1, \ldots, X_n]]$ generated by $X_1, \ldots, X_n$. In particular, $(A[[X_1, \ldots, X_n]], I)$ is complete.   □

LEMMA 5.7.6: *Let A be a complete ring with respect to an ideal $\mathfrak{a}$. Then every $u \in A$ which is a unit of A modulo $\mathfrak{a}$ is a unit of A.*

Proof: By assumption there exists $v \in A$ such that $uv \equiv 1 \bmod \mathfrak{a}$. Thus, $a = 1 - uv \in \mathfrak{a}$. Hence, $\sum_{i=0}^{\infty} a^i$ converges in A and $uv \sum_{i=0}^{n} a^i = 1 - a^{n+1}$ for each positive integer n. Taking n to the limit, we find that $uv \sum_{i=0}^{\infty} a^i = 1$, hence $u \in A^{\times}$.   □

PROPOSITION 5.7.7: *If $(A, \mathfrak{a})$ is a complete domain-ideal pair, then $(A, \mathfrak{a})$ is a Henselian pair and $K = \operatorname{Quot}(A)$ is an ample field.*

Proof: We follow the classical proof that a complete discrete valued field is Henselian.

Let $f \in A[X]$ and $a \in A$ such that (2) holds. In particular $f'(0)$ is a unit modulo $\mathfrak{a}$. By Lemma 5.7.6, $f'(0) \in A^{\times}$. Let $a_0 = 0$ and inductively define a sequence $a_0, a_1, a_2, \ldots$ of elements of K by

$$(3) \qquad\qquad a_{n+1} = a_n - f'(a_n)^{-1} f(a_n).$$

Inductively suppose
(4a) $a_n \in A$, $a_n \equiv 0 \bmod \mathfrak{a}$, and
(4b) $f(a_n) \equiv 0 \bmod \mathfrak{a}^{2^n}$.

Then, $f(a_n) \equiv f(0) \equiv 0 \bmod \mathfrak{a}$ and $f'(a_n) \equiv f'(0) \bmod \mathfrak{a}$ is a unit of A modulo $\mathfrak{a}$, so $f'(a_n)$ is a unit of A (Lemma 5.7.6). Therefore, a_{n+1} is a well defined element of $\mathfrak{a}$. Moreover, by Taylor expansion,

$$f(a_{n+1}) = f(a_n - f'(a_n)^{-1}f(a_n))$$
$$= f(a_n) - f'(a_n)f'(a_n)^{-1}f(a_n) + bf'(a_n)^{-2}f(a_n)^2$$
$$= bf'(a_n)^{-2}f(a_n)^2 \equiv 0 \bmod \mathfrak{a}^{2^{n+1}},$$

for some $b \in A$. Since A is complete with respect to $\mathfrak{a}$, (3) and (4) imply that a_n converges to an element x of A. By (4a), $x \equiv 0 \bmod \mathfrak{a}$. Finally, by (4b), $f(x) \equiv 0 \bmod \mathfrak{a}^{2^n}$ for all n, so $f(x) = 0$. Consequently, $(A, \mathfrak{a})$ is a Henselian pair. By Proposition 5.7.3, K is ample. $\square$

Remark 5.7.8: Completions. Let $(A, \mathfrak{a})$ be a domain-ideal pair and let $(\hat{A}, \hat{\mathfrak{a}})$ be its completion. If $\hat{A}$ is an integral domain, then $\mathrm{Quot}(\hat{A})$ is an ample field (Proposition 5.7.7). This is, for example, the case when $A = K[X_1, \ldots, X_n]$ is the ring of polynomials in $X_1, \ldots, X_n$ over a field K and $\mathfrak{a}$ is the ideal generated by $X_1, \ldots, X_n$. As mentioned in Example 5.7.5, $\hat{A} = K[[X_1, \ldots, X_n]]$ is the ring of formal power series in $X_1, \ldots, X_n$ over K. It is an integral domain and $\mathrm{Quot}(\hat{A}) = K((X_1, \ldots, X_n))$ is the **field of power series** in $X_1, \ldots, X_n$ over K.

More generally, $\hat{A}$ is an integral domain if A is the local ring of an absolutely irreducible variety over a field K at a normal subvariety. In other words, let $K[\mathbf{x}] = K[x_1, \ldots, x_n]$ be a domain such that $K(\mathbf{x})$ is a regular extension of K, let $\mathfrak{p}$ be a prime ideal of $K[\mathbf{x}]$, and let $A = K[\mathbf{x}]_{\mathfrak{p}}$. Suppose A is integrally closed. Then, by a theorem of Zariski [ZaS75, p. 320, Thm. 32], the completion $\hat{A}$ of A with respect to $\mathfrak{p}A$ is an integrally closed domain, hence $\mathrm{Quot}(\hat{A})$ is ample. $\square$

We finish this section with some remarks about the connection between "weakly Henselian pairs" and "Henselian pairs".

PROPOSITION 5.7.9: *The following statements hold for each weakly Henselian pair $(A, \mathfrak{a})$.*
(a) *The ideal $\mathfrak{a}$ is contained in the **Jacobson radical** $J(A)$ of A (defined as the intersection of all maximal ideals of A).*
(b) *If $u \in A$ is a unit of A modulo $\mathfrak{a}$, then u is a unit of A.*
(c) *If $f \in A[X]$ satisfies $f(0) \equiv 0 \bmod \mathfrak{a}$ and $f'(0)$ is a unit of A mod $\mathfrak{a}$, then there exists $x \in A$ such that $f(x) = 0$.*

Proof of (a): Assume $\mathfrak{a} \not\subseteq J(A)$ and choose $a \in \mathfrak{a} \smallsetminus J(A)$. By definition, A has a maximal ideal $\mathfrak{m}$ that does not contain a. Thus, there exist $u \in A$ and $m \in \mathfrak{m}$ such that $m - ua = 1$. The polynomial $f(X) = m(1 + X) - 1$ satisfies $f(0) = m - 1 = ua \equiv 0 \bmod \mathfrak{a}$ and $f'(0) = m = 1 + ua \equiv 1 \bmod \mathfrak{a}$. By definition, f has a root $x \in A$. Thus, $m(1 + x) = 1$. This contradiction to the fact that m is not invertible proves that $\mathfrak{a} \subseteq J(A)$.

Proof of (b): Let u be a unit of A modulo $\mathfrak{a}$. Thus, there exists $b \in A$ with $bu \equiv 1 \bmod \mathfrak{a}$. Assume that u is not a unit of A. Then A has a maximal ideal $\mathfrak{m}$ containing u. By (a), $bu \equiv 1 \bmod \mathfrak{m}$, so $1 \in \mathfrak{m}$. This contradiction proves that u is a unit of A.

Proof of (c): By (b), $f'(0)$ is a unit of A. Hence, the polynomial $g = f'(0)^{-1}f$ satisfies $g(0) \equiv 0 \bmod \mathfrak{a}$ and $g'(0) = f'(0)^{-1}f'(0) = 1$. By definition, there exists $x \in A$ with $g(x) = 0$. Hence, $f(x) = 0$. $\qquad\square$

Remark 5.7.10: On the uniqueness of the root. Let $(A, \mathfrak{a})$ be a ring-ideal pair. Consider a polynomial $f \in A[X]$ such that $f(0) \equiv 0 \bmod \mathfrak{a}$ and $c_1 = f'(0)$ is a unit of A modulo $\mathfrak{a}$. If $(A, \mathfrak{a})$ is a weakly Henselian pair, then there exists $x \in A$ with $f(x) = 0$ (Proposition 5.7.9) while if $(A, \mathfrak{a})$ is a Henselian pair, we may choose x in $\mathfrak{a}$. In both cases f has at most one root in $\mathfrak{a}$.

Indeed, suppose $x, y \in \mathfrak{a}$ and $f(x) = f(y) = 0$. Let $f(X) = \sum_{k=0}^{n} c_k X^k$ with $c_k \in A$. Then

$$0 = f(x) - f(y) = \sum_{k=1}^{n} c_k(x^k - y^k) = (x - y)(c_1 + a),$$

where $a = \sum_{k=2}^{n} c_k \sum_{i=0}^{k-1} x^i y^{k-1-i} \in \mathfrak{a}$. Since c_1 is a unit modulo $\mathfrak{a}$, so is $c_1 + a$. Hence, by Proposition 5.7.9(a), $c_1 + a \in A^\times$. It follows that $x - y = 0$, hence $x = y$. $\qquad\square$

Remark 5.7.11: Henselian closure. An **extension** $(B, \mathfrak{b})$ of $(A, \mathfrak{a})$ is a ring-ideal pair such that $A \subseteq B$ and $\mathfrak{b} \cap A = \mathfrak{a}$. In particular, assuming that $\bigcap_{n=0}^{\infty} \mathfrak{a}^n = 0$, the completion $(\hat{A}, \hat{\mathfrak{a}})$ of $(A, \mathfrak{a})$ is an extension of $(A, \mathfrak{a})$. Let $(A^h, \mathfrak{a}^h)$ be the intersection of all Henselian pairs $(B, \mathfrak{b})$ lying between $(A, \mathfrak{a})$ and $(\hat{A}, \hat{\mathfrak{a}})$. If $f \in A^h[X]$, $f(0) \equiv 0 \bmod \mathfrak{a}^h$, and $f'(0)$ is a unit modulo $\mathfrak{a}^h$, then f satisfies the same conditions with respect to each of the Henselian pairs $(B, \mathfrak{b})$ and in particular with respect to $(\hat{A}, \hat{\mathfrak{a}})$. The unique root x of f in $\hat{\mathfrak{a}}$ (Remark 5.7.10) belongs to each of the ideals $\mathfrak{b}$, hence also to $\mathfrak{a}^h$. Thus, $(A^h, \mathfrak{a}^h)$ is a Henselian pair extending $(A, \mathfrak{a})$, called the **Henselian closure** of $(A, \mathfrak{a})$.

It is possible to reach $(A^h, \mathfrak{a}^h)$ from below. To this end let A_1 be the subring of A^h generated by all of the roots $x \in \mathfrak{a}^h$ of polynomials $f \in A[X]$ such that $f(0) \equiv 0 \bmod \mathfrak{a}$ and $f'(0)$ is a unit of A modulo $\mathfrak{a}$. Let $\mathfrak{a}_1 = \mathfrak{a}^h \cap A_1$. Then $f'(x) \equiv f'(0) \bmod \mathfrak{a}_1$, so $f'(x) \neq 0$. Thus, if both A and $\hat{A}$ are integral domains, then so is A_1 and $\mathrm{Quot}(A_1)$ is a separable algebraic extension of $\mathrm{Quot}(A)$. Now we construct a pair $(A_2, \mathfrak{a}_2)$ out of $(A_1, \mathfrak{a}_1)$ in the same way that $(A_1, \mathfrak{a}_1)$ was constructed from $(A, \mathfrak{a})$. Then we continue inductively to construct $(A_3, \mathfrak{a}_3)$, $(A_4, \mathfrak{a}_4)$, and so on. Finally let $(A', \mathfrak{a}')$ be the union of all pairs $(A_n, \mathfrak{a}_n)$. Then $(A', \mathfrak{a}')$ is a Henselian pair between $(A, \mathfrak{a})$ and $(A^h, \mathfrak{a}^h)$. It follows from the minimality of the last pair, that $(A', \mathfrak{a}') = (A^h, \mathfrak{a}^h)$. If both A and $\hat{A}$ are integral domains, then $\mathrm{Quot}(A^h)$ is a separable algebraic extension of $\mathrm{Quot}(A)$. By Proposition 5.7.7, $\mathrm{Quot}(A^h)$ is an ample field.

The Henselian pair $(A^h, \mathfrak{a}^h)$ has the expected universal property: Let $(C, \mathfrak{c})$ be a Henselian pair and $\alpha\colon (A, \mathfrak{a}) \to (C, \mathfrak{c})$ a homomorphism of ring-ideal pairs, that is, $\alpha\colon A \to C$ is a homomorphism of commutative rings such that $\alpha(\mathfrak{a}) \subseteq \mathfrak{c}$. Then α extends to a homomorphism $\hat{\alpha}\colon (\hat{A}, \hat{\mathfrak{a}}) \to (\hat{C}, \hat{\mathfrak{c}})$ of the completions. Let $B = \hat{\alpha}^{-1}(C)$ and $\mathfrak{b} = \alpha^{-1}(\mathfrak{c})$. Then the uniqueness of the roots proved in Remark 5.7.10 implies that $(B, \mathfrak{b})$ is a Henselian pair between $(A, \mathfrak{a})$ and $(\hat{A}, \hat{\mathfrak{a}})$. By construction, $A^h \subseteq B$. It follows that the restriction of $\hat{\alpha}$ to A^h is a homomorphism $\alpha^h\colon (A^h, \mathfrak{a}^h) \to (C, \mathfrak{c})$ of ring-ideal pairs that extends α. The uniqueness of the roots mentioned above shows that the value of $\hat{\alpha}$ at the element x mentioned at the beginning of the previous paragraph is unique. Therefore, $\hat{\alpha}$ is unique on A_1. Similarly, $\hat{\alpha}$ is uniquely determined on A_2, and so on. This proves that the extension α^h of α is unique. Read more about Henselian pairs in [Laf63], [Ray70, Chap Xi], and [KPR75, Chap. 2]. $\square$

Example 5.7.12: Henselian fields. Let v be a valuation of a field K and (K_v^h, v) a Henselization of (K, v). Let O_v and O_v^h be the corresponding valuation rings and let $\mathfrak{m}_v$ and $\mathfrak{m}_v^h$ be their maximal ideals. Then $(O_v^h, \mathfrak{m}_v^h)$ is a Henselian closure of the ring-pair $(O_v, \mathfrak{m}_v)$. By Remark 5.7.11 or Example 5.6.2, the field K_v^h is ample. $\square$

5.8 Fields with pro-p Absolute Galois Groups

Very rarely has the absolute Galois group of a field K a decisive impact on the diophantine nature of K. If $\mathrm{Gal}(K)$ is trivial, then K is separably closed, if $\mathrm{Gal}(K)$ has order 2, then K is real closed (Artin-Schreier), and if $\mathrm{Gal}(K) \cong \mathrm{Gal}(\mathbb{Q}_p)$, then K is p-adically closed (Efrat-Koenigsmann-Pop). It is therefore surprising to find out that if $\mathrm{Gal}(K)$ is a pro-p group, then K is ample.

Remark 5.8.1: Algebraic function fields of one variable. Let F be a function field of one variable over a field K. We briefly recall the definitions of the main objects attached to F/K and their properties. For a more comprehensive survey see [FrJ08, Sections 3.1-3.2].

(a) Recall that a K-place of F is a place $\varphi\colon F \to \tilde{K} \cup \{\infty\}$ such that $\varphi(a) = a$ for each $a \in K$ (Definition 4.1.2). A prime divisor $\mathfrak{p}$ of F/K is an equivalence class of valuations of F that are trivial on K (Remark 4.1.1), or what amounts to be equivalent, of K-places of F. Let $\varphi_{\mathfrak{p}}$ be a place in that class, $v_{\mathfrak{p}}$ the corresponding discrete valuation of F/K, and $\bar{F}_{\mathfrak{p}}$ the residue field. The latter field is a finite extension of K which is uniquely determined by $\mathfrak{p}$ up to K-conjugation. We set $\deg(\mathfrak{p}) = [\bar{F}_{\mathfrak{p}} : K]$. A **divisor** of F/K is a formal sum $\mathfrak{a} = \sum k_{\mathfrak{p}} \mathfrak{p}$ where $\mathfrak{p}$ ranges over all prime divisors of F/K, for each $\mathfrak{p}$ the coefficient $k_{\mathfrak{p}}$ is an integer, and $k_{\mathfrak{p}} = 0$ for all but finitely many $\mathfrak{p}'s$. The **degree** of $\mathfrak{a}$ is $\deg(\mathfrak{a}) = \sum k_{\mathfrak{p}} \deg(\mathfrak{p})$. The divisor attached to an element $f \in F^\times$ is defined to be $\mathrm{div}(f) = \sum v_{\mathfrak{p}}(f)\mathfrak{p}$, where $\mathfrak{p}$ ranges over all prime divisors of F/K. This makes sense, since $v_{\mathfrak{p}}(f) = 0$ for all but finitely many $\mathfrak{p}$'s. Further, one attaches to f the **divisor of zeros** $\mathrm{div}_0(f) = \sum_{v_\mathfrak{p}(f)>0} v_\mathfrak{p}(f)\mathfrak{p}$ and the **divisor of poles** $\mathrm{div}_\infty(f) = -\sum_{v_\mathfrak{p}(f)<0} v_\mathfrak{p}(f)\mathfrak{p}$. If $f \notin K$, the degrees of both divisors are equal to $[F : K(f)]$. Hence, $\deg(\mathrm{div}(f)) = \deg(\mathrm{div}_0(f)) - \deg(\mathrm{div}_\infty(f)) = 0$. If $\mathfrak{a} = \sum k_\mathfrak{p}\mathfrak{p}$ is a divisor of F/K, we write $v_\mathfrak{p}(\mathfrak{a}) = k_\mathfrak{p}$ for each prime divisor $\mathfrak{p}$ of F/K and note that $v_\mathfrak{p}(\mathrm{div}(f)) = v_\mathfrak{p}(f)$ for each $f \in F^\times$. Given two divisors $\mathfrak{a}, \mathfrak{b}$ of F/K, we write $\mathfrak{a} \leq \mathfrak{b}$ if $v_\mathfrak{p}(\mathfrak{a}) \leq v_\mathfrak{p}(\mathfrak{b})$ for each prime divisor $\mathfrak{p}$ of F/K. Finally, one attaches to each divisor $\mathfrak{a}$ a finitely generated vector space $\mathcal{L}(\mathfrak{a})$ over K consisting of 0 and of all $f \in F^\times$ with $\mathrm{div}(f) + \mathfrak{a} \geq 0$ and write $\dim(\mathfrak{a})$ for $\dim(\mathcal{L}(\mathfrak{a}))$. If $\mathfrak{a} \leq \mathfrak{b}$, then $\mathcal{L}(\mathfrak{a}) \subseteq \mathcal{L}(\mathfrak{b})$.

The inequality $\mathrm{div}(f) + \mathfrak{a} \geq 0$ can be rewritten as $\mathrm{div}_0(f) + \mathfrak{a} \geq \mathrm{div}_\infty(f)$. Since $\mathrm{div}_0(f)$ and $\mathrm{div}_\infty(f)$ have no common prime divisors, the latter inequality is equivalent to $\mathfrak{a} \geq \mathrm{div}_\infty(f)$ if $\mathfrak{a} \geq 0$.

The latter inequality can be obtained for an arbitrary divisor $\mathfrak{a}$ with $\mathcal{L}(\mathfrak{a}) \neq 0$ by shifting it with a principal divisor. Indeed, fix $f \in \mathcal{L}(\mathfrak{a})$, $f \neq 0$ and set $\mathfrak{a}' = \mathrm{div}(f) + \mathfrak{a}$. Then $\mathfrak{a}' \geq 0$, $\deg(\mathfrak{a}') = \deg(\mathfrak{a})$, and the map $x \mapsto fx$ is a K-isomorphism of $\mathcal{L}(\mathfrak{a}')$ onto $\mathcal{L}(\mathfrak{a})$, in particular, $\dim(\mathfrak{a}') = \dim(\mathfrak{a})$.

(b) The set $\mathrm{Div}(F/K)$ of all divisors of F/K is an additive Abelian group freely generated by the set $\mathbb{P}(F/K)$ of all prime divisors of F/K. The map $\deg \colon \mathrm{Div}(F/K) \to \mathbb{Z}$ is a homomorphism whose kernel is the subgroup $\mathrm{Div}_0(F/K)$ of all divisors of degree 0. If in addition F/K has a prime divisor of degree 1, then the short sequence

$$0 \longrightarrow \mathrm{Div}_0(F/K) \longrightarrow \mathrm{Div}(F/K) \xrightarrow{\deg} \mathbb{Z} \longrightarrow 0,$$

is exact.

(c) The Riemann-Roch theorem gives a nonnegative integer g, called the **genus** of F/K, such that if $\deg(\mathfrak{a}) > 2g - 2$, then $\dim(\mathfrak{a}) = \deg(\mathfrak{a}) + 1 - g$. In the general case $\dim(\mathfrak{a}) = \deg(\mathfrak{a}) + 1 - g + \dim(\mathfrak{w} - \mathfrak{a})$, where $\mathfrak{w}$ is a **canonical divisor** of F/K [FrJ08, Thm. 3.2.1]. To this end recall that all canonical divisors of F/K are **linearly equivalent** (i.e. differ from each other by a divisor of an element of $F^\times$), $\deg(\mathfrak{w}) = 2g - 2$, and $\dim(\mathfrak{w}) = g$ [FrJ08, Lemma 3.2.2].

(d) An easy corollary of the Riemann-Roch theorem characterizes rational function fields $K(x)$ as those algebraic function fields of one variables over K of genus zero with a prime divisor of degree 1 [FrJ08, Example 3.2.4].

(e) If L is an algebraic extension of K, then FL/L is an algebraic function field of one variable. We say that a prime divisor $\mathfrak{P}$ of FL/L **lies over** a prime divisor $\mathfrak{p}$ of F/K if $v_\mathfrak{P}$ lies over $v_\mathfrak{p}$. If L/K is separable, then $\mathfrak{P}/\mathfrak{p}$ is unramified [Deu, p. 113]. In the general case, FL/L has only finitely many prime divisors $\mathfrak{P}_1, \ldots, \mathfrak{P}_r$ that lie over $\mathfrak{p}$ [Deu73, p. 96]. Then we identify $\mathfrak{p}$ with the divisor $\sum_{i=1}^r e_{\mathfrak{P}_i/\mathfrak{p}}\mathfrak{P}_i$ of FL/L and extend this identification to an embedding of $\mathrm{Div}(F/K)$ into $\mathrm{Div}(FL/L)$. Then we write $\mathcal{L}_{FL}(\mathfrak{a})$ to denote the linear subspace $\{f \in FL \mid \mathrm{div}(f) + \mathfrak{a} \geq 0\}$ of FL associated with $\mathfrak{a}$. By [Deu73, p. 132, Thm. 1], $\mathrm{genus}(FL/L) \leq \mathrm{genus}(F/K)$ and equality holds if

L/K is separable. In this case $\dim_K(\mathcal{L}_F(\mathfrak{a})) = \dim_L(\mathcal{L}_{FL}(\mathfrak{a}))$ for each divisor $\mathfrak{a}$ of F/K [Deu73, p. 132, Thm. 2].

(f) If F is a finite extension of a field E that contains K, then E is also a function field of one variable over K. If in addition, F/E is separable, then the **Riemann-Hurwitz genus formula** relates the genera of the two function fields:

$$2\text{genus}(F/K) - 2 = [F : E](2\text{genus}(E/K) - 2) + \deg(\text{Diff}(F/E)),$$

where $\text{Diff}(F/E)$ is a nonnegative divisor of F/K whose prime divisors are exactly those that ramify over E [FrJ08, Thm. 3.6.1]. In particular,

$$\text{genus}(E/K) \leq \text{genus}(F/K).$$

Also, if F/E is unramified, then $2\text{genus}(F/K) - 2 = [F : E](2\text{genus}(E/K) - 2)$. Hence, if in addition, $\text{genus}(E/K) = 1$, then also $\text{genus}(F/K) = 1$.

If $\text{genus}(E/K) = 0$ (in particular if $E = K(x)$) and $[F : E] > 1$, then F/E must ramify. Otherwise, $\deg(\text{Diff}(F/E)) = 0$ and we get a contradiction $2\text{genus}(F/K) - 2 = -2[F : E]$. $\qquad\square$

PROPOSITION 5.8.2: *Let K be a perfect field such that $\text{Gal}(K)$ is a pro-p group for some prime number p. Then K is ample.*

Proof: Consider a function field F of one variable over K with a prime divisor $\mathfrak{p}$ of degree 1. Let $\mathfrak{p}_1, \ldots, \mathfrak{p}_m$ be additional prime divisors of F/K of degree 1. By the weak approximation theorem there exists $f \in F$ with $v_{\mathfrak{p}}(f) = 1$ and $v_{\mathfrak{p}_i}(f) = 0$ for $i = 1, \ldots, m$ [FrJ08, Prop. 2.1.1]. Then $\text{div}(f) = \mathfrak{p} + \sum_{j=1}^{n} k_j \mathfrak{q}_j$, for some additional distinct prime divisors $\mathfrak{q}_1, \ldots, \mathfrak{q}_n$. It follows that

$$(1) \qquad\qquad 1 + \sum_{j=1}^{n} k_j \deg(\mathfrak{q}_j) = \deg(\text{div}(f)) = 0.$$

Denote the residue field of F at $\mathfrak{q}_j$ by $\bar{F}_{\mathfrak{q}_j}$. Since K is perfect, $\bar{F}_{\mathfrak{q}_j}/K$ is separable. Since $\deg(\mathfrak{q}_j) = [\bar{F}_{\mathfrak{q}_j} : K]$ and $\text{Gal}(K)$ is a pro-p group, each of the numbers $\deg(\mathfrak{q}_j)$ is a power of p. We conclude from (1) that $\deg(\mathfrak{q}_j) = 1$ for some j between 1 and n. Consequently, F/K has infinitely many prime divisors of degree 1. In other words, K is ample (Lemma 5.3.1(d)). $\qquad\square$

More arguments are needed to prove Proposition 5.8.2 without the condition "K is perfect".

THEOREM 5.8.3: *Let K be a field such that $\text{Gal}(K)$ is a pro-p group for some prime number p. Then K is ample.*

Proof: Each finite field has finite extensions of every degree, in particular its absolute Galois group is not pro-p. It follows that K is infinite.

Let F be a function field of one variable of genus g over K with a prime divisor $\mathfrak{p}$ of degree 1. Set $\mathfrak{p}_0 = \mathfrak{p}$ and let $\mathfrak{p}_1, \ldots, \mathfrak{p}_m$ with $m \geq 1$ be additional prime divisors of F/K of degree 1. Choose a positive multiple k of p such that $k \geq 2g$ and $\operatorname{char}(K)|k$ if $\operatorname{char}(K) \neq 0$. Consider the divisors $\mathfrak{a} = \mathfrak{p} + k \sum_{i=1}^m \mathfrak{p}_i$ and $\mathfrak{a}_i = \mathfrak{a} - \mathfrak{p}_i$, $i = 0, \ldots, m$, of F/K. Then $\deg(\mathfrak{a}) > \deg(\mathfrak{a}_0) \geq k \geq 2g$ and $\deg(\mathfrak{a}) > \deg(\mathfrak{a}_i) \geq k - 1 \geq 2g - 1$ for $i = 1, \ldots, m$. By Riemann-Roch, $\dim(\mathcal{L}(\mathfrak{a})) = \deg(\mathfrak{a}) + 1 - g$ and $\dim(\mathcal{L}(\mathfrak{a}_i)) = \deg(\mathfrak{a}_i) + 1 - g$. Thus, $\mathcal{L}(\mathfrak{a}_i)$ is a proper subspace of $\mathcal{L}(\mathfrak{a})$, $i = 0, \ldots, m$. Since K is infinite, there exists $t \in \mathcal{L}(\mathfrak{a}) \smallsetminus \bigcup_{i=0}^m \mathcal{L}(\mathfrak{a}_i)$. Hence, $\operatorname{div}(t) + \mathfrak{a} \geq 0$ but $\operatorname{div}(t) + \mathfrak{a}_i \not\geq 0$ for each i. It follows that $\operatorname{div}_\infty(t) = \mathfrak{a}$, so $\operatorname{div}_\infty(t - a) = \mathfrak{a}$ for each $a \in K$.

By definition

$$(2) \qquad \deg(\mathfrak{a}) = 1 + k \sum_{i=1}^m \deg(\mathfrak{p}_i).$$

Hence,

$$(3) \qquad [F : K(t - a)] = \deg(\operatorname{div}_\infty(t - a)) = \deg(\mathfrak{a}) \equiv 1 \mod k.$$

In particular, if $\operatorname{char}(K) \neq 0$, then $\operatorname{char}(K) \nmid [F : K(t)]$. Thus, in each case, $F/K(t)$ is a finite separable extension.

Now choose a primitive element x for $F/K(t)$, integral over $K[t]$. Let $f = \operatorname{irr}(x, K(t))$. Then $f(T, X) \in K[T, X]$ is an absolutely irreducible polynomial separable in X [FrJ08, Cor. 10.2.2(b)]. Hence, there exists $a \in K$ such that all roots of $f(a, X)$ are simple. In particular, they belong to K_s. These roots correspond to zeros of $t - a$ (as an element of F). Therefore, $\operatorname{div}_0(t - a) = \sum_{i=1}^r \mathfrak{q}_i$ and for each i, $\mathfrak{q}_i$ is a prime divisor of F/K with residue field $\bar{F}_{\mathfrak{q}_i}$ separable over K. The assumption on K implies that $\deg(\mathfrak{q}_i) = [\bar{F}_{\mathfrak{q}_i} : K]$ is a power of p. By (3),

$$\sum_{i=1}^r \deg(\mathfrak{q}_i) = \deg(\operatorname{div}_0(t - a)) = \deg(\operatorname{div}_\infty(t - a)) \equiv 1 \mod p.$$

Hence, there exists i between 1 and r with $\deg(\mathfrak{q}_i) = 1$. In addition, $\mathfrak{q}_i$ is relatively prime to $\mathfrak{a}$, so $\mathfrak{q}_i$ differs from $\mathfrak{p}, \mathfrak{p}_1, \ldots, \mathfrak{p}_m$. Consequently, K is ample. $\qquad\square$

PROBLEM 5.8.4: *Let K be a field such that the order of $\operatorname{Gal}(K)$ is divisible by only finitely many prime numbers. Is K ample?*

5.9 Embedding Problems over Ample Fields

In this section K/K_0 is an arbitrary finite Galois extension with Galois group Γ and x is transcendental over K. Suppose Γ acts on a finite group G. We look for a rational solution of the constant split embedding problem

$$(1) \qquad \operatorname{pr}: \operatorname{Gal}(K(x)/K_0(x)) \ltimes G \to \operatorname{Gal}(K(x)/K_0(x))$$

over $K_0(x)$. When K_0 is complete under an ultrametric absolute value, this problem reduces to the special case solved in Section 4.4.

Consider also a regular extension $\hat{K}_0$ of K_0 such that x is transcendental over $\hat{K}_0$ and let $\hat{K} = K\hat{K}_0$. Then $\hat{K}_0(x)$ is a regular extension of $K_0(x)$ [FrJ08, Lemma 2.6.8(a)], so $\hat{K}_0(x)$ is linearly disjoint from $K(x)$ over $K_0(x)$. Hence, res: $\text{Gal}(\hat{K}(x)/\hat{K}_0(x)) \to \text{Gal}(K(x)/K_0(x))$ is an isomorphism. This gives rise to a finite split embedding problem over $\hat{K}_0(x)$,

$$(2) \qquad \text{pr: } \text{Gal}(\hat{K}(x)/\hat{K}_0(x)) \ltimes G \to \text{Gal}(\hat{K}(x)/\hat{K}_0(x))$$

such that $\text{pr} \circ (\text{res}_{K(x)} \times \text{id}_G) = \text{res}_{K(x)} \circ \text{pr}$.

We identify each of the groups $\text{Gal}(\hat{K}(x)/\hat{K}_0(x))$, $\text{Gal}(K(x)/K_0(x))$, and $\text{Gal}(\hat{K}/\hat{K}_0)$ with $\Gamma = \text{Gal}(K/K_0)$ via restriction. Moreover, if F (resp. $\hat{F}$) is a solution field of embedding problem (1) (resp. (2)), then we identify $\text{Gal}(F/K_0(x))$ (resp. $\text{Gal}(\hat{F}/\hat{K}_0(x))$) with $\Gamma \ltimes G$ via an isomorphism θ (resp. $\hat{\theta}$) satisfying $\text{pr} \circ \theta = \text{res}$ (resp. $\text{pr} \circ \hat{\theta} = \text{res}$). We say that (F, θ) is a **split rational solution** of (1) if F has a K-rational place φ such that $\Gamma = D_\varphi$. We say that (F, θ) is **unramified** if φ can be chosen to be unramified over $K_0(x)$.

LEMMA 5.9.1: *In the above notation suppose K_0 is ample and existentially closed in $\hat{K}_0$. Let $\hat{F}$ be a solution field to embedding problem (2) with a $\hat{K}$-rational place $\hat{\varphi}$, unramified over $\hat{K}_0(x)$, such that $\hat{\varphi}(x) \in \hat{K}_0$ and $D_{\hat{\varphi}} = \Gamma$. Then embedding problem (1) has a solution field F with a K-rational place φ unramified over $K_0(x)$ such that $\varphi(x) \in K_0$ and $D_\varphi = \Gamma$.*

Proof: We break up the proof into several parts. First we solve embedding problem (1) over $\hat{K}_0(x)$, then we push the solution down to a solution over a function field $K_0(\mathbf{u}, x)$ which is regular over K_0, and finally we specialize the latter solution to a solution over $K_0(x)$ with a place satisfying all of the prescribed conditions.

PART A: *A solution of (1) over $\hat{K}_0(x)$.* By assumption, there exists an isomorphism

$$\hat{\theta}: \text{Gal}(\hat{F}/\hat{K}_0(x)) \to \text{Gal}(\hat{K}(x)/\hat{K}_0(x)) \ltimes G$$

such that $\text{pr} \circ \hat{\theta} = \text{res}_{\hat{K}(x)}$. Let $\hat{F}_0$ be the fixed field in $\hat{F}$ of $D_{\hat{\varphi}}$ ($= \Gamma$). Then, $\hat{F}_0 \cap \hat{K}(x) = \hat{K}_0(x)$ and $\hat{F}_0 \cdot \hat{K}(x) = \hat{F}$, so $m = [\hat{F}_0 : \hat{K}_0(x)] = [\hat{F} : \hat{K}(x)]$. By Lemma 4.4.1, $\hat{\varphi}(\hat{F}_0) = \hat{K}_0 \cup \{\infty\}$. Hence, $\hat{F}_0/\hat{K}_0$ is regular [FrJ08, Lemma 2.6.9(b)].

We choose a primitive element y for the extension $\hat{F}_0/\hat{K}_0(x)$ integral over $\hat{K}_0[x]$. By the preceding paragraph, $\hat{F} = \hat{K}(x, y)$.

By Lemma 5.1.2, there exists an absolutely irreducible polynomial $h \in \hat{K}_0[V, W]$ and elements $v, w \in \hat{F}_0$ such that $\hat{K}_0(v, w) = \hat{F}_0$, $h(v, w) = 0$, $h(0, 0) = 0$, and $\frac{\partial h}{\partial W}(0, 0) \neq 0$.

We also choose a primitive element c for K over K_0, a primitive element z for $\hat{F}$ over $\hat{K}_0(x)$ integral over $\hat{K}_0[x]$, and note that $\hat{F} = \hat{K}_0(c, x, y)$. Then there exist polynomials $f, p_0, p_1 \in \hat{K}_0[X, Z]$, $g, r_0, r_1, r_2 \in \hat{K}_0[X, Y]$, $q_0, q_1 \in \hat{K}_0[T, X, Y]$, and $s_0, s_1, s_2 \in \hat{K}_0[V, W]$ such that the following conditions hold:

(3a) $\hat{F} = \hat{K}_0(x, z)$ and $f(x, Z) = \mathrm{irr}(z, \hat{K}_0(x))$; in particular $\mathrm{discr}(f(x, Z)) \in \hat{K}_0(x)^\times$.

(3b) $g(x, Y) = \mathrm{irr}(y, \hat{K}_0(x)) = \mathrm{irr}(y, \hat{K}(x))$; since $\hat{F}_0/\hat{K}_0$ is regular (by the first paragraph of Part A), $g(X, Y)$ is absolutely irreducible [FrJ08, Cor. 10.2.2(b)].

(3c) $y = \frac{p_1(x, z)}{p_0(x, z)}$, $z = \frac{q_1(c, x, y)}{q_0(c, x, y)}$, $p_0(x, z) \neq 0$, and $q_0(c, x, y) \neq 0$.

(3d) $v = \frac{r_1(x, y)}{r_0(x, y)}$, $w = \frac{r_2(x, y)}{r_0(x, y)}$, $x = \frac{s_1(v, w)}{s_0(v, w)}$, $y = \frac{s_2(v, w)}{s_0(v, w)}$, $r_0(x, y) \neq 0$, and $s_0(v, w) \neq 0$.

PART B: *Pushing down.* The polynomials introduced in Part A depend on only finitely many parameters from $\hat{K}_0$. Thus, there are $u_1, \ldots, u_n \in \hat{K}_0$ with the following properties:

(4a) The coefficients of $f, g, h, p_0, p_1, q_0, q_1, r_0, r_1, r_2, s_0, s_1, s_2$ are in $K_0[\mathbf{u}]$.

(4b) $F_\mathbf{u} = K_0(\mathbf{u}, x, z)$ is a Galois extension of $K_0(\mathbf{u}, x)$, $f(x, Z) = \mathrm{irr}(z, K_0(\mathbf{u}, x))$, and $\mathrm{discr}(f(x, Z)) \in K_0(\mathbf{u}, x)^\times$.

(4c) $g(x, Y) = \mathrm{irr}(y, K_0(\mathbf{u}, x)) = \mathrm{irr}(y, K(\mathbf{u}, x))$; we set $F_{0,\mathbf{u}} = K_0(\mathbf{u}, x, y)$.

It follows that restriction maps the groups $\mathrm{Gal}(\hat{F}/\hat{K}_0(x))$, $\mathrm{Gal}(\hat{F}/\hat{F}_0)$, and $\mathrm{Gal}(\hat{F}/\hat{K}(x))$ isomorphically onto the groups $\mathrm{Gal}(F_\mathbf{u}/K_0(\mathbf{u}, x))$, $\mathrm{Gal}(F_\mathbf{u}/F_{0,\mathbf{u}})$, and $\mathrm{Gal}(F_\mathbf{u}/K(\mathbf{u}, x))$, respectively. Therefore, restriction transfers $\hat{\theta}$ to an isomorphism

$$(5) \qquad \theta \colon \mathrm{Gal}(F_\mathbf{u}/K_0(\mathbf{u}, x)) \to \mathrm{Gal}(K(\mathbf{u}, x)/K_0(\mathbf{u}, x)) \ltimes G$$

satisfying $\mathrm{pr} \circ \theta = \mathrm{res}_{F_\mathbf{u}/K(\mathbf{u}, x)}$.

PART C: *Specialization.* Since K_0 is existentially closed in $\hat{K}_0$, the field $\hat{K}_0$ and therefore also $K_0(\mathbf{u})$ are regular extensions of K_0 (Lemma 5.2.6). Thus, $\mathbf{u}$ generates an absolutely irreducible variety $U = \mathrm{Spec}(K_0[\mathbf{u}])$ over K_0 [FrJ08, Cor. 10.2.2]. The variety U has a nonempty Zariski-open subset U' that contains $\mathbf{u}$ such that for each $\mathbf{u}' \in U'$ the K_0-specialization $\mathbf{u} \to \mathbf{u}'$ extends to a $K(x)$-homomorphism $' \colon K(x)[\mathbf{u}, v, w, y, z] \to K(x)[\mathbf{u}', v', w', y', z']$ such that the following conditions, derived from (3) and (4), hold:

(6a) The coefficients of $f', g', h', p_0', p_1', q_0', q_1', r_0', r_1', r_2', s_0', s_1', s_2'$ belong to $K_0[\mathbf{u}']$.

(6b) $F = K_0(\mathbf{u}', x, z')$ is a Galois extension of $K_0(\mathbf{u}', x)$, $f'(x, z') = 0$, and $\mathrm{discr}(f'(x, Z)) \in K_0(\mathbf{u}', x)^\times$.

(6c) $y' = \frac{p_1'(x, z')}{p_0'(x, z')}$, $z' = \frac{q_1'(c, x, y')}{q_0'(c, x, y')}$, $p_0'(x, z') \neq 0$, and $q_0'(c, x, y') \neq 0$; we set $F_0 = K_0(\mathbf{u}', x, y')$ and find that $F = F_0 K$.

(6d) $g'(X, Y)$ is absolutely irreducible, $\deg_Y(g'(x, Y)) = \deg_Y(g(x, Y))$, $g'(x, y') = 0$, and so $g'(x, Y) = \mathrm{irr}(y', K_0(\mathbf{u}', x)) = \mathrm{irr}(y', K(\mathbf{u}', x))$;

(6e) $h'(V, W)$ is absolutely irreducible, $h'(0, 0) = 0$, and $\frac{\partial h'}{\partial W}(0, 0) \neq 0$.

(6f) $v' = \frac{r_1'(x, y')}{r_0'(x, y')}$, $w' = \frac{r_2'(x, y')}{r_0'(x, y')}$, $x = \frac{s_1'(v', w')}{s_0'(v', w')}$, $y' = \frac{s_2'(v', w')}{s_0'(v', w')}$, $r_0'(x, y') \neq 0$, and $s_0'(v', w') \neq 0$; thus $F_0 = K_0(\mathbf{u}', v', w')$.

To achieve the absolute irreducibility of g' and h' we have used the Bertini-Noether theorem [FrJ08, Prop. 9.4.3].

PART D: *Choosing* $\mathbf{u}' \in \hat{K}_0^n$. Since K_0 is existentially closed in $\hat{K}_0$ and since $\mathbf{u} \in U'(\hat{K}_0)$, we can choose $\mathbf{u}' \in U'(K_0)$. Then $K_0[\mathbf{u}'] = K_0$, $K_0(\mathbf{u}', x) = K_0(x)$, $F_0 = K_0(x, y') = K_0(v', w')$, and $F = K_0(x, z')$. Since $\mathrm{discr}(f'(x, Z)) \neq 0$ (by (6b)) the homomorphism $'$ induces an embedding

(7) $$\psi^*\colon \mathrm{Gal}(F/K_0(x)) \to \mathrm{Gal}(F_{\mathbf{u}}/K_0(\mathbf{u}, x))$$

such that $(\psi^*(\sigma)(s))' = \sigma(s')$ for all $\sigma \in \mathrm{Gal}(F/K_0(x))$ and $s \in F_{\mathbf{u}}$ with $s' \in F$ [Lan93, p. 344, Prop. 2.8]. Each $s \in K(x)$ is fixed by $'$, hence $\psi^*(\sigma)(s) = \sigma(s)$ for each $\sigma \in \mathrm{Gal}(F/K_0(x))$. It follows that ψ^* commutes with restriction to $K(x)$.

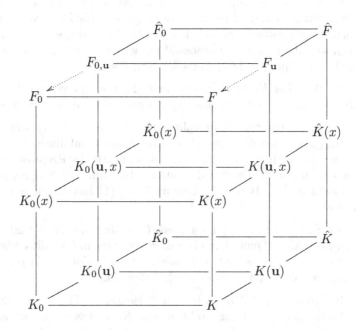

By (6c), $F = K(x, y') = F_0 K$. By (6d) and [FrJ08, Cor. 10.2.2(b)], F_0/K_0 is a regular extension, so F_0 is linearly disjoint from K over K_0. Therefore, F_0 is linearly disjoint from $K(x)$ over $K_0(x)$, hence $F_0 \cap K(x) =$

$K_0(x)$ and $[F_0 : K_0(x)] = [F : K(x)]$. It follows from (6d) that

$$
\begin{aligned}
|\mathrm{Gal}(F/K_0(x))| &= [F : K_0(x)] \\
&= [F : K(x)][K(x) : K_0(x)] \\
&= \deg_Y g'(x,Y)[K : K_0] \\
&= \deg_Y g(x,Y)[K : K_0] \\
&= [F_{\mathbf{u}} : K(\mathbf{u},x)][K(\mathbf{u},x) : K_0(\mathbf{u},x)] \\
&= [F_{\mathbf{u}} : K_0(\mathbf{u},x)] = |\mathrm{Gal}(F_{\mathbf{u}}/K_0(\mathbf{u},x))|.
\end{aligned}
$$

Therefore ψ^* is an isomorphism. Let

$$
\rho\colon \mathrm{Gal}(K(\mathbf{u},x)/K_0(\mathbf{u},x)) \ltimes G \to \mathrm{Gal}(K(x)/K_0(x)) \ltimes G
$$

be the isomorphism whose restriction to $\mathrm{Gal}(K(\mathbf{u},x)/K_0(\mathbf{u},x))$ is the restriction map and to G is the identity map. Then, $\theta' = \rho \circ \theta \circ \psi^*$ satisfies $\mathrm{pr} \circ \theta' = \mathrm{res}_{F/K(x)}$ (by (5)). This means that θ' is a solution of embedding problem (1).

PART E: *Rational place.* Finally, by (6e) and (6f), the curve defined by $h'(X,Y) = 0$ is a model of F_0/K_0 and $(0,0)$ is a K_0-rational simple point of it. Therefore, by Lemma 5.1.4(b), F_0 has a K_0-rational place $\varphi_0 \colon F_0 \to K_0 \cup \{\infty\}$. Since K_0 is ample, F_0 has infinitely many K_0-places (Lemma 5.3.1). Only finitely many of them are ramified over $K_0(x)$. Hence, we may choose φ_0 to be unramified over $K_0(x)$. Using the linear disjointness of F_0 and K over K_0, we extend φ_0 to a K-rational place $\varphi \colon F \to K \cup \{\infty\}$ unramified over $K_0(x)$. Consequently, by Lemma 4.4.1(c), $\Gamma = D_\varphi$. $\square$

THEOREM 5.9.2: *Let K_0 be an ample field. Then each constant finite split embedding problem over $K_0(x)$ has a split unramified rational solution.*

Proof: Consider a constant finite split embedding problem (1) over $K_0(x)$. Let $\hat{K}_0 = K_0((t))$. Then $\hat{K}_0$ is complete under a nontrivial discrete ultrametric absolute value with prime element t. Consequently, by Proposition 4.4.2, (2) has a split unramified rational solution. By Lemma 5.3.1, K_0 is existentially closed in $\hat{K}_0$. Hence, by Lemma 5.9.1, (1) has a split unramified rational solution. $\square$

COROLLARY 5.9.3: *Let $(A, \mathfrak{a})$ be a weak Henselian pair such that A is a domain and set $K_0 = \mathrm{Quot}(A)$. Then each constant finite split embedding problem over $K_0(x)$ has a split unramified rational solution. In particular, this conclusion holds if $(A, \mathfrak{a})$ is a complete domain.*

Proof: By Proposition 5.7.3, K_0 is ample. Hence, by Theorem 5.9.2, each constant finite split embedding problem over $K_0(x)$ has a split unramified rational solution. In particular this conclusion holds if $(A, \mathfrak{a})$ is a complete domain (by Proposition 5.7.7). $\square$

5.10 PAC Hilbertian Fields are ω-Free

The statement of the title was a major open problem of Field Arithmetic. Theorem 5.10.3 settles that problem.

Recall that the **rank** of a profinite group G is the least cardinality of a system of generators of G that converges to 1. If G is not finitely generated, then rank(G) is also the cardinality of the set of all open normal subgroups of G [FrJ08, Prop. 17.1.2]. We denote the free profinite group of rank m by $\hat{F}_m$.

An **embedding problem** for a profinite group G is a couple

$$(1) \qquad\qquad (\varphi\colon G \to A,\, \alpha\colon B \to A),$$

of homomorphisms of profinite groups with φ and α surjective. The embedding problem is said to be **finite** if B is finite. If there exists a homomorphism $\alpha'\colon A \to B$ such that $\alpha \circ \alpha' = \mathrm{id}_A$, we say that (1) **splits**. A **weak solution** to (1) is a homomorphism $\gamma\colon G \to B$ such that $\alpha \circ \gamma = \varphi$. If γ is surjective, we say that γ is a **solution** to (1). We say that G is **projective** if every finite embedding problem for G has a weak solution.

An **embedding problem** over a field K is an embedding problem (1), where $G = \mathrm{Gal}(K)$. If L is the fixed field of $\mathrm{Ker}(\varphi)$, we may identify A with $\mathrm{Gal}(L/K)$ and φ with $\mathrm{res}_{K_s/L}$ and then consider $\alpha\colon B \to \mathrm{Gal}(L/K)$ as the given embedding problem. This shows that our present definition generalizes the one given in Section 1.2. Note that if $\gamma\colon \mathrm{Gal}(K) \to B$ is a solution of (1) and F is the fixed field in K_s of $\mathrm{Ker}(\gamma)$, then F is a solution field of the embedding problem $\alpha\colon B \to \mathrm{Gal}(L/K)$ and γ induces an isomorphism $\bar{\gamma}\colon \mathrm{Gal}(F/K) \to B$ such that $\alpha \circ \bar{\gamma} = \mathrm{res}_{F/L}$.

The first statement of the following proposition is due to Gruenberg [FrJ08, Lemma 22.3.2], the second one is a result of Iwasawa [FrJ08, Cor. 24.8.2].

PROPOSITION 5.10.1: *Let G be a projective group. If each finite split embedding problem for G is solvable, then every finite embedding problem for G is solvable. If in addition rank(G) $\leq \aleph_0$, then $G \cong \hat{F}_\omega$.*

We say that a field K is ω-**free** if every finite embedding problem over K (that is, finite embedding problem for $\mathrm{Gal}(K)$) is solvable.

THEOREM 5.10.2: *Let K be an ample field.*
(a) *If K is Hilbertian, then each finite split embedding problem over K is solvable.*
(b) *If in addition, $\mathrm{Gal}(K)$ is projective, then K is ω-free.*
(c) *If in addition, $\mathrm{Gal}(K)$ has countably many generators, and in particular, if K is countable, then $\mathrm{Gal}(K) \cong \hat{F}_\omega$.*

Proof of (a): Every finite split embedding problem over K gives a finite split constant embedding problem over $K(x)$. The latter is solvable by Theorem 5.9.2. Now use the Hilbertianity and specialize to get a solution of the original embedding problem over K [FrJ08, Lemma 13.1.1].

Proof of (b): By (a), every finite split embedding problem over K is solvable. Hence, by Proposition 5.10.1, every finite embedding problem over K is solvable.

Proof of (c): Use (b) and Proposition 5.10.1.   □

The following special case of Theorem 5.10.2 is a solution of [FrJ86, Prob. 24.41].

THEOREM 5.10.3: *Let K be a PAC field. Then K is ω-free if and only if K is Hilbertian.*

Proof: That 'K is ω-free' implies 'K is Hilbertian' is a result of Roquette [FrJ08, Cor. 27.3.3]. Conversely, if K is PAC, then $\mathrm{Gal}(K)$ is projective [FrJ08, Thm. 11.6.2]. By Example 5.6.1, K is ample. Hence, if K is Hilbertian, then by Theorem 5.10.2(b), K is ω-free.   □

Remark 5.10.4: It is a major problem of Galois theory whether statements (a) and (b) of Theorem 5.10.2 are true without the assumption 'K is ample'. For the first statement, this is a conjecture of Débes and Deschamps [DeD97, Section 2.1, Conjecture (Split EP/$_{K\ hilb.}$)], the second it is a conjecture of Fried and Völklein [FrJ08, Conjecture 24.8.6]. As a matter of fact Débes and Deschamps make a stronger conjecture than the former one: If K is an arbitrary field and x is a variable, then every constant split embedding problem over $K(x)$ has a regular solution [DeD97, Section 2.1, Conjecture (Split EP/$_{K(T)}$)]. By Theorem 5.9.2, the latter conjecture holds if K is ample. On the other hand, we don't know of any nonample field K for which this conjecture holds.   □

Example 5.10.5: The maximal Abelian extension of $\mathbb{Q}$. We denote that field by $\mathbb{Q}_{\mathrm{ab}}$. It is an Abelian extension of $\mathbb{Q}$. Since $\mathbb{Q}$ is Hilbertian [FrJ08, Thm. 13.4.2], a result of Kuyk asserts that $\mathbb{Q}_{\mathrm{ab}}$ is also Hilbertian [FrJ08, Thm. 16.11.3]. Moreover, $\mathrm{Gal}(\mathbb{Q}_{\mathrm{ab}})$ is projective. Indeed, since $\sqrt{-1} \in \mathbb{Q}_{\mathrm{ab}}$, the field $\mathbb{Q}_{\mathrm{ab}}$ has no embedding into $\mathbb{R}$. Moreover, for each prime number p the maximal unramified extension $\mathbb{Q}_{p,\mathrm{ur}}$ of the field of p-adic numbers $\mathbb{Q}_p$ is an Abelian extension with Galois group $\hat{\mathbb{Z}}$. Since $\hat{\mathbb{Z}} \cong \prod_l \mathbb{Z}_l$, we have $l^\infty | [\mathbb{Q}_{\mathrm{ab}}\mathbb{Q}_p : \mathbb{Q}_p]$ for each prime number l. In addition, $\mathbb{Q}_{\mathrm{ab}}$ has a cyclic extension of degree l (e.g. because $\mathbb{Q}_{\mathrm{ab}}$ is Hilbertian and contains ζ_l). It follows by class field theory that the l-th cohomological dimension of $\mathrm{Gal}(\mathbb{Q}_{\mathrm{ab}})$ is 1 [Rib70, p. 303, Thm. 8.8]. Since this holds for each l, the cohomological dimension of $\mathrm{Gal}(\mathbb{Q}_{\mathrm{ab}})$ is 1. Consequently, $\mathrm{Gal}(\mathbb{Q}_{\mathrm{ab}})$ is projective (Subsection 9.3.16).

It is not known whether $\mathbb{Q}_{\mathrm{ab}}$ is ample. But if it is, it will follow from the preceding paragraph and Theorem 5.10.2(c) that $\mathrm{Gal}(\mathbb{Q}_{\mathrm{ab}}) \cong \hat{F}_\omega$. This will prove a well known conjecture of Shafarevich.   □

Example 5.10.6: The maximal pro-solvable extension of $\mathbb{Q}$. We denote the compositum of all finite solvable extensions of $\mathbb{Q}$ by $\mathbb{Q}_{\mathrm{solv}}$. It is an infinite extension of $\mathbb{Q}$. Moreover, $\mathrm{Gal}(\mathbb{Q}_{\mathrm{solv}}) \leq \mathrm{Gal}(\mathbb{Q}_{\mathrm{ab}})$, so by Example 5.10.5, $\mathrm{Gal}(\mathbb{Q}_{\mathrm{solv}})$ is projective. If N is a finite solvable extension of $\mathbb{Q}_{\mathrm{solv}}$, then there exists a finite Galois extension N_0 of $\mathbb{Q}$ such that $\mathbb{Q}_{\mathrm{solv}} N_0 = N$. Then $\mathrm{Gal}(N/\mathbb{Q}_{\mathrm{solv}}) \cong \mathrm{Gal}(N_0/K_0)$, where $K_0 = \mathbb{Q}_{\mathrm{solv}} \cap N_0$. Both $K_0/\mathbb{Q}$ and N_0/K_0 are solvable extensions, hence so is $N_0/\mathbb{Q}$. Therefore, $N_0 \subseteq \mathbb{Q}_{\mathrm{solv}}$, so $N = \mathbb{Q}_{\mathrm{solv}}$. In other words, $\mathbb{Q}_{\mathrm{solv}}$ has no proper solvable extensions. In particular, $\mathbb{Q}_{\mathrm{solv}}$ has no quadratic extensions. It follows that $\mathbb{Q}_{\mathrm{solv}}$ is not Hilbertian. However, by Weissauer, each proper finite extension K of $\mathbb{Q}_{\mathrm{solv}}$ is Hilbertian [FrJ08, Thm. 13.9.1(b)]. In addition, since closed subgroups of projective groups are projective [FrJ08, Prop. 22.4.7], $\mathrm{Gal}(K)$ is projective. It follows that if $\mathbb{Q}_{\mathrm{solv}}$ is ample, then so is K (Lemma 5.5.1(b)), hence by Theorem 5.10.2(c), $\mathrm{Gal}(K) \cong \hat{F}_\omega$.

Of course, the question if $\mathbb{Q}_{\mathrm{solv}}$ is ample is at this time far from being settled. However, we feel that there are more chances for an affirmative answer to that question than for the correspondent question for $\mathbb{Q}_{\mathrm{ab}}$.   □

Example 5.10.7: Let K be a countable Hilbertian field. Suppose K has an embedding into $\mathbb{R}$. Then this embedding defines a real local prime $\mathfrak{p}$ on K. Let $S = \{\mathfrak{p}\}$. By Example 5.6.6, for almost all $\boldsymbol{\sigma} \in \mathrm{Gal}(K)^e$ the field $K_{\mathrm{tot},S}[\boldsymbol{\sigma}]$ is P$\mathcal{S}$C, that is $K_{\mathrm{tot},S}[\boldsymbol{\sigma}]$ is P$\mathcal{K}$C, where $\mathcal{K}$ is the family of all real closures of K. By Example 5.6.4, $M = K_{\mathrm{tot},S}[\boldsymbol{\sigma}](\sqrt{-1})$ is P$\mathcal{K}'$C, where $\mathcal{K}' = \{\bar{K}(\sqrt{-1}) \mid \bar{K} \in \mathcal{K}\}$. Since each $\bar{K}$ in $\mathcal{K}$ is real closed, $\mathcal{K}' = \{\tilde{K}\}$. Hence, M is PAC. In addition, as a finite proper extension of a Galois extension of a Hilbertian field, M is Hilbertian [FrJ08, Weissauer's theorem 13.9.1]. Consequently, by Theorem 5.10.3, $\mathrm{Gal}(M) \cong \hat{F}_\omega$ (note that M itself is countable).

The case where $K = \mathbb{Q}$ and $e = 0$ is especially attractive. In this case $K_{\mathrm{tot},S}$ is the **maximal totally real extension of** $\mathbb{Q}$ and is usually denoted by $\mathbb{Q}_{\mathrm{tr}}$. The preceding paragraph asserts that $\mathbb{Q}_{\mathrm{tr}}(\sqrt{-1})$ is a PAC, Hilbertian field with absolute Galois group isomorphic to $\hat{F}_\omega$.   □

Example 5.10.8: Let K be a Hilbertian field algebraic over either $\mathbb{Q}$ or $\mathbb{F}_p(x)$, where x is an indeterminate (e.g. K is a global field [FrJ08, Thm. 13.4.2]). Let K_{cycl} be the extension of K obtained by adjoining all roots of unity in $\tilde{K}$ and let M be an algebraic extension of K_{cycl}. If $\mathbb{Q} \subseteq M$, then $\sqrt{-1} \in M$ and $l^\infty \| [M\mathbb{Q}_p : \mathbb{Q}_p] = \infty$ for all prime numbers l, p and every embedding M into the algebraic closure of $\mathbb{Q}_p$. Hence, the cohomological dimension of $\mathrm{Gal}(M)$ is at most 1 [Rib70, p. 303, Thm. 8.8]. Thus, $\mathrm{Gal}(M)$ is projective [Ser79, p. 75, Prop. 3 or Subsection 9.3.16]. If M is an algebraic extension of $\mathbb{F}_p(x)_{\mathrm{cycl}}$, then M is an algebraic extension of $\tilde{\mathbb{F}}_p(x)$, so $\mathrm{Gal}(M)$ is projective, by [Ser79, p. 75, Prop. 3].

We consider a finite set S of local primes of K. By Example 5.6.6, for almost all $\boldsymbol{\sigma} \in \mathrm{Gal}(K)^e$ the field $K_{\mathrm{tot},S}[\boldsymbol{\sigma}]$ is ample. Hence, so is $K_{\mathrm{tot},S}[\boldsymbol{\sigma}]_{\mathrm{cycl}} = K_{\mathrm{tot},S}[\boldsymbol{\sigma}] \cdot K_{\mathrm{cycl}}$ (Lemma 5.5.1(b)). Omitting the complex local primes of K from S does not effect $K_{\mathrm{tot},S}$ nor $K_{\mathrm{tot},S}[\boldsymbol{\sigma}]$. Thus, we assume that S contains no complex local prime.

Next observe that for almost all $\boldsymbol{\sigma} \in \mathrm{Gal}(K)^e$ the field $K_{\mathrm{tot},S}[\boldsymbol{\sigma}]_{\mathrm{cycl}}$ is Hilbertian. Indeed, if $S \neq \emptyset$, we choose $\mathfrak{p} \in S$ and observe that $K_{\mathrm{tot},S}[\boldsymbol{\sigma}]$ is contained in the completion $\hat{K}_{\mathfrak{p}}$ of K at $\mathfrak{p}$ and $\hat{K}_{\mathfrak{p}}$, as a non-complex local field, contains only finitely many roots of unity. Thus, $K_{\mathrm{cycl}} \not\subseteq K_{\mathrm{tot},S}[\boldsymbol{\sigma}]$. Therefore, we may choose a finite extension K' of K in K_{cycl} with $K' \not\subseteq K_{\mathrm{tot},S}[\boldsymbol{\sigma}]$. Then, $K_{\mathrm{tot},S}[\boldsymbol{\sigma}]K'$ is a finite proper extension of $K_{\mathrm{tot},S}[\boldsymbol{\sigma}]$. Since the latter field is Galois over K and K is Hilbertian, $K_{\mathrm{tot},S}[\boldsymbol{\sigma}]K'$ is Hilbertian [FrJ08, Weissauer's theorem 13.9.1]. Since $K_{\mathrm{tot},S}[\boldsymbol{\sigma}]_{\mathrm{cycl}}$ is an Abelian extension of $K_{\mathrm{tot},S}[\boldsymbol{\sigma}]K'$, a theorem of Kuyk implies that $K_{\mathrm{tot},S}[\boldsymbol{\sigma}]_{\mathrm{cycl}}$ is Hilbertian [FrJ08, 16.11.3].

If $S = \emptyset$, then $K_{\mathrm{tot},S}[\boldsymbol{\sigma}] = K_s[\boldsymbol{\sigma}]$ and the latter field is Hilbertian by [Jar97, Thm. 2.7]. Hence, by Kuyk, $K_{\mathrm{tot},S}[\boldsymbol{\sigma}]_{\mathrm{cycl}}$ is Hilbertian.

It follows that for almost all $\boldsymbol{\sigma} \in \mathrm{Gal}(K)^{\sigma}$, the field $K_{\mathrm{tot},S}[\boldsymbol{\sigma}]_{\mathrm{cycl}}$ is ample, Hilbertian, and $G = \mathrm{Gal}(K_{\mathrm{tot},S}[\boldsymbol{\sigma}]_{\mathrm{cycl}})$ is projective. Consequently, by Theorem 5.10.2(c), $G \cong \hat{F}_{\omega}$. $\square$

5.11 Krull Domains

Krull domains give rise to a large family of ample Hilbertian fields.

Definition 5.11.1: An integral domain A with a quotient field K is said to be a **Krull domain** if there exists a family $\mathcal{P}$ of prime ideals of A satisfying the following conditions:
(1a) $A_{\mathfrak{p}}$ is a discrete valuation ring of K for each $\mathfrak{p} \in \mathcal{P}$.
(1b) Every nonzero $a \in A$ belongs to only finitely many $\mathfrak{p} \in \mathcal{P}$.
(1c) $A = \bigcap_{\mathfrak{p} \in \mathcal{P}} A_{\mathfrak{p}}$. $\square$

Remark 5.11.2: Examples of Krull domains.

(a) Definition 5.11.1 is a reformulation of the definition of a Krull domain in [Mats94, p. 87] (where it is called a **Krull ring**) and also to the one given in [Bou89, p. 480]. It is also equivalent to the definition given in [FrJ08, beginning of Sec. 15.5]. In particular, by [Mats94, p. 87, Thm. 12.3], each nonzero minimal prime ideal of A belongs to $\mathcal{P}$. Moreover, the family of nonzero minimal prime ideals of A satisfies Condition (1).

(b) If S is a multiplicative subset of a Krull domain, then the localization A_S is also a Krull domain [Mats94, Thm. 12.1].

(c) Let $A_1, \ldots, A_n$ be Krull subdomains of a field K. Then $A = \bigcap_{i=1}^n A_n$ is a Krull domain [Bou89, p. 480].

(d) Let A be a Krull domain, L a finite extension of $K = \mathrm{Quot}(A)$, and B the integral closure of A in L. Then B is a Krull domain.

(e) Every Noetherian integrally closed domain is a Krull domain [Mats94, Thm. 12.4(i)]. In particular, every Dedekind domain is a Krull domain.

(f) If A is a Krull domain, then so is each of the rings $A[X_1, \ldots, X_n]$ and $A[[X_1, \ldots, X_n]]$ [Mats94, Thm. 12.4(iii)].

(g) Every unique factorization domain is a Krull domain, the prime ideals appearing in Definition 5.11.1 being those generated by the prime elements.

In particular, if K is a field, then the ring of polynomials $B = K[X_i \mid i \in I]$ is a Krull domain. Note however that if I is infinite, then B is not Noetherian. More generally, if A is a Krull domain, then so is $A[X_i \mid i \in I]$ [Bou89, p. 547, Exer. 8].

(h) Deeper than the previous results in this remark is the following theorem of Mori-Nagata: If A is a Noetherian domain, then its integral closure A' is a Krull domain [Nag62, Thm. 33.10]. Note that A' need not be Noetherian [Nag62, p. 207, Example 5]. □

THEOREM 5.11.3: *Let A be a Krull domain and set $K = \mathrm{Quot}(A)$.*
(a) *If $A \neq K$, then $F = \mathrm{Quot}(A[[X_1, \ldots, X_n]])$ is an ample Hilbertian field for each $n \geq 1$ (Pop-Weissauer).*
(b) *$F = K((X_1, \ldots, X_n))$ is an ample Hilbertian field for each $n \geq 2$ (Pop-Weissauer).*
(c) *In each case, every finite split embedding problem over F is solvable (Paran).*

Proof: For each positive integer n the integral domain $R = A[[X_1, \ldots, X_n]]$ is complete with respect to the ideal $\mathfrak{a}$ generated by $X_1, \ldots, X_n$ (Example 5.7.5). Hence, by Proposition 5.7.7, $F = \mathrm{Quot}(R)$ is ample.

On the other hand, by Remark 5.11.2(f), R is a Krull domain. If $A \neq K$ and $n \geq 1$, then A has a nonzero prime ideal $\mathfrak{p}$. Then $P_1 = \sum_{i=1}^{n} RX_i$ is a nonzero prime ideal of R properly contained in the prime ideal $P_2 = \sum_{i=1}^{n} RX_i + R\mathfrak{p}$. If $A = K$ and $n \geq 2$, then $RX_1 \subset RX_1 + RX_2$ are nonzero prime ideals of R. Thus, in both cases $\dim(R) \geq 2$. It follows from Weissauer that F is Hilbertian [FrJ08, Thm. 15.4.6]. This concludes the proof of (a) and (b). Statement (c) now follows from Theorem 5.10.2(a). □

Remark 5.11.4: In the case where $n = 1$ in Theorem 5.11.3(b), $F = K((X_1))$ is a complete discrete valued field, hence ample (Example 5.6.2). However, F is Henselian, so F is not Hilbertian [FrJ08, Lemma 15.5.4]. □

5.12 Lifting

Let K/K_0 be a finite Galois extension of fields with Galois group Γ and x an indeterminate. The **Beckmann-Black Problem** for (K_0, K, Γ) asks whether $K_0(x)$ has a Galois extension F_0 such that $\mathrm{Gal}(F_0/K_0(x)) \cong \Gamma$, F_0/K_0 is a regular extension, and there is a prime divisor $\mathfrak{p}$ of F_0/K_0 with decomposition field $K_0(x)$ and residue field K. Dèbes solved the Beckmann-Black problem for PAC fields [FrJ08, Thm. 24.2.2]. Here we apply Theorem 5.9.2 to solve the problem when K_0 is ample.

THEOREM 5.12.1: *Let K_0 be an ample field, G a finite group, Γ a subgroup, K a Galois extension of K_0 with Galois group Γ, and x an indeterminate. Then $K_0(x)$ has a Galois extension E with the following properties:*
(1a) *$\mathrm{Gal}(E/K_0(x)) \cong G$.*
(1b) *E/K_0 is a regular extension.*

(1c) E/K_0 has infinitely many prime divisors $\mathfrak{p}$ with decomposition group Γ over $K_0(x)$ and residue field K.

Proof: Consider the action of Γ on G by conjugation and write the corresponding semidirect product as $\Gamma \ltimes G = \{(\gamma, g) \mid \gamma \in \Gamma,\ g \in G\}$ with the multiplication rule $(\gamma_1, g_1)(\gamma_2, g_2) = (\gamma_1\gamma_2, g_1^{\gamma_2} g_2)$ (Compare with Section 1.2). The projection pr: $\Gamma \ltimes G \to \Gamma$ on the first factor maps the subgroup $\Gamma' = \{(\gamma, 1) \mid \gamma \in \Gamma\}$ of $\Gamma \ltimes G$ isomorphically onto Γ.

Theorem 5.9.2 gives a Galois extension F of $K_0(x)$ containing $K(x)$ such that $\mathrm{Gal}(F/K_0(x)) = \Gamma \ltimes G$ and res: $\mathrm{Gal}(F/K_0(x)) \to \mathrm{Gal}(K(x)/K_0(x))$ is the map pr: $\Gamma \ltimes G \to \Gamma$. Moreover, F/K has a prime divisor $\mathfrak{q}$ unramified over $K_0(x)$ with residue field K and decomposition group Γ' over $K_0(x)$. In particular, F/K is regular [FrJ08, Lemma 2.6.9(b)].

Note that the map $(\gamma, g) \mapsto (\gamma, \gamma g)$ is an isomorphism $\varphi \colon \Gamma \ltimes G \to \Gamma \times G$. Indeed,

$$\varphi((\gamma_1, g_1)(\gamma_2, g_2)) = \varphi(\gamma_1\gamma_2, g_1^{\gamma_2} g_2) = (\gamma_1\gamma_2, \gamma_1 g_1 \gamma_2 g_2)$$
$$= (\gamma_1, \gamma_1 g_1)(\gamma_2, \gamma_2 g_1) = \varphi(\gamma_1, g_1)\varphi(\gamma_2, g_2).$$

Composing φ with the projection on G gives rise to an epimorphism $\rho \colon \Gamma \ltimes G \to G$. Explicitly, we have $\rho(\gamma, g) = \gamma g$ for $\gamma \in \Gamma$ and $g \in G$. Let $N = \mathrm{Ker}(\rho)$ and denote the fixed field of N in F by E. Then we may identify $\mathrm{Gal}(E/K_0(x))$ with G and ρ with res: $\mathrm{Gal}(F/K_0(x)) \to \mathrm{Gal}(E/K_0(x))$.

It follows that $E \cap K(x) = K_0(x)$ and $[E : K_0(x)] = |G| = [F : K(x)]$. Hence, E is linearly disjoint from $K(x)$ over $K_0(x)$, so E is linearly disjoint from K over K_0. Since F is regular over K, this implies that E is regular over K_0.

Let $\mathfrak{p}$ be the restriction of $\mathfrak{q}$ to E. Since ρ maps Γ' isomorphically onto Γ, we have $\Gamma = D_\mathfrak{p}$ [Ser79, p. 22, Prop. 22(b)]. The residue field $\bar{E}_\mathfrak{p}$ of E at $\mathfrak{p}$ is contained in the residue field $\bar{F}_\mathfrak{q}$, which is K. In addition, $\mathrm{Gal}(\bar{E}_\mathfrak{p}/K_0) \cong D_\mathfrak{p}$, because $\mathfrak{q}$, hence also $\mathfrak{p}$, are unramified over $K_0(x)$. Hence, $|\Gamma| = |D_\mathfrak{p}| = [\bar{E}_\mathfrak{p} : K_0] \le [K : K_0] = |\Gamma|$. It follows that $\bar{E}_\mathfrak{p} = K$.

Finally, denote the fixed field of $D_\mathfrak{q}$ in F by F_0 and let $\mathfrak{q}_0$ be the restriction of $\mathfrak{q}$ to F_0. Then $\mathfrak{q}_0$ is unramified over $K_0(x)$ and its residue field is K_0. Since K_0 is ample, F_0 has infinitely many prime divisors $\mathfrak{q}_0'$ unramified over $K_0(x)$ and with residue field K_0 (Lemma 5.3.1(d)). Extend each $\mathfrak{q}_0'$ to a prime divisor $\mathfrak{q}'$ of F/K unramified over $K_0(x)$ with residue field K. Then, as in the preceding paragraph, the restriction $\mathfrak{p}'$ of $\mathfrak{q}'$ to E is a prime divisor of E/K_0 unramified over $K_0(x)$ with residue field K. Since over each prime divisor of $K_0(x)/K_0$ there lie only finitely many prime divisors of F/K, infinitely many of the prime divisors $\mathfrak{p}'$ are distinct. $\square$

Remark 5.12.2: In case of $\Gamma = G$, Theorem 5.12.1 says that an ample field K_0 has the so-called **arithmetic lifting property** of Beckmann-Black [Bla99]. $\square$

Remark 5.12.3: When K_0 is PAC, Pierre Dèbes strengthens Theorem 5.12.1 considerably [FrJ08, Theorem 24.2.2]:

Let $E = K_0(x)$ with an indeterminate x and G a finite group. For $i = 1, \ldots, n$ let Γ_i be a subgroup of G and K_i a Galois extension of K_0 with Galois group Γ_i. Then E has a Galois extension F such that

(2a) $\mathrm{Gal}(F/E) \cong G$.

(2b) F/K_0 is a regular extension.

(2c) For each i there exists a prime divisor $\mathfrak{p}_i$ of F/K_0 with decomposition group over E equal to Γ_i and with residue field K_i. Moreover, $\mathfrak{p}_1, \ldots, \mathfrak{p}_n$ are distinct. □

Notes

Pop's observation mentioned in Roquette's letter to Geyer (Notes to Chapter 4) that each finite group G has a K-regular realization over $K(x)$ whenever K is PAC or Henselian is based on the common property that both types of fields have: in each case K is existentially closed in $K((t))$. This property is equivalent to some other seemingly unrelated properties which are collected into Lemma 5.3.1. The source of that lemma is [Pop96, Prop. 11]. In that paper Pop calls the fields sharing the equivalent properties of Lemma 5.3.1 'large fields' . Unfortunately, the adjective 'large' in the naive sense has been attached to algebraic extensions of Hilbertian fields in several papers (e.g. [Jar72], [FyJ74], [Jar75], [GeJ78], [Jar79], [Jar82], [Jar97], [ImB06]). Thus, in order not to create confusion we have replaced 'large' by the adjective 'ample' with similar attributes. See also [Hrb03, Remark 3.3.12] for other alternatives.

The idea appearing in the first paragraph of the proof of Lemma 5.4.2 (which is a special case of [Feh10, Lemma 4]) to find two points (u_1, v_1) and (u_2, v_2) on a given absolutely irreducible affine plane K-curve C over an ample field K such that $\frac{u_1}{u_2}$ is a given element c of K is due to Koenigsmann (Proposition 6.1.5). Pop applied that idea to prove Proposition 5.4.3(b).

In [HaS05] Harbater and Stevenson call a field K satisfying the conclusion of Proposition 5.4.3(b) **very large**. We prefer the expression **very ample**. Question 4.5 of [HaS05] asks whether every ample field is very ample. This question is positively answered by Proposition 5.4.3(b). The proof is due to Pop [Har09, Prop. 3.3]. Proposition 5.4.3(a) and Corollary 5.4.5 are proved by Fehm in [Feh10].

Pop uses restriction of scalars to prove Lemma 5.5.1(b). Our proof is due to Haran.

Pop proves in [Pop96] that $K_{\mathrm{tot},S}$ is PSC if S is a finite set of local primes of a Hilbertian field K. He notices that the field $K_{\mathrm{tot},S}(\mu_\infty)$ obtained by adjoining all roots of unity is ample and has a projective absolute Galois group, so $\mathrm{Gal}(K_{\mathrm{tot},S}(\mu_\infty)) \cong \hat{F}_\omega$. It turns out that that in the special case where S consists of the infinite prime of $\mathbb{Q}$, the field $\mathbb{Q}_{\mathrm{tr}}(\sqrt{-1})$ is PAC with absolute Galois group $\hat{F}_\omega$. This is perhaps the most explicit example for fields of that kind.

Harbater-Stevenson prove in [HaS05] that if K is a field, then every finite split embedding problem over $K((X_1, X_2))$ can be lifted from a problem over $K((X_1))(X_2)$ and the latter problem has a solution that can be lifted to a solution of the original problem.

Harbater and Stevenson noticed in the introduction to [HaS05] that their result would follow from previous results if they knew that $K((X_1, X_2))$ was ample. They suspected it was not.

Paran [Par08] generalizes the method of algebraic patching over complete fields to algebraic patching over domains A with a prime ideal $\mathfrak{p}$ such that the function $v_{\mathfrak{p}}(x) = \sup(n \in \mathbb{N} \mid x \in \mathfrak{p}^n)$ extends to a discrete valuation of $K = \mathrm{Quot}(A)$ and $(A, \mathfrak{p})$ is complete. If in addition, A is a Noetherian integrally closed domain, Paran's Main Theorem in [Par09] states that every constant finite split embedding problem over $K(x)$ has a rational solution. In particular, if K is Hilbertian, then every finite split embedding problem over K is solvable [Par09, Thm. B]. This proves [Par09, Cor. C] that every finite split embedding problem over $K((X_1, \ldots, X_n))$ is solvable if $n \geq 2$ (and generalizes the theorem of Harbater-Stevenson.) It also proves that every finite split embedding problem over $\mathrm{Quot}(A[[X_1, \ldots, X_n]])$ is solvable if $n \geq 1$ and A is a Noetherian integrally closed domain which is not a field. In particular, it follows that every finite split embedding problem over $\mathrm{Quot}(\mathbb{Z}[[X]])$ is solvable.

Contrary to the suspicion that $K((X_1, X_2))$ is not ample, Pop realized that the quotient field of each Henselian domain is ample [Pop10, Thm. 1.1], in particular, so is $K((X_1, X_2))$. That showed that all of the results of Harbater-Stevenson and Paran mentioned above about the solvability of finite split embedding problems follow already from the corresponding result over ample field (Theorem 5.9.2).

Although Pop's result has far reaching consequences, the insight behind the proof is very simple: $K((X_1, X_2))$ is the quotient field of a Henselian ring. In fact our proof of Proposition 5.7.3, which is an adjustment of a proof of Fehm-Geyer (private communication) that every Henselian field is ample, was written upon hearing the announcement of Pop's result. Moreover, it has turned out that our proof is essentially the same that of Pop.

Proposition 5.8.2 saying that every field of characteristic 0 with a pro-p absolute Galois group is ample is due to Colliot-Thélène [CoT00, paragraph preceding Thm. 1]. Theorem 5.8.3, extends that result to arbitrary characteristic [Jar03].

Theorem 5.11.3(c) is proved by Paran in [Par08, Thm. B] in a stronger form that we partially recapitulate in even a stronger form in Example 12.4.4.

Colliot-Thélène [CoT00] uses technique of Kollár, Miyaoka, and Mori to prove Theorem 5.12.1 in characteristic 0. Moret-Bailly [MoB01, Thm. 1.1] generalizes the theorem to arbitrary characteristic. We follow [HaJ00b].

Theorem 5.2.3 of [Hrb03] due to Harbater and Pop generalizes both Lemma 5.9.1 and 5.12.1. It solves finite split embedding problems over function fields of one variable over ample fields with information on decomposition groups.

Chapter 6.
Non-Ample Fields

It is sometimes more difficult to give examples of objects that do not have a certain property P than examples of objects that have that property. A standard method to do that is to prove that each object having the property P has another property P' and then to look for an object that does not have the property P'. For example, by Corollary 5.3.3, every ample field is infinite. Hence, finite fields are not ample. More sophisticated examples of nonample fields are function fields of several variables over arbitrary fields (Theorem 6.1.8(a)). Likewise we prove that if E/K is a function field of one variable and F is the compositum of a directed family of finite extensions of E of bounded genus, then F is nonample (Theorem 6.1.8(b)). The proof uses elementary methods like the Riemann-Hurwitz genus formula. We have not been able to prove that number fields are nonample by elementary means. We have rather used in Proposition 6.2.5 the deep theorem of Faltings (formerly, Mordell's conjecture).

Section 6.3 surveys concepts and results on Abelian varieties, Jacobian varieties, and homogeneous spaces (the latter is applied only in 11.5). Likewise, Section 6.4 surveys the very deep Mordell-Lang conjecture proved by Faltings and others. As a consequence we prove that the rational rank of every nonzero Abelian variety over an ample field of characteristic zero is infinite (Theorem 6.5.2). That result combined with a result of Kato-Rohrlich (Example 6.5.5) leads to examples of infinite algebraic extensions of number fields that are nonample. Finally, we prove that for each positive integer d there is a linearly disjoint sequence $K_1, K_2, K_3, \ldots$ of extensions of $\mathbb{Q}$ of degree d whose compositum is nonample (Example 6.8.9). The proof is based on the concept of the "gonality" of a function field of one variable that we establish in Sections 6.6 and 6.7 as well as on a result of Frey (Lemma 6.8.7) based on the Mordell-Lang conjecture.

6.1 Nonample Fields — Elementary Methods

The nonampleness of function fields depends on the definability of ample fields in function fields.

LEMMA 6.1.1: *Let K be a field of positive characteristic p, F a function field of one variable over K, and n a positive integer. Then:*
(a) $[F : KF^{p^n}] = p^n$.
(b) *If $K \subset E \subseteq F$ and F/E is a purely inseparable extension of degree p^n, then $E = KF^{p^n}$.*

Proof: Statement (b) follows from the inclusion $KF^{p^n} \subseteq E \subseteq F$ and from (a).

M. Jarden, *Algebraic Patching*, Springer Monographs in Mathematics,
DOI 10.1007/978-3-642-15128-6_6, © Springer-Verlag Berlin Heidelberg 2011

In order to prove (a) recall that our convention of a function field implies that F/K is a regular extension. In particular F/K has a separating transcendence element x. In particular $F/K(x)$ is a separable extension. Hence, so is $F^{p^n}/K(x)^{p^n}$. Therefore, $KF^{p^n}/K(x^{p^n})$ is also separable. In addition, $K(x)/K(x^{p^n})$ is a purely inseparable extension of degree p^n. Hence, $K(x)$ is linearly disjoint from KF^{p^n} over $K(x^{p^n})$. This implies that $[K(x)F^{p^n} : KF^{p^n}] = [K(x) : K(x^{p^n})] = p^n$. Finally, F is both a separable and a purely inseparable extension of $K(x)F^{p^n}$. Hence, $F = K(x)F^{p^n}$.

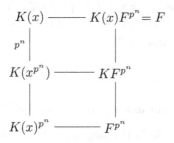

Consequently, $[F : KF^{p^n}] = p^n$. $\qquad\qquad\qquad\qquad\qquad\qquad\qquad\square$

LEMMA 6.1.2: *Let K be a perfect field of positive characteristic p, F a function field of one variable over K, and n a positive integer. Then:*
(a) $[F^{1/p^n} : F] = p^n$.
(b) *If F' is a purely inseparable extension of F of degree p^n, then $F' = F^{1/p^n}$ and the map $x \mapsto x^{p^n}$ is an isomorphism of F'/K onto F/K. In particular, F'/K is a function field of one variable and genus$(F'/K) = $ genus(F/K).*

Proof: Since K is perfect, $KF^{p^n} = K^{p^n}F^{p^n} = F^{p^n}$. Hence, by Lemma 6.1.1, $[F : F^{p^n}] = p^n$. It follows that $[F^{1/p^n} : F] = p^n$.

Now let F' be a purely inseparable extension of F of degree p^n. Then $F' \subseteq F^{1/p^n}$, so by (a), $F' = F^{1/p^n}$. This implies all of the other statements of (b). $\qquad\qquad\qquad\qquad\qquad\qquad\qquad\square$

LEMMA 6.1.3: *Let E and F be function fields of one variable over a field K such that $E \subseteq F$. Then genus$(E/K) \leq$ genus(F/K).*

Proof: By assumption, F/E is a finite algebraic extension. Denote the maximal separable extension of E in F by F_0. By Riemann-Hurwitz, genus$(E/K) \leq$ genus(F_0/K) (Remark 5.8.1). Moreover, F/F_0 is purely inseparable. Thus, we may assume that K has a positive characteristic p and that F/E is a purely inseparable extension of degree p^n for some n. Arguing inductively, we may assume that $n = 1$.

By Lemma 6.1.1, $E = KF^p$. Now note that

$$\text{genus}(F/K) = \text{genus}(F^p/K^p)$$

and genus$(F^p/K^p) \geq$ genus(F^pK/K) (Remark 5.8.1). Hence, genus$(E/K) \leq$ genus(F/K).   □

Giving an absolutely irreducible K-variety V, we denote the function field of V over $\tilde{K}$ (or, in the language of schemes, of $V \times_K \tilde{K}$) by $\tilde{K}(V)$.

LEMMA 6.1.4: *Let F/K be a function field of one variable, C an absolutely irreducible K-curve. Suppose* genus$(F\tilde{K}/\tilde{K}) <$ genus$(\tilde{K}(C)/\tilde{K})$. *Then $C(F) = C(K)$.*

Proof: Assume there exists $\mathbf{x} \in C(F) \smallsetminus C(K)$. Then $\mathbf{x} \in C(F\tilde{K}) \smallsetminus C(\tilde{K})$, because $F \cap \tilde{K} = K$. Therefore, $\mathbf{x}$ is a generic point of C over $\tilde{K}$, so $\tilde{K}(\mathbf{x}) \cong_{\tilde{K}} \tilde{K}(C)$. Hence, by Lemma 6.1.3,

$$\text{genus}(\tilde{K}(C)/\tilde{K}) = \text{genus}(\tilde{K}(\mathbf{x})/\tilde{K}) \leq \text{genus}(F\tilde{K}/\tilde{K}).$$

This contradiction to our assumption proves that $C(F) = C(K)$.   □

If $\varphi(X)$ is a formula in the language of rings in the free variable X, K is a field, and $x \in K$, we write $K \models \varphi(x)$ to denote that the statement $\varphi(x)$ holds in K.

PROPOSITION 6.1.5: *For each prime number q there exists an existential formula $\varphi_q(Z)$ in the language of rings (Definition 5.2.4) with the following property: For each ample field K with* char$(K) \neq q$ *and for each function field F of one variable over K with* genus$(F\tilde{K}/\tilde{K}) < \frac{1}{2}(q-1)(q-2)$ *we have $K = \{z \in F \mid F \models \varphi_q(z)\}$.*

Proof: Consider the Fermat polynomial $f(X,Y) = X^q + Y^q - 1$ and let $\varphi_q(Z)$ be the formula $(\exists X_1)(\exists X_2)(\exists Y_1)(\exists Y_2)[X_1 = ZX_2 \wedge X_2 \neq 0 \wedge f(X_1, Y_1) = 0 \wedge f(X_2, Y_2) = 0]$. Now let K be an ample field with char$(K) \neq q$ and let F be a function field of one variable over K with genus$(F\tilde{K}/\tilde{K}) < \frac{1}{2}(q-1)(q-2)$. Since $Y^q - 1$ is a separable polynomial, Eisenstein's criterion implies that $f(X,Y)$ is irreducible in $\tilde{K}[X,Y]$, that is $f(X,Y)$ is absolutely irreducible. Let C be the absolutely irreducible K-curve defined by the equation $f(X,Y) = 0$. It has a smooth K-rational point $(0,1)$. Explicitly, $f(0,1) = 0$ and $\frac{\partial f}{\partial Y}(0,1) \neq 0$. Moreover, the projective completion C^* of C is defined in $\mathbb{P}^2$ by the equation $f^*(X,Y,Z) = X^q + Y^q - Z^q = 0$. Since the only common zero of the polynomials $f^*(X,Y,Z)$, $\frac{\partial f^*}{\partial X} = qX^{q-1}$, $\frac{\partial f^*}{\partial Y} = qY^{q-1}$, $\frac{\partial f^*}{\partial Z} = -qZ^{q-1}$ is $(0,0,0)$ the curve C^* is smooth. It follows from [FrJ08, Cor. 5.3.6] that the genus of C^* (and of C) over $\tilde{K}$ is $\frac{1}{2}(q-1)(q-2)$. Therefore, genus$(F\tilde{K}/\tilde{K}) <$ genus$(\tilde{K}(C)/\tilde{K})$. By Lemma 6.1.4, $C(F) = C(K)$. We prove that an element $z \in F$ is in K if and only if $F \models \varphi_q(z)$.

First suppose $F \models \varphi_q(z)$. Then there exist $x_1, x_2, y_1, y_2 \in F$ such that $x_1 = zx_2$, $x_2 \neq 0$, $f(x_1, y_1) = 0$, and $f(x_2, y_2) = 0$. Then $(x_i, y_i) \in C(F) = C(K)$ for $i = 1, 2$. In particular, $x_1, x_2 \in K$. Hence, $z = \frac{x_1}{x_2} \in K$.

Conversely, let $z \in K$ and $z \neq 0$. Then each of the equations $f(t, Y) = 0$ and $f(tz, Y) = 0$ has a solution in $K((t))$ because 1 is a simple root of $f(0, Y)$ and $K((t))$ is Henselian with respect to the t-adic valuation. It follows that the existential sentence

(1)
$$(\exists X_1)(\exists X_2)(\exists Y_1)(\exists Y_2)[X_1 = zX_2 \wedge X_2 \neq 0 \wedge f(X_1, Y_1) = 0 \wedge f(X_2, Y_2) = 0]$$

of $\mathcal{L}(\text{ring}, K)$ holds in $K((t))$. Since K is ample, K is existentially closed in $K((t))$ (Lemma 5.3.1(f)), hence (1) is true also in K. In other words, there exist $x_1, x_2, y_1, y_2 \in K$ such that $x_1 = zx_2$, $x_2 \neq 0$ and $f(x_i, y_i) = 0$ for $i = 1, 2$. Thus, $K \models \varphi_q(z)$ and also $F \models \varphi_q(z)$.

Finally, for $z = 0$, we use Lemma 5.3.1(b) to choose $x_2 \in K^\times$ and $y_2 \in K$ such that $f(x_2, y_2) = 0$. In addition, we let $x_1 = 0$ and $y_1 = 1$ to conclude that $\varphi_q(0)$ holds in both K and F.   □

Given a function field F/K of one variable, we refer to genus$(F\tilde{K}/\tilde{K})$ as the **absolute genus** of F/K.

We generalize Proposition 6.1.5 to certain infinite algebraic extensions of function fields of one variable. To this end we say that a field F is a **generalized function field of one variable over a field** K if trans.deg$(F/K) = 1$ and F/K is a regular extension. In particular, every function field F of one variable over K is a generalized function field of one variable over K. We define the **genus** of a generalized function field F/K as the supremum of genus(E/K), where E ranges over all function fields of one variable over K in F. By Lemma 6.1.3, this definition coincides with the usual definition of genus(F/K) if F/K is a function field of one variable.

PROPOSITION 6.1.6 (Fehm): *For every prime number q there exists an existential formula $\varphi_q(X)$ in $\mathcal{L}(\text{ring})$ with the following property: For each ample field K with* char$(K) \neq q$ *and for each generalized function field F of one variable over K with absolute genus $g < \frac{1}{2}(q-1)(q-2)$ we have $K = \{x \in F \mid F \models \varphi_q(x)\}$.*

Proof: We write $F = \bigcup_{i \in I} F_i$, where F_i ranges over all function fields of one variable over K in F. Then genus$(F_i\tilde{K}/\tilde{K}) \leq$ genus$(F_i/K) \leq g$ (Remark 5.8.1(e)). Hence, by Proposition 6.1.5, $K = \{x \in F_i \mid F_i \models \varphi_q(x)\}$. Since $\varphi_q(X)$ is an existential formula,

$$\{x \in F \mid F \models \varphi_q(x)\} = \bigcup_{i \in I} \{x \in F_i \mid F_i \models \varphi_q(x)\}.$$

Consequently, $K = \{x \in F \mid F \models \varphi_q(x)\}$.   □

COROLLARY 6.1.7: *Let F be a generalized function field of one variable over an ample field K. Suppose $g =$ genus(F/K) is finite. Then:*
(a) *Every automorphism of F leaves K invariant.*

(b) If K' is an ample subfield of F and F is a generalized function field of one variable over K' such that $g' = \text{genus}(F/K') < \infty$, then $K' = K$.

Proof: We choose a prime number $q \neq \text{char}(K)$ such that $g < \frac{1}{2}(q-1)(q-2)$. By Proposition 6.1.6, there exists a formula $\varphi_q(X)$ of $\mathcal{L}(\text{ring})$ such that for all $x \in F$ the statement $x \in K$ is equivalent to $F \models \varphi_q(x)$. If $\alpha \in \text{Aut}(F)$, then $F \models \varphi_q(x)$ if and only if $F \models \varphi_q(\alpha(x))$. Hence, $\alpha(K) = K$.

If K' is as in (b), enlarge q such that $g' < \frac{1}{2}(q-1)(q-2)$. Then $\text{char}(K') = \text{char}(K) \neq q$. Hence, $K' = \{x \in F \mid F \models \varphi_q(x)\} = K$.   □

PROPOSITION 6.1.8: Let F be a transcendental extension of a field K. Then F is nonample in each of the following cases:
(a) F is finitely generated over K.
(b) $\text{trans.deg}(F/K) = 1$ and $\text{genus}(F\tilde{K}/\tilde{K})$ is finite (Fehm).

Proof: Assume F is ample. Replacing K by a bigger field in F shows that we may assume that $\text{trans.deg}(F/K) = 1$ also in Case (a). Replacing K by $\tilde{K}$ and F by $F\tilde{K}$, we may use Lemma 5.5.1 to assume that K is algebraically closed. Thus, in both cases, F is a generalized function field of one variable over K of finite genus. Let $t \in F$ be transcendental over K and choose an element t' transcendental over F. Then the map $t \mapsto t'$ extends to a K-isomorphism of F onto a field F'. The field F' is a generalized function field of one variable over K, F' is ample, and $F' \neq F$. By [FrJ08, Cor. 2.6.8], $N = FF'$ is a regular extension of transcendence degree 1 over both F and F'. Moreover, by definition, $F = \bigcup_{i \in I} F_i$, where F_i/K is a function field of one variable with $\text{genus}(F_i/K) \leq \text{genus}(F/K)$. Then $N = \bigcup_{i \in I} F_i F'$ and $F_i F'/F'$ is a function field of one variable with $\text{genus}(F_i F'/F') = \text{genus}(F_i/K) \leq \text{genus}(F/K)$ (Remark 5.8.1). Thus, N/F' is a generalized function field of one variable of a finite genus. Similarly, N/F is a generalized function field of one variable of a finite genus. This contradiction to Corollary 6.1.7(b) proves that F is not ample.   □

Example 6.1.9: We give three examples of nonample generalized function fields of one variable that are not function fields of one variable.

(a) (Fehm) Let K be a field and x an indeterminate. For each positive integer n with $\text{char}(K) \nmid n$ we choose $x_n \in K(x)_s$ such that $x_1 = x$ and $x_{mn}^n = x_m$ for all m, n. Then $F = \bigcup_{\text{char}(K) \nmid n} K(x_n)$ is a generalized function field of one variable over K of genus 0. By Proposition 6.1.8, F is nonample. Moreover, $[F : K(x)] = \infty$, so F/K is not a function field of one variable.

(b) Let E be a function field of one variable over a field K with $\text{genus}(E\tilde{K}/\tilde{K}) = 1$ (e.g. E is the function field of an elliptic curve defined over K). If E' is a finite unramified extension of E and E' is regular over K, then $E'\tilde{K}$ is unramified over $E\tilde{K}$. Hence, by Riemann-Hurwitz, $\text{genus}(E'\tilde{K}/\tilde{K}) = 1$ (Remark 5.8.1). Thus, if we denote the compositum of all finite unramified extensions of E by E_{ur}, then $\text{trans.deg}(E_{\text{ur}}/K) = 1$ and $\text{genus}(E_{\text{ur}}\tilde{K}/\tilde{K}) = 1$. By Proposition 6.1.8(b), E_{ur} is nonample.

Note that $E_{\mathrm{ur}}\tilde{K}/\tilde{K}$ is not a function field of one variable. Indeed, even more is true, namely that $K_s \subseteq E_{\mathrm{ur}}$, $[E_{\mathrm{ur}} : EK_s] = \infty$, and E_{ur} is linearly disjoint from $E\tilde{K}$ over EK_s, so $[E_{\mathrm{ur}}\tilde{K} : E\tilde{K}] = \infty$. However, in order to keep our references as elementary as possible, we assume that K is algebraically closed and prove that $[E_{\mathrm{ur}} : E] = \infty$.

Indeed, E has a K-rational point, so E is the function field of an elliptic K-curve C [Sil86, Sec. III3]. Thus, $E = K(\mathbf{x})$, where $\mathbf{x}$ is a generic point of C over K. For each positive integer n with $\mathrm{char}(K) \nmid n$ we choose a point $\mathbf{x}_n \in C(\tilde{E})$ such that $n\mathbf{x}_n = \mathbf{x}$ (a consequence of [Sil86, p. 71, Prop. 4.2]). Then $\mathbf{x}_n$ is also a generic point of C over K and $K(\mathbf{x}_n)$ is a Galois extension of $K(\mathbf{x})$ of degree n^2 (a consequence of [Sil86, p.89, Cor. 6.4(b)]). In particular, $\mathrm{genus}(K(\mathbf{x}_n)/K) = 1$, so by Riemann-Hurwitz, $K(\mathbf{x}_n)$ is an unramified extension of $K(\mathbf{x})$. It follows that $[E_{\mathrm{ur}} : E] = \infty$, which proves our claim.

(c) The scope of the method that led to the preceding examples is limited. Indeed, if E is a function field of one variable over a field K, $\mathrm{genus}(E/K) \geq 2$, and F is an infinite separable algebraic extension of E regular over K, then E has finite extensions E' in F of unbounded degree. By Riemann-Hurwitz, $2\mathrm{genus}(E'/K) - 2 \geq [E' : E](2\mathrm{genus}(E/K) - 2)$. Hence, $\mathrm{genus}(E'/K)$ is unbounded, so $\mathrm{genus}(F/K) = \infty$ and Proposition 6.1.8 does not apply. In the following section we use more powerful tools to get more examples of nonample fields. $\square$

For each prime number p there exists an example of Hrushovski of a countable non-PAC field F of characteristic p such that F_{ins} is PAC [FrJ08, Thm. 11.7.8]. Problem 11.7.9 of [FrJ08] then asks if that F can be finitely generated over $\mathbb{F}_p$. The following result gives a negative answer to that question.

PROPOSITION 6.1.10 (Fehm): *Let K be a field of positive characteristic and F a finitely generated transcendental extension of K. Then F_{ins} is not ample (hence also not PAC).*

Proof: We break up the proof into two parts.

PART A: *K is perfect and F is a function field of one variable over K.* Let $g = \mathrm{genus}(F/K)$. For each positive integer n the map $x \mapsto x^{p^n}$ is an isomorphism of F^{1/p^n} onto F that leaves K invariant, so F^{1/p^n} is also a function field of one variable over K of genus g. It follows that $F_{\mathrm{ins}} = \bigcup_n F^{1/p^n}$ is a generalized function field of one variable over K of genus g. By Proposition 6.1.8(b), F_{ins} is nonample.

PART B: *The general case.* Replacing K by a bigger field in F, if necessary, we may assume that $\mathrm{trans.deg}(F/K) = 1$ and K is algebraically closed in F. Then K has a finite extension K' such that $F' = FK'$ is a regular extension of K', that is F'/K' is a function field of one variable. Moreover, F'_{ins} is an algebraic extension of F_{ins}. If F_{ins} is ample, then so is F'_{ins} (Lemma 5.5.1(b)). Thus, without loss of generality, we may assume that F is a function field of one variable over K. But then FK_{ins} is a function field of one variable over

K_{ins}. Since K_{ins} is perfect and $F_{\text{ins}} = (FK_{\text{ins}})_{\text{ins}}$, it follows from Part A that F_{ins} is not ample. $\qquad\square$

6.2 Nonample Fields — Advanced Methods

We call a field K **small** if K is finite, a number field, or a finitely generated transcendental extension of another field. In each of these cases, K is nonample. We have already proved this statement for the first and the third types of fields (Corollary 5.3.3 and Theorem 6.1.8). To prove it for the second type (and reprove it for the third type), we need examples of absolutely irreducible K-curves with finitely many K-rational points. These are given by two deep results of arithmetic geometry, Faltings' theorem and the Grauert-Manin theorem. Each of these theorems speaks about curves of genus at least 2. Thus, we start by establishing examples of such curves.

Let K/K_0 be an extension of fields. An absolutely irreducible K-curve C is said to be **nonconstant** over K/K_0 if C is not birationally equivalent over $\tilde{K}$ to any $\tilde{K}_0$-curve C_0.

LEMMA 6.2.1 ([FrJ08, Remark 11.7.7 and Proposition 3.8.4]): *Let K be a field of characteristic p and $u_0, u_1, u_2, u_3, u_4, u_5$ distinct elements of K. Then the K-curve*

$$C: \begin{cases} Y^2 = (X-u_1)(X-u_2)(X-u_3)(X-u_4)(X-u_5) & \text{if } p \neq 2 \\ Y^2 + Y = \frac{X-u_0}{(X-u_1)(X-u_2)(X-u_3)(X-u_4)(X-u_5)} & \text{if } p = 2 \end{cases}$$

is absolutely irreducible (and is hyperelliptic). Its genus over K and over $\tilde{K}$ is 2 if $p \neq 2$ and 4 if $p = 2$. Moreover, if K_0 is a subfield of K and $u_4 \notin \tilde{K}_0(u_1, u_2, u_3)$, then C is nonconstant over K/K_0.

Remark 6.2.2: Genus of a curve. Let C be an absolutely irreducible curve defined over a field K and let F be the function field of C over K. We define the **genus** (resp. **absolute genus**) of C to be the genus of F/K (resp. $F\tilde{K}/\tilde{K}$). By Remark 5.8.1, the genus of C is greater or equal to its absolute genus. Moreover, equality holds if K is perfect. $\qquad\square$

The following result is known as the "Mordell's Conjecture for function fields" and is due to Grauert and Manin.

PROPOSITION 6.2.3 ([Sam66, pp. 107 and 118]): *Let K be a finitely generated regular extension of a field K_0 and C an absolutely irreducible K-curve nonconstant over K/K_0. Suppose the absolute genus of C is at least 2. Then $C(K)$ is a finite set.*

We also need Faltings' theorem, formerly known as "Mordell's Conjecture".

PROPOSITION 6.2.4 ([Fal83]): *Let K be a number field and C an absolutely irreducible K-curve of genus at least 2. Then $C(K)$ is finite.*

PROPOSITION 6.2.5: *In each of the following cases K is a nonample field:*
(a) *K is a number field.*
(b) *K is a finitely generated transcendental extension of a subfield K_0.*

Proof: We consider the absolutely irreducible K-curve C of genus ≥ 2 defined in Lemma 6.2.1. In case (b) we may replace K_0 by a finite algebraic extension K_0' and K by KK_0', if necessary, to assume that K is a regular extension of K_0. If char$(K) \neq 2$, then $(u_1, 0) \in C_{\text{simp}}(K)$. If char$(K) = 2$, then $(u_0, 0) \in C_{\text{simp}}(K)$. Indeed, in each case, the corresponding point does not satisfy the equation obtained by taking the derivative with respect to X of the defining equation. Moreover, in case (b), C is nonconstant over K/K_0. It follows from Propositions 6.2.3 and 6.2.4 that $C(K)$ is finite. By Lemma 5.3.1(a), K is nonample. $\square$

Note that Case (b) of Proposition 6.2.5 is proved in Theorem 6.1.8 by much simpler means. However, the theorem of Grauert-Manin comes in handy in the proof of a generalization of the following result:

THEOREM 6.2.6 (Fehm): *Let K be a field of infinite cardinality m, n a positive integer, $x_1, \ldots, x_n$ algebraically independent elements over K, and F a compositum of less than m finite extensions of $K(x_1, \ldots, x_n)$. Then F is nonample.*

Proof: The case where $m = \aleph_0$ is covered by Theorem 6.1.8(a) and also by Proposition 6.2.5(b). Therefore we assume that $m > \aleph_0$. We may also assume that $n = 1$ and set $F_0 = K(x_1)$. Lemma 6.2.1 gives an absolutely irreducible F_0-curve C nonconstant over F_0/K of genus ≥ 2. Moreover, $(0, u_1)$ (resp. u_0) is a simple F_0-rational point of C if char$(F_0) \neq 2$ (resp. if char$(F_0) = 2$). By assumption $F = \bigcup_{i \in I} F_i$, where I is a set of cardinality less than m and each F_i is a finite extension of F_0. By Grauert-Manin, $C(F_i)$ is finite, hence $C(F_i)$ is finite for each i. Hence, card$(C(F)) \leq$ card$(I)\aleph_0 < m$. It follows from Proposition 5.4.3(b) that F is nonample. $\square$

6.3 Abelian Varieties and Jacobian Varieties

We survey in this section elements of the theory of Abelian varieties and Jacobian varieties needed in this book. Our sources are Lang's book [Lan59], Mumford's book [Mum74], and Milne's two chapters in [CoS86].

6.3.1 ABELIAN VARIETIES.

A **group variety** over a field K is an absolutely irreducible K-variety G with two morphisms $G \times G \to G$ (multiplication), $G \to G$ (inverse), and an element $e \in G(K)$ (the unit element) defined over K such that the structure on $G(\tilde{K})$ defined by the multiplication and the inverse is that of a group with identity element e. In particular, G is **smooth**, that is each point of $G(\tilde{K})$ is simple [CoS86, p. 104].

We say that a group variety A is an **Abelian variety** if A is complete. In that case $A(L)$ is an Abelian group for every field extension L of K [CoS86,

p. 105, Cor. 2.4]. We write the operation of A as addition and use $\mathbf{o}$ (or $\mathbf{o}_A$) for the zero element of A. One can embed A into a projective space $\mathbb{P}_K^r$ for some positive integer r [CoS86, p. 113, Thm. 7.1].

A **homomorphism** $\alpha\colon A \to B$ of Abelian varieties over K is a morphism that respects addition. In particular, for each extension L of K, α gives rise to a homomorphism $\alpha\colon A(L) \to B(L)$ of Abelian groups. One says that α is an **isogeny**, if α is surjective with a finite kernel and let $\deg(\alpha) = [K(A) : K(B)]$ be the **degree** of α. In this case, $\dim(A) = \dim(B)$ [CoS86, p. 114, Prop. 8.1].

Multiplication of elements of an Abelian variety A by a positive integer n is an isogeny of degree $n^{2\dim(A)}$ denoted by n_A (or just by n) [CoS86, p. 115, Thm. 8.2]. Thus, $A(\tilde{K})$ is a **divisible group**, i.e. for each $\mathbf{a} \in A(\tilde{K})$ and every positive integer n there exists $\mathbf{b} \in A(\tilde{K})$ such that $n\mathbf{b} = \mathbf{a}$. The isogeny n_A is étale (in particular separable) if $\mathrm{char}(K) \nmid n$. Writing $A_n = \mathrm{Ker}(n_A)$, we get in the latter case that $A_n(\tilde{K}) \cong (\mathbb{Z}/n\mathbb{Z})^{2\dim(A)}$ [Mum74, p. 64].

Now consider a subgroup Γ of $A(\tilde{K})$ and an extension L of K. We write $\Gamma^{\mathrm{div}}(L) = \{\mathbf{a} \in A(\tilde{L}) \mid n\mathbf{a} \in \Gamma \text{ for some } n \in \mathbb{N}\}$ for the divisible hull of Γ in $A(\tilde{L})$. For each $n \in \mathbb{N}$ and every $\mathbf{b} \in \Gamma$ the set $\{\mathbf{a} \in A(\tilde{L}) \mid n\mathbf{a} = \mathbf{b}\}$ is a coset of $A_n(\tilde{K})$, hence it is a finite algebraic set defined over $\tilde{K}$, so it is contained in $A(\tilde{K})$. It follows that $\Gamma^{\mathrm{div}} = \Gamma^{\mathrm{div}}(L) \leq A(\tilde{K})$ is independent of L.

Finally let α be a rational map from an absolutely irreducible variety V into an Abelian variety A defined over K. Then α is defined at each simple point of V [Lan59, p. 20, Thm. 2]. Thus, if V is smooth, the α is a morphism [CoS86, p. 106, Thm. 3.1]. Moreover, if V is an Abelian variety and $\alpha(\mathbf{o}_V) = \mathbf{o}_A$ then α is a homomorphism (see [Lan59, p. 24, Thm. 4] or [CoS86, p. 107, Cor. 3.6]).

6.3.2 JACOBIAN VARIETIES.

Let C be a complete smooth absolutely irreducible curve of positive genus g defined over a field K. We assume that C has a K-rational point $\mathbf{o}$. Then there exists an Abelian variety $J = J_C$ of dimension g and a rational map $\gamma\colon C \to J$ defined over K such that $\gamma(\mathbf{o}) = \mathbf{o}_J$ with the following universal property: If $\alpha\colon C \to A$ is a rational map into an Abelian variety A defined over K such that $\alpha(\mathbf{o}) = \mathbf{o}_A$, then there is a unique homomorphism $\alpha'\colon J \to A$ such that $\alpha' \circ \gamma = \alpha$ [Lan59, p. 35, Thm. 9]. Note that by the final remark of Subsection 6.3.1, both maps γ, α are actually morphisms.

By [Lan59, p. 35, Thm. 8], the image of the morphism $C^g \to J$ given by

$$(\mathbf{p}_1, \ldots, \mathbf{p}_g) \to \sum_{i=1}^{g} \gamma(\mathbf{p}_i)$$

is Zariski-dense in J. Since C^g is complete and the image of morphisms of complete varieties is Zariski-closed, each point of J is the sum of g points of $\gamma(C)$ (see also [CoS86, p. 182]). Thus, $\gamma(C)$ generates J. By [Lan59, p. 49, Prop. 4], the map $\gamma\colon C \to \gamma(C)$ is an isomorphism.

In particular, if C happens to be also an Abelian variety, then by the final remark of Subsection 6.3.1, γ is a homomorphism. Hence, by the preceding paragraph, $\gamma(C(\tilde{K})) = J(\tilde{K})$. Therefore, $\gamma \colon C \to J$ is surjective and $1 \leq g = \dim(J) \leq \dim(C) = 1$. Consequently, $g = 1$, i.e. C is an elliptic curve.

In the general case, we choose a generic point $\mathbf{x}$ of C over K. Let $F = K(\mathbf{x})$ be the function field of C over K. Recall (Remark 5.8.1) that $\mathbb{P}(F\tilde{K}/\tilde{K})$ denotes the set of prime divisors of $F\tilde{K}/\tilde{K}$. Each $\mathfrak{p} \in \mathbb{P}(F\tilde{K}/\tilde{K})$ is represented by a unique place $\varphi_{\mathfrak{p}} \colon F\tilde{K} \to \tilde{K} \cup \{\infty\}$ such that $\varphi_{\mathfrak{p}}(a) = a$ for each $a \in \tilde{K}$. Since C is complete $\mathbf{p} = \varphi_{\mathfrak{p}}(\mathbf{x})$ is a point of $C(\tilde{K})$. Since C is smooth, each point $\mathbf{p} \in C(\tilde{K})$ is uniquely obtained in this way. Thus, the map $\mathfrak{p} \mapsto \varphi_{\mathfrak{p}}(\mathbf{x})$ from $\mathbb{P}(F\tilde{K}/\tilde{K})$ into $C(\tilde{K})$ is bijective. Composing this map with γ and extending linearly to $\mathrm{Div}(F\tilde{K}/\tilde{K})$ we get a homomorphism $\beta \colon \mathrm{Div}(F\tilde{K}/\tilde{K}) \to J(\tilde{K})$, namely if $\mathfrak{a} = \sum_{i=1}^{n} k_i \mathfrak{p}_i$, then $\beta(\mathfrak{a}) = \sum_{i=1}^{n} k_i \gamma(\varphi_{\mathfrak{p}_i}(\mathbf{x}))$. By the second paragraph of the present subsection, β is surjective.

Now let $\mathfrak{o}$ be the prime divisor of F/K with $\varphi_{\mathfrak{o}}(\mathbf{x}) = \mathbf{o}$. Then $\mathfrak{o}$ remains prime over $\tilde{K}$ and $\beta(\mathfrak{o}) = \gamma(\mathbf{o}) = \mathbf{o}_J$. Thus, $\deg(\mathfrak{a} - \deg(\mathfrak{a})\mathfrak{o}) = 0$, and $\beta(\mathfrak{a} - \deg(\mathfrak{a})\mathfrak{o}) = \beta(\mathfrak{a})$. It follows that the restriction β_0 of β to $\mathrm{Div}_0(F\tilde{K}/\tilde{K})$ is also surjective. By Abel, $\mathrm{Ker}(\beta_0)$ is the group $\mathrm{div}((F\tilde{K})^{\times})$ of principal divisors [Lan59, p. 36, Thm. 10]. It follows that the following short sequence of Abelian groups is exact:

$$(1) \qquad 0 \to \mathrm{div}((F\tilde{K})^{\times}) \to \mathrm{Div}_0(F\tilde{K}/\tilde{K}) \xrightarrow{\beta_0} J(\tilde{K}) \to 0$$

Since γ is defined over K, it commutes with the action of $\mathrm{Gal}(K)$. On the other hand, the equality $\varphi_{\mathfrak{p}^{\sigma}}(\mathbf{x}) = \varphi_{\mathfrak{p}}(\mathbf{x})^{\sigma^{-1}}$ implies that the map

$$\mathbb{P}(F\tilde{K}/\tilde{K}) \to C(\tilde{K})$$

commutes with the action of $\mathrm{Gal}(K)$ up to the involution $\sigma \mapsto \sigma^{-1}$. One way to fix this problem is to let $\mathrm{Gal}(K)$ act both on $\mathbb{P}(F\tilde{K}/\tilde{K})$ and on $C(\tilde{K})$ (and then on $J(\tilde{K})$) from the left. Then $\varphi_{\sigma\mathfrak{p}}(\mathbf{x}) = \sigma\varphi_{\mathfrak{p}}(\mathbf{x})$ and the map $\mathbb{P}(F\tilde{K}/\tilde{K}) \to C(\tilde{K})$ commutes with the action of $\mathrm{Gal}(K)$. It follows that (1) yields for $\mathrm{Pic}_0(F\tilde{K}/\tilde{K}) = \mathrm{Div}_0(F\tilde{K}/\tilde{K})/\mathrm{div}((F\tilde{K})^{\times})$ a $\mathrm{Gal}(K)$-isomorphism

$$(2) \qquad \mathrm{Pic}_0(F\tilde{K}/\tilde{K}) \cong J(\tilde{K}).$$

Like for any other Abelian variety, the group $J(\tilde{K})$ is divisible (Subsection 6.3.1). By (2) the quotient group, $\mathrm{Div}_0(F\tilde{K}/\tilde{K})/\mathrm{div}((F\tilde{K})^{\times})$ is also divisible. Since $\tilde{K}$ is algebraically closed, the degree of each prime divisor $\mathfrak{o}$ of $F\tilde{K}/\tilde{K}$ is 1. Hence, for each $\mathfrak{a} \in \mathrm{Div}(F\tilde{K}/\tilde{K})$ and for each positive integer n there exists $f \in (F\tilde{K})^{\times}$ such that $\mathfrak{a} - \deg(\mathfrak{a})\mathfrak{o} \equiv \mathrm{div}(f) \bmod n\mathrm{Div}(F\tilde{K}/\tilde{K})$.

6.3.3 HOMOGENEOUS SPACES.

Let A be an Abelian variety defined over a field K. A **principal homogeneous space** over A (also called a **torsor**) is an absolutely irreducible K-variety V together with a right simply transitive action of A on V. In other words, there is a K-morphism $V \times A \to V$ mapping $(\mathbf{v}, \mathbf{a})$ onto $\mathbf{va}$, such that for all $\mathbf{v}, \mathbf{v}_1, \mathbf{v}_2 \in V(\tilde{K})$ and $\mathbf{a}, \mathbf{b} \in A(\tilde{K})$ the following holds:

(3a) $\mathbf{v} \cdot 0 = \mathbf{v}$.

(3b) $\mathbf{v}(\mathbf{a} + \mathbf{b}) = (\mathbf{va})\mathbf{b}$.

(3c) For every $\mathbf{v}_1, \mathbf{v}_2 \in V(\tilde{K})$ there is a unique $\mathbf{a} \in \tilde{K}$ with $\mathbf{v}_1\mathbf{a} = \mathbf{v}_2$.

(3d) The map $\nu \colon V \times V \to A$ mapping $(\mathbf{v}_1, \mathbf{v}_2)$ onto $\mathbf{a}$ of (3c) is a K-morphism.

Two principal homogeneous spaces V and V' are K-**isomorphic** if there exists a K-isomorphism $\varphi \colon V \to V'$ of absolutely irreducible varieties such that $\varphi(\mathbf{va}) = \varphi(\mathbf{v})\mathbf{a}$ for all $\mathbf{v} \in V(\tilde{K})$ and $\mathbf{a} \in A(\tilde{K})$. We denote the set of K-isomorphism classes of principal homogeneous spaces over A by $P(A)$.

The Abelian variety A is a principal homogeneous space over itself. The action is given by addition: $(\mathbf{x}, \mathbf{a}) \mapsto \mathbf{x} + \mathbf{a}$.

If L is an algebraic extension of K and $\mathbf{v}_0 \in V(L)$, then the map $\mathbf{a} \mapsto \mathbf{v}_0\mathbf{a}$ is an L-isomorphism of A onto V. If $\mathbf{v}_0 \in V(K)$, then that map is even a map of principal homogeneous spaces over A.

By definition, if K is a PAC field, then every absolutely irreducible K-variety has a K-rational point. Hence, in this case $P(A)$ consists of one element, namely the class of A.

The main interest in principal homogeneous spaces arises from a bijective correspondence between $P(A)$ and the first cohomology group $H^1(\mathrm{Gal}(K), A(K_s))$. The class of A corresponds to the zero element of $H^1(\mathrm{Gal}(K), A(K_s))$ [LaT58, p. 667, Prop. 4]. By the preceding paragraph, $H^1(\mathrm{Gal}(K), A(K_s)) = 0$ if K is PAC.

6.4 The Mordell-Lang Conjecture

The Mordell-Lang conjecture (proved by Faltings and others) is a far reaching generalization of the Mordell conjecture proved previously also by Faltings (and others). We quote the result and draw several special cases of it.

We start by quoting Theorem 4.2 of [Fal94].

PROPOSITION 6.4.1: *Let K be a number field, A an Abelian variety defined over K, and V a Zariski-closed subset of A defined over K. Then there exist $\mathbf{a}_1, \ldots, \mathbf{a}_n \in A(K)$ and Abelian subvarieties $B_1, \ldots, B_n$ of A defined over K such that $\mathbf{a}_i + B_i \subseteq V$ for each i and $V(K) = \bigcup_{i=1}^n [\mathbf{a}_i + B_i(K)]$.*

Remark 6.4.2: Rational rank. Let Γ be an additive Abelian group. Elements $c_1, \ldots, c_n$ of Γ are said to be **linearly independent over** $\mathbb{Z}$ if for all $k_1, \ldots, k_n \in \mathbb{Z}$ the equality $\sum_{i=1}^n k_i c_i = 0$ implies $k_1 = \cdots = k_n = 0$. An arbitrary subset of Γ is **linearly independent** if every finite subset is linearly independent. The **rational rank** of Γ is the supremum of the cardinalities of all linearly independent subsets of Γ over $\mathbb{Z}$. It is denoted by $\mathrm{rr}(\Gamma)$. For example, $\mathrm{rr}(\Gamma) = 0$ if and only if Γ is a torsion group. By [Bou89, p. 437], $\mathrm{rr}(\Gamma)$ is also the dimension of the $\mathbb{Q}$-vector space $\mathbb{Q} \otimes_{\mathbb{Z}} \Gamma$. If Γ is finitely generated, then $\Gamma \cong \bigoplus_{i=1}^{m} \mathbb{Z}/q_i\mathbb{Z} \oplus \mathbb{Z}^r$ for some prime powers $q_1, \ldots, q_m$ and an integer $r \geq 0$. Then $\mathrm{rr}(\Gamma) = r$.

Now suppose Γ is contained in a divisible Abelian group A. Then $\mathrm{rr}(\Gamma) < \infty$ if and only if Γ has a finitely generated subgroup Γ_0 such that $\Gamma \leq \Gamma_0^{\mathrm{div}} = \{a \in A \mid na \in \Gamma_0 \text{ for some } n \in \mathbb{Z}\}$.

Indeed, if there exists Γ_0 as in the preceding paragraph and $c_1, \ldots, c_n$ are linearly independent elements of Γ over $\mathbb{Z}$, then there exists $m \in \mathbb{N}$ such that $mc_1, \ldots, mc_n \in \Gamma_0$. By definition, also $mc_1, \ldots, mc_n$ are linearly independent over $\mathbb{Z}$. Hence, $m \leq \mathrm{rr}(\Gamma_0)$. It follows that $\mathrm{rr}(\Gamma) \leq \mathrm{rr}(\Gamma_0) < \infty$.

Conversely, if $r = \mathrm{rr}(\Gamma) < \infty$, we choose linearly independent elements $c_1, \ldots, c_r \in \Gamma$ over $\mathbb{Z}$ and let Γ_0 be the subgroup of Γ generated by $c_1, \ldots, c_r$. If c is an arbitrary element of Γ, then $c, c_1, \ldots, c_r$ are linearly dependent over $\mathbb{Z}$. Thus, there exist $k, k_1, \ldots, k_r \in \mathbb{Z}$ with $k \neq 0$ such that $kc + k_1 c_1 + \cdots + k_r c_r = 0$. Therefore, $c \in \Gamma_0^{\mathrm{div}}$. $\qquad\square$

PROPOSITION 6.4.3 (The Mordell-Lang Conjecture): *Let K be a field of characteristic 0, A an Abelian variety defined over $\tilde{K}$, V a Zariski-closed subset of A, and Γ a subgroup of A of a finite rational rank. Then there exist elements $\mathbf{a}_1, \ldots, \mathbf{a}_n \in \Gamma$ and Abelian subvarieties $B_1, \ldots, B_n$ of A defined over $\tilde{K}$ such that $\mathbf{a}_i + B_i(\tilde{K}) \subseteq V(\tilde{K})$ for each i and $V(\tilde{K}) \cap \Gamma = \bigcup_{i=1}^{n}[\mathbf{a}_i + (B_i(\tilde{K}) \cap \Gamma)]$.*

References to the proof: Following Raynaud [Ray83], Hindry reduces the proposition to the case where Γ is finitely generated rather than of finite rational rank [Hin88]. Then comes a reduction to the case where K is a number field, A and V are defined over K, and $\Gamma = A(K)$ [Maz00, §7]. The latter case is due to Faltings (Proposition 6.4.1), who applies the technique of diophantine approximation due to Vojta. $\qquad\square$

COROLLARY 6.4.4: *Let K be a field of characteristic 0, A an Abelian variety defined over K, C an absolutely irreducible K-subcurve of A with genus$(C) \geq 2$, and Γ a subgroup of $A(\tilde{K})$ of finite rational rank. Then:*
(a) *$C(\tilde{K}) \cap \Gamma$ is a finite set.*
(b) *If $\mathrm{rr}(A(K)) < \infty$, then $C(K)$ is a finite set.*
(c) *If K is a number field, then $C(K)$ is a finite set.*

Proof of (a): By Proposition 6.4.3 there exist points $\mathbf{a}_1, \ldots, \mathbf{a}_n \in \Gamma$ and Abelian subvarieties $B_1, \ldots, B_n$ of A defined over $\tilde{K}$ such that $\mathbf{a}_i + B_i(\tilde{K}) \subseteq C(\tilde{K})$ for each i and $C(\tilde{K}) \cap \Gamma = \bigcup_{i=1}^{n}[\mathbf{a}_i + (B_i(\tilde{K}) \cap \Gamma)]$. If $C(\tilde{K}) \cap \Gamma$ is infinite, then there is an i such that $\mathbf{a}_i + (B_i(\tilde{K}) \cap \Gamma)$ is infinite. Hence, $\mathbf{a}_i + B_i(\tilde{K})$ is infinite. Since $\mathbf{a}_i + B_i$ is a closed subset of C and C is irreducible, $C = \mathbf{a}_i + B_i$. In particular, B_i is isomorphic to C. Hence, B_i is a curve of genus at least 2. This contradicts the consequence of the third paragraph of Section 6.3.2 and proves that $C(\tilde{K}) \cap \Gamma$ is finite.

Proof of (b): Taking $\Gamma = A(K)$, it follows from (a) that $C(K) = C(\tilde{K}) \cap A(K)$ is finite.

Proof of (c): By Mordell-Weil, $A(K)$ is in our case finitely generated. Hence, by (b), $C(K)$ is finite.   □

6.5 Consequences of the Mordell-Lang Conjecture

We use the Mordell-Lang conjecture in order to prove that if K is an ample field of characteristic 0 and A is an Abelian variety of positive dimension defined over K, then $\mathrm{rr}(A(K)) = \infty$ (Theorem 6.5.2). Using another deep theorem of Kato-Rohrlich, we give examples of infinite algebraic extensions of $\mathbb{Q}$ that are nonample (Example 6.5.5).

In the following result we construct the subcurve C of A required in Corollary 6.4.4 under the condition that $\dim(A) \geq 2$.

LEMMA 6.5.1: *Let A be an Abelian variety of dimension ≥ 2 defined over an infinite perfect field K. Then there exists an absolutely irreducible curve C of genus ≥ 2 defined over K, lying in A, and having a simple K-rational point.*

Proof: Recall that an Abelian variety B defined over K is said to be K-**simple** if 0 and B are the only Abelian subvarieties of B defined over K. By Poincaré, A is K-isogeneous to a product $B = B_1 \times \cdots \times B_r$ of K-simple Abelian subvarieties [Lan59, p. 28, Thm. 6]. Thus, there exists a morphism $\alpha \colon A \to B$, defined over K, such that $\alpha(A(\tilde{K})) = B(\tilde{K})$ and $\mathrm{Ker}(\alpha)(\tilde{K})$ is finite. In particular, $\dim(B) = \dim(A) \geq 2$. For each i we choose a nonzero point $\mathbf{q}_i \in B_i(\tilde{K})$ and a point $\mathbf{p}_i \in A(\tilde{K})$ with $\alpha(\mathbf{p}_i) = \mathbf{q}_i$. By Lemma 5.1.3, there exists an absolutely irreducible projective curve C defined over K that lies on A and goes through the zero point $\mathbf{o}_A$ of A and through each of the points $\mathbf{p}_i$. Moreover, since $\mathbf{o}_A$ and each of the points $\mathbf{p}_i$ are simple on A, we may choose C such that $\mathbf{o}_A$ and each of the points $\mathbf{p}_i$ are simple on C.

The proof of [JaR98, Lemma 10.1] (on which Lemma 5.1.3 is based) can be refined to make C smooth (This may be used to remove the condition on K to be perfect.) But since this is not explicitly stated there, we consider the K-normalization $\pi \colon C' \to C$ of C. Since K is perfect, C' is smooth. Moreover, there exist unique points $\mathbf{o}' \in C'(K)$ and $\mathbf{p}'_1, \ldots, \mathbf{p}'_r \in C'(\tilde{K})$ such that $\pi(\mathbf{o}') = \mathbf{o}_A$ and $\pi(\mathbf{p}'_i) = \mathbf{p}_i$ for $i = 1, \ldots, r$.

Since C' is a smooth projective curve defined over K and $\mathbf{o}' \in C'(K)$, the Jacobian variety J of C' is defined over K. Let $\gamma' \colon C' \to J$ be the unique morphism such that $\gamma'(\mathbf{o}')$ is the zero point $\mathbf{o}_J$ of J (Subsection 6.3.2). It yields a homomorphism $\varphi \colon J \to A$ defined over K making the following diagram commutative:

$$
\begin{array}{ccc}
C' & \xrightarrow{\gamma'} & J \\
{\scriptstyle \pi}\downarrow & & \downarrow{\scriptstyle \varphi} \\
C & \xrightarrow[\text{incl.}]{} A & \xrightarrow{\alpha} B = B_1 \times \cdots \times B_r
\end{array}
$$

For each i we have $\alpha(\varphi(\gamma'(\mathbf{p}'_i))) = \alpha(\pi(\mathbf{p}'_i)) = \mathbf{q}_i \in B_i(\tilde{K})$. Hence, $\alpha(\varphi(J)) \cap B_i$ is a nonzero Abelian subvariety of B_i defined over K. Since B_i is K-simple, $B_i \subseteq \alpha(\varphi(J))$. It follows that $\alpha \circ \varphi \colon J \to B$ is surjective. Therefore, $\mathrm{genus}(C) = \mathrm{genus}(C') = \dim(J) \geq \dim(B) \geq 2$, as desired. $\qquad\square$

THEOREM 6.5.2 (Fehm-Petersen): *Let A be an Abelian variety of dimension ≥ 1 defined over an ample field K of characteristic 0. Then $\mathrm{rr}(A(K)) = \infty$.*

Proof: If $\mathrm{rr}(A(K) \times A(K)) = \infty$, then $\mathrm{rr}(A(K)) = \infty$. Therefore, replacing A by $A \times A$, if necessary, we may assume that $\dim(A) \geq 2$. Let C be an absolutely irreducible curve of genus at least 2 lying on A that Lemma 6.5.1 gives. In particular, the zero point $\mathbf{o}_A$ of A is a simple K-rational point of C. Since K is ample, $C(K)$ is infinite (Lemma 5.3.1(b)). It follows from Corollary 6.4.4(b) that $\mathrm{rr}(A(K)) = \infty$. $\qquad\square$

It is somewhat surprising that the latest result can be proved for ample fields of infinite transcendence degree without the restriction on the characteristic by much simpler means. One of the ingredients of that proof seems to be folklore:

LEMMA 6.5.3: *Let K be a field with an infinite transcendece degree over its prime field K_0. Then $\mathrm{card}(K) = \mathrm{trans.deg}(K/K_0)$.*

Proof: Let T be a transcendence basis of K/K_0. By assumption $m = \mathrm{card}(T) = \mathrm{trans.deg}(K/K_0) \geq \aleph_0$. Hence, the cardinality of the set $\mathcal{T}$ of all finite subsets of T is also m. Since $K_0(T) = \bigcup_{T_0 \in \mathcal{T}} K_0(T_0)$ and $\mathrm{card}(K_0(T_0)) \leq \aleph_0$, we have $\mathrm{card}(T) \leq \mathrm{card}(K_0(T)) \leq \aleph_0 \cdot m = m$. Thus, $\mathrm{card}(K_0(T)) = m$. Since K is an algebraic extension of $K_0(T)$, the cardinality of K is also m. $\qquad\square$

THEOREM 6.5.4: *Let A be an Abelian variety of dimension ≥ 1 defined over an ample field K. Let K_0 be the prime field of K with $\mathrm{trans.deg}(K/K_0) = \infty$. Then, $\mathrm{rr}(A(K)) = \mathrm{card}(K)$.*

Proof: Assume $\mathrm{rr}(A(K)) < \mathrm{card}(K)$. Since $\mathrm{trans.deg}(K/K_0) = \infty$, we have $\mathrm{trans.deg}(K/K_0) = \mathrm{card}(K)$. Let $(\mathbf{a}_i)_{i \in I}$ be a maximal linearly independent set elements of $A(K)$ with $\mathrm{card}(I) = \mathrm{rr}(A(K))$. By our assumption, $\mathrm{card}(I) < \mathrm{card}(K)$.

We choose a finitely generated subfield K_1 in K over which A is defined and let $K_2 = K_1(\mathbf{a}_i \mid i \in I)$. If $\mathrm{card}(K) = \aleph_0$, then I is finite. Hence, K_2/K_1 is finitely generated, so $\mathrm{trans.deg}(K_2/K_0) < \aleph_0 = \mathrm{card}(K)$. If $\mathrm{card}(K) > \aleph_0$, then trans.deg$(K_2/K_0) \leq$ trans.deg$(K_1/K_0) + \aleph_0 \cdot$ card$(I) <$ card(K). Thus,

(1)   trans.deg$(K_2/K_0) <$ card(K)

in each case.

For each $\mathbf{a} \in A(K)$ there exists a positive integer n such that $n\mathbf{a}$ belongs to the group generated by the $\mathbf{a}_i$'s, hence to $A(K_2)$. Since $n_A^{-1}(A(K_2)) \subseteq A(\tilde{K}_2)$, we have $K_2(A(K)) \subseteq \tilde{K}_2$. On the other hand, $K_2(A(K)) = K$, by Lemma 5.4.3(b), hence $K \subseteq \tilde{K}_2$. It follows from (1) that trans.deg$(\tilde{K}_2/K_0) =$ trans.deg$(K_2/K_0) <$ card$(K) =$ trans.deg$(K/K_0) \leq$ trans.deg$(\tilde{K}_2/K_0)$. This contradiction proves that rr$(A(K)) =$ card(K). $\square$

Example 6.5.5: Examples of infinite algebraic extensions of $\mathbb{Q}$ that are non-ample.

(a) Let S be a finite set of prime numbers, E a modular elliptic curve defined over $\mathbb{Q}$, and L the maximal Abelian extension of $\mathbb{Q}$ that is at most ramified in S. By a deep theorem of Kato and Rohrlich [Gre01, Thm. 1.5], $E(L)$ is finitely generated. Hence, by Theorem 6.5.2, L is nonample. By Lemma 5.5.1(b), every subfield of L is nonample. Of course, we know now by the theorem of Breuil-Conrad-Diamond (that generalizes the Taylor-Wiles theorem) that every elliptic $\mathbb{Q}$-curve is modular. However, for our purpose it suffices to choose only one such modular curve. For example, we may choose $X_0(11)$ [Sil86, p. 355]. Recall that the latter curve parametrizes all equivalence classes of pairs (E, C), where E is an elliptic curve defined over $\mathbb{Q}$ with a subgroup C of order 11 invariant under the action of Gal$(\mathbb{Q})$ [Sil86, p. 354, Thm. 13.1(a)].

(b) Let p be a prime number and set $F = \mathbb{Q}(\zeta_{p^i})_{i=1,2,3,\ldots}$. Then Gal$(F/\mathbb{Q}) \cong A \times \mathbb{Z}_p$, where A is a finite group (isomorphic to $\mathbb{Z}/2\mathbb{Z}$ if $p = 2$ or to $\mathbb{Z}/(p-1)\mathbb{Z}$ if $p \neq 2$). Let L be the fixed field of A in F. Then, Gal$(L/\mathbb{Q}) \cong \mathbb{Z}_p$. For each number field K, Gal(LK/K) is isomorphic to an open subgroup of $\mathbb{Z}_p$, hence to $\mathbb{Z}_p$ [FrJ08, Lemma 1.4.2(e)]. The field LK is called the **cyclotomic $\mathbb{Z}_p$-extension** of K.

The only prime number that ramifies in L is p [CaF67, p. 87, Lemma 4]. In addition, only finitely many prime numbers ramify in K. Hence, only finitely many prime numbers ramify in LK. If, in addition, $K/\mathbb{Q}$ is Abelian, then by (a), LK is nonample. $\square$

Our next goal is to prove that the inverse of Theorem 6.5.2 is not true.

LEMMA 6.5.6: *Let K be a field, C an absolutely irreducible K-curve of absolute genus at least 2, and A an Abelian variety defined over K. Denote the function field of A over K by F. Then $C(K) = C(F)$.*

Proof: Assume there is a point $\mathbf{p} \in C(F) \smallsetminus C(K)$. Then $\mathbf{p} \in C(\tilde{F}) \smallsetminus C(\tilde{K})$ (because K is algebraically closed in F). Hence, making a base change from K to $\tilde{K}$, if necessary, we may assume that K is algebraically closed.

Next let C' be a projective closure of C. Then $\mathbf{p} \in C'(F) \smallsetminus C'(K)$ (because $K(\mathbf{p})$ is a transcendental extension of K). Finally, let $\pi: C'' \to C'$

be the projective normalization of C'. Since $\mathbf{p}$ is a generic point of C over K, there exists a unique point $\mathbf{p}'' \in C''(F) \smallsetminus C''(K)$ with $\pi(\mathbf{p}'') = \mathbf{p}$. Replacing C by C'', if necessary, we may assume that C is a smooth projective curve over K.

The inclusion $K(\mathbf{p}) \subseteq F$ defines a dominant rational map $\alpha\colon A \to C$ defined over K. It is defined at some point $\mathbf{a} \in A(K)$. Replacing α by α' defined by $\alpha'(\mathbf{x}) = \alpha(\mathbf{a} + \mathbf{x})$, if necessary, we may assume that α is defined at the zero point $\mathbf{o}_A$ of A.

Set $\mathbf{o} = \alpha(\mathbf{o}_A)$. Let J be the Jacobian variety of C over K and let $\gamma\colon C \to J$ be the canonical map satisfying $\gamma(\mathbf{o}) = \mathbf{o}_J$. Then $\theta = \gamma \circ \alpha$ is a rational map from A to J with $\theta(\mathbf{o}_A) = \mathbf{o}_J$. By Subsection 6.3.1, θ is a homomorphism of Abelian varieties. In particular, $B = \theta(A)$ is an Abelian subvariety of J defined over K. In addition, B is birationally equivalent to $\gamma(C)$. On the other hand, by the second paragraph of Subsection 6.3.2, $\gamma(C)$ is isomorphic to C. It follows that B is an absolutely irreducible curve over K of genus at least 2 as well as an Abelian variety. This contradicts a consequence of Subsection 6.3.2 and proves that $C(F) = C(K)$, as contended. $\square$

PROPOSITION 6.5.7: *Let K be either a number field or a finite field. Then K has an extension F such that F is nonample but $\mathrm{rr}(A(F)) = \infty$ for every Abelian variety A of positive dimension defined over F.*

Proof: We choose an absolutely irreducible K-curve C of absolute genus at least ≥ 2 with a K-rational simple point. For example we may take C to be a Fermat curve appearing in the proof of Proposition 6.1.5. We construct for every extension L of K and for every Abelian variety A of positive dimension defined over L an extension L' such that $C(L) = C(L')$ and $\mathrm{rr}(A(L')) = \infty$. Moreover, if L is countable, then so is L'.

By induction we choose a sequence $\mathbf{x}_1, \mathbf{x}_2, \mathbf{x}_3, \ldots$ of algebraically independent generic points of A over L. For each $n \geq 0$ let $L_n = L(\mathbf{x}_1, \ldots, \mathbf{x}_n)$. Then we set $L' = L(\mathbf{x}_1, \mathbf{x}_2, \mathbf{x}_3, \ldots)$. By Lemma 6.5.6, $C(L_{n-1}) = C(L_n)$ for each $n \geq 1$. Hence, $C(L) = C(L')$. Moreover, the sequence $\mathbf{x}_1, \mathbf{x}_2, \mathbf{x}_3, \ldots$ of points of $A(L')$ is linearly independent. Otherwise, there exist $k_1, \ldots, k_n \in \mathbb{Z}$ such that $k_n \neq 0$ and $k_1\mathbf{x}_1 + \cdots + k_{n-1}\mathbf{x}_{n-1} = k_n\mathbf{x}_n$. By Subsection 6.3.1, $\mathbf{x}_n \in A(\tilde{L}_{n-1})$. Since $L(\mathbf{x}_n)$ is a regular extension of L and $L(\mathbf{x}_n)$ is algebraically independent from $\tilde{L}_{n-1}$ over L, the extensions $L(\mathbf{x}_n)$ and $\tilde{L}_{n-1}$ are linearly disjoint over L. It follows that $\mathbf{x}_n \in A(L)$. This contradicts the fact that $\dim(A) > 0$.

Next we list all of the Abelian varieties of positive dimension defined over K as $A_1, A_2, A_3, \ldots$. Using the construction of the preceding paragraph, we construct an ascending sequence of fields $K = K_0 \subset K_1 \subset K_2 \subset \cdots$ such that $C(K_n) = C(K_{n-1})$ and $\mathrm{rr}(A_n(K_n)) = \infty$ for each $n \geq 1$. The field $K^{(1)} = \bigcup_{n=0}^{\infty} K_n$ satisfies $C(K^{(1)}) = C(K)$ and $\mathrm{rr}(A(K^{(1)})) = \infty$ for each Abelian variety A of positive dimension defined over K.

Finally, we continue by induction to construct an ascending sequence of fields $K = K^{(0)} \subset K^{(1)} \subset K^{(2)} \cdots$ such that $C(K^{(n)}) = C(K^{(n-1)})$ and $\mathrm{rr}(A(K^{(n)})) = \infty$ for every Abelian variety A of positive dimension defined over $K^{(n-1)}$. Let $F = \bigcup_{n=0}^{\infty} K^{(n)}$. Then $C(F) = C(K)$ is a finite set (Proposition 6.2.4) containing a simple K-rational (hence F-rational) point. Thus, F is nonample. Moreover, every Abelian variety A of positive dimension defined over F is already defined over $K^{(n)}$ for some n. Hence, $\mathrm{rr}(A(K^{(n+1)})) = \infty$, so $\mathrm{rr}(A(F)) = \infty$, as desired.   □

CONJECTURE 6.5.8: *Let K be a field such that $\mathrm{Gal}(K)$ is finitely generated. Then:*
(a) *If K is infinite, then K is ample.*
(b) *If K is not an algebraic extension of a finite field and A is a nonzero Abelian variety defined over K, then $\mathrm{rr}(A(K)) = \infty$.*

Remark 6.5.9: The reason for making Conjecture 6.5.8(a) is that each of the known fields of the form $\tilde{\mathbb{Q}}(\sigma)$, with $\sigma \in \mathrm{Gal}(\mathbb{Q})^e$ belongs to one of the families of ample fields appearing in Section 5.6.

Conjecture 6.5.8(b) is stated as [Lar03, Question 1]. It is proved in [ImL08] in the case where $\mathrm{Gal}(K)$ is procyclic and $\mathrm{char}(K) \neq 2$. In the general case Conjecture 6.5.8(b) follows from Conjecture 6.5.8(a) and Theorem 6.5.2 in the case where $\mathrm{char}(K) = 0$ and Remark in general.   □

The following conjecture generalizes Example 6.5.5(b).

CONJECTURE 6.5.10: *The maximal pro-p extension of a number field is nonample.*

6.6 On the Gonality of a Function Field of One Variable

The **gonality** of a function field F/K of one variable is defined to be the minimal number $\mathrm{gon}(F/K)$ of all degrees $[F : K(x)]$ with $x \in F \smallsetminus K$. We prove in this section that if K is perfect and F/K has a prime divisor of degree 1, then $\mathrm{gon}(F\tilde{K}/\tilde{K}) \geq \sqrt{\mathrm{gon}(F/K)}$ (Proposition 6.6.12(c)) and use this inequality to construct for each perfect field K and each positive integer d a function field F over K such that $\mathrm{gon}(F\tilde{K}/\tilde{K}) \geq d$.

LEMMA 6.6.1: *The following holds for a function field F/K of one variable.*
(a) $\mathrm{gon}(F/K) = 1$ *if and only if F is a rational function field over K.*
(b) $\mathrm{gon}(F/K) \leq d$ *if and only if F/K has a divisor $\mathfrak{a}$ with $\dim(\mathfrak{a}) \geq 2$ and $\deg(\mathfrak{a}) \leq d$.*

Proof of (a): We have $\mathrm{gon}(F/K) = 1$ if and only if $F = K(x)$ for some $x \in F$.

Proof of (b): If $[F : K(x)] \leq d$, then $\mathfrak{a} = \mathrm{div}_{\infty}(x)$ satisfies $\deg(\mathfrak{a}) = [F : K(x)] \leq d$ (Remark 5.8.1) and $1, x$ are K-linearly independent elements of $\mathcal{L}(\mathfrak{a})$, so $\dim(\mathfrak{a}) \geq 2$.

Conversely, suppose $\mathfrak{a}$ is a divisor of F/K with $\dim(\mathfrak{a}) \geq 2$ and $\deg(\mathfrak{a}) \leq d$. Replacing $\mathfrak{a}$ by a linearly equivalent divisor, we may assume that $\mathfrak{a} \geq 0$

(Remark 5.8.1(a)). Then $\mathcal{L}(\mathfrak{a})$ contains an element $x \notin K$ and this element satisfies $\mathrm{div}_\infty(x) \leq \mathfrak{a}$ (again by Remark 5.8.1(a)). Hence, $[F : K(x)] = \deg(\mathrm{div}_\infty(x)) \leq \deg(\mathfrak{a}) \leq d$, as desired. $\qquad\square$

LEMMA 6.6.2: *Suppose a function field F/K of one variable has a prime divisor of degree 1. Then $\mathrm{gon}(F/K) = 2$ if and only if F/K is elliptic or hyperelliptic.*

Proof: By definition, F/K is **elliptic** (resp. **hyperelliptic**) if $\mathrm{genus}(F/K) = 1$ (resp. $\mathrm{genus}(F/K) \geq 2$), and F has a subfield E containing K such that $[F : E] = 2$ and $\mathrm{genus}(E/K) = 0$. In this case, the restriction of the prime divisor of degree 1 of F/K to E is a prime divisor of degree 1 of E/K [FrJ08, Example 3.2.4], hence E/K is a field of rational functions. Thus, $\mathrm{gon}(F/K) = 2$.

Conversely, if $\mathrm{gon}(F/K) = 2$, then F is nonrational (so, $\mathrm{genus}(F/K) \geq 1$) and there exists $x \in F$ with $[F : K(x)] = 2$. Consequently, F/K is either elliptic or hyperelliptic. $\qquad\square$

LEMMA 6.6.3 (Monotony): *Let $K \subset E \subseteq F$ be a tower of fields such that F/K is a function field of one variable and $[F : E] < \infty$. Then*

$$\mathrm{gon}(E/K) \leq \mathrm{gon}(F/K) \leq [F : E]\mathrm{gon}(E/K).$$

Proof: We choose an element $z \in E \smallsetminus K$ with $[E : K(z)] = \mathrm{gon}(E/K)$. Then, $\mathrm{gon}(F/K) \leq [F : K(z)] = [F : E][E : K(z)] = [F : E]\mathrm{gon}(E/K)$, which proves the right inequality.

In order to prove the left inequality we choose a $y \in F$ with $[F : K(y)] = \mathrm{gon}(F/K)$. Set $x = \mathrm{norm}_{F/E}(y)$. Then, by [Deu73, p. 109, (3)], $\mathrm{gon}(E/K) \leq [E : K(x)] = \deg(\mathrm{div}_{E/K,\infty}(x)) = \deg(\mathrm{div}_{F/K,\infty}(y)) = [F : K(y)] = \mathrm{gon}(F/K)$. $\qquad\square$

LEMMA 6.6.4 (Constant field extension): *Let F/K be a function field of one variable and L a finite extension of K. Then $\mathrm{gon}(FL/L) \leq \mathrm{gon}(F/K) \leq [L : K]\mathrm{gon}(FL/L)$.*

Proof: We choose an $x \in F$ with $[F : K(x)] = \mathrm{gon}(F/K)$. Since F is a regular extension of K, F is linearly disjoint from $L(x)$ over $K(x)$. Hence, $\mathrm{gon}(FL/L) \leq [FL : L(x)] = [F : K(x)] = \mathrm{gon}(F/K)$.

In order to prove the right hand inequality we choose a $y \in FL$ with $\mathrm{gon}(FL/L) = [FL : L(y)]$ and set $x = \mathrm{norm}_{FL/F}(y)$. Then, by [Che51, p. 66, Lemma 2],

$$\mathrm{gon}(F/K) \leq [F : K(x)] = \deg(\mathrm{div}_{F/K,\infty}x) = [L : K]\deg(\mathrm{div}_{FL/L,\infty}y)$$
$$= [L : K][FL : L(y)] = [L : K]\mathrm{gon}(FL/L),$$

as claimed. $\qquad\square$

We use Riemann-Roch to give some basic upper bounds for $\mathrm{gon}(F/K)$.

LEMMA 6.6.5: *Let F/K be a function field of one variable of genus g.*
(a) *If $g = 0$, then $\mathrm{gon}(F/K) \leq 2$.*
(b) *If $g \geq 2$, then $\mathrm{gon}(F/K) \leq 2g - 2$.*
(c) *If F/K has a prime divisor $\mathfrak{p}$ of degree 1, then $\mathrm{gon}(F/K) \leq g + 1$.*
(d) *If F/K has a prime divisor $\mathfrak{p}$ of degree 1 and $g \geq 2$, then $\mathrm{gon}(F/K) \leq g$.*

Proof: Let $\mathfrak{w}$ be a canonical divisor of F/K. Then $\deg(\mathfrak{w}) = 2g - 2$ and $\dim(\mathfrak{w}) = g$ (Remark 5.8.1).

Proof of (a): In this case, $\deg(-\mathfrak{w}) = -\deg(\mathfrak{w}) = 2$. Hence, by Riemann-Roch, $\dim(-\mathfrak{w}) = 3$. It follows from Lemma 6.6.1(b), that $\mathrm{gon}(F/K) \leq 2$.

Proof of (b): In this case $\dim(\mathfrak{w}) \geq 2$. Hence, by Lemma 6.6.1(b), $\mathrm{gon}(F/K) \leq 2g - 2$.

Proof of (c): We have $\deg((g+1)\mathfrak{p}) = g+1$ and $\dim((g+1)\mathfrak{p}) \geq g+1+1-g = 2$. Hence, by Lemma 6.6.1(b), $\mathrm{gon}(F/K) \leq g + 1$.

Proof of (d): We consider the divisor $\mathfrak{a} = \mathfrak{w} - (g-2)\mathfrak{p}$. It satisfies $\deg(\mathfrak{a}) = 2g - 2 - (g-2) = g$ and $\mathfrak{w} - \mathfrak{a} = (g-2)\mathfrak{p}$. Since $1 = \dim(0) \leq \dim((g-2)\mathfrak{p})$, Riemann-Roch asserts that $\dim(\mathfrak{a}) = g + 1 - g + \dim((g-2)\mathfrak{p}) \geq 2$. Hence, by Lemma 6.6.1(b), $\mathrm{gon}(F/K) \leq g$. □

Remark 6.6.6: The results stated in the following comments are much deeper than those proven so far in this section, but they will not be used in this book.

(a) In contrast to the cases $g = 0$ and $g \geq 2$, the gonality of function fields of one variable with genus 1 is not bounded from above. Indeed, at the end of the introduction to [LaT58] the authors mention that for each positive integer d there is a number field K and a function field F of one variable over K with genus 1 such that the degree of each divisor of F/K is a multiple of d. Thus, by Lemma 6.6.1(b), $\mathrm{gon}(F/K) \geq d$.

(b) A theorem of Kempf-Kleiman-Laksov improves Lemma 6.6.5(c) in the case where K is algebraically closed. It says that in this case, $\mathrm{gon}(F/K) \leq \frac{g+3}{2}$ [KlL72].

(c) For each positive integer g there exists a variety M_g called the "moduli space of curves of genus g".

A theorem named after Brill-Noether-Kleiman-Laksov gives a lower bound for the gonality in the case where K is algebraically closed. Given a positive integer g, there exists a nonempty Zariski-open subset U of the moduli space M_g of curves of genus g such that for each curve Γ corresponding to a point of $U(K)$ the function field F of Γ over K satisfies $\mathrm{gon}(F/K) \geq \frac{1}{2}g+1$. For a proof see Laksov's appendix to [Kle76]. □

We could use Remark 6.6.6(c) in order to find function fields of one variable over number fields with high gonalities. But we prefer to present a result of Poonen proved by a much more elementary arguments that serves the same purpose.

PROPOSITION 6.6.7 (Castelnuovo-Severi inequality [Sti93, p. 130, Thm. III.10.3]): *Let F, F_1, F_2 be function fields of one variable over a perfect field K with genera g, g_1, g_2, respectively, such that $F = F_1 F_2$. Then*

$$g \leq [F : F_1]g_1 + [F : F_2]g_2 + ([F : F_1] - 1)([F : F_2] - 1).$$

Definition 6.6.8: Let F/K be a function field of one variable and d a positive integer. A function field of one variable E over K with $E \subseteq F$ is d-**controlled by** F if d has a divisor e such that $[F : E] = \frac{d}{e}$ and $\mathrm{genus}(E/K) \leq (e-1)^2$. $\square$

LEMMA 6.6.9: *In the notation of Definition 6.6.8 suppose that K is perfect and the function field E is d-controlled by F. Then $E(y)$ is also d-controlled by F for each $y \in F$ with $[F : K(y)] = d$.*

Proof: We consider an e as in Definition 6.6.8 and set $f = [E(y) : E]$. Then $[F : E(y)] = \frac{d}{ef}$ and $[E(y) : K(y)] = ef$. Hence, by Proposition 6.6.7,

$$\mathrm{genus}(E(y)/K) \leq ef \cdot 0 + f \cdot \mathrm{genus}(E/K) + (ef - 1)(f - 1)$$
$$\leq f(e-1)^2 + (ef - 1)(f - 1) \leq (ef - 1)^2,$$

where the latter inequality is simplified to $0 \leq (e-1)(f-1)$. Thus, $E(y)$ is d-controlled by F. $\square$

LEMMA 6.6.10: *Let F be a function field of one variable over a perfect field K. Then $K(y_1, \ldots, y_n)$ is d-controlled by F for all $y_1, \ldots, y_n \in F$ with $[F : K(y_i)] = d$, $i = 1, \ldots, n$.*

Proof: For $n = 1$ we have $[F : K(y_1)] = \frac{d}{1}$ and $\mathrm{genus}(K(y_1)/K) = 0 \leq (1-1)^2$. Hence, $K(y_1)$ is d-controlled by F. The case where $n > 1$ follows now by induction from Lemma 6.6.9. $\square$

LEMMA 6.6.11: *Let F be a function field of one variable over a perfect field K, L a Galois extension of K, and $d = \mathrm{gon}(FL/L)$. Then F has a d-controlled subfield.*

Proof: By assumption, FL has an element y with $[FL : L(y)] = d$. Let $y_1, \ldots, y_n$ be all of the conjugates of y over F. Thus, for each i there exists $\sigma \in \mathrm{Gal}(FL/F)$ with $y_i = y^\sigma$. Since L is Galois over K, we have $L(y)^\sigma = L(y_i)$, so $[FL : L(y_i)] = [FL : L(y)] = d$. Hence, by Lemma 6.6.10, the field $E' = L(y_1, \ldots, y_n)$ is d-controlled by FL. Thus, d has a divisor e such that $[FL : E'] = \frac{d}{e}$ and $\mathrm{genus}(E'/L) \leq (e-1)^2$.

Let $G = \mathrm{Gal}(FL/F)$. By construction, E' is G-invariant. Let E be the fixed field of E' under G. Then $F \cap E' = E$ and $FE' = FL$, so $[F : E] = [FL : E'] = \frac{d}{e}$. Since F/K is regular, the restriction of G to L is a bijection onto $\mathrm{Gal}(L/K)$. Hence, $K \subseteq E$ and $EL = E'$. Therefore, $\mathrm{genus}(E/K) = \mathrm{genus}(E'/L)$ (Remark 5.8.1). By the preceding paragraph, $\mathrm{genus}(E/K) \leq (e-1)^2$. Consequently, E is d-controlled by F. $\square$

All of these lemmas lead to the inequality for the gonality under constant field extensions.

PROPOSITION 6.6.12 (Poonen): *Let F be a function field of one variable over a perfect field K with a prime divisor of degree 1. Consider an algebraic extension L of K and set $d = \mathrm{gon}(FL/L)$.*
(a) *If $d \le 2$, then $\mathrm{gon}(F/K) = d$.*
(b) *If $d > 2$, then $\mathrm{gon}(F/K) \le (d-1)^2$.*
(c) *In any case, $\mathrm{gon}(FL/L) \ge \sqrt{\mathrm{gon}(F/K)}$.*

Proof of (a): If $d = 1$, then FL is a rational function field over L, so $\mathrm{genus}(F/K) = \mathrm{genus}(FL/L) = 0$. Since F/K has a prime divisor of degree 1, F is also a rational function field over K [FrJ08, Example 3.2.4], so $\mathrm{gon}(F/K) = 1$.

If $d = 2$, then by Lemma 6.6.2, FL/L is elliptic or hyperelliptic.

If FL/L is elliptic, then $\mathrm{genus}(F/K) = \mathrm{genus}(FL/L) = 1$. Since F/K has a prime divisor $\mathfrak{p}$ of degree 1, it is also elliptic. Indeed, by Riemann-Roch, $\dim(\mathfrak{p}) = 1$ and $\dim(2\mathfrak{p}) = 2$, so we may choose an $x \in \mathcal{L}(2\mathfrak{p}) \smallsetminus \mathcal{L}(\mathfrak{p})$. Then $\mathrm{div}_\infty(x) = 2\mathfrak{p}$, so $[F : K(x)] = 2$. Consequently $\mathrm{gon}(F/K) = 2$.

Now we suppose FL/L is hyperelliptic. Again, $g = \mathrm{genus}(F/K) = \mathrm{genus}(FL/L)$. We choose a canonical divisor $\mathfrak{w}$ of F/K. Then $\deg(\mathfrak{w}) = 2g - 2$ and $\dim(\mathfrak{w}) = g$ (Remark 5.8.1). Since L/K is a separable extension, $\dim_{FL/L}(\mathfrak{w}) = \dim_{F/K}(\mathfrak{w}) = g$. Hence, $\mathfrak{w}$ is also a canonical divisor of FL/L [Sti93, p. 30, Prop. I.6.2]. Moreover, a base $x_1, \ldots, x_g$ of $\mathcal{L}_{F/K}(\mathfrak{w})$ over K is also a base of $\mathcal{L}_{FL/L}(\mathfrak{w})$ over L. By [FrJ08, Prop. 3.7.4], $E' = L(\frac{x_2}{x_1}, \ldots, \frac{x_g}{x_1})$ is the only subfield of FL containing L with $[FL : E'] = 2$ and $\mathrm{genus}(E'/L) = 0$. Set $E = K(\frac{x_2}{x_1}, \ldots, \frac{x_g}{x_1})$. Then $EL = E'$. Hence, $\mathrm{genus}(E/K) = \mathrm{genus}(E'/L) = 0$. Since F/K has a prime divisor of degree 1, so does E/K, hence E is a rational function field over K. Finally, since F/K is regular, we have $[F : E] = [FL : E'] = 2$. Consequently, $\mathrm{gon}(F/K) = 2$.

Proof of (b): Let $\hat{L}$ be the Galois closure of L/K. By Lemma 6.6.4, $\mathrm{gon}(F\hat{L}/\hat{L}) \le \mathrm{gon}(FL/L)$. We may therefore replace L by $\hat{L}$, if necessary, to assume that L/K is Galois.

By Lemma 6.6.11, K has an extension E in F and d has a divisor e such that $[F : E] = \frac{d}{e}$ and $g = \mathrm{genus}(E/K) \le (e-1)^2$. Since F has a prime divisor of degree 1, so has E. Thus, if $g = 0$, then E is rational and $\mathrm{gon}(F/K) \le \frac{d}{e} \le d < (d-1)^2$.

If $g = 1$, then $e \ge 2$. Moreover, E/K is elliptic and $\mathrm{gon}(E/K) = 2$ (Lemma 6.6.2). Hence, by Lemma 6.6.3, $\mathrm{gon}(F/K) \le \frac{d}{e}\mathrm{gon}(E/K) \le \frac{d}{2}2 = d < (d-1)^2$.

If $g \ge 2$, then $\mathrm{gon}(E/K) \le g$ (Lemma 6.6.5(d)). Hence, by Lemma 6.6.3, $\mathrm{gon}(F/K) \le [F : E]\mathrm{gon}(E/K) \le \frac{d}{e}(e-1)^2 \le (d-1)^2$, where the latter inequality holds because $1 \le e \le d$.

Proof of (c): If $d \le 2$, then by (a), $\mathrm{gon}(F/K) = d \le d^2$. If $d > 2$, then by (b), $\mathrm{gon}(F/K) \le (d-1)^2 < d^2$. In both cases $\mathrm{gon}(F/K) \le \mathrm{gon}(FL/L)^2$. $\square$

6.7 Gonality under Constant Reduction

The aim of this section is to construct explicit function fields over $\mathbb{F}_p$ with large gonalities and then to lift them to function fields over $\mathbb{Q}$ with large gonalities.

Definition 6.7.1: Let F/K be a function field of one variable and v a valuation of F. We use a bar to denote the reduction of objects at v and say that F/K has a **regular constant reduction at** v if $\bar{F}/\bar{K}$ is also a function field of one variable. □

LEMMA 6.7.2: *Let F/K be a function field of one variable with a regular constant reduction at a valuation v of F. Then $\mathrm{gon}(F/K) \geq \mathrm{gon}(\bar{F}/\bar{K})$.*

Proof: We choose an $x \in F$ with $\mathrm{gon}(F/K) = [F : K(x)]$. By assumption, $\bar{F}/\bar{K}$ is a function field of one variable. Since $[\bar{F} : \overline{K(x)}] \leq [F : K(x)] < \infty$, there exist elements $y \in K(x) \smallsetminus K$ such that $\bar{y} \in \overline{K(x)} \smallsetminus \bar{K}$. Let y_0 be an element among them with $n = [K(x) : K(y_0)]$ minimal. We write $y_0 = \frac{f}{g}$ with $f = \sum_{i=0}^r a_i x^i$, $g = \sum_{j=0}^s b_j x^j$, where $a_i, b_j \in K$ for all i, j and $a_r, b_s \neq 0$. By [FrJ08, Example 3.2.4], $n = \max(r, s)$. We define an element $y \in K(y_0)$ by cases and check that in each case
(1a) $K(y) = K(y_0)$, $v(y) = 0$, $\bar{y} \in \overline{K(x)} \smallsetminus \bar{K}$, and
(1b) x is a root of a monic polynomial $p \in K[y][X]$ of degree n.

CASE A: If $r > s$, we set $y = y_0$. Then $r = n$ and $\sum_{i=0}^s (a_i - y b_i) x^i + \sum_{i=s+1}^n a_i x^i = 0$. Dividing both sides by a_r, we verify (1).

CASE B: If $r < s$, we set $y = y_0^{-1}$. Exchanging the roles of f and g, we find again that (1) holds.

CASE C: If $r = s$, then $f = cg + h$, where $c = \frac{a_n}{b_n} \in K^\times$ and $h \in K[x]$ has degree less than n. This case splits into two subcases.

CASE C1: If $v(c) \geq 0$, we set $y = \frac{g}{h} = \frac{1}{y_0 - c}$. Then $K(y) = K(y_0)$ and $\bar{y}_0 \neq \bar{c}$, because $\bar{y}_0 \notin \bar{K}$. Thus, $\bar{y} \in \overline{K(x)} \smallsetminus \bar{K}$. As in case A, also (1b) holds.

CASE C2: If $v(c) < 0$, we set $y = \frac{cf}{h} = \left(c^{-1} - y_0^{-1}\right)^{-1}$ and note that $\bar{c}^{-1} \neq \bar{y}_0^{-1}$, hence $\bar{y} \in \overline{K(x)} \smallsetminus \bar{K}$. The verification of (1b) proceeds as in Case A.

Having chosen y as in (1), we may present each $q \in K[X]$ as $q(X) = q_1(X)p(X) + q_2(X)$ with $q_1, q_2 \in K[y][X]$ and $\deg_X(q_2) < n$. Substituting x for X in the latter equation, we conclude that each element of $K[x]$ is a polynomial in x of degree at most $n - 1$ with coefficients in $K[y]$. The latter expression may be rewritten as a polynomial in y with coefficients in $K[x]$ of degree at most $n - 1$.

Now we consider an element $z \in K(x)^\times$. By the preceding paragraph there exist $u_0, \ldots, u_m, w_0, \ldots, w_m \in K[x]$ of degree at most $n - 1$ such that

$$(2) \qquad z = \frac{u_0 + u_1 y + \cdots + u_m y^m}{w_0 + w_1 y + \cdots + w_m y^m}.$$

Dividing the numerator and the denominator of the right hand side of (2) by an element having the least value among the coefficients, we may assume that $v(u_i), v(w_j) \geq 0$ for all i, j. Moreover, $v(u_i) = 0$ for at least one i or $v(w_j) = 0$ for at least one j. Since $[K(x) : K(u_i)] = \deg(u_i) \leq n - 1$, the minimality of n implies that $\bar{u}_i \in \bar{K}$ for each i. Similarly, $\bar{w}_j \in \bar{K}$ for each j. Since $\bar{y}$ is transcendental over $\bar{K}$, this implies that $\bar{u}_0 + \bar{u}_1 \bar{y} + \cdots + \bar{u}_m \bar{y}^m$ is not zero unless all of the coefficients are zero. Similarly, $\bar{w}_0 + \bar{w}_1 \bar{y} + \cdots + \bar{w}_m \bar{y}^m$ is not zero unless each of the coefficients is zero. Therefore, in the expression

$$\bar{z} = \frac{\bar{u}_0 + \bar{u}_1 \bar{y} + \cdots + \bar{u}_m \bar{y}^m}{\bar{w}_0 + \bar{w}_1 \bar{y} + \cdots + \bar{w}_m \bar{y}^m}$$

both the numerator and the denominator belong to $\bar{K}[\bar{y}]$ and at least one of them is not zero. It follows that $\bar{z} \in \bar{K}(\bar{y}) \cup \{\infty\}$. Consequently, $\overline{K(x)} = \bar{K}(\bar{y})$.

By the latter consequence

$$\mathrm{gon}(\bar{F}/\bar{K}) \leq [\bar{F} : \bar{K}(\bar{y})] = [\bar{F} : \overline{K(x)}] \leq [F : K(x)] = \mathrm{gon}(F/K),$$

as claimed. $\square$

We supply an explicit computation of the gonality over $\mathbb{F}_p$ and over $\mathbb{Q}$ but first we need a general lemma.

LEMMA 6.7.3: *Let K be a field and $f, g \in K[X]$ polynomials with relatively prime degrees. Let x, y be elements in a field extension of K such that x is transcendental over K and $g(y) = f(x)$.*
(a) *The infinite valuation $v_{x,\infty}$ of $K(x)/K$ is totally ramified in $K(x,y)$, $[K(x,y) : K(x)] = \deg(g)$, $K(x,y)/K$ is a function field of one variable, and the polynomial $g(Y) - f(X)$ is absolutely irreducible.*
(b) *Let v be a valuation of $K(x)$ such that $\bar{x}$ is transcendental over $\bar{K}$ (where we use a bar to denote reduction at v). Then $v(a_m x^m + \cdots + a_0) = \min(v(a_m), \ldots, v(a_0))$ for all $a_m, \ldots, a_0 \in K$. Moreover, $\overline{K(x)} = \bar{K}(\bar{x})$.*
(c) *Let v be a valuation of $K(x,y)$ such that $\bar{x}$ is transcendental over $\bar{K}$, $\deg(f) = \deg(\bar{f})$ and $\deg(g) = \deg(\bar{g})$ (where we use a bar to denote reduction at v). Then $\bar{K}(\bar{x}, \bar{y})/\bar{K}$ is a function field of one variable, $\overline{K(x,y)} = \bar{K}(\bar{x}, \bar{y})$, and $\mathrm{gon}(\bar{K}(\bar{x}, \bar{y})/\bar{K}) \leq \mathrm{gon}(K(x,y)/K)$.*

Proof of (a): Let w be an extension of $v_{x,\infty}$ to $K(x,y)$ and e the ramification index of w over $K(x)$. Then $v_{x,\infty}(x) = -1$, $w(f(x)) = -\deg(f)e$, and $w(y) < 0$, so $w(g(y)) = \deg(g)w(y)$. Thus, $\deg(g)w(y) = -\deg(f)e$. Since $\gcd(\deg(f), \deg(g)) = 1$, we have $\deg(g)|e$. On the other hand, $e \leq [K(x,y) : K(x)] \leq \deg(g)$. Therefore, $e = \deg(g) = [K(x,y) : K(x)]$. In particular $w/v_{x,\infty}$ is totally ramified.

The same argument, now applied to $\tilde{K}$ rather than to K, proves that $[\tilde{K}(x,y) : \tilde{K}(x)] = \deg(g)$, hence $[\tilde{K}(x,y) : \tilde{K}(x)] = [K(x,y) : K(x)]$. This implies that $K(x,y)/K$ is a regular extension, hence a function field of one variable. In addition, we find that $g(Y) - f(X)$ is absolutely irreducible.

Proof of (b): Let $a_0, \ldots, a_m \in K$ and set $f(X) = a_m X^m + \cdots + a_0$. Let a be one of the a_i's with a minimal value. For each i we set $b_i = \frac{a_i}{a}$. Then $v(b_i) \geq 0$ for each i and one of the b_i's is 1. Since $\bar{x}$ is transcendental over $\bar{K}$ this implies that $\bar{b}_m \bar{x}^m + \cdots + \bar{b}_0 \neq 0$, hence $v(b_m x^m + \cdots + b_0) = 0$. It follows from the relation $f(x) = a(b_m x^m + \cdots + b_0)$ that $v(f(x)) = v(a)$, as claimed.

Now we consider an element

$$(3) \qquad z = \frac{c_k x^k + \cdots + c_0}{d_l x^l + \cdots + d_0}$$

of $K(x)$ with $c_i, d_j \in K$ and $d_j \neq 0$ for at least one j. Assuming that $\bar{z} \neq \infty$, we conclude from the preceding paragraph that $\min(v(d_l), \ldots, v(d_0)) \leq \min(v(c_k), \ldots, v(c_0))$. Dividing the numerator and denominator of (3) by one of the d_j's with the least value, we may assume that $v(c_i), v(d_j) \geq 0$ for all i, j and one of the d_j's is 1. Thus,

$$\bar{z} = \frac{\bar{c}_k \bar{x}^k + \cdots + \bar{c}_0}{\bar{d}_l \bar{x}^l + \cdots + \bar{d}_0} \in \bar{K}(\bar{x}).$$

Since the inclusion $\bar{K}(\bar{x}) \subseteq \overline{K(x)}$ is clear, we conclude that $\overline{K(x)} = \bar{K}(\bar{x})$.

Proof of (c): By (b), $\overline{K(x)} = \bar{K}(\bar{x})$. Since $\bar{K}(\bar{x}, \bar{y}) \subseteq \overline{K(x,y)}$, we have, by (a) applied both to K and to $\bar{K}$, that

$$\deg(\bar{g}) = [\bar{K}(\bar{x}, \bar{y}) : \bar{K}(\bar{x})] \leq [\overline{K(x,y)} : \overline{K(x)}] \leq [K(x,y) : K(x)] = \deg(g).$$

Using the assumption $\deg(\bar{g}) = \deg(g)$, we find that $\bar{K}(\bar{x}, \bar{y}) = \overline{K(x,y)}$. In addition, by (a) applied to $\bar{K}$ rather than to K, $\bar{K}(\bar{x}, \bar{y})/\bar{K}$ is a function field of one variable. Hence, by Lemma 6.7.2, $\mathrm{gon}(\bar{K}(\bar{x}, \bar{y})/\bar{K}) \leq \mathrm{gon}(K(x,y)/K)$. □

LEMMA 6.7.4 (Fehm): *Let p be a prime number and n a positive integer. Let K be either $\mathbb{F}_p$ or $\mathbb{Q}$. Consider elements x, y in a field extension of K such that x is transcendental over K and $y^{p^n} - y - x^{p^n - 1} + 1 = 0$. Then $K(x,y)/K$ is a function field of one variable and $\mathrm{gon}(K(x,y)/K) = p^n - 1$.*

Proof: By Lemma 6.7.3(a), $K(x,y)/K$ is a function field of one variable. Moreover, $\mathrm{gon}(K(x,y)/K) \leq [K(x,y) : K(y)] \leq p^n - 1$. Thus, it suffices to prove that $p^n - 1 \leq \mathrm{gon}(K(x,y)/K)$.

CASE A: $K = \mathbb{F}_p$. Let $E = \mathbb{F}_{p^n}(x)$ and $F = \mathbb{F}_{p^n}(x,y)$. By Lemma 6.7.3(a), $F/\mathbb{F}_{p^n}$ is a function field of one variable and $[F : E] = p^n$. For each $a \in \mathbb{F}_{p^n}^\times$, the specialization $x \to a$ gives rise to a prime divisor $\mathfrak{p}_{x,a}$ of $E/\mathbb{F}_{p^n}$ of degree 1. For each $b \in \mathbb{F}_{p^n}$ we have $b^{p^n} - b = 0 = a^{p^n - 1} - 1$. Moreover, the partial derivative of $Y^{p^n} - Y - X^{p^n-1} + 1$ with respect to Y at (a,b) is -1. Hence, $\mathfrak{p}_{x,a}$ extends to a prime divisor of $F/\mathbb{F}_{p^n}$ of degree 1 in p^n ways. Thus, the number of prime divisors of $F/\mathbb{F}_{p^n}$ of degree 1 is at least $(p^n - 1)p^n$.

On the other hand, let $d = \mathrm{gon}(F/\mathbb{F}_{p^n})$ and let z be an element of F with $[F : \mathbb{F}_{p^n}(z)] = d$. Since $\mathbb{F}_{p^n}(z)/\mathbb{F}_{p^n}$ has exactly $p^n + 1$ prime divisors of degree 1 and each of them extends to F in at most d distinct ways, the number of prime divisors of $F/\mathbb{F}_{p^n}$ of degree 1 is at most $d(p^n + 1)$. Combining this estimate with the conclusion of the preceding paragraph, we get $p^n(p^n - 1) \leq d(p^n + 1)$. It follows that $p^n - 2 < \frac{p^n(p^n-1)}{p^n+1} \leq d$. Since d is an integer, $p^n - 1 \leq d$.

Finally, by Lemma 6.6.4, $\mathrm{gon}(\mathbb{F}_p(x,y)/\mathbb{F}_p) \geq \mathrm{gon}(\mathbb{F}_{p^n}(x,y)/\mathbb{F}_{p^n}) = d \geq p^n - 1$. Therefore, $\mathrm{gon}(\mathbb{F}_p(x,y)/\mathbb{F}_p) = p^n - 1$.

CASE B: $K = \mathbb{Q}$. Let v be the p-adic valuation of $\mathbb{Q}$. Applying Lemma 6.7.3(c) on the polynomials $f(X) = X^{p^n-1} - 1$ and $g(Y) = Y^{p^n} - Y$, we conclude from Case A that $\mathrm{gon}(\mathbb{Q}(x,y)/\mathbb{Q}) \geq \mathrm{gon}(\mathbb{F}_p(\bar{x},\bar{y})/\mathbb{F}_p) = p^n - 1$, where $\bar{x}$ and $\bar{y}$ are zeros of f and g considered as polynomials over $\mathbb{F}_p$. Therefore, $\mathrm{gon}(\mathbb{Q}(x,y)/\mathbb{Q}) = p^n - 1$. $\qquad\square$

We combine Lemma 6.7.4 with Proposition 6.6.12(c) to conclude that the absolute gonality of function fields over $\mathbb{F}_p$ and over $\mathbb{Q}$ is unbounded.

PROPOSITION 6.7.5: *Let p be a prime number and n a positive integer. Let K be either $\mathbb{F}_p$ or $\mathbb{Q}$. Consider elements x, y in a field extension of K such that x is transcendental over K and $y^{p^n} - y - x^{p^n-1} + 1 = 0$. Then for each algebraic extension L of K, $L(x,y)/L$ is a function field of one variable and $\mathrm{gon}(L(x,y)/L) \geq \sqrt{p^n - 1}$.*

Let $F/\mathbb{Q}$ be a function field of one variable. We prove in Lemma 8.3.1 that for almost all prime numbers p, reduction modulo p preserves the genus of $F/\mathbb{Q}$. Lemma 6.7.4 gives an example where the same phenomenon happens with the gonality. We do not know if this is true for an arbitrary function field $F/\mathbb{Q}$.

PROBLEM 6.7.6 (Fehm): *Let $F/\mathbb{Q}$ be a function field of one variable with a prime divisor of degree 1. Is $\mathrm{gon}(\bar{F}/\mathbb{F}_p) = \mathrm{gon}(F/\mathbb{Q})$ for almost all prime numbers p?*

6.8 Points of Degree d

We give more examples of infinite nonample extensions of $\mathbb{Q}$. We prove that for each positive integer d there exist infinitely many linearly disjoint extensions of $\mathbb{Q}$ of degree d whose compositum is nonample. The main ingredient of the proof is Faltings' theorem (Corollary 6.4.4(c)).

We consider a field K of characteristic 0 and an absolutely irreducible smooth projective curve C defined over K. Let F be the function field of C over K. Following Subsection 6.3.2, we identify the set $C(\tilde{K})$ with the set of prime divisors of $F\tilde{K}/\tilde{K}$. This identification is compatible with the action of $\mathrm{Gal}(K)$ from the left and preserves the degrees over K. It identifies $\mathrm{Div}(F\tilde{K}/\tilde{K})$ with the free Abelian additive group $\mathrm{Div}(C)$ generated by $C(\tilde{K})$. We say that a divisor $\mathbf{a}$ of C is K-**rational** if $\sigma\mathbf{a} = \mathbf{a}$ for all $\sigma \in \mathrm{Gal}(K)$.

Now we assume that C has a K-rational point $\mathbf{o}$ and let $\gamma\colon C \to J$ be a morphism such that $\gamma(\mathbf{o}) = \mathbf{o}_J$. Then γ extends to an epimorphism $\beta\colon \mathrm{Div}(C) \to J(\tilde{K})$ such that $\beta(\sum_{i=1}^{n} k_i\mathbf{p}_i) = \sum_{i=1}^{n} k_i\gamma(\mathbf{p}_i)$. The restriction β_0 of β to the subgroup $\mathrm{Div}_0(C)$ of divisors of degree 0 is also an epimorphism onto $J(\tilde{K})$ whose kernel is the group of principal divisors (see (1) of Section 6.3).

LEMMA 6.8.1: *Let $\mathbf{a}, \mathbf{b}$ be distinct nonnegative divisors of C of the same degree such that $\beta(\mathbf{a}) = \beta(\mathbf{b})$. Suppose $\mathbf{a}$ is K-rational. Then there exists $f \in F$ such that $[F : K(f)] \le \deg(\mathbf{a})$.*

Proof: By assumption, $\deg(\mathbf{b} - \mathbf{a}) = 0$ and $\beta_0(\mathbf{b} - \mathbf{a}) = 0$. By the remarks preceding this lemma, there exists $g \in (F\tilde{K})^{\times}$ such that $\mathbf{b} - \mathbf{a} = \mathrm{div}(g)$. Thus, $g \in \mathcal{L}_{F\tilde{K}}(\mathbf{a})$. Since $\mathbf{b} - \mathbf{a} \ne 0$, we have $\mathrm{div}(g) \ne 0$, so $g \notin \tilde{K}$. In addition, since $\mathbf{a} \ge 0$, each element of $\tilde{K}$ belongs to $\mathcal{L}_{F\tilde{K}}(\mathbf{a})$. Hence, $\dim(\mathcal{L}_{F\tilde{K}}(\mathbf{a})) \ge 2$. By Remark 5.8.1(e), $\dim(\mathcal{L}_F(\mathbf{a})) = \dim(\mathcal{L}_{F\tilde{K}}(\mathbf{a}))$. Hence, there exists $f \in \mathcal{L}_F(\mathbf{a}) \smallsetminus K$. Therefore, $\mathrm{div}_{\infty}(f) \le \mathbf{a}$ (Remark 5.8.1(a)), so $[F : K(f)] = \deg(\mathrm{div}_{\infty}(f)) \le \deg(\mathbf{a})$, as asserted. $\square$

Definition 6.8.2: *Symmetric products.* We fix a positive integer d and consider the cartesian product C^d of d copies of C. The dth **symmetric product** of C is the quotient $C^{(d)}$ of C^d by the symmetric group S_d of degree d. By [Ser88, p. 53], each point of $C^{(d)}(\tilde{K})$ can be identified with a nonnegative divisor $\sum_{i=1}^{d} \mathbf{p}_i$ of degree d. Under this identification there is a surjective morphism $\pi\colon C^d \to C^{(d)}$ such that $\pi(\mathbf{p}_1, \ldots, \mathbf{p}_d) = \sum_{i=1}^{d} \mathbf{p}_i$. This morphism commutes with the action of $\mathrm{Gal}(K)$ from the left, so $C^{(d)}$ and π are defined over K. A point $\sum_{i=1}^{d} \mathbf{p}_i$ belongs to $C^{(d)}(K)$ if and only if $\sum_{i=1}^{d} \sigma\mathbf{p}_i = \sum_{i=1}^{d} \mathbf{p}_i$ for each $\sigma \in \mathrm{Gal}(K)$, that is the divisor $\sum_{i=1}^{d} \mathbf{p}_i$ of C is K-rational. Finally we note, that since C is projective, absolutely irreducible, and smooth, so is $C^{(d)}$ [Ser88, p. 53]. $\square$

We denote the set of all elements of $\tilde{K}$ of degree at most d over K by $K^{(d)}$ and let $C(K^{(d)})$ be the union of all $C(L)$ with L ranging over all algebraic extensions of K of degree at most d.

LEMMA 6.8.3: *The set $C(K^{(d)})$ is infinite if and only if $C^{(d)}(K)$ is infinite.*

Proof: We define a map $\lambda\colon C(K^{(d)}) \to C^{(d)}(K)$ in the following way. Given $\mathbf{p} \in C(K^{(d)})$, let $k = [K(\mathbf{p}) : K]$. Then $k \le d$ and $K(\mathbf{p})$ has exactly k K-embeddings $\sigma_1, \ldots, \sigma_k$ into $\tilde{K}$. We set $\lambda(\mathbf{p}) = \sigma_1\mathbf{p} + \cdots + \sigma_k\mathbf{p} + (d - k)\mathbf{o}$. Then $\lambda(\mathbf{p})$ is fixed under each $\sigma \in \mathrm{Gal}(K)$, so $\lambda(\mathbf{p}) \in C^{(d)}(K)$. Since each $\mathbf{p} \in C(K^{(d)})$ has at most d conjugates over K, the fibers of λ have at most d elements. Therefore, if $C(K^{(d)})$ is infinite, so is $C^{(d)}(K)$.

Conversely, if $\sum_{i=1}^{d} \mathbf{p}_i \in C^{(d)}(K)$, then each $\mathbf{p}_i$ has at most d conjugates over K, namely $\mathbf{p}_1, \ldots, \mathbf{p}_d$. It follows that $\mathbf{p}_i \in C(K^{(d)})$. Moreover, if $C^{(d)}(K)$ is infinite, so is $C(K^{(d)})$. $\square$

The morphism $C^d \to J$ given by $(\mathbf{p}_1, \ldots, \mathbf{p}_d) \mapsto \sum_{i=1}^{d} \gamma(\mathbf{p}_i)$ has a unique value on each S_d-class, so it defines a morphism $\varphi_d \colon C^{(d)} \to J$ over K satisfying $\varphi_d(\sum_{i=1}^{d} \mathbf{p}_i) = \sum_{i=1}^{d} \gamma(\mathbf{p}_i)$. Thus, φ_d coincides with $\beta|_{C^{(d)}}$. We denote the image of $C^{(d)}$ under φ_d by W_d. Since $C^{(d)}$ is projective and absolutely irreducible, W_d is a closed absolutely irreducible subvariety of J. In particular, $\varphi_d(C^{(d)}(\tilde{K})) = W_d(\tilde{K})$ and $\varphi_d(C^{(d)}(K)) \subseteq W_d(K)$.

We denote the restriction of φ_d to $C^{(d)}(K)$ by φ.

LEMMA 6.8.4: *If the map $\varphi \colon C^{(d)}(K) \to J(K)$ is not injective, then there exists $f \in F$ such that $[F : K(f)] \leq d$.*

Proof: We assume that C has distinct nonnegative K-rational divisors $\sum_{i=1}^{d} \mathbf{p}_i$ and $\sum_{i=1}^{d} \mathbf{q}_i$ of degree d such that $\varphi(\sum_{i=1}^{d} \mathbf{p}_i) = \varphi(\sum_{i=1}^{d} \mathbf{q}_i)$. Then, by Lemma 6.8.1, there exists $f \in F$ such that $[F : K(f)] \leq d$. $\square$

The conclusion of Lemma 6.8.4 can be achieved if φ is injective under additional assumptions. To this end we need a certain combinatorial argument.

We denote the number of d-tuples $(\varepsilon_1, \ldots, \varepsilon_d) \in \{0, 1, 2\}^d$ such that

$$\text{(1)} \qquad\qquad\qquad \sum_{j=1}^{d} \varepsilon_j = d$$

by $n(d)$.

LEMMA 6.8.5: *Let $x \colon \{1, \ldots, d\} \to X$ be a surjective map of sets. Set $x_i = x(i)$ for $i = 1, \ldots, d$ and let A the free additive Abelian monoid generated by X. Denote the set of all sums $\sum_{j=1}^{d} y_j$ with $y_1, \ldots, y_d \in X$ by $X^{(d)}$; in particular $x = \sum_{j=1}^{d} x_j \in X^{(d)}$. Then the number of all $y \in X^{(d)}$ for which there exists $z \in A$ such that*

$$\text{(2)} \qquad\qquad\qquad y + z = 2x$$

is at most $n(d)$.

Proof: First we note that if $y \in X^{(d)}$ and $z \in A$ solves (2), then $z \in X^{(d)}$.

CASE A: $x_1, \ldots, x_d$ *are distinct.*   Then there is a bijective correspondence between the set of all $y \in X^{(d)}$ such that (2) is solvable and the set of solutions $(\varepsilon_1, \ldots, \varepsilon_d) \in \{0, 1, 2\}^d$ of (1). Thus, in this case, the number of the y's is $n(d)$.

Indeed, let y, z be elements of $X^{(d)}$ such that (2) holds and choose $y_1, \ldots, y_d \in X$ with $y = \sum_{j=1}^{d} y_j$. For each j let ε_j be the number of occurrences of x_j in the d-tuple $(y_1, \ldots, y_d)$. Since $x_1, \ldots, x_d$ are distinct, $\varepsilon_j \in \{0, 1, 2\}$ and $\sum_{j=1}^{d} \varepsilon_j = d$. Moreover, $(\varepsilon_1, \ldots, \varepsilon_d)$ depends only on y.

Conversely, let $(\varepsilon_1, \ldots, \varepsilon_d)$ be a solution of (1). Choose $(y_1, \ldots, y_d)$, $(z_1, \ldots, z_d) \in X^d$ such that for each j the number of occurrences of x_j in $(y_1, \ldots, y_d)$ is ε_j and in $(z_1, \ldots, z_d)$ is $2 - \varepsilon_j$. Then $y = \sum_{j=1}^{d} y_j$ and $z = \sum_{j=1}^{d} z_j$ solve (2).

CASE B: $x_1, \ldots, x_d$ are *not necessarily distinct.* We choose distinct elements $x_1', \ldots, x_d'$, set $X' = \{x_1', \ldots, x_d'\}$, let A' be the free additive Abelian monoid generated by X', let $(X')^{(d)}$ be the set of all sums $\sum_{j=1}^{d} y_j'$ with $y_1', \ldots, y_d' \in X'$, and let $x' = \sum_{j=1}^{d} x_j'$. By Case A, the number of $y' \in (X')^{(d)}$ for which there exists $z' \in A'$ with

(2') $$y' + z' = 2x'$$

is $n(d)$.

The map $(x_1', \ldots, x_d') \mapsto (x_1, \ldots, x_d)$ defines an epimorphism α of A' onto A and $\alpha((X')^{(d)}) = X^{(d)}$. If $y', z' \in (X')^{(d)}$ solve (2'), then $y = \alpha(y'), z = \alpha(z')$ solve (2).

Conversely, let $y = \sum_{j=1}^{d} y_j$ and $z = \sum_{j=1}^{d} z_j$ be elements of $X^{(d)}$ satisfying (2). Then there exists $k(1)$ between 1 and d such that $y_1 = x_{k(1)}$. Canceling y_1 and $x_{k(1)}$ from both sides of (2), we find a $k(2)$ between 1 and d such that $y_2 = x_{k(2)}$. Continuing in this way, we find a permutation $(k(1), \ldots, k(d), l(1), \ldots, l(d))$ of $(1, \ldots, d, 1, \ldots, d)$ such that $y_j = x_{k(j)}$ and $z_j = x_{l(j)}$ for $j = 1, \ldots, d$. Now we set $y_j' = x_{k(j)}'$ and $z_j' = x_{l(j)}'$ for $j = 1, \ldots, d$, $y' = \sum_{j=1}^{d} y_j'$, and $z' = \sum_{j=1}^{d} z_j'$. Then $y', z' \in (X')^{(d)}$ satisfy (2'), $\alpha(y') = y$, and $\alpha(z') = z$.

We conclude that the number of $y \in X^{(d)}$ for which (2) is solvable is at most $n(d)$, as claimed. $\qquad\square$

LEMMA 6.8.6: *Suppose there exist a point* $\mathbf{p} = \sum_{j=1}^{d} \mathbf{p}_j \in C^{(d)}(K)$ *and at least* $n(d) + 1$ *points* $\mathbf{b}_0, \ldots, \mathbf{b}_{n(d)} \in J(\tilde{K})$ *such that* $\beta(\mathbf{p}) \pm \mathbf{b}_i \in W_d(\tilde{K})$, $i = 0, \ldots, n(d)$. *Then there exists* $f \in F$ *such that* $[F : K(f)] \leq 2d$.

Proof: For each $0 \leq i \leq n(d)$ there exist $\mathbf{q}_{i1}, \ldots, \mathbf{q}_{id}, \mathbf{r}_{i1}, \ldots, \mathbf{r}_{id} \in C(\tilde{K})$ such that with $\mathbf{q}_i = \sum_{j=1}^{d} \mathbf{q}_{ij}$ and $\mathbf{r}_i = \sum_{j=1}^{d} \mathbf{r}_{ij}$ we have

(3) $$\begin{aligned} \beta(\mathbf{p}) + \mathbf{b}_i &= \beta(\mathbf{q}_i) \\ \beta(\mathbf{p}) - \mathbf{b}_i &= \beta(\mathbf{r}_i). \end{aligned}$$

Summing up the equalities in (3), we have

(4) $$\beta(\mathbf{q}_i + \mathbf{r}_i) = \beta(2\mathbf{p}).$$

By definition, $\mathrm{Div}(C)$ is the free Abelian group generated by the set $C(\tilde{K})$. Let A be the submonoid of $\mathrm{Div}(C)$ generated by $\mathbf{p}_1, \ldots, \mathbf{p}_d$. Then A is the free Abelian monoid generated by $X = \{\mathbf{p}_1, \ldots, \mathbf{p}_d\}$. If $\mathbf{q}_{ij} \notin X$ for some i, j, then $\mathbf{q}_i + \mathbf{r}_i \neq 2\mathbf{p}$. Similarly, if $\mathbf{r}_{ij} \notin X$ for some i, j, then $\mathbf{q}_i + \mathbf{r}_i \neq 2\mathbf{p}$. If $\mathbf{q}_{ij}, \mathbf{r}_{ij} \in X$ for all i, j, then by Lemma 6.8.5, there exists i between 0 and $n(d)$ such that $\mathbf{q}_i + \mathbf{r}_i \neq 2\mathbf{p}$. Since $\mathbf{p}$ is K-rational and both $\mathbf{q}_i + \mathbf{r}_i$ and $\mathbf{p}$ are nonnegative, (4) and Lemma 6.8.1 give an $f \in F$ such that $[F : K(f)] \leq \deg(2\mathbf{p}) = 2d$. $\qquad\square$

LEMMA 6.8.7: *Suppose K is a number field and $C(K^{(d)})$ is infinite. Then there exists $f \in F$ such that $[F : K(f)] \leq 2d$.*

Proof: If the map $\varphi \colon C^{(d)}(K) \to J(K)$ is not injective, then by Lemma 6.8.4, there exists $f \in F$ such that $[F : K(f)] \leq d$. We may therefore assume that φ is injective.

By Faltings (Proposition 6.4.1), there exist $\mathbf{a}_1, \ldots, \mathbf{a}_n \in J(K)$ and Abelian subvarieties $B_1, \ldots, B_n$ defined over K such that $W_d(K) = \bigcup_{i=1}^{n}[\mathbf{a}_i + B_i(K)]$ and $\mathbf{a}_i + B_i(\tilde{K}) \subseteq W_d(\tilde{K})$ for each i. By Lemma 6.8.3, $C^{(d)}(K)$ is infinite. Hence, $\varphi(C^{(d)}(K))$ is an infinite subset of $W_d(K)$, hence of $\bigcup_{i=1}^{n}[\mathbf{a}_i + B_i(K)]$. Therefore, there exists i such that $\varphi(C^{(d)}(K)) \cap [\mathbf{a}_i + B_i(K)]$ is infinite. This allows us to choose $\mathbf{p} \in C^{(d)}(K)$ and $\mathbf{b} \in B_i(K)$ such that $\varphi(\mathbf{p}) = \mathbf{a}_i + \mathbf{b}$. In addition, it follows that $B_i(K)$ is infinite, so we choose distinct points $\mathbf{b}_0, \ldots, \mathbf{b}_{n(d)} \in B_i(K)$. For each of them we have $\beta(\mathbf{p}) \pm \mathbf{b}_j = \varphi(\mathbf{p}) \pm \mathbf{b}_j = \mathbf{a}_i + (\mathbf{b} \pm \mathbf{b}_j) \in \mathbf{a}_i + B_i(K) \subseteq W_d(\tilde{K})$. We conclude from Lemma 6.8.6 that there exists $f \in F$ such that $[F : K(f)] \leq 2d$, as desired. $\qquad\square$

PROPOSITION 6.8.8 (Corvaja): *Let K be a number field, d a positive integer, and $K_1, K_2, K_3, \ldots$ a linearly disjoint sequence of extensions of degree d over K. Then there exists a subsequence $K_{i(1)}, K_{i(2)}, K_{i(3)}, \ldots$ whose compositum $K_{i(1)}K_{i(2)}K_{i(3)} \cdots$ is a nonample field.*

Proof: We choose a prime number p such that $\sqrt{p-1} > 2d$ and consider the affine absolutely irreducible $\mathbb{Q}$-curve C defined by the equation $Y^p - Y - X^{p-1} + 1 = 0$ (Lemma 6.7.3(a)). Let x be a transcendental element over $\mathbb{Q}$ and y an element of $\widetilde{\mathbb{Q}(x)}$ such that $y^p - y - x^{p-1} + 1 = 0$. Since $(1, 0)$ is a simple $\mathbb{Q}$-rational point of C, for each number field L, $L(C)/L$ is a function field of one variable with a prime divisor of degree 1. By Proposition 6.7.5, $\mathrm{gon}(L(C)/L) \geq \sqrt{p-1} > 2d$. Hence, by Lemma 6.8.7, $C(L^{(d)})$ is finite. Also, by Lemma 6.6.5, $\mathrm{genus}(L(C)/L) + 1 \geq \mathrm{gon}(L(C)/L) > 2d$, so $\mathrm{genus}(L(C)/L) \geq 2$.

By Faltings (Proposition 6.2.4), $C(K)$ is finite. Suppose by induction we have already found $i(1) < i(2) < \cdots < i(m)$ such that

$$(5) \qquad\qquad C(K') = C(K),$$

where $K' = K_{i(1)}K_{i(2)} \cdots K_{i(m)}$. By [FrJ08, Lemma 2.5.7] there exists $n > m$ such that the fields $K'_j = K'K_j$, $j = n, n+1, n+2, \ldots$, are linearly disjoint over K'. By the preceding paragraph, $C((K')^{(d)})$ is finite. Since $[K'_j : K'] \leq [K_j : K] = d$, it follows that the set $\bigcup_{j=n}^{\infty} C(K'_j)$ is finite.

Now we consider integers $k > j \geq n$. By the linear disjointness, $C(K'_j) \cap C(K'_k) = C(K')$, so $[C(K'_j) \smallsetminus C(K')] \cap [C(K'_k) \smallsetminus C(K')] = \emptyset$. Therefore, there exists m such that $i(m+1) > n$ and $C(K'_{i(m+1)}) = C(K')$. It follows from (5) that $C(K_{i(1)} \cdots K_{i(m)} K_{i(m+1)}) = C(K)$. This concludes the induction.

Let M be the compositum of all the fields $K_{i(k)}$. Then $C(M) = C(K)$. Since $(1,0)$ is a simple M-rational point of C, M is nonample. $\qquad\square$

Example 6.8.9: Linearly disjoint extensions. For each positive integer d there exists an infinite sequence $p_1, p_2, p_3, \ldots$ of prime numbers satisfying $p_i \equiv 1 \bmod d$. This special case of Dirichlet's theorem on primes in arithmetic progressions has an elementary proof that uses arguments including roots of unity [Nag51, p. 168, Thm. 96]. The fields $\mathbb{Q}(\zeta_{p_1}), \mathbb{Q}(\zeta_{p_2}), \mathbb{Q}(\zeta_{p_3}), \cdots$ are linearly disjoint over $\mathbb{Q}$ and $\mathrm{Gal}(\mathbb{Q}(\zeta_{p_i})/\mathbb{Q}) \cong \mathbb{Z}/(p_i - 1)\mathbb{Z}$ [Lan93, p. 278, Thms. 3.2 and 3.1]. In particular, for each i, the field $\mathbb{Q}(\zeta_{p_i})$ contains a unique cyclic extension K_i of $\mathbb{Q}$ of degree d. Moreover, the fields $K_1, K_2, K_3, \ldots$ are linearly disjoint over $\mathbb{Q}$. By Proposition 6.8.8, there exist $i(1) < i(2) < i(3) < \ldots$ such that $M = K_{i(1)} K_{i(2)} K_{i(3)} \cdots$ is nonample. Note that in this case M is an Abelian extension of $\mathbb{Q}$.

A more general method to construct linearly disjoint field extensions applies Hilbert's irreducibility theorem. One starts from the general polynomial $f(t_1, \ldots, t_d, X)$ of degree d and successively specializes the $(t_1, \ldots, t_d)$ to d-tuples $(a_1, \ldots, a_n)$ with coordinates in a given number field K such that the root field (and even the decomposition field) of $f(a_1, \ldots, a_d, X)$ is linearly disjoint from the previously constructed field extensions of K (see for example the proof of [FrJ08, Lemma 16.2.6]). Applying Proposition 6.8.8, this gives many more examples of infinite extensions of K that are nonample. $\quad\square$

Notes

Proposition 6.1.5 is due to Koenigsmann [Koe02, Thm. 2].

Theorem 6.5.2 is due to Fehm and Petersen [FeP10]. This theorem has a history that goes back at least to a paper of Lutz [Lut37] from 1937. In that paper Lutz proves that if E is an elliptic curve defined over a p-adic field K, then $E(K)$ contains a subgroup isomorphic to the additive group K^+ of K. Mattuck [Mat55] generalizes Lutz's result to an arbitrary Abelian variety A defined over a complete field K under a complete nonarchimedean absolute value having characteristic 0. This implies that $\mathrm{rr}(A(K)) = \infty$, because K has infinite degree over its prime field. This also implies that $\mathrm{rr}(A(K)) = \infty$ if K is an algebraically closed field of characteristic 0. The latter result was generalized in [FyJ74, Thm. 10.1] to arbitrary algebraically closed fields that are not algebraic extensions of finite fields. Moreover, in that paper, the authors prove that if K is an infinite finitely generated field over its prime field, then for almost all $\boldsymbol{\sigma} \in \mathrm{Gal}(K)^e$ and for each Abelian variety A defined over $K_s(\boldsymbol{\sigma})$, we have $\mathrm{rr}(A(K_s(\boldsymbol{\sigma})) = \infty$. The same result holds for the smaller fields $K_s[\boldsymbol{\sigma}]$, again for almost all $\boldsymbol{\sigma} \in \mathrm{Gal}(K)^e$, by [GeJ06]. In an appendix to [Pet06] the author of this book uses methods of Petersen to generalize the latter result to the case where K is a countable Hilbertian field. He also considers a finite set S of local primes of K in the sense of Example 5.6.5 and a nonnegative integer e. By Example 5.6.6, $K_{\mathrm{tot},S}[\boldsymbol{\sigma}]$ is an ample field for almost all $\boldsymbol{\sigma} \in \mathrm{Gal}(K)^e$. If in addition, $\mathrm{char}(K) = 0$, then by Theorem 6.5.2, $\mathrm{rr}(A(K_{\mathrm{tot},S}[\boldsymbol{\sigma}])) = \infty$ for almost all $\boldsymbol{\sigma} \in \mathrm{Gal}(K)^e$ and for all Abelian varieties A defined over $K_{\mathrm{tot},S}[\boldsymbol{\sigma}]$. Fehm and Petersen remove the assumption that $\mathrm{char}(K) = 0$ from the latter result [FeP09, Thm. 2.2]. Finally, as the book goes to press, Fehm and Petersen announced a generalization of Theorem 6.5.2 to an arbitrary ample field that is not an algebraic extension of a finite field.

Corollary 6.1.7, Proposition 6.1.8, Example 6.1.9(a), Proposition 6.1.10, and Theorem 6.2.6, are taken from Fehm's work [Feh11]. Lemma 6.7.4 is also due to Fehm (private communication).

The results of Section 6.6 are taken from the work of Poonen [Poo07].

Lemma 6.7.2 originated from [Ohm83].

Most of Section 6.8 follows Frey's article [Fre94]. Proposition 6.8.8 is due to Corvaja.

Chapter 7.
Split Embedding Problems over Complete Fields

Let K_0 be a complete field with respect to an ultrametric absolute value. In Proposition 4.4.2 we considered a finite Galois extension K of K_0 with Galois group Γ acting on a finite group G and let x be an indeterminate. We constructed a finite Galois extension F of $K_0(x)$ that contains K and with Galois group $\Gamma \ltimes G$ that solves the constant embedding problem $\Gamma \ltimes G \to \mathrm{Gal}(K(x)/K_0(x))$. Using an appropriate specialization we have been then able to prove the same result in the case where K_0 was an arbitrary ample field (Theorem 5.9.2). This was sufficient for the proof that each Hilbertian PAC field is ω-free (Theorem 5.10.3).

In this chapter we lay the foundation to the proof of the third major result of this book: Giving a function field E of one variable over an ample field K of cardinality m, each finite split embedding problem over E has m linearly disjoint solution fields (Theorem 11.7.1).

Here we let K_0 be as in the first paragraph, and consider a finite Galois extension E' of $K_0(x)$ (where E' is not necessarily of the form $K(x)$ with K/K_0 Galois) acting on a finite group H. We prove that the finite split embedding problem $\mathrm{Gal}(E'/K_0(x)) \ltimes H \to \mathrm{Gal}(E'/K_0(x))$ has a solution field F'. Moreover, if H is generated by finitely many cyclic subgroups G_j, then for each j there is a branch point b_j with G_j as an inertia group.

7.1 Total Decomposition

An auxiliary goal in solving regular finite split embedding problems over a field K is to achieve a solution field F with a K-rational place φ which is unramified over $K(x)$. In other words, the place $\mathfrak{p}$ of $K(x)/K$ which corresponds to $\varphi|_{K(x)}$ **totally decomposes** in F. This means that there are $[F : K(x)]$ prime divisors of F/K lying over $\mathfrak{p}$. This extra condition ensures that F is regular over K. Replacing x by another generator of $K(x)$ over K, we may assume that $\mathfrak{p}$ is the pole $\mathfrak{p}_{x,\infty}$ of x. The following result gives a necessary and sufficient condition for $\mathfrak{p}_{x,\infty}$ to totally decompose in F in terms of the irreducible polynomial of an appropriate primitive element of $F/K(x)$.

LEMMA 7.1.1: *Let K be an arbitrary field and consider a Galois extension F of $E = K(x)$ of degree d which is regular over K. Let $\mathfrak{p}_{x,\infty}$ be the pole of x in $K(x)/K$ and $\mathfrak{p}$ a prime divisor of F/K over $\mathfrak{p}_{x,\infty}$.*

(a) Suppose $\mathfrak{p}_{x,\infty}$ totally decomposes in F and let y be an element of F with $\mathrm{div}_{F,\infty}(y) = k\mathfrak{p}$ for some positive integer k. Then y is integral over $K[x]$,

M. Jarden, *Algebraic Patching*, Springer Monographs in Mathematics,
DOI 10.1007/978-3-642-15128-6_7, © Springer-Verlag Berlin Heidelberg 2011

$F = E(y)$, and $f = \mathrm{irr}(y, E)$ has the form

$$(1) \qquad f(x, Y) = Y^d + a_{d-1}(x)Y^{d-1} + \cdots + a_0(x).$$

with $a_i \in K[x]$ and $\deg(a_i(x)) \le \deg(a_{d-1}(x)) = k$, $i = 0, \ldots, d - 1$.

(b) Suppose $\mathfrak{p}_{x,\infty}$ totally decomposes in F. Then there exists y as in (a) with $k \le \mathrm{genus}(F/K) + 1$.

(c) Conversely, suppose $y \in F$ and $f = \mathrm{irr}(y, E)$ is given by (1) such that $a_i(x) \in K[x]$, $\deg(a_{d-1}(x)) > 0$, and $\deg(a_i(x)) \le \deg(a_{d-1}(x))$, $i = 0, \ldots, d - 1$. Then $\mathfrak{p}_{x,\infty}$ totally decomposes in F.

Proof of (a): We denote the normalized valuation of F/K that corresponds to $\mathfrak{p}$ by $v_{\mathfrak{p}}$. Then $v_{\mathfrak{p}}(y) = -k$ and $w(y) \ge 0$ for each other valuation w of F/K. If $\sigma \in G = \mathrm{Gal}(F/E)$ and $\sigma \ne 1$, then $v_{\mathfrak{p}}^{\sigma} \ne v_{\mathfrak{p}}$ (because $\mathfrak{p}_{x,\infty}$ totally decomposes on F), so $v_{\mathfrak{p}}(y^{\sigma^{-1}}) = v_{\mathfrak{p}}^{\sigma}(y) \ge 0$. Hence, $y^{\sigma^{-1}} \ne y$ for each $\sigma \ne 1$. Therefore $F = K(x, y)$.

In addition $w(y) \ge 0$ if $w(x) \ge 0$. Hence, y is integral over $K[x]$. In particular $f(x, Y) = \mathrm{irr}(y, K(x)) \in K[x, Y]$ is a monic polynomial in Y. Since F/E is a Galois extension, $f(x, Y)$ decomposes into distinct linear factors over F:

$$(2) \qquad f(x, Y) = \prod_{\sigma \in G} (Y - y^{\sigma}).$$

A comparison of (1) and (2) gives:

$$(3) \qquad a_i(x) = (-1)^{d-i} \sum_{S \in \mathcal{P}_i} \prod_{\sigma \in S} y^{\sigma}$$

where $\mathcal{P}_i$ is the collection of all subsets of G of cardinality $d - i$. Note that $v_{\mathfrak{p}}$ is unramified over E. Hence the restriction of $v_{\mathfrak{p}}$ to E coincides with the valuation $v_{x,\infty}$ that corresponds to $\mathfrak{p}_{x,\infty}$. Since $\sigma = 1$ appears at most once in each of the summands $\prod_{\sigma \in S} y^{\sigma}$, and since $v_{\mathfrak{p}}(y^{\sigma}) \ge 0$ for $\sigma \ne 1$, this gives

$$- \deg(a_i(x)) = v_{x,\infty}(a_i(x)) = v_{\mathfrak{p}}(a_i(x)) \ge \min_{S \in \mathcal{P}_i} \sum_{\sigma \in S} v_{\mathfrak{p}}(y^{\sigma}) \ge -k.$$

Finally, $- \deg(a_{d-1}(x)) = v_{x,\infty}(a_{d-1}(x)) = v_{\mathfrak{p}}(-y - \sum_{\sigma \ne 1} y^{\sigma}) = v_{\mathfrak{p}}(y) = -k$, as desired.

Proof of (b): Let $g = \mathrm{genus}(F/K)$. Since $\mathfrak{p}_{x,\infty}$ decomposes in F into d prime divisors, $\deg(\mathfrak{p}) = 1$. We claim that there exists k between 1 and $g + 1$ with $\dim((k - 1)\mathfrak{p}) = 1$ and $\dim(k\mathfrak{p}) = 2$. Indeed, by Riemann-Roch (Remark 5.8.1(c)), if $g = 0$, then $\dim(0 \cdot \mathfrak{p}) = 1$ and $\dim(1 \cdot \mathfrak{p}) = 2$. If $g = 1$, then $\dim(1 \cdot \mathfrak{p}) = 1$ and $\dim(2 \cdot \mathfrak{p}) = 2$. If $g \ge 2$, we choose a canonical divisor $\mathfrak{w}$ of F/K. Then $\mathcal{L}(\mathfrak{w} - (k+1)\mathfrak{p}) \subseteq \mathcal{L}(\mathfrak{w} - k\mathfrak{p})$. By Riemann-Roch,

$\dim(k\mathfrak{p}) = k + 1 - g + \dim(\mathfrak{w} - k\mathfrak{p})$. Hence, $\dim((k+1)\mathfrak{p}) - \dim(k\mathfrak{p}) \leq 1$. In addition, $\dim(0 \cdot \mathfrak{p}) = 1$. If there exists no k as above, then $\dim((g+1)\mathfrak{p}) = 1$, hence $\dim((2g - 1)\mathfrak{p}) \leq g - 1$. This will contradict Riemann-Roch which predict that $\dim((2g - 1)\mathfrak{p}) = g$. This concludes the proof of our claim.

Let $y \in \mathcal{L}(k\mathfrak{p}) \smallsetminus \mathcal{L}((k - 1)\mathfrak{p})$. Then $\mathrm{div}_{F,\infty}(y) = k\mathfrak{p}$, as contended.

Proof of (c): Let $k = \deg(a_{d-1}(x))$. Then $z = y/x^k$ satisfies

$$(4) \qquad z^d + b_{d-1}(x)z^{d-1} + b_{d-2}(x)z^{d-2} + \cdots + b_0(x) = 0,$$

where $b_i(x) = a_i(x)/x^{(d-i)k}$. As in (a), we choose a prime divisor $\mathfrak{p}$ of F/K lying over $\mathfrak{p}_{x,\infty}$, let $v_\mathfrak{p}$ be the normalized valuation of F/K corresponding to $\mathfrak{p}$, and set $e = e_{v_\mathfrak{p}/v_{x,\infty}}$. Then $v_\mathfrak{p}(b_{d-1}(x)) = e(v_{x,\infty}(a_{d-1}(x)) - kv_{x,\infty}(x)) = e(-\deg(a_{d-1}(x)) + k) = 0$ and $v_\mathfrak{p}(b_i(x)) = e(-\deg(a_i(x)) + (d - i)k)) > 0$ for $i = 0, \ldots, d - 2$. Hence, reduction of (4) modulo $\mathfrak{p}$ gives $\bar{z}^{d-1}(\bar{z} + c) = 0$ with $c = \overline{b_{d-1}(x)} \in K^\times$. By Hensel's Lemma, $h(Z) = Z^d + b_{d-1}(x)Z^{d-1} + \cdots + b_0(x)$ has a root in the completion $K((x^{-1}))$ of E with respect to $\mathfrak{p}_{x,\infty}$. Since $F = E(z)$ is Galois over E, all of the roots of $h(Z)$ are in $K((x^{-1}))$. Consequently, $\mathfrak{p}_{x,\infty}$ totally decomposes in F. $\qquad\square$

7.2 Ramification

The most effective way by which we can distinguish between two finite Galois extensions F_1 and F_2 of $E = K(x)$, where K is a field and x is an indeterminate, is through ramification. That is, $F_1 \neq F_2$ if $\mathrm{Ram}(F_1/E) \neq \mathrm{Ram}(F_2/E)$ or, equivalently, if $\mathrm{Branch}(F_1/E) \neq \mathrm{Branch}(F_2/E)$. A more refined separation of F_1 and F_2 is achieved, when they are linearly disjoint over E (Lemma 7.4.1 and Proposition 7.4.4). One way to do it is to use inertia groups. The basic information on inertia groups we need is contained in the following result.

LEMMA 7.2.1: *Let (E, v) be a discrete valued field, $(\hat{E}, \hat{v})$ its completion, and F, F' finite Galois extensions of E. Suppose there is an embedding of $\tilde{E}$ into the algebraic closure of $\hat{E}$ such that $\hat{F} = \hat{E}F = \hat{E}F'$. Denote the unique extension of $\hat{v}$ to $\hat{F}$ by $\hat{w}$ and let w and w' be the restrictions of $\hat{w}$ to F and F', respectively. Then v is ramified in F if and only if v is ramified in F'. Moreover, the restriction maps $\mathrm{res}_{\hat{F}/F}\colon \mathrm{Gal}(\hat{F}/\hat{E}) \to \mathrm{Gal}(F/E)$ and $\mathrm{res}_{\hat{F}/F'}\colon \mathrm{Gal}(\hat{F}/\hat{E}) \to \mathrm{Gal}(F'/E)$ map $I_{\hat{w}/\hat{v}}$ isomorphically onto $I_{w/v}$ and $I_{w'/v}$, respectively. In particular, if v is totally ramified in F, then $I_{w'/v} = \mathrm{Gal}(F'/\hat{E} \cap F')$.*

Proof: The valuation group and the residue field of a discrete valuation coincide with the valuation group and the residue field, respectively, of its completion. Hence, v is ramified in F if and only if v is ramified in $\hat{F}$ if and only if v is ramified in F'.

To prove the statement about the inertia groups we note that if $\hat\sigma \in I_{\hat w/\hat v}$, then $\hat w(\hat x^{\hat\sigma} - \hat x) > 0$ for each $\hat x \in O_{\hat w}$. Hence, $\sigma = \hat\sigma|_F$ satisfies $w(x^\sigma - x) > 0$ for each $x \in O_w = F \cap O_{\hat w}$. Therefore, $\sigma \in I_{w/v}$.

Conversely, we extend each $\sigma \in I_{w/v}$ to an element $\hat\sigma$ of $\mathrm{Gal}(\hat F/\hat E)$. To this end we present each $\hat x \in \hat F$ as the limit of a sequence of elements $x_i \in F$. In particular, the x_i's form a w-Cauchy sequence. Being in $I_{w/v}$ the automorphism σ belongs also to the decomposition group, i.e. $O_w^\sigma = O_w$. Hence, the x_i^σ's also form a w-Cauchy sequence and we denote its unique limit in $\hat F$ by $\hat x^{\hat\sigma}$. One observes that $\hat x^{\hat\sigma}$ is independent of the approximation of $\hat x$ by elements of F. This implies that $\hat\sigma$ is indeed a lifting of σ to an element of $\mathrm{Gal}(\hat F/\hat E)$.

Now we suppose $\hat x \in O_{\hat w}$ and choose $x \in O_w$ with $\hat w(\hat x - x) > 0$. Since $\hat w^{\hat\sigma^{-1}} = \hat w$, this implies $\hat w(\hat x^{\hat\sigma} - x^{\hat\sigma}) > 0$. Hence, $\hat w(\hat x^{\hat\sigma} - \hat x) = \hat w(\hat x^{\hat\sigma} - x^\sigma + x^\sigma - x + x - \hat x) \geq \min\left(\hat w(\hat x^{\hat\sigma} - x^\sigma), w(x^\sigma - x), \hat w(x - \hat x)\right) > 0$. It follows that $\hat\sigma \in I_{\hat w/\hat v}$. Since $\hat F = \hat E F$, the restriction map $\mathrm{res}_{\hat F/F}$ is injective, so it maps $I_{\hat w/\hat v}$ isomorphically onto $I_{w/v}$.

Similarly, $\mathrm{res}_{\hat F/F'}$ maps $I_{\hat w/\hat v}$ isomorphically onto $I_{w'/v}$. In particular, if v totally ramifies in F, then $I_{w/v} = \mathrm{Gal}(F/E)$. Hence, $I_{\hat w/\hat v} = \mathrm{Gal}(\hat F/\hat E)$ and therefore $I_{w'/v} = \mathrm{Gal}(F'/\hat E \cap F')$. $\qquad\square$

For the rest of this section we assume that K is complete under a nontrivial ultrametric absolute value $|\ |$ and extend $|\ |$ to $\tilde K$ in the unique possible way. As in Chapter 3 we consider a nonempty set I, and for each $i \in I$ an element $c_i \in K$ and $r \in K^\times$ such that

$$(1) \qquad\qquad |r| \leq |c_i - c_j| \qquad \text{if} \quad i \neq j.$$

For each $i \in I$ let $w_i = \frac{r}{x - c_i}$. We extend $|\ |$ to a norm of the ring $R_0 = K[w_i \mid i \in I]$ by the rule

$$\left\| \sum_{i=0}^n a_i x^i \right\| = \max_n(|a_0|, \ldots, |a_n|)$$

and let $R = K\{w_i \mid i \in I\}$ be the completion of R_0 (Lemma 2.1.5). By Lemma 3.2.1, each $f \in R$ has a unique presentation as a multiple power series:

$$f = a_0 + \sum_{i \in I} \sum_{n=1}^\infty a_{in} w_i^n,$$

where $a_0, a_{in} \in K$ and $|a_{in}| \to 0$ as $n \to \infty$. Let $Q = \mathrm{Quot}(R)$.

LEMMA 7.2.2: Let $\mathfrak{p}_{x,\alpha}$ be the prime divisor of E/K associated with an element $\alpha \in \tilde K \cup \{\infty\}$ and let $v = v_{x,\alpha}$ be the correspondent discrete valuation of E/K.

(a) If $|\alpha - c_i| \geq |r|$ for all $i \in I$, then v extends to a discrete valuation v_Q of Q such that the completion of Q at v_Q is a completion of E at v.

(b) Let F/E be a finite Galois extension such that $F \subseteq Q$. If $\alpha \in \mathrm{Branch}(F/E)$, then there is an $i \in I$ such that $|\alpha - c_i| < |r|$.

Proof of (a): Since K is complete under $|\ |$, so is $K(\alpha)$. Hence, we may apply Lemma 2.2.1(d) to consider the evaluation homomorphism $\varphi \colon R \to K(\alpha)$ given by

$$a_0 + \sum_{i \in I} \sum_{n=1}^{\infty} a_{in} w_i^n \quad \mapsto \quad a_0 + \sum_{i \in I} \sum_{n=1}^{\infty} a_{in} \left(\frac{r}{\alpha - c_i}\right)^n.$$

Since $\varphi(w_i) = \frac{r}{\alpha - c_i}$, we have $\varphi(x) = \alpha$.

Now fix an $i \in I$. By Proposition 3.2.9, R is a principal ideal domain and the ideal $\mathrm{Ker}(\varphi)$ of R is generated by an element $q \in K[w_i]$ such that $\mathrm{Ker}(\varphi) \cap K[w_i] = qK[w_i]$.

Since q is irreducible in R, the localization R_{qR} is a discrete valuation ring, hence φ uniquely extends to a place $\varphi \colon Q \to K(\alpha) \cup \{\infty\}$. The corresponding discrete valuation v_Q of Q extends v. The residue fields of both v and v_Q is $K(\alpha)$ and q is a common uniformizer. This implies that E is v_Q-dense in Q.

Indeed, let $f \in Q$ and assume without loss that $v_Q(f) \geq 0$. Inductively assume that there exist $p_0, \ldots, p_{n-1} \in E$ such that $v_Q(f - \sum_{i=0}^{n-1} p_i q^i) \geq n$. Then there exists $f_n \in Q$ with $f - \sum_{i=0}^{n-1} p_i q^i = f_n q^n$ and there exists $p_n \in E$ with $v(f_n - p_n) \geq 1$. Therefore, $v(f_n q^n - p_n q^n) \geq n+1$, hence $v_Q(f - \sum_{i=0}^{n} p_i q^i) \geq n+1$, and the induction is complete.

Consequently, the completion of Q at v_Q is also a completion of E at v.

Note that the above proof works also for $\alpha = \infty$. In this case $|\alpha - c_i| \geq |r|$, $\frac{r}{\alpha - c_i} = 0$, and $w_i = \frac{rx^{-1}}{1 - c_i x^{-1}}$ for each $i \in I$, so $q = x^{-1}$. Alternatively, $w_i \in K((x^{-1}))$ for each $i \in I$, so $K((x^{-1}))$ is the completion of both Q and E at v.

Proof of (b): Assume that $|\alpha - c_i| \geq |r|$ for each $i \in I$. By (a), v extends to a discrete valuation v_Q of Q such that the completion $\hat{Q}$ of Q at v_Q is also a completion of E at v. In particular, v is unramified in $\hat{Q}$, hence in F. Consequently, $\alpha \notin \mathrm{Branch}(F/E)$. $\qquad\square$

We assume from now till the end of this section that $|I| \geq 2$. Let G be a finite group and for each $i \in I$ let G_i be a subgroup of G such that $G = \langle G_i \mid i \in I \rangle$. As in Section 3.3 let $P_i = K\{w_j \mid j \neq i\}$ and $P_i' = K\{w_i\}$. Finally let F_i be a Galois extension of E in P_i' with $\mathrm{Gal}(F_i/E) \cong G_i$. Then $\mathcal{E} = (E, F_i, P_i, Q; G_i, G)_{i \in I}$ is patching data (Construction B in the proof of Proposition 4.4.2).

LEMMA 7.2.3: *Let F be the compound of $\mathcal{E} = (E, F_i, P_i, Q; G_i, G)_{i \in I}$.*

(a) *Let $i \in I$. If $\alpha \in \mathrm{Branch}(F_i/E)$, then $|\alpha - c_i| < |r|$. In particular, the sets $\mathrm{Branch}(F_i/E)$, for $i \in I$, are disjoint.*

(b) $\mathrm{Branch}(F/E) = \bigcup_{i \in I} \mathrm{Branch}(F_i/E)$.

(c) *Suppose the set I contains the symbol 1 and $G = G_1 \ltimes H$, where $H = \langle G_i \mid i \in I \setminus \{1\} \rangle \lhd G$. Then $F^H = F_1$ and $\bigcup_{\substack{i \in I \\ i \neq 1}} \mathrm{Branch}(F_i/E) = \mathrm{Branch}(F^{G_1}/E)$.*

(d) *Under the assumptions of (c) let $i \in I \setminus \{1\}$ and $\alpha \in \mathrm{Branch}(F_i/F)$. Suppose $\mathfrak{p} = \mathfrak{p}_{x,\alpha}$ totally ramifies in F_i. Then there exists a prime divisor $\mathfrak{q}$ of F/E over $\mathfrak{p}$ such that $I_{\mathfrak{q}/\mathfrak{p}}$ is the subgroup G_i of H.*

Proof of (a): By assumption, $F_i \subseteq P_i'$. By Lemma 7.2.2(b), with $I = \{i\}$, each $\alpha \in \mathrm{Branch}(F_i/E)$ satisfies $|\alpha - c_i| < |r|$. If α also belongs to $\mathrm{Branch}(F_j/E)$ for some $j \neq i$, then $|\alpha - c_j| < |r|$, so $|c_i - c_j| < |r|$, contradicting (1). It follows that $\mathrm{Branch}(F_i/E) \cap \mathrm{Branch}(F_j/E) = \emptyset$ if $i \neq j$.

Proof of (b): Let $\alpha \in \tilde{K} \cup \{\infty\}$ and let $\mathfrak{p} = \mathfrak{p}_{x,\alpha}$ be the corresponding prime divisor of E/K and $v = v_{x,\alpha}$ the associated normalized discrete valuation.

PART A: *First assume that $\alpha \in \mathrm{Branch}(F_i/E)$.* Then v is ramified in F_i. By (a), $|\alpha - c_i| < |r|$. Hence, by (1), $|\alpha - c_j| \geq |r|$ for each $j \neq i$. By Lemma 7.2.2(a), applied to $I \setminus \{i\}$ rather than to I, P_i is contained in the completion $\hat{E}$ of E at v. Now we embed $\tilde{E}$ in the algebraic closure of $\hat{E}$. By Lemma 1.1.7(b), $P_i F = P_i F_i$, so $\hat{E}F = \hat{E}F_i$. By Lemma 7.2.1, v ramifies in F, so $\alpha \in \mathrm{Branch}(F/E)$.

PART B: *Conversely, assume that v is ramified in F.* If there is an $i \in I$ such that $|\alpha - c_i| < |r|$, then, by (1),

(2) $|\alpha - c_j| \geq |r|$ for all $j \neq i$.

Otherwise, $|\alpha - c_j| \geq |r|$ for all j.

Thus, in each case we may fix an i such that (2) holds. As in Part A, P_i is contained in a completion $\hat{E}$ of E at v and $\hat{E}F_i = \hat{E}F$. By Lemma 7.2.1, v is ramified in F_i, so $\alpha \in \mathrm{Branch}(F_i/E)$.

Proof of (c): We have $F^H = F_1$, by Lemma 1.3.1(a). It follows that $F^{G_1} \cap F_1 = E$ and $F^{G_1} F_1 = F$. Hence, $\mathrm{Branch}(F/E) = \mathrm{Branch}(F_1/E) \cup \mathrm{Branch}(F^{G_1}/E)$ (Remark 4.1.1).

Let $\alpha \in \bigcup_{i \neq 1} \mathrm{Branch}(F_i/E)$. By (b), $\alpha \in \mathrm{Branch}(F/E)$; but, by (a), $\alpha \notin \mathrm{Branch}(F_1/E)$. Hence $\alpha \in \mathrm{Branch}(F^{G_1}/E)$.

Conversely, if $\alpha \in \mathrm{Branch}(F^{G_1}/E)$, then $\alpha \in \mathrm{Branch}(F/E)$. If $\alpha \in \mathrm{Branch}(F_1/E)$, then by Part A of the proof of (b), P_1 is contained in the completion of E at $v_{x,\alpha}$. By Lemma 1.1.7(b), $F^{G_1} \subseteq P_1$, so α is unramified in F^{G_1}. This contradiction to our assumption implies that $\alpha \notin \mathrm{Branch}(F_1/E)$. Hence, by (b), $\alpha \in \bigcup_{i \neq 1} \mathrm{Branch}(F_i/E)$. We conclude that $\bigcup_{i \neq 1} \mathrm{Branch}(F_i/E) = \mathrm{Branch}(F^{G_1}/E)$.

Proof of (d): We continue the proof of Part A of (b) under the assumptions of (d) and find, by Lemma 7.2.1, a prime divisor $\mathfrak{q}$ of F/K lying over $\mathfrak{p}$ such that $I_{\mathfrak{q}/\mathfrak{p}} = \mathrm{Gal}(F/\hat{E} \cap F)$. Since $\mathfrak{p}$ totally ramifies in F_i, $\hat{E} \cap F_i = E$. Hence, $\hat{E} \cap P_i F = \hat{E} \cap P_i F_i = P_i$. Therefore, by Lemma 1.1.7, $G_i = \mathrm{Gal}(F/\hat{E} \cap F)$, so $G_i = I_{\mathfrak{q}/\mathfrak{p}}$, as claimed. $\square$

7.3 Solution of Embedding Problems over Complete Fields

We generalize Proposition 4.4.2 and prove that if $\hat{K}_0$ is a complete absolute valued field, then each finite split embedding problem over $\hat{K}_0(x)$ is solvable. Moreover, if $\hat{K}_0$ is an extension of a subfield K_0 with $\mathrm{trans.deg}(\hat{K}_0/K_0) = \infty$, then we may find a solution field with a branch point which is transcendental over K_0. This allows us to find as many solution fields as the cardinality of $\hat{K}_0$ with branch points having given inertia groups.

PROPOSITION 7.3.1: *Consider a diagram of fields and Galois groups*

$$
\begin{array}{ccccc}
E_0 & \xrightarrow{\ \Gamma\ } & E & \xrightarrow{\ G_1\ } & E' \\
\big| & & \big| & & \\
K_0 & \text{------} & \hat{K}_0 & \xrightarrow{\ \Gamma\ } & \hat{K}
\end{array}
$$

satisfying the following assumptions:
(1a) $\hat{K}/\hat{K}_0$ *is a finite Galois extension of complete fields under an ultrametric absolute value* $|\ |$;
(1b) $E_0 = \hat{K}_0(x)$, $E = \hat{K}(x)$, *where x is an indeterminate;*
(1c) E'/E_0 *is a finite Galois extension;*
(1d) E' *has a $\hat{K}$-rational place φ unramified over E with $\varphi(x) \in \hat{K}_0 \cup \{\infty\}$;*
(1e) $\mathrm{Gal}(E'/E_0)$ *acts on a finite group H.*

Then E_0 has a finite Galois extension F with the following properties:
(a) *F contains E' and there exists an isomorphism*

$$\alpha\colon \mathrm{Gal}(F/E_0) \to \mathrm{Gal}(E'/E_0) \ltimes H$$

such that $\mathrm{pr} \circ \alpha = \mathrm{res}_{F/E'}$;
(b) *F has a $\hat{K}$-rational place φ' that extends φ and is unramified over E.*
(c) *Let $\{G_j \mid j \in J\}$ be a finite family of cyclic subgroups of H of prime power orders that generate H. Then for each $j \in J$ there exists a point $b_j \in \mathrm{Branch}(F/E) \cap \hat{K}_{0,s}$ with G_j as an inertia group.*
(d) *If $\mathrm{trans.deg}(\hat{K}_0/K_0) = \infty$, then the b_j's can be chosen to be algebraically independent over K_0.*

Proof: We may assume that $H \neq 1$. Let $\Gamma = \mathrm{Gal}(\hat{K}/\hat{K}_0) = \mathrm{Gal}(E/E_0)$ and $F_1 = E'$.

We break up the proof into several parts. The idea of the proof is to extend (E, F_1, G_1) to patching data $\mathcal{E} = (E, F_i, P_i, Q; G_i, G)_{i \in I}$ with $1 \in I$ on which Γ properly acts. The compound F of $\mathcal{E}$ will be the required solution field.

PART A: *The completion of* $(E, |\ |)$. We extend $|\ |$ to an ultrametric absolute value of E by the formula $|\sum_{i=0}^{n} a_i x^i| = \max(|a_0|, \ldots, |a_n|)$ for $a_0, \ldots, a_n \in \hat{K}$. Then let $(\hat{E}_0, |\ |)$ be the completion of $(E_0, |\ |)$ and set $\hat{E} = \hat{E}_0 E$. By Remark 2.3.2(f), $\hat{E}$ is the completion of E at $|\ |$, the map res: $\mathrm{Gal}(\hat{E}/\hat{E}_0) \to \mathrm{Gal}(E/E_0)$ is an isomorphism, and $|y^\gamma| = |y|$ for each $y \in \hat{E}$ and $\gamma \in \Gamma$.

PART B: *Construction of the* P_i's. We assume without loss $1 \notin J$. We put $I_2 = J \times \Gamma$ and let Γ act on I_2 by $(j, \gamma)^{\gamma'} = (j, \gamma\gamma')$. For each $j \in J$ we identify $(j, 1) \in I_2$ with j and notice that
(2) every $i \in I_2$ can be uniquely written as $i = j^\gamma$ with $j \in J$ and $\gamma \in \Gamma$.
We let $I = \{1\} \cup I_2$ and extend the action of Γ on I_2 to an action on I by $1^\gamma = 1$ for each $\gamma \in \Gamma$.

CLAIM: *There exists a subset* $\{c_i \mid i \in I\}$ *of* $\hat{K}$ *such that* $c_i \neq c_j$ *and* $c_i^\gamma = c_{i^\gamma}$ *for all distinct* $i, j \in I$ *and* $\gamma \in \Gamma$.
It suffices to choose $c_1 \in \hat{K}_0$ and $c_j \in \hat{K}$ for $j \in J$ (and then define c_i, for $i = j^\gamma \in I_2$, as c_j^γ) such that $c_1 \neq c_j^\varepsilon$ and $c_j^\delta \neq c_j^\varepsilon$ for all $j \in J$ and all distinct $\delta, \varepsilon \in \Gamma$, and $c_j^\delta \neq c_k^\varepsilon$ for all distinct $j, k \in J$ and all $\delta, \varepsilon \in \Gamma$.

The first condition says that c_j is a primitive element for $\hat{K}/\hat{K}_0$ and $c_j \neq c_1$ for $j \in J$ if $\hat{K}_0 = \hat{K}$; the second condition means that distinct c_j, c_k are not conjugate over $\hat{K}_0$. Thus, it suffices to show that there are infinitely many primitive elements for $\hat{K}/\hat{K}_0$. But if $c \in \hat{K}^\times$ is primitive, then so is $c+a$, for each $a \in \hat{K}_0$. Since $\hat{K}_0$ is complete, hence infinite, the claim follows.

Having proved the claim we choose $r \in \hat{K}_0^\times$ such that $|r| \leq |c_i - c_j|$ for all distinct $i, j \in I$. For each $i \in I$ we put $w_i = \frac{r}{x - c_i} \in \hat{K}(x)$. As in Section 3.2, we consider the closure $R = \hat{K}\{w_i \mid i \in I\}$ of $\hat{K}[w_i \mid i \in I]$ in $\hat{E}$ and let Q be its quotient field. For each $i \in I$ we set

$$P_i = P_{I \smallsetminus \{i\}} = \mathrm{Quot}(\hat{K}\{w_j \mid j \neq i\}) \quad \text{and} \quad P_i' = P_{\{i\}} = \mathrm{Quot}(\hat{K}\{w_i\})$$

(we use the notation of Section 3.3).

By the Claim, each $\gamma \in \Gamma$ satisfies $w_i^\gamma = w_{i^\gamma}$, hence maps $R_0 = \hat{K}[w_i \mid i \in I]$ onto itself. Since the action of γ on $\hat{E}$ is continuous, γ leaves R, hence also Q, invariant. We identify Γ with its image in $\mathrm{Aut}(Q/E_0)$. In addition, $P_i^\gamma = P_{i^\gamma}$ and $(P_i')^\gamma = P_{i^\gamma}'$ for each $i \in I$.

PART C: *Without loss of generality* $F_1 \subset P_1'$ *and* $\varphi(w_1) = 0$. To show this it suffices to construct a $\hat{K}$-embedding $\theta \colon F_1 \to P_1'$ such that $\theta(E_0) = E_0$, $\theta(E) = E$, and $\varphi \circ \theta^{-1}(w_1) = 0$. Then the assumptions and the conclusions of our proposition hold for (F_1, φ) if and only if they hold for $(\theta(F_1), \varphi \circ \theta^{-1})$.

We construct θ as above in two steps.

Since φ maps w_1 into $\hat{K}_0 \cup \{\infty\}$, there is a $\hat{K}_0$-automorphism ω of $E_0 = \hat{K}_0(w_1)$ such that $\varphi \circ \omega^{-1}(w_1) = 0$. We extend ω to a $\hat{K}$-automorphism of E and then to an isomorphism of fields $F_1 \to F_1'$. Applying the first paragraph of Part C, we assume that $\varphi(w_1) = 0$, so that $F_1 \subseteq \hat{K}((w_1))$.

Let y_1 be a primitive element for F_1/E. Since y_1 is algebraic over E, and since for each $c \in \hat{K}^\times$ there are elements in $\hat{K}_0^\times$ of smaller absolute value, there is a $c \in \hat{K}_0$ at which y_1 converges (Proposition 2.4.5). In the notation of Remark 4.3.1, this implies that $\mu_c(y_1) \in \mathrm{Quot}(\hat{K}\{w_1\}) = P_1'$. Therefore, the automorphism μ_c of $\hat{K}((w_1))$ leaves $E_0 = \hat{K}_0(w_1)$ and $E = \hat{K}(w_1)$ invariant and maps F_1 into P_1'. In particular, $\varphi(\mu_c^{-1}(w_1)) = \varphi(c^{-1} w_1) = c^{-1}\varphi(w_1) = 0$, as needed.

PART D: *Groups.* Since $F_1 \subset Q$ is a Galois extension of E_0, it is Γ-invariant. We identify $\Gamma \le \mathrm{Aut}(Q/E_0)$ with its image in $\mathrm{Gal}(F_1/E_0)$. Then $\mathrm{Gal}(F_1/E_0) = \Gamma \ltimes G_1$, where Γ acts on G_1 by conjugation in $\mathrm{Gal}(F_1/E_0)$. Thus,

(3) $(a^\tau)^\gamma = (a^\gamma)^{\tau^\gamma}$ for all $\gamma \in \Gamma$, $a \in F_1$, and $\tau \in G_1$.

The given action of $\mathrm{Gal}(F_1/E_0)$ on H induces an action of its subgroups G_1 and Γ on H. Let $G = G_1 \ltimes H$ with respect to this action. Then

$$\mathrm{Gal}(F_1/E_0) \ltimes H = (\Gamma \ltimes G_1) \ltimes H = \Gamma \ltimes (G_1 \ltimes H) = \Gamma \ltimes G.$$

By assumption $H = \langle G_j \mid j \in J \rangle$. For each $j \in J$ we choose a generator τ_j of G_j. For each $i \in I_2$ we use (2) to write $i = j^{\gamma'}$ with unique $j \in J$ and $\gamma' \in \Gamma$. Then we define $G_i = G_j^{\gamma'}$, $\tau_i = \tau_j^{\gamma'}$ (so $G_i = \langle \tau_i \rangle$), and observe that

(4a) $G_i^\gamma = G_{i^\gamma}$ for all $i \in I_2$ and $\gamma \in \Gamma$.
(4b) $H = \langle G_i \mid i \in I_2 \rangle$ and $G = \langle G_i \mid i \in I \rangle$.
(4c) $|I| \ge 2$.

PART E: *Patching data.* Let π_0 be an element of $\hat{K}_0^\times$ with $|\pi_0| < 1$ and set $\Pi(\pi_0, \hat{K}_0) = \{\pi_0^k \zeta_n^l \mid k = 0, 1, 2, \ldots; \; l = 0, 1; \; n \in \mathbb{N}, \; \mathrm{char}(\hat{K}_0) \nmid n\}$. Then, $\Pi(\pi_0, \hat{K}_0) \subseteq \hat{K}_{0,s}$ and for each $j \in J$ and each $a_j \in \hat{K}_0$ Lemma 4.3.5 gives a cyclic extension F_j/E with Galois group isomorphic to G_j and a $\pi_j \in \Pi(\pi_0, \hat{K}_0)$ such that $F_j \subset P_j'$, $b_j = \frac{\pi_j a_j c_j + r}{\pi_j a_j} \in \mathrm{Branch}(F_j/E)$, and $\mathfrak{p}_{E,x,b_j}$ is totally ramified in F_j, so that $I_{b_j} = G_j$ (here we are using that $|G_j|$ is a power of a prime). If $\mathrm{trans.deg}(\hat{K}_0/K_0) = \infty$, we may choose the a_j's such that they are algebraically independent over $K_1 = K_0(r, c_j, \pi_0)_{j \in J}$. Since $\pi_0 \in K_1$, the elements b_j are algebraically independent over K_1, hence also over K_0.

For an arbitrary $i \in I_2$ there exist unique $j \in J$ and $\gamma \in \Gamma$ such that $i = j^\gamma$ (by (2)). Let $F_i = F_j^\gamma$. Since γ acts on Q and leaves E invariant, F_i is a Galois extension of E and $F_i = F_j^\gamma \subseteq (P_j')^\gamma = P_i'$ (by Part B).

The isomorphism $\gamma \colon F_j \to F_i$ yields an isomorphism $\mathrm{Gal}(F_j/E) \cong \mathrm{Gal}(F_i/E)$ that maps each $\tau \in \mathrm{Gal}(F_j/E)$ onto $\gamma^{-1} \circ \tau \circ \gamma \in \mathrm{Gal}(F_i/E)$.

We can therefore identify G_i with $\mathrm{Gal}(F_i/E)$ such that τ_i coincides with $\gamma^{-1} \circ \tau_j \circ \gamma$. This means that $(a^\tau)^\gamma = (a^\gamma)^{\tau^\gamma}$ for all $a \in F_j$ and $\tau \in G_j$.

It follows that for all $i \in I$ and $\gamma \in \Gamma$ we have $F_i^\gamma = F_{i\gamma}$. Moreover, $(a^\tau)^\gamma = (a^\gamma)^{\tau^\gamma}$ for all $a \in F_i$ and $\tau \in G_i$; this generalizes (3).

By Proposition 3.4.5, $\mathcal{E} = (E, F_i, P_i, Q; G_i, G)_{i \in I}$ is patching data. By construction Γ acts properly acts on $\mathcal{E}$ (Definition 1.2.1). Let F be the compound of $\mathcal{E}$. Lemma 1.3.1(d) gives an identification $\alpha \colon \mathrm{Gal}(F/E_0) \to \mathrm{Gal}(E'/E_0) \ltimes H$ such that $\mathrm{pr} \circ \alpha = \mathrm{res}_{F/E'}$, as claimed. By Lemma 7.2.3(b), $\mathrm{Branch}(F/E) = \bigcup_{i \in I} \mathrm{Branch}(F_i/E)$. In particular, $b_j \in \mathrm{Branch}(F/E) \cap \hat{K}_{0,s}$ for each $j \in J$. Moreover, by Lemma 7.2.3(d), G_j is an inertia group of $\mathfrak{p}_{E,x,b_j}$ in $\mathrm{Gal}(F/E)$ for each $j \in J$.

PART F: *Extension of* φ. Let $b \in \hat{K}_0$ such that $|b| > |r|$ and $|b| > |c_i|$ for each $i \in I$. We set $z = \frac{b}{x}$ and let $\hat{K}\{z\}$ be the ring of convergent power series in z over $\hat{K}$ with respect to the absolute value $|\ |_z$ given by $|\sum_{n=0}^\infty a_n z^n|_z = \max(|a_n|)$. Then,

$$(5) \qquad w_i = \frac{r}{x - c_i} = \frac{rz}{b - c_i z} = \frac{rz}{b} \cdot \frac{1}{1 - \frac{c_i}{b} z} = \frac{rz}{b} \sum_{n=0}^\infty \left(\frac{c_i}{b}\right)^n z^n \in \hat{K}\{z\}$$

for each $i \in I$. Thus, $R_0 \subset \hat{K}\{z\}$. Moreover, $|w_i|_z = \frac{|r|}{|b|} < 1$. By Lemma 3.1.1, every $f \in R_0$ is of the form $f = a_0 + \sum_{i \in I} \sum_{n=1}^\infty a_{in} w_i^n$, where $a_0, a_{in} \in \hat{K}$ and almost all of them are 0. Hence, $|f|_z \le |f|$, so the inclusion $R_0 \subset \hat{K}\{z\}$ is a continuous R_0-embedding. Since R is the completion of R_0 with respect to $|\ |$ (second paragraph of Section 3.2), this inclusion induces a continuous R_0-homomorphism $\kappa \colon R \to \hat{K}\{z\}$. By Proposition 3.2.9, there is a $p \in R_0$ such that $\mathrm{Ker}(\kappa) = Rp$. Therefore, $p = 0$ and κ is injective.

We identify R with its image under κ to assume that $R \subset \hat{K}\{z\} \subset \hat{K}[[z]]$. Then $F \subseteq Q = \mathrm{Quot}(R) \subseteq \hat{K}((z))$. The specialization $z \to 0$ extends to a $\hat{K}$-rational place of $\hat{K}((z))$ unramified over $E = \hat{K}(z)$. Its restriction to F is a $\hat{K}$-rational place ψ of F unramified over $E = \hat{K}(z)$.

Since $\psi(w_1) = 0 = \varphi(w_1)$ (by (5)) and $E = \hat{K}(w_1)$, we have $\psi|_E = \varphi|_E$. We may therefore replace ψ by $\psi \circ \sigma$ for a suitable $\sigma \in \mathrm{Gal}(F/E)$, if necessary, to assume that $\psi|_{F_1} = \varphi$. $\qquad \square$

7.4 Linearly Disjoint Solutions

Consider a finite embedding problem

$$(1) \qquad (\mathrm{res} \colon \mathrm{Gal}(E) \to \mathrm{Gal}(E'/E), \ \alpha \colon H \to \mathrm{Gal}(E'/E))$$

over a field E and a set $\mathcal{F} = \{F_i \mid i \in I\}$ of solution fields of (1). In particular, each F_i is a finite Galois extension of E that contains E'. We say that $\mathcal{F}$ is **linearly disjoint** if the fields F_i, $i \in I$, are linearly disjoint over E'. The following lemma applies simple valuation theoretic principles to achieve linear disjointness of a solution field from a given Galois extension of E containing E'.

LEMMA 7.4.1: *Let $E \subseteq E' \subseteq F$ and $E \subseteq E' \subseteq N$ be towers of fields such that E'/E, F/E, and N/E are Galois extensions and F/E is finite. For each j in a set J let G_j be a subgroup of $\mathrm{Gal}(F/E)$, v_j a discrete valuation of E, and w_j an extension of v_j to F. Suppose that $\mathrm{Gal}(F/E') = \langle G_j \mid j \in J \rangle$, $G_j \leq I_{w_j/v_j}$, and v_j is unramified in N for each $j \in J$. Then $F \cap N = E'$.*

Proof: We denote the restriction of w_j to $F \cap N$ by v'_j. By assumption v'_j is unramified over E, hence $I_{v'_j/v_j} = 1$ (Remark 4.2.3). Since $I_{v'_j/v_j}$ is the image of I_{w_j/v_j} under the map res: $\mathrm{Gal}(F/E) \to \mathrm{Gal}(F \cap N/E)$, we have $I_{w_j/v_j} \leq \mathrm{Gal}(F/F \cap N)$. By assumption,

$$\mathrm{Gal}(F/E') = \langle G_j \mid j \in J \rangle \leq \langle I_{w_j/v_j} \mid j \in J \rangle \leq \mathrm{Gal}(F/F \cap N),$$

so $F \cap N \subseteq E'$. Since $E' \subseteq F \cap N$, we conclude that $F \cap N = E'$, as claimed. $\square$

LEMMA 7.4.2: *If K/K_0 is an extension of fields such that $\mathrm{card}(K_0) < \mathrm{card}(K)$ and $\mathrm{card}(K) > \aleph_0$, then $\mathrm{trans.deg}(K/K_0) = \infty$.*

Proof: The cardinality of a field does not change or becomes at most $\aleph_0$ by algebraic extensions or by finitely generated purely transcendental extensions. Hence, $\mathrm{trans.deg}(K/K_0) = \infty$. $\square$

LEMMA 7.4.3: *Let $\hat{K}$ be a complete field under an ultrametric absolute value $|\ |$. Then $\mathrm{card}(\hat{K}) \geq 2^{\aleph_0}$.*

Proof: By assumption there exists $\pi \in \hat{K}$ with $0 < |\pi| < 1$. For each sequence $\mathbf{a} = (a_0, a_1, a_2, \ldots)$ consisting of 0's and 1's the infinite series $\sum_{n=0}^{\infty} a_n \pi^n$ converges in $\hat{K}$ to an element, say $s(\mathbf{a})$. If $\mathbf{b} = (b_0, b_1, b_2, \ldots)$ is different similar sequence, then there is an $r \geq 0$ such that $a_i = b_i$ for $i = 0, \ldots, r - 1$ and, say, $a_r = 0$ and $b_r = 1$. Then $s(\mathbf{b}) - s(\mathbf{a}) = \pi^r + \sum_{n=r+1}^{\infty} (b_n - a_n) \pi^n$. Hence, $|s(\mathbf{b}) - s(\mathbf{a})| = |\pi|^r \neq 0$, so $s(\mathbf{b}) \neq s(\mathbf{a})$. Since there are $2^{\aleph_0}$ sequences $\mathbf{a}$ of that type, we conclude that $\mathrm{card}(\hat{K}) \geq 2^{\aleph_0}$. $\square$

PROPOSITION 7.4.4: *Consider a diagram of fields and Galois groups*

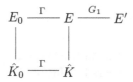

satisfying the following assumptions:

(2a) $\hat{K}/\hat{K}_0$ is a finite Galois extension of complete fields under an ultrametric absolute value $|\ |$;

(2b) $E_0 = \hat{K}_0(x)$, $E = \hat{K}(x)$, where x is an indeterminate;

(2c) E'/E_0 is a finite Galois extension;

(2d) E' has a $\hat{K}$-rational place φ unramified over E with $\varphi(x) \in \hat{K}_0 \cup \{\infty\}$; and

(2e) $\mathrm{Gal}(E'/E_0)$ acts on a finite group H.

Then, with $m = \mathrm{card}(\hat{K}_0)$, the embedding problem

$$(3)\qquad\qquad \mathrm{Gal}(E'/E_0) \ltimes H \to \mathrm{Gal}(E'/E_0)$$

over E_0 has m linearly disjoint solution fields, each equipped with a $\hat{K}$-rational place that extends φ and is unramified over E.

Proof: We use transfinite induction to construct a transfinite sequence $(F_\alpha)_{\alpha < m}$ of solution fields of (3) with the desired properties. Suppose $\beta < m$ is an ordinal number and for each $\alpha < \beta$ we have constructed a solution field F_α of (3) equipped with a $\hat{K}$-rational place φ_α unramified over E that extends φ. Then $N = \prod_{\alpha < \beta} F_\alpha$ is a Galois extension of E_0 that contains E'. Moreover, $\mathrm{Branch}(N/E_0) = \bigcup_{\alpha < \beta} \mathrm{Branch}(F_\alpha/E_0)$ (Remark 4.1.1). Since $\mathrm{Branch}(F_\alpha/E_0)$ is a finite set, $\mathrm{card}(\mathrm{Branch}(N/E_0)) < m$.

Let K_0 be the prime field of $\hat{K}_0$ and set $K_1' = K_0(\mathrm{Branch}(N/E_0))$. By Lemma 7.4.3, $m \ge 2^{\aleph_0}$. Since $\mathrm{card}(K_0) \le \aleph_0$, this implies that $\mathrm{card}(K_1') < m$. Moreover, each element y of K_1' is algebraic over $\hat{K}_0$. Let K_1 be the field generated over K_0 by the coefficients of $\mathrm{irr}(y, \hat{K}_0)$, where y ranges on K_1'. Then, $K_1' \subseteq \hat{K}_1$ and $\mathrm{card}(K_1) \le \mathrm{card}(K_1') < m = \mathrm{card}(\hat{K}_0)$. Hence, by Lemma 7.4.2, $\mathrm{trans.deg}(\hat{K}_0/K_1) = \infty$.

Since each element of H is a product of elements of prime power order, we may find a finite set $\{G_j \mid j \in J\}$ of cyclic subgroups of H of prime power order such that $H = \langle G_j \mid j \in J \rangle$. Proposition 7.3.1 then gives a solution field F_β of embedding problem (3) and for each j a point $b_j \in \mathrm{Branch}(F_\beta/E_0)$ with G_j as an inertia group such that the b_j's are algebraically independent over K_1. Moreover, φ extends to a $\hat{K}$-rational place φ_β of F_β unramified over E. In particular, $b_j \notin \mathrm{Branch}(N/E_0)$, so $\mathfrak{p}_{E_0, x, b_j}$ is unramified in N for all $j \in J$. It follows from Lemma 7.4.1 that $F_\beta \cap N = E'$. Consequently, the transfinite sequence of solutions $(F_\alpha)_{\alpha \le \beta}$ is linearly disjoint. This completes the transfinite induction.   □

Proposition 8.6.3 generalizes Proposition 7.4.4 to all ample fields.

Notes

The results of section 7.1 appear in [GeJ98, Sec. 9].

Lemma 7.2.1 already appears in [HaV96, Lemma 3.6(e)].

Lemma 7.2.2 is based on [HaJ00a, Lemma 1.2] and, in its present form, on [Har05, Lemma 3.12].

The first three parts of Lemma 7.2.3 are taken from [HaJ00a, Lemma 1.4].

Proposition 7.3.1 is the main result of this chapter. One of its main ingredients, namely information about inertia groups appears already in [Pop94].

In contrast to Proposition 4.4.2, where only constant split embedding problems are considered, Proposition 7.3.1 solves arbitrary finite split embedding problems over $K_0(x)$, where K_0 is a complete field with respect to an ultrametric absolute value. This parts appears for the first time in [Pop03, Lemma 2.8]. Our proof follows that of [HaJ98b, Prop. 4.1]. It adds the information about the branch points and their inertia groups. That feature appears already in [Pop94].

Chapter 8.
Split Embedding Problems over Ample Fields

We generalize Theorem 5.9.2 and prove that if K_0 is an ample field, then not only constant finite split embedding problems over $K_0(x)$ are solvable but every finite split embedding problem $\mathrm{Gal}(E/K_0(x)) \ltimes H \to \mathrm{Gal}(E/K_0(x))$ has as many linearly disjoint solution fields F_α, with $\alpha < \mathrm{card}(K)$ (Proposition 8.6.3). Moreover, let K be the algebraic closure of K_0 in E. Then each K-rational place φ of E unramified over $K_0(x)$ with $\varphi(x) \in K_0 \cup \{\infty\}$ extends to a K-rational place of F_α unramified over $K_0(x)$ (Lemma 8.6.1).

The construction of the solutions for general finite split embedding problems over $K_0(x)$ in the case where K_0 is an ample field relies on Proposition 7.3.1, where K_0 is assumed to be complete under an ultrametric absolute value. For an arbitrary ample field K_0, we first lift the embedding problem to one over $K_0((t))(x)$, and apply Proposition 7.3.1 to solve it with additional information on the branch points (in particular they should be algebraically independent over K_0) and on their inertia groups. Then, we use Bertini-Noether as in Lemma 5.9.1 to reduce that solution to one of the original problem. In order to achieve many linearly disjoint solutions the reduction has to keep track of the branch points and their inertia groups. This can be done once the reduction is normal in the sense of Section 8.1.

8.1 Normal Reduction

Let (K, w) be a valued field and x a variable. The **Gauss extension** of w to a valuation of the field $K(x)$ bearing the same notation is defined by

$$(1) \qquad w\Big(\sum_{i=0}^{n} a_i x^i \Big) = \min(w(a_0), \ldots, w(a_n))$$

for $a_0, \ldots, a_n \in K$ [FrJ08, Example 2.3.3].

We denote the residue field of K by $\bar{K}$ and the residue of each element $y \in K(x)$ by $\bar{y}$. Definition (1) implies that $\bar{x}$ is transcendental over $\bar{K}$. Indeed, suppose there are $\bar{a}_0, \ldots, \bar{a}_n \in \bar{K}$ such that $\sum_{i=0}^{n} \bar{a}_i \bar{x}^i = 0$. Let a_i be a lifting of $\bar{a}_i$ to K. Then $w(\sum_{i=0}^{n} a_i x^i) > 0$. Hence, by (1), $w(a_i) > 0$, so $\bar{a}_i = 0$ for each i.

We denote the valuation rings of w in K and in $K(x)$ by $O_{w,K}$ and $O_{w,K(x)}$, respectively. By definition, $O_{w,K(x)} \cap K = O_{w,K}$. Each element of $O_{w,K(x)}$ has the form

$$y = \frac{\sum a_i x^i}{\sum b_j x^j},$$

where i and j range over finite index sets, $a_i, b_j \in K$, and $\min(w(a_i)) \geq \min(w(b_j))$ and not all b_j are zero. We choose an index l with $w(b_l) =$

M. Jarden, *Algebraic Patching*, Springer Monographs in Mathematics,
DOI 10.1007/978-3-642-15128-6_8, © Springer-Verlag Berlin Heidelberg 2011

$\min(w(b_j))$ and divide the numerator and the denominator of y by b_l, if necessary, to assume that $\min(w(a_i)), \min(w(b_j)) \geq 0$ and one of the b_j's is 1. Taking residues, we find that

$$\bar{y} = \frac{\sum \bar{a}_i \bar{x}^i}{\sum \bar{b}_j \bar{x}^j} \in \bar{K}(\bar{x}).$$

It follows that

(2) $$\overline{K(x)} = \bar{K}(\bar{x}).$$

In the case where the denominator of y is 1, the previous arguments imply that

(3) $$O_{w,K(x)} \cap K[x] = O_{w,K}[x] \text{ and } \overline{K[x]} = \bar{K}[\bar{x}].$$

Now we consider a finite extension F of $K(x)$, extend w to a valuation of F having the same notation w, and write $O_{w,F}$ for the valuation ring of w in F. Let $\bar{F}$ be the residue field of F at w. By (2), $[\bar{F} : \bar{K}(\bar{x})] \leq [F : K(x)]$. We say that F has a **degree preserving constant reduction** at w over K with respect to x if the latter inequality is an equality, i.e.

(4) $$[F : K(x)] = [\bar{F} : \bar{K}(\bar{x})].$$

LEMMA 8.1.1: *Under Assumption (4), w extends uniquely from $K(x)$ to F, $w(F^\times) = w(K^\times)$, and $O_{w,F}$ is the integral closure of $O_{w,K(x)}$ in F. If in addition, $\bar{F}/\bar{K}(\bar{x})$ is separable, then w is unramified over $K(x)$.*

Proof: We list the extensions of w from $K(x)$ to F as $w_1, \ldots, w_m$. By the fundamental inequality for valuations, the ramification indices and the residue fields of F at these valuations satisfy $\sum_{i=1}^m e_i[\bar{F}_{w_i} : \bar{K}(\bar{x})] \leq [F : K(x)]$ [Efr06, Thm. 17.1.5]. It follows from (4) that $m = 1$, w has a unique extension from $K(x)$ to F, and the ramification index is 1. In particular, $w(F^\times) = w(K(x)^\times)$. Hence, by (1), $w(F^\times) = w(K^\times)$. Since the integral closure of $O_{w,K(x)}$ in F is the intersection of all valuation rings of F that lie over $O_{w,K(x)}$ [Lan58, p. 14, Prop. 5], $O_{w,F}$ is the integral closure of $O_{w,K(x)}$ in F. Finally, if $\bar{F}/\bar{K}(\bar{x})$ is a separable extension, then w is unramified over $K(x)$. $\qquad\square$

Note that $\bar{F}$ is the set of all residues $\bar{z}$ with $z \in O_{w,F}$. More generally, for a subset A of F we write $\bar{A} = \{\bar{a} \in \bar{F} \mid a \in A \cap O_{w,F}\}$ and observe that $\bar{A} = \overline{A \cap O_{w,F}}$. Moreover, $A \subseteq B \subseteq F$ implies $\bar{A} \subseteq \bar{B} \subseteq \bar{F}$. If R is a subring of F, then $\bar{R}$ is a subring of $\bar{F}$; and if I is an ideal of R, then $\bar{I}$ is an ideal of $\bar{R}$.

DEFINITION 8.1.2: *Let F/K be a function field of one variable, x a separating transcendental element, and w a valuation of $F\tilde{K}$. We say that F/K has a **degree preserving regular constant reduction** at w with respect to x if the following conditions hold:*
(5a) *The element $\bar{x}$ is a separating transcendental element of $\bar{F}/\bar{K}$.*
(5b) *The extensions $F/K(x)$ and $\bar{F}/\bar{K}(\bar{x})$ have the same degree n.*
(5c) *$\bar{F}/\bar{K}$ is a regular extension.*   □

LEMMA 8.1.3: *In the notation of Definition 8.1.2, let L be an algebraic extension of K. Then FL/L has a degree preserving regular constant reduction at w with respect to x. Moreover, $\overline{FL} = \bar{F}\bar{L}$.*

Proof: By (2) applied to L rather than to K, we have $\overline{L(x)} = \bar{L}(\bar{x})$. Since both F/K and $\bar{F}/\bar{K}$ are regular extensions, so are FL/L and $\bar{F}\bar{L}/\bar{L}$. Moreover, $[FL : L(x)] = [F : K(x)]$ and $[\bar{F}\bar{L} : \bar{L}(\bar{x})] = [\bar{F} : \bar{K}(\bar{x})]$. Also, note that $\bar{L}(\bar{x}) \subseteq \bar{F}\bar{L} \subseteq \overline{FL}$. Hence, by (5b),

$$[\overline{FL} : \bar{L}(\bar{x})] \geq [\bar{F}\bar{L} : \bar{L}(\bar{x}] = [\bar{F} : \bar{K}(\bar{x})]$$
$$= [F : K(x)] = [FL : L(x)] \geq [\overline{FL} : \bar{L}(\bar{x})],$$

so $[\overline{FL} : \bar{L}(\bar{x})] = [\bar{F}\bar{L} : \bar{L}(\bar{x})]$. This proves that $\overline{FL} = \bar{F}\bar{L}$ and also (5a), (5b), and (5c) for L rather than for K, as claimed.   □

Definition 8.1.4: In the notation of Definition 8.1.2, let R (resp. R') be the integral closure of $K[x]$ (resp. $K[x^{-1}]$) in F. We say that F/K has a **normal reduction** at w with respect to x if in addition to (5a), (5b), and (5c) also the following condition holds:
(6) $\bar{R}$ and $\bar{R'}$ are integrally closed.   □

Remark 8.1.5: *Good reduction.* We say that F/K has a **good reduction** at w with respect to x if (5a), (5b), (5c), and
(6') genus(F/K) = genus$(\bar{F}/\bar{K})$
hold. In this book we use "normal reduction". However, Green and Roquette prove in the forthcoming book [GrR] that "normal reduction" is equivalent to "good reduction" of F/K at w with respect to x, if K is perfect.   □

By [Ser79, Prop. I.4.8], the rings R and R' mentioned in Definition 8.1.4 are finitely generated over K. Combined with the following result, this gives a convenient criterion for Condition (6) to hold.

LEMMA 8.1.6: *Let F be a function field of one variable over an algebraically closed field L and let $y_1, \ldots, y_n \in F$. Let $f_1, \ldots, f_m \in L[\mathbf{Y}]$, with $\mathbf{Y} = (Y_1, \ldots, Y_n)$ such that $f_i(\mathbf{y}) = 0$ for $i = 1, \ldots, m$, and consider the **Jacobian matrix***

$$J(\mathbf{Y}) = \left(\frac{\partial f_i}{\partial Y_j}\right)_{1 \leq i \leq m, \, 1 \leq j \leq n}.$$

(a) *If the ring $L[\mathbf{y}]$ is integrally closed and $f_1, \ldots, f_m$ generate the ideal of all polynomials in $L[\mathbf{Y}]$ that vanish at $\mathbf{y}$, then 1 is a linear combination of all $(n-1) \times (n-1)$ subdeterminants of $J(\mathbf{y})$ with coefficients in $L[\mathbf{y}]$.*

(b) *If 1 is a linear combination of all $(n-1) \times (n-1)$ subdeterminants of $J(\mathbf{y})$ with coefficients in $L[\mathbf{y}]$, then $L[\mathbf{y}]$ is integrally closed.*

Proof: Let $d_1, \ldots, d_r \in L[\mathbf{Y}]$ be the determinants of the $(n-1) \times (n-1)$ submatrices of $J(\mathbf{Y})$.

Proof of (a): The assumption that $L[\mathbf{y}]$ is integrally closed implies that the affine curve $C = \operatorname{Spec}(L[\mathbf{y}])$ generated by the point $\mathbf{y}$ over L is smooth, so $\operatorname{rank}(J(\mathbf{a})) = n-1$ for each $\mathbf{a} \in C(L)$ [Sha77, p. 112, Cor]. Hence, there is a k such that $d_k(\mathbf{a}) \neq 0$. It follows that $f_1, \ldots, f_m, d_1, \ldots, d_r$ have no common zero in L^n. By Hilbert Nullstellensatz [FrJ08, Prop. 9.4.1], 1 is a linear combination of $f_1, \ldots, f_m, d_1, \ldots, d_r$ with coefficients in $L[\mathbf{Y}]$. Substituting $\mathbf{y}$ for $\mathbf{Y}$, we conclude that 1 is a linear combination of $d_1(\mathbf{y}), \ldots, d_r(\mathbf{y})$ with coefficients in $L[\mathbf{y}]$, as claimed.

Proof of (b): We add $f_{m+1}, \ldots, f_{m'} \in L[\mathbf{Y}]$ to $f_1, \ldots, f_m$ such that $f_1, \ldots, f_{m'}$ generate the ideal of all polynomials in $L[\mathbf{Y}]$ that vanish at $\mathbf{y}$ and consider the Jacobian matrix $J'(\mathbf{Y}) = \left(\frac{\partial f_i}{\partial Y_j}\right)_{1 \leq i \leq m', 1 \leq j \leq n}$. By assumption, 1 is a linear combination of all $(n-1) \times (n-1)$ subdeterminants of $J(\mathbf{Y})$, hence of $J'(\mathbf{Y})$, with coefficients in $L[\mathbf{y}]$. Therefore, for each point $\mathbf{a} \in C(L)$ there is a subdeterminant of $J'(\mathbf{Y})$ whose value at $\mathbf{a}$ is nonzero. This implies that $\operatorname{rank}(J'(\mathbf{a})) \geq n-1$. By [Mum88, Cor. III.4.1], $\mathbf{a}$ is a simple point of C. We conclude from [Sha77, p. 110, Thm. 1] that the coordinate ring $L[\mathbf{y}]$ of C is integrally closed. $\square$

8.2 Inertia Groups

Our main application of reduction theory is to make sure that inertia groups in appropriate Galois extensions of algebraic function fields of one variable are mapped under normal reductions into inertia groups of the reduced Galois extensions. Unfortunately, we can not always produce normal reductions. However, we can produce degree preserving regular constant reductions that become normal reductions after algebraic extensions of the base field.

Remark 8.2.1: Inertia groups of extensions of prime ideals. Let K be a field, x a variable, and F a finite Galois extension of $E = K(x)$. Let $b \in \tilde{K}$, set $p_{x,b} = \operatorname{irr}(b, K)$, let $P_{x,b} = p_{x,b}(x)K[x]$ be the corresponding prime ideal of $K[x]$, and let v be the valuation of E/K with $v(p_{x,b}(x)) = 1$. Next we denote the integral closure of $K[x]$ in F by R and choose a valuation v' of F lying over v. Let $P = \{y \in R \mid v'(y) > 0\}$ be the center of v' at R.

$O_{v'}$ is the local ring R_P of R at P.

The **inertia group** of $P/P_{x,b}$ is defined as

$$I_{P/P_{x,b}} = \{\sigma \in \operatorname{Gal}(F/E) \mid y^\sigma \equiv y \bmod P \text{ for each } y \in R\}.$$

It contains the inertia group

$$I_{v'/v} = \{\sigma \in \mathrm{Gal}(F/E) \mid v'(z^\sigma - z) > 0 \text{ for each } z \in O_{v'}\}.$$

Conversely, suppose $\sigma \in I_{P/P_{x,b}}$. Then $R^\sigma = R$ and $P^\sigma = P$. Each $z \in O_{v'}$ can be written as $z = \frac{y}{u}$ with $y \in R$ and $u \in R \smallsetminus P$. In particular, $v'(y), v'(y^\sigma) \geq 0$ and $v'(u) = v'(u^\sigma) = 0$. Hence,

$$\begin{aligned}
v'(z^\sigma - z) &= v'(uy^\sigma - u^\sigma y) - v'(u^\sigma) - v'(u) \\
&= v'(uy^\sigma - uy + uy - u^\sigma y) = v'(u(y^\sigma - y) + (u - u^\sigma)y) \\
&\geq \min(v'(u) + v'(y^\sigma - y), v'(u - u^\sigma) + v'(y)) > 0,
\end{aligned}$$

so $\sigma \in I_{v'/v}$. Consequently, $I_{P/P_{x,b}} = I_{v'/v}$.

By Remark 4.1.1, b is a branch point of F/E with respect to x if and only if v'/v is ramified. By Remark 4.2.3, v'/v is ramified if and only if $I_{v'/v} \neq 1$. It follows from the preceding paragraph that b is a branch point of F/E if and only if $I_{P/P_{x,b}} \neq 1$. Moreover, in this case $I_{P/P_{x,b}}$ is an inertia group of b in $\mathrm{Gal}(F/E)$.

As usual, we define the inertia groups in F of ∞ with respect to x as the inertia group in F of 0 with respect to x^{-1}.   □

LEMMA 8.2.2: *Let F/K be a function field of one variable, x a separating transcendental element, and w a valuation of F. Suppose F/K has a degree preserving regular constant reduction at w with respect to x and $F/K(x)$ is Galois.*

Then there exists an isomorphism $\sigma \mapsto \bar\sigma$ of $\mathrm{Gal}(F/K(x))$ onto $\mathrm{Gal}(\bar F/\bar K(\bar x))$ such that $\bar y^{\bar\sigma} = \overline{y^\sigma}$ for all $\sigma \in \mathrm{Gal}(F/K(x))$ and $y \in O_{w,F}$.

Proof: By (2) of Section 8.1, $\overline{K(x)} = \bar K(\bar x)$, so $\bar F/\bar K(\bar x)$ is normal. By (5a) of Section 8.2, $\bar F/\bar K(\bar x)$ is separable, hence $\bar F/\bar K(\bar x)$ is Galois, Therefore, by Lemma 8.1.1, w is unramified over $K(x)$. Since $[F : K(x)] = [\bar F : \bar K(\bar x)]$, the decomposition group of w over $K(x)$ is $\mathrm{Gal}(F/K(x))$. Thus, the lemma is a special case of [FrJ08, Lemma 6.1.1(b)].   □

LEMMA 8.2.3: *Let E/K be a function field of one variable and F/E a finite Galois extension with F/K regular. Let K' be an algebraic extension of K and set $E' = EK'$ and $F' = FK'$. Let w' be a valuation of F'/K' and denote the restriction of w' to E, F, E', respectively, by v, w, v'. Then the isomorphism res: $\mathrm{Gal}(F'/E') \to \mathrm{Gal}(F/E)$ maps $I_{w'/v'}$ onto $I_{w/v}$.*

Proof: First we observe that we may assume that K'/K is finite and then also that F'/F is normal. Secondly we may assume that either K'/K is separable or K'/K is purely inseparable.

If K'/K is separable, then K'/K is Galois, hence F'/E is a finite Galois extension. By Remark 5.8.1(e), v'/v is unramified, so $I_{w'/v} = I_{w'/v'}$. By Remark 4.2.3, $\mathrm{res}(I_{w'/v'}) = \mathrm{res}(I_{w'/v}) = I_{w/v}$.

If char$(K) > 0$ and K'/K is purely inseparable, then there exists a power q of char(K) with $(K')^q \subseteq K$, so $(F')^q \subseteq F$. Each $\sigma \in I_{w/v}$ can be lifted to an element $\sigma' \in \mathrm{Gal}(F'/E')$. Consider $x \in O_{w'}$. It satisfies $x^q \in O_w$, hence $w((x^q)^\sigma - x^q) > 0$. Therefore, with $e = e_{w'/w}$ we have $eq \cdot w'(x^\sigma - x) > 0$. It follows that $w'(x^\sigma - x) > 0$, hence $\sigma \in I_{w'/v}$, as claimed. $\square$

LEMMA 8.2.4: *Let F/K be a function field of one variable, x a separating transcendental element, and w a valuation of $F\tilde{K}$. Suppose F/K has a degree preserving regular constant reduction at $w|_F$ with respect to x and $F/K(x)$ is Galois. Suppose $F\tilde{K}/\tilde{K}$ has a normal reduction at w with respect to x.*

Then for each $b \in \tilde{K}$ there exists an inertia group I_b of b in $\mathrm{Gal}(F/K(x))$ and there exists an inertia group $I_{\bar{b}}$ of $\bar{b}$ in $\mathrm{Gal}(\bar{F}/\bar{K}(\bar{x}))$ such that, under the isomorphism of Lemma 8.2.2, $\bar{I}_b \leq I_{\bar{b}}$. In particular, if b is a branch point of $F/K(x)$, then $\bar{b}$ is a branch point of $\bar{F}/\bar{K}(\bar{x})$.

Proof: As in the proof of Lemma 8.2.2, $\bar{F}/\bar{K}(\bar{x})$ is Galois, so the second paragraph of our lemma makes sense. First we prove the lemma over $\tilde{K}$ and then we descend to K.

PART A: *Working over $L = \tilde{K}$.*   Exchanging x with x^{-1}, if necessary, we assume that $\bar{b} \in \bar{L}$, so $w(b) \geq 0$.

Let S be the integral closure of $L[x]$ in FL. We consider the maximal ideal $\mathfrak{q} = (x - b)L[x]$ of $L[x]$ and choose a maximal ideal Q of S over $\mathfrak{q}$. Then we set $L[x]_w = L[x] \cap O_{w,FL}$, $\mathfrak{q}_w = \mathfrak{q} \cap O_{w,FL}$, $S_w = S \cap O_{w,FL}$, and $Q_w = Q \cap O_{w,FL}$. Then $\mathfrak{q}_w$ and Q_w are prime ideals of the rings $L[x]_w$ and S_w, respectively. By (3) of Section 8.1, $L[x]_w = O_{w,L}[x]$.

CLAIM A1: $L[x]_w/\mathfrak{q}_w = O_{w,L}$.   Indeed, if $(x - b)f(x) \in L[x]_w$ with $f \in L[X]$, then $w(x - b) + w(f(x)) \geq 0$. Since by (1) of Section 8.1, $w(x - b) = \min(w(1), w(b)) = 0$, we have $w(f(x)) \geq 0$, so $f(x) \in L[x]_w$. Thus, $\mathfrak{q}_w = (x - b)L[x]_w$. It follows that $L[x]_w/\mathfrak{q}_w = O_{w,L}[x]/(x - b)O_{w,L}[x] = O_{w,L}$. Moreover, $\bar{\mathfrak{q}} = (\bar{x} - \bar{b})\bar{L}[\bar{x}]$.

CLAIM A2: S_w *is the integral closure of $L[x]_w$ in FL.*   Indeed, consider $y \in S_w$. Then $y \in S$, so y is integral over $L[x]$. Hence, irr$(y, L(x)) \in L[x][Y]$ and the coefficients of irr$(y, L(x))$ are the fundamental symmetric polynomials in the conjugates of y over $L(x)$. Since w is equal to each of its conjugates over $L(x)$ (Lemma 8.1.1), each of the conjugates of y belongs to $O_{w,FL}$. Hence, the coefficients of irr$(y, L(x))$ belong to $L[x]_w$. It follows that

$\operatorname{irr}(y, L(x)) \in L[x]_w[Y]$, so y is integral over $L[x]'_w$.

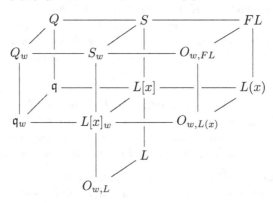

Conversely, let $y \in FL$ be integral over $L[x]_w$. Then y is integral over $L[x]$. So, $y \in S$. Also, y is integral over $O_{w,L(x)}$, so $w(y) \geq 0$. Hence, $y \in S_w$.

CLAIM A3: $S_w/Q_w = O_{w,L}$. Indeed, $Q_w = Q \cap S_w$ is a prime ideal of S_w that lies over $\mathfrak{q}_w$. Moreover, by Claim A2, S_w/Q_w is an integral extension of $L[x]_w/\mathfrak{q}_w = O_{w,L}$ (Claim A1). Since L is algebraically closed, $S_w/Q_w \subseteq L$. Finally, since $O_{w,L}$, as a valuation ring, is integrally closed in L, we have $S_w/Q_w = O_{w,L}$.

Following the convention used above, we use a bar for the reduction at w and recall that $\bar{S} = \overline{S_w}$ and $\bar{L} = \overline{O_{w,L}}$. Let M_w and $M_{w,L}$ be the kernels of the maps $S_w \to \bar{S}$ and $O_{w,L} \to \bar{L}$, respectively. Thus, $\bar{S} \cong S_w/M_w$ and $\bar{L} \cong O_{w,L}/M_{w,L}$.

CLAIM A4: We have $(Q_w + M_w)/Q_w = M_{w,L}$ under the identification $S_w/Q_w = O_{w,L}$ of Claim A3. Let $\pi \colon S_w \to O_{w,L}$ be the quotient map of Claim A3. Each $c \in M_{w,L}$ belongs to M_w. Since $Q_w \cap L = 0$, we have $\pi(c) = c$. Thus, $M_{w,L} \subseteq \pi(Q_w + M_w)$.

Assume $M_{w,L} \neq \pi(Q_w + M_w)$. Since $O_{w,L}/M_{w,L} = \bar{L}$ is a field, $M_{w,L}$ is a maximal ideal of $O_{w,L}$. Hence, $\pi(Q_w + M_w) = O_{w,L}$. Thus, there exist $q \in Q_w$ and $m \in M_w$ such that $1 = q + m$. Taking the norm from FL to $L(x)$ of both sides and using that S_w is integral over $O_{w,L}[x]$ (Claim A2), we get

$$1 = \prod_{\sigma \in \operatorname{Gal}(FL/L(x))} (q^\sigma + m^\sigma) = q_0 + m_0,$$

where

$$q_0 = \prod_{\sigma \in \operatorname{Gal}(FL/L(x))} q^\sigma \in Q_w \cap O_{w,L}[x] = \mathfrak{q}_w = (x - b)O_{w,L}[x]$$

(by Claim A1) and

$$m_0 = \sum_A \prod_{\substack{\sigma \in A \\ \tau \notin A}} q^\tau m^\sigma \in O_{w,L}[x]$$

with A ranging over all nonempty subsets of $\mathrm{Gal}(FL/L(x))$. Note that $w(m_0) > 0$ satisfies, because $w(q^\tau) = w(q) \geq 0$ and $w(m^\sigma) = w(m) > 0$ for all $\sigma, \tau \in \mathrm{Gal}(FL/L(x)))$. Thus, there exists $g \in O_{w,L}[X]$ such that $1 = (x - b)g(x) + m_0$. Taking reduction on both sides, we get $1 = (\bar{x} - \bar{b})\bar{g}(\bar{x})$, which is a contradiction. We conclude from this contradiction that $\pi(Q_w + M_w) = M_{w,L}$, as claimed.

CLAIM A5: $\bar{S}/\bar{Q} = \bar{L}$. Note that $\bar{Q} = \overline{Q_w} \cong Q_w/(Q_w \cap M_w) \cong (Q_w + M_w)/M_w$.

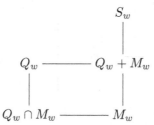

It follows from Claims A3 and A4 that

$$\bar{S}/\bar{Q} \cong (S_w/M_w)/((Q_w + M_w)/M_w) \cong S_w/(Q_w + M_w)$$
$$\cong (S_w/Q_w)/((Q_w + M_w)/Q_w)$$
$$\cong O_{w,L}/M_{w,L} \cong \bar{L},$$

as claimed.

CLAIM A6: $\overline{I_{Q/\mathfrak{q}}} \subseteq I_{\bar{Q}/\bar{\mathfrak{q}}}$. By Claim A5, $\bar{Q}$ is a maximal ideal of $\bar{S}$. Since $\bar{\mathfrak{q}} \subseteq \bar{Q}$ and $\bar{\mathfrak{q}}$ is a maximal ideal of $\bar{L}[\bar{x}]$, it follows that $\bar{Q} \cap \bar{L}[\bar{x}] = \bar{\mathfrak{q}}$.

Let $\sigma \in I_{Q/\mathfrak{q}}$. Then $y^\sigma \equiv y \mod Q$ for each $y \in S$, hence for each $y \in S_w$. Therefore, in the notation of Lemma 8.2.2, $\bar{y}^{\bar{\sigma}} \equiv \bar{y} \mod \bar{Q}$. Since the reduction at w is normal with respect to x, the ring $\bar{S}$ is the integral closure of $\bar{L}[\bar{x}]$ in $\bar{F}\bar{L}$. Since the map $S_w \to \bar{S}$ is surjective, the latter congruence implies that $\bar{\sigma} \in I_{\bar{Q}/\bar{\mathfrak{q}}}$, as claimed.

PART B: *Descending from $\tilde{K}$ to K.* Let R be the integral closure of $K[x]$ in F. In the notation of Part A, $\mathfrak{q} = (x - b)L[x]$, $P_{x,b} = K[x] \cap \mathfrak{q}$ is the prime ideal of all polynomials in $K[x]$ that vanish at b and $P = R \cap Q$ is a prime ideal of R lying over $P_{x,b}$. By Remark 8.2.1, $I_b = I_{P/P_{x,b}}$ is an inertia group of b in $\mathrm{Gal}(F/K(x))$.

Again, by Part A, $\bar{\mathfrak{q}} = (\bar{x} - \bar{b})\bar{L}[\bar{x}]$ and $\bar{Q}$ is a prime ideal of $\bar{S}$ lying over $\bar{\mathfrak{q}}$. Thus, $P_{\bar{x},\bar{b}} = \bar{K}(\bar{x}) \cap \bar{\mathfrak{q}}$ is the prime ideal of $\bar{K}[\bar{x}]$ of all polynomials that vanish at $\bar{b}$.

Since $\bar{S}$ is the integral closure of $\bar{L}[\bar{x}]$ in $\bar{F}\bar{L}$ (end of Part A), the ring $R' = \bar{F} \cap \bar{S}$ is the integral closure of $\bar{K}[\bar{x}]$ in $\bar{F}$. Then $P' = R' \cap \bar{Q}$ is a prime ideal of R' that lies over $P_{\bar{x},\bar{b}}$. By Remark 8.2.1, $I_{\bar{b}} = I_{P'/P_{\bar{x},\bar{b}}}$ is an inertia group of $\bar{b}$ in $\mathrm{Gal}(\bar{F}/\bar{K}(\bar{x}))$.

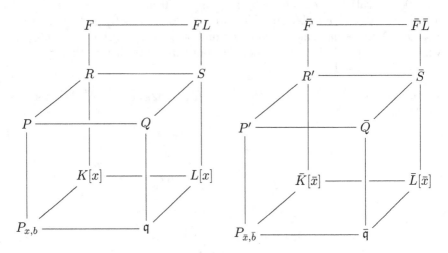

The diagrams of fields give rise to a diagram of Galois groups.

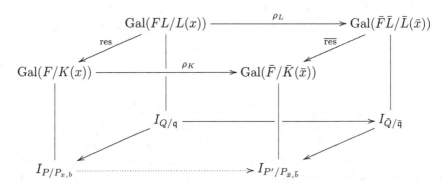

The maps ρ_K and ρ_L are the reduction maps and res and $\overline{\mathrm{res}}$ are the restriction maps. Since F/K and FL/L have degree preserving regular constant reduction at w with respect to x, both ρ_K and ρ_L are isomorphisms, by Lemma 8.2.2. Moreover, since F/K and $\bar{F}/\bar{K}$ are regular extensions, res and $\overline{\mathrm{res}}$ are isomorphisms. Since the reduction of Galois groups given by Lemma 8.2.2 commutes with restrictions, the upper side of the diagram is commutative.

The vertical lines in the diagram are inclusions. By Part A, $\rho_L(I_{Q/\mathfrak{q}}) \leq I_{\bar{Q}/\bar{\mathfrak{q}}}$. By Lemma 8.2.3, $\mathrm{res}(I_{Q/\mathfrak{q}}) = I_{P/P_{x,b}}$ and $\overline{\mathrm{res}}(I_{\bar{Q}/\bar{\mathfrak{q}}}) = I_{P'/P_{\bar{x},\bar{b}}}$. It follows from the commutativity of the lower part of the diagram that $\rho_K(I_{P/P_{x,b}}) \leq I_{P'/P_{\bar{x},\bar{b}}}$, as claimed. $\qquad\square$

8.3 Almost All Reductions

We consider a field K, a function field of one variable F over K, and a separating transcendental element x for F/K such that $F/K(x)$ is Galois. Let W be a set of valuations of K. We extend each $w \in W$ to a valuation w of $\tilde{K}$. Each $w \in W$ has a unique Gauss extension to $\tilde{K}(x)$ and then finitely many extensions to $F\tilde{K}(x)$, conjugate to each other over $\tilde{K}(x)$. We use w again to denote one of these extensions. As in previous sections, we use a bar to denote reduction of elements of $F\tilde{K}$, residue fields at w, and reduction of sets at w in the sense of Section 8.1. In particular, $\bar{K}$ and $\bar{F}$ are the residue fields of K and F, respectively, and the residue $\bar{x}$ of x at w is transcendental over $\bar{K}$. We say that a possible property P of elements and fields with respect to a valuation holds for **almost all** $w \in W$ if there exist $c_1, \ldots, c_n \in \tilde{K}^{\times}$ such that P holds for each $w \in W$ with $w(c_i) \geq 0$, $i = 1, \ldots, n$. It follows that if each one of finitely many possible properties $P_1, \ldots, P_m$ holds for almost all $w \in W$, then their conjunction $P_1 \wedge \cdots \wedge P_m$ holds for almost all $w \in W$.

LEMMA 8.3.1: *In the above notation suppose K is infinite, F/K is a function field of one variable, and $F/K(x)$ is a Galois extension. Then, for almost all $w \in W$, F/K has a degree preserving regular constant reduction at w with respect to x such that $\bar{F}/\bar{K}(\bar{x})$ is Galois. Moreover, $F\tilde{K}/\tilde{K}$ has a normal reduction at w with respect to x.*

Proof: We break the proof into two parts.

PART A: *Proof of Conditions (5a)–(5c) of Definition 8.1.2.* We choose a primitive element y for the extension $F/K(x)$ that is integral over $K[x]$. Then $\mathrm{irr}(y, K(x))$ is a monic irreducible polynomial $f(x, Y)$ in $K[x, Y]$. By [FrJ08, Cor. 10.2.2], $f(X, Y)$ is absolutely irreducible. Hence, by Bertini-Noether, $\bar{f}(X, Y)$ is absolutely irreducible for almost all $w \in W$ [FrJ08, Prop. 9.4.3]. In particular, $\bar{f}(\bar{x}, Y)$ is a monic irreducible polynomial in $\bar{K}[\bar{x}, Y]$ and $\bar{f}(\bar{x}, \bar{y}) = 0$. By (2) of Section 8.1, $\bar{K}(\bar{x}) = \overline{K(x)}$, hence $[F : K(x)] = \deg(f(x, Y)) = \deg(\bar{f}(\bar{x}, Y)) = [\bar{K}(\bar{x}, \bar{y}) : \bar{K}(\bar{x})] \leq [\bar{F} : \bar{K}(\bar{x})] = [\bar{F} : \overline{K(x)}] \leq [F : K(x)]$. Therefore, $[F : K(x)] = [\bar{F} : \bar{K}(\bar{x})]$ and $\bar{F} = \bar{K}(\bar{x}, \bar{y})$. By [FrJ08, Cor. 10.2.2(b)], this implies that $\bar{F}/\bar{K}$ is a regular extension. Moreover, by [FrJ08, Lemma 13.1.1], $\bar{F}/\bar{K}(\bar{x})$ is a Galois extension for almost all $w \in W$.

PART B: *Normal reduction.* We set $L = \tilde{K}$ and use Lemma 8.1.3 to observe that Conditions (5a), (5b), and (5c) of Definition 8.1.2 for FL/L follow from the same conditions for F/K. In order to prove also Condition (6) of Definition 8.1.4 we denote the integral closure of $L[x]$ in FL by S. By [Ser79, Prop. I.4.8], $S = L[y_1, \ldots, y_n]$ for some $y_1, \ldots, y_n \in FL$. For each i we have $h_i = \mathrm{irr}(y_i, L(x)) \in L[x, Y]$.

Let $f_1, \ldots, f_m$ be generators of the ideal of all polynomials in $L[\mathbf{Y}]$ that vanish at $\mathbf{y}$ and consider the Jacobian matrix

$$J(\mathbf{Y}) = \left(\frac{\partial f_i}{\partial Y_j} \right)_{1 \leq i \leq m, \, 1 \leq j \leq n}.$$

Let $d_1, \ldots, d_r \in L[\mathbf{Y}]$ be the $(n-1) \times (n-1)$ subdeterminants of $J(\mathbf{Y})$. By Lemma 8.1.6(a), there exist $c_1, \ldots, c_r \in L[\mathbf{Y}]$ such that $\sum_{k=1}^r c_k(\mathbf{y}) d_k(\mathbf{y}) = 1$.

For almost all $w \in W$ the elements $y_1, \ldots, y_n$ belong to $O_{w,FL}$, and the coefficients of $f_1, \ldots, f_m, h_1, \ldots, h_n, c_1, \ldots, c_r, d_1, \ldots, d_r$ belong to $O_{w,L}$. Hence, reduction at w gives that $\bar{h}_i \in \bar{L}[x, Y]$ is a monic polynomial in Y satisfying $\bar{h}(\bar{x}, \bar{y}_i) = 0$, so that $\bar{y}_i$ is integral over $\bar{L}[\bar{x}]$. Therefore, $\bar{L}[\bar{\mathbf{y}}]$ is integral over $\bar{L}[\bar{x}]$. Moreover, $\bar{J}(\mathbf{Y}) = \left(\frac{\partial \bar{f}_i}{\partial Y_j}\right)_{1 \le i \le m, 1 \le j \le n}$ is the Jacobian matrix of $\bar{f}_1, \ldots, \bar{f}_m$, and $\bar{d}_1, \ldots, \bar{d}_r$ are the $(n-1) \times (n-1)$ subdeterminants of $\bar{J}(\mathbf{Y})$. Finally, $\sum_{k=1}^r \bar{c}_k(\bar{\mathbf{y}}) \bar{d}_k(\bar{\mathbf{y}}) = 1$. Hence, by Lemma 8.1.6(b), $\bar{L}[\bar{\mathbf{y}}] = \overline{L[\mathbf{y}]}$ is integrally closed. It follows that $\bar{L}[\bar{\mathbf{y}}]$ is the integral closure of $\bar{L}[\bar{x}]$.

Similarly, for almost all $w \in W$, the reduction at w of the integral closure of $L[x^{-1}]$ is the integral closure of $\bar{L}[\bar{x}^{-1}]$. This completes the verification of Condition (6) of Section 8.1. $\qquad\qquad\Box$

Next we assume that K is a finitely generated regular extension of a field K_0 and let $\mathbf{u} = (u_1, \ldots, u_n)$ with $K = K_0(\mathbf{u})$. Let $U = \mathrm{Spec}(K_0[\mathbf{u}])$ be the affine absolutely irreducible variety defined over K_0 in $\mathbb{A}^n$ with generic point $\mathbf{u}$ [FrJ08, Cor. 10.2.2(a)]. For each $\mathbf{a} \in U_{\mathrm{simp}}(\tilde{K}_0)$ we choose a K_0-place $\varphi_{\mathbf{a}}$ of K with residue field $K_0(\mathbf{a})$ such that $\varphi_{\mathbf{a}}(\mathbf{u}) = \mathbf{a}$ [JaRo80, Cor. A2]. Let $w_{\mathbf{a}}$ be a valuation of K that corresponds to $\varphi_{\mathbf{a}}$ and set $W = \{w_{\mathbf{a}} \mid \mathbf{a} \in U_{\mathrm{simp}}(\tilde{K}_0)\}$. As above, we fix a valuation of $K\tilde{K}_0$ extending $w_{\mathbf{a}}$ and call it $w_{\mathbf{a}}$ too. In addition, we consider a variable x, a finite Galois extension F of $K(x)$, and fix an extension of $w_{\mathbf{a}}$ to a valuation of $F\tilde{K}$ with the same notation such that $w_{\mathbf{a}}|_{\tilde{K}(x)}$ is the Gauss extension of $w_{\mathbf{a}}|_{\tilde{K}}$. We say that a property P of elements and fields with respect to a valuation holds for **almost all** $\mathbf{a} \in U(\tilde{K}_0)$ if there exists a nonempty Zariski-open subset U_0 of U_{simp} such that P holds for $w_{\mathbf{a}}$ for all $\mathbf{a} \in U_0(\tilde{K}_0)$. Note that if $c \in \tilde{K}^\times$, then $\mathrm{irr}(c, K) = X^r + \sum_{i=0}^{r-1} \frac{f_i(\mathbf{u})}{g_i(\mathbf{u})} X^i$ with $f_i, g_i \in K_0[X_1, \ldots, X_n]$ such that $g_i(\mathbf{u}) \ne 0$, $i = 0, \ldots, r-1$. If $g_i(\mathbf{u}) \ne 0$, $i = 0, \ldots, r-1$, then $w_{\mathbf{a}}(c) \ge 0$. It follows that the property P holds for almost all $\mathbf{a} \in U(\tilde{K}_0)$ if it holds for almost all $w \in W$. Conversely, if P holds for almost all $\mathbf{a} \in U(\tilde{K}_0)$, then it holds for almost all $w \in W$.

PROPOSITION 8.3.2: *Let K_0 be a field, $K = K_0(\mathbf{u})$ a finitely generated regular extension of K_0, and $U = \mathrm{Spec}(K_0[\mathbf{u}])$ the absolutely irreducible affine variety defined over K_0 with generic point $\mathbf{u}$. Let x be a transcendental element over K and F a finite Galois extension of $K(x)$ which is regular over K. Then the following statements hold for almost all $\mathbf{a} \in U_{\mathrm{simp}}(\tilde{K}_0)$:*

(a) *The valuation $w_{\mathbf{a}}$ of $F\tilde{K}$ chosen above gives rise to a regular constant reduction with residue fields $\bar{F}$ and $\bar{K}$ of F and K, respectively, such that $\bar{F}/\bar{K}(\bar{x})$ is a Galois extension, and an isomorphism $\sigma \mapsto \bar{\sigma}$ of $\mathrm{Gal}(F/K(x))$ onto $\mathrm{Gal}(\bar{F}/\bar{K}(\bar{x}))$ such that $\bar{y}^{\bar{\sigma}} = \overline{y^\sigma}$ for all $\sigma \in \mathrm{Gal}(F/K(x))$ and $y \in F$ with $\bar{y} \in \bar{F}$.*

(b) *For each $b \in \tilde{K}$ there exists an inertia group I_b of b in $\mathrm{Gal}(F/K(x))$ and* *there exists an inertia group $I_{\bar{b}}$ of $\bar{b}$ in $\mathrm{Gal}(\bar{F}/\bar{K}(\bar{x}))$ such that under the isomorphism of (a), $\overline{I_b} \leq I_{\bar{b}}$.*

Again, we use a bar over elements to denote reduction at $w_{\mathbf{a}}$.

Proof: By Lemma 8.3.1, F/K has a degree preserving regular constant reduction at $w_{\mathbf{a}}$ with respect to x for almost all $\mathbf{a} \in U(\tilde{K}_0)$. Moreover, $F\tilde{K}/\tilde{K}$ has a normal reduction at $w_{\mathbf{a}}$ with respect to x for almost all $\mathbf{a} \in U(\tilde{K}_0)$. For almost all $\mathbf{a} \in U(\tilde{K}_0)$, the reduction of F at $w_{\mathbf{a}}$ is a Galois extension of the reduction of $K(\mathbf{x})$ at $w_{\mathbf{a}}$. Now we apply Lemma 8.2.2 and then Lemma 8.2.4.   □

8.4 Embedding Problems under Existentially Closed Extensions

We generalize Lemma 5.9.1 from constant finite split embedding problems to arbitrary finite split embedding problems.

Let $\hat{K}_0/K_0$ be a field extension such that K_0 is existentially closed in $\hat{K}_0$ and let x be an indeterminate. Consider a finite split embedding problem

$$(1) \qquad \mathrm{Gal}(E/K_0(x)) \ltimes H \xrightarrow{\ \mathrm{pr}\ } \mathrm{Gal}(E/K_0(x)) \ .$$

Let K be the algebraic closure of K_0 in E. Then E is a Galois extension of $K(x)$, so E is a separable extension of K, hence E/K is a regular extension. A solution field F of (1) is **regular** if F is regular over K.

We set $\hat{K} = K\hat{K}_0$ and $\hat{E} = E\hat{K}_0$. By Lemma 5.2.6, $\hat{K}_0$ is a regular extension of K_0. In addition, $\hat{K}_0$ is algebraically independent from E over K_0. Hence, $\hat{K}_0$ is linearly disjoint from E over K_0 [FrJ08, Lemma 2.6.7], so $\hat{K}_0(x)$ is linearly disjoint from E over $K_0(x)$. Therefore, $\mathrm{res}_{\hat{E}/E}\colon \mathrm{Gal}(\hat{E}/\hat{K}_0(x)) \to \mathrm{Gal}(E/K_0(x))$ is an isomorphism. Thus, $\mathrm{Gal}(\hat{E}/\hat{K}_0(x))$ acts on H via $\mathrm{res}_{\hat{E}/E}$. This gives rise to a finite split $\hat{K}_0$-embedding problem

$$(2) \qquad \mathrm{Gal}(\hat{E}/\hat{K}_0(x)) \ltimes H \xrightarrow{\ \mathrm{pr}\ } \mathrm{Gal}(\hat{E}/\hat{K}_0(x)).$$

Suppose E has a K-rational place φ unramified over $K(x)$. Since $\hat{K}$ and E are linearly disjoint over K, the place φ extends to a $\hat{K}$-rational place $\hat{\varphi}$ of $\hat{E}$, unramified over $\hat{K}(x)$ [FrJ08, Lemma 2.5.11].

In this setup we prove:

LEMMA 8.4.1: *Suppose (2) has a solution field $\hat{F}$ such that $\hat{\varphi}$ extends to a $\hat{K}$-rational place of $\hat{F}$ unramified over $\hat{K}(x)$. Then (1) has a regular solution*

field F such that φ extends to a K-rational place of F unramified over $K(x)$.

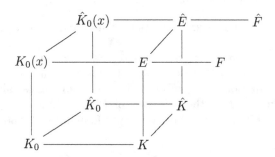

If in addition, K_0 is ample, $\hat{K}_0 = K_0((t))$, H is generated by finitely many subgroups G_j, $j \in J$, we identify H with $\mathrm{Gal}(\hat{F}/\hat{E})$, and for each $j \in J$ there exists a point $b_j \in \mathrm{Branch}(\hat{F}/\hat{K}_0(x))$ transcendental over K_0 such that G_j is an inertia group of b_j, then for each ordinal $\beta < \mathrm{card}(K)$ there is a regular solution field F_β to (1) such that φ extends to a K-place of F_β unramified over $K(x)$ and there exists

$$b_{\beta,j} \in \big(\mathrm{Branch}(F_\beta/K_0(x)) \cap K_s\big) \smallsetminus \bigcup_{\alpha < \beta} \mathrm{Branch}(F_\alpha/K_0(x))$$

such that G_j is contained in an inertia group of $b_{\beta,j}$ in $\mathrm{Gal}(F_\beta/K_0(x))$.

Proof: We may assume that $\varphi(x) = \infty$, so that the place $\mathfrak{p}_{x,\infty}$ of $\hat{K}(x)/\hat{K}$ totally decomposes in $\hat{F}$; otherwise $\varphi(x) = a \in K$ and we may replace x by $\frac{1}{x-a}$. In particular, $\hat{F}$ is a regular extension of $\hat{K}$ [FrJ08, Lemma 2.6.9(b)].

PART A: *Solution of (2).* By assumption, there exists an isomorphism

$$\gamma\colon \mathrm{Gal}(\hat{F}/\hat{K}_0(x)) \to \mathrm{Gal}(\hat{E}/\hat{K}_0(x)) \ltimes H$$

such that $\mathrm{pr} \circ \gamma = \mathrm{res}_{\hat{F}/\hat{E}}$. Thus, there exist polynomials $f \in \hat{K}_0[X, Z]$, $g \in \hat{K}[X, Y]$, and elements $z, y \in \hat{F}$ such that the following conditions hold:
(3a) $\hat{F} = \hat{K}_0(x, z)$, $f(x, Z) = \mathrm{irr}(z, \hat{K}_0(x))$; we identify $\mathrm{Gal}(f(x, Z), \hat{K}_0(x))$
with $\mathrm{Gal}(\hat{F}/\hat{K}_0(x))$.
(3b) $\hat{F} = \hat{K}(x, y)$, $g(x, Y) = \mathrm{irr}(y, \hat{K}(x))$. Therefore, $g(X, Y)$ is absolutely
irreducible [FrJ08, Lemma 10.2.2(b)]. By Lemma 7.1.1, we may assume
that $g(X, Y) = Y^d + a_{d-1}(X)Y^{d-1} + \cdots + a_0(X)$ with $a_i \in \hat{K}[X]$ and
$\deg(a_i(X)) \le \deg(a_{d-1}(X)) \ge 1$, for $i = 0, \ldots, d - 1$.

PART B: *Pushing the tower over $\hat{K}_0$ down toward K_0.* All of the objects in
(3) together depend on only finitely many parameters from $\hat{K}_0$. Let $u_1, \ldots, u_n$
be elements of $\hat{K}_0$ satisfying the following conditions:

(4a) $F = K_0(\mathbf{u}, x, z)$ is a Galois extension of $K_0(\mathbf{u}, x)$, the coefficients of $f(X, Z)$ lie in $K_0[\mathbf{u}]$, $f(x, Z) = \mathrm{irr}(z, K_0(\mathbf{u}, x))$, and

$$\mathrm{Gal}(f(x, Z), K_0(\mathbf{u}, x)) = \mathrm{Gal}(f(x, Z), \hat{K}_0(x));$$

again we identify $\mathrm{Gal}(f(x, Z), K_0(\mathbf{u}, x))$ with $\mathrm{Gal}(F/K_0(\mathbf{u}, x))$.

(4b) $F = K(\mathbf{u}, x, y)$ and the coefficients of g lie in $K[\mathbf{u}]$; hence $g(x, Y) = \mathrm{irr}(y, K(\mathbf{u}, x))$.

By (4a) we may consider γ as an isomorphism of

(5) $\gamma \colon \mathrm{Gal}(F/K_0(\mathbf{u}, x)) \to \mathrm{Gal}(E(\mathbf{u})/K_0(\mathbf{u}, x)) \ltimes H$

satisfying $\mathrm{pr} \circ \gamma = \mathrm{res}_{F/E(\mathbf{u})}$.

PART C: *Descending to K_0.* Since $\hat{K}_0$ is regular over K_0, so is $K_0(\mathbf{u})$. Thus, $\mathbf{u}$ generates an absolutely irreducible variety $U = \mathrm{Spec}(K_0[\mathbf{u}])$ in $\mathbb{A}^n$ over K_0. The variety U has a nonempty Zariski-open subset U' such that for each $\mathbf{u}' \in U'(\tilde{K})$ the K_0-specialization $(\mathbf{u}, x) \to (\mathbf{u}', x)$ extends to an E-homomorphism $\ '\colon E[\mathbf{u}, x, z, y] \to E[\mathbf{u}', x, z', y']$ such that the following conditions hold:

(6a) $f'(x, z') = 0$, $\mathrm{discr}(f'(x, Z), K_0(\mathbf{u}', x)) \neq 0$, $F' = K_0(\mathbf{u}', x, z')$ is the splitting field of $f'(x, Z)$ over $K_0(\mathbf{u}', x)$; in particular $F'/K_0(\mathbf{u}', x)$ is Galois and we may identify $\mathrm{Gal}(f'(x, Z), K_0(\mathbf{u}', x))$ with $\mathrm{Gal}(F'/K_0(\mathbf{u}', x))$.

(6b) $g'(X, Y)$ is absolutely irreducible and $g'(x, y') = 0$, so

$$g'(x, Y) = \mathrm{irr}(y', K(\mathbf{u}', x)).$$

Furthermore, $g'(X, Y) = Y^d + a'_{d-1}(X)Y^{d-1} + \cdots + a'_0(X)$ with $a'_i \in K[\mathbf{u}', X]$ and $\deg(a'_i(X)) \leq \deg(a'_{d-1}(X)) \geq 1$, for $i = 0, \ldots, d-1$.

Condition (6a) follows from a Lemma of Hilbert [FrJ08, Lemma 13.1.1]. To achieve the absolute irreducibility of g' we use Bertini-Noether [FrJ08, Prop. 9.4.3]. Since K_0 is existentially closed in $\hat{K}_0$ and since $\mathbf{u} \in U'(\hat{K}_0)$, we can choose $\mathbf{u}' \in U'(K_0)$. By (6a), the homomorphism $'$ induces an embedding

$$\varphi^* \colon \mathrm{Gal}(f'(x, Z), K_0(x)) \to \mathrm{Gal}(f(x, Z), K_0(\mathbf{u}, x))$$

that commutes with the restriction to $\mathrm{Gal}(E/K_0(x))$ [FrJ08, Lemma 16.1.1].

Observe that $K(x)$ is linearly disjoint from $K_0(\mathbf{u})$ over K_0.

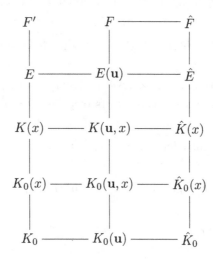

Hence, by (6b),

$$
\begin{aligned}
|\mathrm{Gal}(f'(x,Z), K_0(x))| &= [F' : K_0(x)] = \deg(g'(x,Y))[K(x) : K_0(x)] \\
&= \deg(g(x,Y))[K(\mathbf{u},x) : K_0(\mathbf{u},x)] \\
&= [F : K_0(\mathbf{u},x)] = |\mathrm{Gal}(f(x,Z), K_0(\mathbf{u},x))|.
\end{aligned}
$$

It follows that φ^* is an isomorphism. Hence, $\gamma \circ \varphi^*$ solves embedding problem (1).

PART D: *Extending φ.* We extend φ to a place φ' of F'. Then φ' extends the specialization $x \to \infty$. By Lemma 7.1.1 and (6b), φ' totally decomposes in $F'/K(x)$, that is, φ' is unramified and K-rational.

PART E: *Branch points.* Finally we assume that K_0 is ample and for each j in a finite set J there exists $b_j \in \mathrm{Branch}(\hat{F}/\hat{K}_0(x)) \cap K_s$ transcendental over K_0 that has G_j (when H is identified with $\mathrm{Gal}(\hat{F}/\hat{E})$) as an inertia group. We may choose $u_1, \ldots, u_n$ such that $b_j = h_j(\mathbf{u})$, for some $h_j \in K_s[X_1, \ldots, X_n]$. By Proposition 8.3.2, we may make U' smaller, if necessary, to assume that for each $\mathbf{u}' \in U'(K_0)$, the isomorphism $\mathrm{Gal}(F/K_0(\mathbf{u},x)) \to \mathrm{Gal}(F'/K_0(x))$ maps each G_j (which is an inertia group of b_j) into an inertia group of $b'_j = h_j(\mathbf{u}')$.

Let β be an ordinal less than $\mathrm{card}(K)$. Assume by transfinite induction that for each $\alpha < \beta$ there exists $\mathbf{u}^{(\alpha)} \in U'(K_0)$ and an E-homomorphism $\psi^{(\alpha)} \colon E[\mathbf{u}, x, z, y] \to E[\mathbf{u}^{(\alpha)}, x, z^{(\alpha)}, y^{(\alpha)}] = E[x, z^{(\alpha)}, y^{(\alpha)}]$ such that (6a) and (6b) hold with $'$ replaced by (α), so that $F^{(\alpha)} = K_0(x, z^{(\alpha)})$ is a solution field of embedding problem (1), φ extends to a K-rational place of $F^{(\alpha)}$ unramified over $K(x)$, and for each $j \in J$, the group G_j (now considered as a subgroup of H identified with $\mathrm{Gal}(F_\beta/E)$) is contained in an inertia group of $b_j^{(\alpha)} = h_j(\mathbf{u}^{(\alpha)})$, and $b_j^{(\alpha)} \notin \bigcup_{\alpha'<\alpha} \mathrm{Branch}(F^{(\alpha')}/K_0(x))$. Since each of the sets $\mathrm{Branch}(F^{(\alpha)}/K_0(x))$ is finite, $D = \bigcup_{\alpha<\beta} \mathrm{Branch}(F^{(\alpha)}/K_0(x))$ is of cardinality less than $\mathrm{card}(K)$. Since b_j is transcendental over K_0, $h_j(\mathbf{u})$ is non-constant on U'. It follows from Corollary 5.4.4 that there exists $\mathbf{u}^{(\beta)} \in U'(K_0)$ such that $b_j^{(\beta)} = h_j(\mathbf{u}^{(\beta)}) \notin D$. The corresponding field $F^{(\beta)}$ is a solution of (1) and the points $b_j^{(\beta)}$ will have all of the above properties. This completes the transfinite induction. $\square$

8.5 The Transcendence Degree of the Field of Formal Power Series

In Proposition 7.3.1 we consider a field K_0 and an extension $\hat{K}_0$ complete under an ultrametric valuation and solve a finite split embedding problem over $\hat{K}_0(x)$ under appropriate assumptions. In Part (d) of that proposition we further assume that $\mathrm{trans.deg}(\hat{K}_0/K_0) = \infty$ in order to enhance the solution of the embedding problem with branch points that are algebraically independent over K_0. In this section we prove the extra assumption for fields of formal power series. Indeed, we even compute the exact transcendence degree in that case.

LEMMA 8.5.1: *Let K be an uncountable field and K_0 the prime field of K. Then:*
(a) $\mathrm{trans.deg}(K/K_0) = \mathrm{card}(K)$, *and*
(b) $\mathrm{trans.deg}(K((t))/K) \geq \mathrm{card}(K)$.

Proof: We choose a transcendence base T for K/K_0. Since K_0 is at most countable and K is uncountable, T is infinite, so $\mathrm{card}(K) = \mathrm{card}(K_0(T)) = \mathrm{card}(T) = \mathrm{trans.deg}(K/K_0)$.

To prove (b) we set $\kappa = \mathrm{card}(K)$. Since $\kappa \cdot \aleph_0 = \kappa$, we can choose a transcendence base $(x_{\alpha k})_{\alpha<\kappa,\ k<\aleph_0}$ of K/K_0 indexed by $\kappa \times \aleph_0$. For each $\alpha < \kappa$ we consider the element

$$y_\alpha = \sum_{k=0}^\infty x_{\alpha k} t^k$$

of $K[[t]]$. We claim that the family $(y_\alpha)_{\alpha<\kappa}$ is algebraically independent over K. It suffices to prove that the y_α's are algebraically independent over $K_0(x_{\alpha k})_{\alpha<\kappa,\ k<\aleph_0}$, equivalently, that the set $\{x_{\alpha k}, y_\alpha \mid \alpha < \kappa,\ k < \aleph_0\}$ is algebraically independent over K_0.

It suffices to prove for distinct $\alpha_1, \ldots, \alpha_m < \kappa$, $k_1, \ldots, k_m < \aleph_0$, and distinct $\beta_1, \ldots, \beta_n, \beta < \kappa$ that $x_{\alpha_1 k_1}, \ldots, x_{\alpha_m k_m}, y_{\beta_1}, \ldots, y_{\beta_n}, y_\beta$ are algebraically independent over K_0. Arguing inductively, it suffices to prove that $y_\beta = \sum_{k=0}^\infty x_{\beta k} t^k$ is transcendental over

$$K_1 = K_0(x_{\alpha_1 k_1}, \ldots, x_{\alpha_m k_m}, y_{\beta_1}, \ldots, y_{\beta_n}),$$

under the assumption that $x_{\alpha_1 k_1}, \ldots, x_{\alpha_m k_m}, y_{\beta_1}, \ldots, y_{\beta_n}$ are algebraically independent over K_0.

Otherwise, there is a nonzero polynomial

$$f \in K_0[X_1, \ldots, X_m, Y_1, \ldots, Y_n, Y]$$

such that $f(x_{\alpha_1 k_1}, \ldots, x_{\alpha_m k_m}, y_{\beta_1}, \ldots, y_{\beta_n}, y_\beta) = 0$ and Y occurs in f. Thus,

$$f(X_1, \ldots, X_m, Y_1, \ldots, Y_n, Y) = \sum_{j=0}^{l} f_j(X_1, \ldots, X_m, Y_1, \ldots, Y_n) Y^j$$

with $f_0, \ldots, f_l \in K_0[X_1, \ldots, X_m, Y_1, \ldots, Y_n]$, $l \geq 1$, and $f_l \neq 0$ and

$$(1) \qquad \sum_{j=0}^{l} f_j(x_{\alpha_1 k_1}, \ldots, x_{\alpha_m k_m}, y_{\beta_1}, \ldots, y_{\beta_n}) y_\beta^j = 0.$$

The left hand side of (1) is an element of the ring $K_0[x_{\alpha k} \mid \alpha < \kappa,\ k < \aleph_0][[t]]$. Since the $x_{\alpha k}$'s are algebraically independent over K_0, we may specialize some of the $x_{\alpha k}$ to 0 leaving the elements of K_0 and all the other $x_{\alpha k}$'s unchanged. In particular, we may set $r = \max(k_1, \ldots, k_m)$ and specialize each $x_{\beta_i, s}$ with $1 \leq i \leq n$ and $s > r$ to 0. Then for each $1 \leq i \leq n$ the power series $y_{\beta_i} = \sum_{k=0}^{\infty} x_{\beta_i k} t^k$ is mapped onto the polynomial $\bar{y}_{\beta_i} = \sum_{k=0}^{r} x_{\beta_i k} t^k$. For each $0 \leq j \leq l$, the power series $f_j(x_{\alpha_1 k_1}, \ldots, x_{\alpha_m k_m}, y_{\beta_1}, \ldots, y_{\beta_n})$ is mapped onto the polynomial

$$\bar{f}_j(t) = f_j(x_{\alpha_1 k_1}, \ldots, x_{\alpha_m k_m}, \bar{y}_{\beta_1}, \ldots, \bar{y}_{\beta_n}) \in K_0[x_{\alpha k} \mid \alpha < \kappa,\ k < \aleph_0][t].$$

By the induction hypothesis $f_l(x_{\alpha_1 k_1}, \ldots, x_{\alpha_m k_m}, y_{\beta_1}, \ldots, y_{\beta_n}) \neq 0$. The left hand side is a power series in t with coefficients in

$$K_0[x_{\alpha_1 k_1}, \ldots, x_{\alpha_m k_m}, x_{\beta_1 l_1}, \ldots, x_{\beta_n l_n}],$$

where $k_1, \ldots, k_m$ are as above and $l_j = 0, 1, 2, \ldots$. One of these coefficients is nonzero. It involves only finitely many of the variables $x_{\alpha_i k_i}, x_{\beta_j l_j}$. We may therefore enlarge r such that none of those finitely many variables is changed under the specialization of the $x_{\alpha k}$'s. This implies that

$$(2) \qquad f_l(x_{\alpha_1 k_1}, \ldots, x_{\alpha_m k_m}, \bar{y}_{\beta_1}, \ldots, \bar{y}_{\beta_n}) \neq 0.$$

Now we let $d_0 = \max(\deg(\bar{f}_0(t)), \ldots, \deg(\bar{f}_l(t)))$, set $d = \max(r, d_0)$, and specialize each $x_{\beta s}$ with $s > d + 1$ to 0 and keep all of the other $x_{\alpha k}$'s unchanged. Then the power series $y_\beta = \sum_{k=0}^{\infty} x_{\beta k} t^k$ specializes to a polynomial

$\bar{y} = \sum_{k=0}^{d+1} x_{\beta k} t^k$ of degree $d + 1$. The elements $x_{\alpha_i k_i}$, $i = 1, \ldots, m$, are fixed under that specialization, so (1) becomes

$$(3) \qquad \sum_{j=0}^{l} f_j(x_{\alpha_1 k_1}, \ldots, x_{\alpha_m k_m}, \bar{y}_{\beta_1}, \ldots, \bar{y}_{\beta_n}) \bar{y}^j = 0.$$

It follows from (2) that the degree of $f_l(x_{\alpha_1 k_1}, \ldots, x_{\alpha_m k_m}, \bar{y}_{\beta_1}, \ldots, \bar{y}_{\beta_n}) \bar{y}^l$ in t is at least the degree of $\bar{y}^l$, that is $(d + 1)l$. On the other hand, for each $j < l$, the degree of $f_j(x_{\alpha_1 k_1}, \ldots, x_{\alpha_m k_m}, \bar{y}_{\beta_1}, \ldots, \bar{y}_{\beta_n}) \bar{y}^j$ is at most $d + (d + 1)j < (d + 1)l$. Therefore, the left hand side of (3), which is actually $f(x_{\alpha_1 k_1}, \ldots, x_{\alpha_m k_m}, \bar{y}_{\beta_1}, \ldots, \bar{y}_{\beta_n}, \bar{y})$, is nonzero. This contradiction completes our recursion and proves that the set $(y_\alpha)_{\alpha < \kappa}$ is algebraically independent over K. Consequently, $\mathrm{trans.deg}(K((t))/K) \geq \kappa = \mathrm{card}(K)$, as claimed. $\square$

PROPOSITION 8.5.2 (Fehm): *Let K be a field. Then* $\mathrm{trans.deg}(K((t))/K) = \mathrm{card}(K)^{\aleph_0}$.

Proof: As in Lemma 8.5.1 we denote the prime field of K by K_0, set $\kappa = \mathrm{card}(K)$, and observe that $\mathrm{card}(K((t))) = \mathrm{card}(K[[t]]) = \kappa^{\aleph_0} \geq 2^{\aleph_0} > \aleph_0$. Hence, by Lemma 8.5.1(a), $\mathrm{trans.deg}(K((t))/K_0) = \mathrm{card}(K((t))) = \kappa^{\aleph_0}$. We distinguish between two cases.

CASE A: $\kappa^{\aleph_0} > \kappa$. We use the inequality $\mathrm{trans.deg}(K/K_0) \leq \mathrm{card}(K) = \kappa$ and the equality

$$\kappa^{\aleph_0} = \mathrm{trans.deg}(K((t))/K_0) = \mathrm{trans.deg}(K((t))/K) + \mathrm{trans.deg}(K/K_0)$$

to conclude that $\mathrm{trans.deg}(K((t))/K) = \kappa^{\aleph_0}$.

CASE B: $\kappa^{\aleph_0} = \kappa$. Then $\kappa > \aleph_0$, because $\kappa^{\aleph_0} \geq 2^{\aleph_0} > \aleph_0$. Hence, by Lemma 8.5.1, $\mathrm{trans.deg}(K((t))/K) \geq \mathrm{card}(K)$, so

$$\kappa = \mathrm{card}(K) \leq \mathrm{trans.deg}(K((t))/K) \leq \mathrm{trans.deg}(K((t))/K_0) = \kappa^{\aleph_0} = \kappa.$$

Consequently, $\mathrm{trans.deg}(K((t))/K) = \kappa^{\aleph_0}$, as claimed. $\square$

8.6 Solution of Embedding Problems over Ample Fields

Let K_0 be an ample field. Combining the results of the previous sections, we prove that every finite split embedding problem over $K_0(x)$ has as many linearly disjoint solutions as the cardinality of K_0.

LEMMA 8.6.1: *Let K_0 be an ample field and x an indeterminate. Consider a finite split embedding problem over $K_0(x)$*

$$(1) \qquad \mathrm{Gal}(E/K_0(x)) \ltimes H \xrightarrow{\ \mathrm{pr}\ } \mathrm{Gal}(E/K_0(x))$$

with $H \neq 1$. Let $\{G_j \mid j \in J\}$ be a finite family of cyclic groups of prime power orders that generate H. Let K be the algebraic closure of K_0 in E. Suppose E has a K-rational K-place φ unramified over $K(x)$ such that $\varphi(x) \in K_0 \cup \{\infty\}$. Then:

(a) (1) has for each ordinal $\alpha < \mathrm{card}(K)$ a regular solution field F_α.

(b) For each $\beta < \mathrm{card}(K)$ and every $j \in J$ there exists

$$b_{\beta,j} \in \mathrm{Branch}(F_\beta/K_0(x)) \smallsetminus \bigcup_{\alpha<\beta} \mathrm{Branch}(F_\alpha/K_0(x))$$

such that, when H is identified with $\mathrm{Gal}(F_\beta/E)$, G_j is contained in an inertia group of $b_{\beta,j}$.

(c) The solutions fields F_α, $\alpha < \mathrm{card}(K)$, of (1) are linearly disjoint over E.

(d) The place φ extends to a K-rational K-place of F_α unramified over $K(x)$.

Proof: Let t be an indeterminate. Set $\hat{K}_0 = K_0((t))$, $\hat{K} = K((t))$, and $\hat{E} = E\hat{K}$. Then $\hat{K}/\hat{K}_0$ is a finite Galois extension of complete fields under the t-adic absolute value. Since the extension $\hat{K}/K$ is regular [FrJ08, Example 3.5.1] and algebraically independent from E/K, the fields $\hat{K}$ and E are linearly disjoint over K. Hence, φ extends to a $\hat{K}$-rational place $\hat{\varphi}$ of $\hat{E}$, so $\hat{K}$ is the algebraic closure of $\hat{K}_0$ in $\hat{E}$ [FrJ08, Lemma 2.6.9]. Furthermore, $\hat{\varphi}$ is unramified over $\hat{K}(x)$. Finally, $\mathrm{Gal}(\hat{E}/\hat{K}_0(x))$ is isomorphic to $\mathrm{Gal}(E/K_0(x))$ and acts on H via the restriction map. Thus,

$$(2) \qquad \mathrm{Gal}(\hat{E}/\hat{K}_0(x)) \ltimes H \xrightarrow{\mathrm{pr}} \mathrm{Gal}(\hat{E}/\hat{K}_0(x))$$

is a finite split embedding problem over $\hat{K}_0(x)$.

By Proposition 8.5.2, $\mathrm{trans.deg}(\hat{K}_0/K_0) = \infty$. By Proposition 7.3.1, (2) has a solution field $\hat{F}$ such that $\hat{\varphi}$ extends to a $\hat{K}$-rational place of $\hat{F}$ unramified over $\hat{K}(x)$. Moreover, there exist $b_j \in \mathrm{Branch}(\hat{F}/\hat{K}_0(x))$, $j \in J$, algebraically independent over K such that G_j is an inertia group of b_j. Lemma 8.4.1 therefore gives for each ordinal $\beta < \mathrm{card}(K)$ a regular solution field F_β of (1), an extension of φ to a K-rational K-place of F_β unramified over $K(x)$, and points $b_{\beta,j} \in \mathrm{Branch}(F_\beta/K_0(x)) \smallsetminus \bigcup_{\alpha<\beta} \mathrm{Branch}(F_\alpha/K_0(x))$ such that G_j is contained in an inertia group of $b_{\beta,j}$ over $K_0(x)$. In particular, the valuation $v_{x,b_{\beta,j}}$ of $K_0(x)$ is unramified in the field N generated by F_α for all $\alpha < \beta$. By Lemma 7.4.1, $F_\beta \cap N = E$. Consequently, the fields F_α with $\alpha < \mathrm{card}(K)$ are linearly disjoint over E, as claimed in (c). $\square$

Remark 8.6.2: Fiber products. Let $\bar{\varphi}\colon \hat{A} \to A$ and $\alpha\colon B \to A$ be epimorphisms of profinite groups. Then the closed subgroup $\hat{B} = B \times_A \hat{A} = \{(b, \hat{a}) \in B \times \hat{A} \mid \alpha(b) = \bar{\varphi}(\hat{a})\}$ of $B \times \hat{A}$ is called the **fiber product** of B and $\hat{A}$ over A. Let $\beta\colon \hat{B} \to B$ and $\hat{\alpha}\colon \hat{B} \to \hat{A}$ be the projections on the coordinates.

$$(3)$$

$$\begin{array}{ccc} \hat{B} & \xrightarrow{\hat{\alpha}} & \hat{A} \\ {\scriptstyle\beta}\big\downarrow & & \big\downarrow{\scriptstyle\bar{\varphi}} \\ B & \xrightarrow{\alpha} & A \end{array}$$

Then the fiber product is characterized by the following universal property: If $\hat{\varphi}\colon G \to \hat{A}$ and $\gamma\colon G \to B$ are homomorphisms from a profinite group G satisfying $\alpha \circ \gamma = \bar{\varphi} \circ \hat{\varphi}$, then there exists a unique homomorphism $\hat{\gamma}\colon G \to \hat{B}$ such that $\beta \circ \hat{\gamma} = \gamma$ and $\hat{\alpha} \circ \hat{\gamma} = \hat{\varphi}$. In that case

(4) $\operatorname{Ker}(\beta) \cap \operatorname{Ker}(\hat{\alpha}) = 1$ and $\operatorname{Ker}(\beta) \times \operatorname{Ker}(\hat{\alpha}) = \operatorname{Ker}(\alpha \circ \beta) = \operatorname{Ker}(\varphi \circ \hat{\alpha})$

[FrJ08, Lemma 22.2.4] and we say that diagram (3) is **cartesian**.

In particular, suppose a profinite group A acts continuously on a profinite group C, let $B = A \ltimes C$ and let $\alpha\colon A \ltimes C \to A$ be the projection on the first coordinate. Consider an epimorphism $\bar{\varphi}\colon \hat{A} \to A$. Then $\hat{A}$ acts on C via $\bar{\varphi}$ and we can construct the semidirect product $\hat{B} = \hat{A} \ltimes C$ with the maps $\beta\colon \hat{A} \ltimes C \to A \ltimes C$ and $\hat{\alpha}\colon \hat{A} \ltimes C \to \hat{A}$ defined by $\hat{\alpha}(\hat{a}c) = \hat{a}$ and $\beta(\hat{a}c) = \bar{\varphi}(\hat{a})c$ with $\hat{a} \in \hat{A}$ and $c \in C$. We define a homomorphism $\theta\colon \hat{B} \to B \times_A \hat{A}$ by $\theta(\hat{b}) = (\beta(\hat{b}), \hat{\alpha}(\hat{b}))$ and note that θ followed by the projections on the coordinates is β and $\hat{\alpha}$, respectively. If $\theta(\hat{b}) = 1$, then $\hat{\alpha}(\hat{b}) = 1$, so $\hat{b} \in C$. Since β is injective on C and $\beta(\hat{b}) = 1$, we have $\hat{b} = 1$. Thus, θ is injective. If $(b, \hat{a}) \in B \times_A \hat{A}$, then $b = ac$ with $a = \bar{\varphi}(\hat{a})$ and $c \in C$. Thus, $\hat{b} = \hat{a}c \in \hat{B}$ satisfies $\theta(\hat{b}) = (b, \hat{a})$. Therefore, θ is an isomorphism. In particular, relations (4) hold.   □

We omit the assumption and the conclusion about the place from Lemma 8.6.1.

PROPOSITION 8.6.3: *Let K_0 be an ample field and x an indeterminate. Consider a finite split embedding problem*

(5) $\operatorname{Gal}(E/K_0(x)) \ltimes H \xrightarrow{\ \text{pr}\ } \operatorname{Gal}(E/K_0(x))$

over $K_0(x)$ with $H \neq 1$. Let $\{G_j \mid j \in J\}$ be a finite family of cyclic groups of prime power orders that generate H. Let K be the algebraic closure of K_0 in E. Then:

(a) (5) has for each ordinal $\alpha < \operatorname{card}(K)$ a regular solution field F_α.

(b) For each $\beta < \operatorname{card}(K)$ and every $j \in J$ there exists

$$b_{\beta,j} \in \operatorname{Branch}(F_\beta/K_0(x)) \smallsetminus \bigcup_{\alpha<\beta} \operatorname{Branch}(F_\alpha/K_0(x))$$

such that when H is identified with $\operatorname{Gal}(F_\beta/E)$, G_j is contained in an inertia group of $b_{\beta,j}$.

(c) The solutions fields F_α, $\alpha < \operatorname{card}(K)$, of (5) are linearly disjoint.

Proof: Only finitely many K_0-places of E are ramified over $K_0(x)$ or inseparable [Deu73, p. 111, Thm]. Since K_0 is infinite, we may choose a separable K_0-place φ of E unramified over $K_0(x)$, such that $\varphi(x) \in K_0$. Composing φ

with an automorphism of E over $K_0(x)$, we may assume that the restriction of φ to $K(x)$ is a K-place. Let K' be a finite Galois extension of K_0 that contains the residue field of φ. Set $E' = EK'$. Then φ extends to a K'-rational place φ' of E', unramified over $K'(x)$ [Deu73, p. 128, Thm]. Furthermore, $E'/K_0(x)$ is a Galois extension and its Galois group $\mathrm{Gal}(E'/K_0(x))$ acts on H via the restriction to $\mathrm{Gal}(E/K_0(x))$.

$$
\begin{array}{ccc}
E & \text{------} & E' \\
| & & | \\
K_0(x) \text{------} & K(x) \text{------} & K'(x)
\end{array}
$$

The existence of φ' implies that E'/K' is regular.

By (b) of Lemma 8.6.1, the split embedding problem

$$(6) \qquad \mathrm{Gal}(E'/K_0(x)) \ltimes H \xrightarrow{\ \mathrm{pr}\ } \mathrm{Gal}(E'/K_0(x))$$

has for each ordinal $\beta < \mathrm{card}(K)$ a regular solution field F'_β and for each $j \in J$ there exists

$$b_{\beta,j} \in \mathrm{Branch}(F'_\beta/K'(x)) \smallsetminus \bigcup_{\alpha<\beta} \mathrm{Branch}(F'_\alpha/K'(x))$$

such that G_j is contained in an inertia group of $b_{\beta,j}$. The field F'_β satisfies $\mathrm{Gal}(F'_\beta/K_0(x)) \cong \mathrm{Gal}(E'/K_0(x)) \ltimes H$ and it fits into a commutative diagram

$$(7) \quad
\begin{array}{ccccccc}
1 \longrightarrow & \mathrm{Gal}(F'_\beta/E') & \longrightarrow & \mathrm{Gal}(F'_\beta/K_0(x)) & \xrightarrow{\ \mathrm{res}\ } & \mathrm{Gal}(E'/K_0(x)) & \longrightarrow 1 \\
& \downarrow & & \downarrow & & \downarrow{\scriptstyle\mathrm{res}} & \\
1 \longrightarrow & H & \longrightarrow & \mathrm{Gal}(E/K_0(x)) \ltimes H & \longrightarrow & \mathrm{Gal}(E/K_0(x)) & \longrightarrow 1
\end{array}
$$

where the horizontal sequences are exact and the left vertical arrow is an isomorphism. By Remark 8.6.2, the right square of (7) is cartesian. Thus, if we denote the fixed field in F'_β of the kernel of the middle vertical map in (7) by F_β, then, by (4), $\mathrm{Gal}(F'_\beta/F_\beta) \cap \mathrm{Gal}(F'_\beta/E') = 1$ and

$$\mathrm{Gal}(F'_\beta/F_\beta)\mathrm{Gal}(F'_\beta/E') = \mathrm{Gal}(F'_\beta/E).$$

Therefore, $F_\beta E' = F'_\beta$, hence $F_\beta K' = F'_\beta$ and $F_\beta \cap E' = E$. This implies that F_β is a solution field of (5), $\mathrm{Branch}(F'_\beta/K_0(x)) = \mathrm{Branch}(F_\beta/K_0(x))$, and $\mathrm{Branch}(F'_\alpha/K_0(x)) = \mathrm{Branch}(F_\alpha/K_0(x))$ for each $\alpha < \beta$. Therefore,

$$b_{\beta,j} \in \mathrm{Branch}(F_\beta/K_0(x)) \smallsetminus \bigcup_{\alpha<\beta} \mathrm{Branch}(F_\alpha/K_0(x)).$$

Since G_j as a subgroup of $\mathrm{Gal}(F'_\alpha/K_0(x))$ is mapped onto G_j as a subgroup of $\mathrm{Gal}(F_\alpha/K_0(x))$, and each inertia group of $b_{\beta,j}$ in $\mathrm{Gal}(F'_\alpha/K_0(x))$ is mapped onto an inertia group of $b_{\beta,j}$ in $\mathrm{Gal}(F_\alpha/K_0(x))$, an inertia group of $b_{\beta,j}$ in $\mathrm{Gal}(F_\alpha/K_0(x))$ contains G_j. It follows from Lemma 7.4.1 that the fields F_α, $\alpha < \mathrm{card}(K)$, are linearly disjoint over E, as asserted in (c).

Finally, since F_β is linearly disjoint from K' over K, F'_β is linearly disjoint from $\tilde{K}$ over K', and $F_\beta K' = F'_\beta$, the field F_β is linearly disjoint from $\tilde{K}$ over K, that is F_β is a regular extension of K. □

Notes

The main result of that work asserts that if K is an ample field and E is a finite extension of $K(x)$, then each finite split embedding problem $\mathcal{E}$ over E has a regular solution. The first form of this result appears in in the unpublished paper [Pop93, Thm. 2.7], using methods of rigid analytic geometry. Our presentation follows [HaJ98b]. Using "formal patching", Harbater and Stevenson sharpen those results by constructing as many solutions to $\mathcal{E}$ as the cardinality of K, if the kernel of $\mathcal{E}$ is nontrivial and K is perfect [HaS05, Cor. 4.4]. Actually, Harbater and Stevenson state their corollary only in the case where K is very ample (see Notes to Chapter 5). However, since Proposition 5.4.3(b) asserts that every ample field is very ample, their result applies to ample fields.

The notion of "linearly disjoint solutions of an embedding problem" appears for the first time in [BHH10]. We develop that notion in Chapter 10.

The main result of this Chapter (Proposition 8.6.3) applies our method of algebraic patching to improve the results of Pop, Harbater-Stevenson, and [HaJ98b] over an arbitrary ample field K in the case where $E = K(x)$. We construct many linearly disjoint solutions to each finite split embedding problem over $K(x)$ rather than just supplying many solutions, as is done in the above mentioned papers. The case where E is an arbitrary finite extension of $K(x)$ is given by Theorem 11.7.1.

We achieve linearly disjoint soltions by keeping track of inertia groups of branch points during the reduction of functions fields of one variable (Lemma 8.3.1). An earlier version of the proof of that lemma used an improved version of Deuring's theory of reduction of divisors of function fields. The present version, due to Peter Roquette, is simpler and shorter. It replaces reduction of divisors by reduction of prime ideals in integrally closed subrings of function fields of one variable.

Lemma 8.4.1 enhances [HaJ98b, Lemma 2.1] with many solution fields and information about branch points and their inertia groups.

The proof of Lemma 8.6.1 uses that $\mathrm{trans.deg}(K((t))/K) = \infty$. That result has a simpler proof than the stronger result $\mathrm{trans.deg}(K((t))/K) = \mathrm{card}(K)^{\aleph_0}$ (Proposition 8.5.2) that we prove. The stronger result is due to Arno Fehm (private communication).

Chapter 9.
The Absolute Galois Group of $C(t)$

Let C be an algebraically closed field of cardinality m, x an indeterminate, E a finite extension of $C(x)$ of genus g, and S a set of prime divisors of E/C. We denote the maximal extension of E ramified at most over S by E_S. If X is a smooth projective model of E/C, then we interpret S as a subset of $X(C)$, call $\mathrm{Gal}(E_S/E)$ the **fundamental group of** $X \smallsetminus S$, and denote it by $\pi_1(X \smallsetminus S)$. Starting from the fundamental group of the corresponding Riemann surface and applying the Riemann existence theorem, one proves that when $r = \mathrm{card}(S) < \infty$, $\mathrm{Gal}(E_S/E)$ is the free profinite group generated by $r + 2g$ elements $\sigma_1, \ldots, \sigma_r, \tau_1, \tau_1', \ldots, \tau_g, \tau_g'$ with the unique defining relation $\sigma_1 \cdots \sigma_r [\tau_1, \tau_1'] \cdots [\tau_g, \tau_g'] = 1$ (Proposition 9.1.2). Using Grothendieck's specialization theorem, we generalize that result to an arbitrary algebraically closed field C of characteristic 0 (Proposition 9.1.5). In particular, if $r \geq 1$, then $\mathrm{Gal}(E_S/E) \cong \hat{F}_{r+2g-1}$. When $m = \mathrm{card}(S)$ is infinite, we take the limit on all finite subsets of S to conclude that $\mathrm{Gal}(E_S/E) \cong \hat{F}_m$ (Corollary 9.1.9). In particular, if S is all of the prime divisors of E/C, then $\mathrm{card}(S) = \mathrm{card}(C)$ and we find that $\mathrm{Gal}(E) \cong \hat{F}_m$ (Corollary 9.1.10). In particular, $\mathrm{Gal}(E)$ is projective (Corollary 9.1.11).

The situation is quite different when $\mathrm{char}(C)$ is a positive prime number p. We can not use the Riemann existence theorem to determine the structure of $\mathrm{Gal}(E_S/E)$. Indeed, if S is nonempty and of cardinality less than that of C, then $\mathrm{Gal}(E_S/E)$ is even not a free profinite group (Proposition 9.9.4) as is the case in characteristic 0. What we do know is the structure of the Galois group $\mathrm{Gal}(E_{S,p'}/E)$, where $E_{S,p'}$ is the maximal Galois extension of E ramified at most over S and of degree not divisible by p. Using Grothendieck's lifting to characteristic 0, one proves that the latter group is just the maximal quotient of order not divisible by p of the corresponding group in characteristic 0 (Proposition 9.2.1). But, this does not help us to compute $\mathrm{Gal}(E)$. Instead, we prove by algebraic means that $\mathrm{Gal}(E)$ is a free profinite group of cardinality m. This proof works over every algebraically closed field and does not use the Riemann existence theorem.

The first step is to prove that $\mathrm{Gal}(E)$ is projective (Proposition 9.4.6). Our proof applies some basic properties of the cohomology of profinite groups. Then we use that every finite split embedding problem for $\mathrm{Gal}(E)$ has m solutions (Proposition 8.6.3) to conclude that $\mathrm{Gal}(E) \cong \hat{F}_m$ (Corollary 9.4.9).

Interesting enough, the same arguments work if E is a finite extension of $K(x)$, where K is a field of cardinality m of positive characteristic p and $\mathrm{Gal}(K)$ is a pro-p group. Thus, even in this case $\mathrm{Gal}(E) \cong \hat{F}_m$ (Theorem 9.4.8).

Next we prove for each nonempty set of prime divisors of E/C that $\mathrm{Gal}(E_S/E)$ is projective (Corollary 9.5.8). In addition to the projectivity of

M. Jarden, *Algebraic Patching*, Springer Monographs in Mathematics, DOI 10.1007/978-3-642-15128-6_9, © Springer-Verlag Berlin Heidelberg 2011

Gal(E), the main tool used in the proof is the Jacobian variety of a smooth projective model Γ of E/C. The same tool helps us to prove that Gal(E_S/E) is not projective if S is empty (Proposition 9.6.1). The latter group can be interpreted as the fundamental group of Γ.

Finally we consider the case where $E = C(x)$ and apply algebraic patching to solve each split embedding problems m times in E_S, first in the case that C is complete under an ultrametric absolute value and then when C is an arbitrary algebraically closed field. This proves that Gal(E_S/E) $\cong \hat{F}_m$ if card(S) $= m$ (Theorem 9.8.5). This is an optimal result in characteristic p. In that case, Gal(E_S/E) is not free if card(S) $< m$ (Proposition 9.9.4).

9.1. The Fundamental Group of a Riemann Surface

Algebraic topology teaches us that the fundamental group of a sphere punctured in r points is generated by r elements $\sigma_1, \ldots, \sigma_r$ with the single relation $\sigma_1 \cdots \sigma_r = 1$. The theory of Riemann surfaces and in particular Riemann existence theorem translates this result to a theorem about finite Galois groups over $\mathbb{C}(x)$ (Proposition 9.1.1) and more generally over algebraic function fields E of one variable over $\mathbb{C}$ (Proposition 9.1.2). Using Grothendieck's specialization theorem, it is possible to generalize these results to arbitrary algebraically closed field C of characteristic 0 (Proposition 9.1.5). Taking the limit over the sets of prime divisors that we allow to ramify in the extensions prove the main result of this section: Let S be a set of prime divisors of E/C of infinite cardinality m. Denote the maximal Galois extension of E ramified at most over S by E_S. Then Gal(E_S/E) $\cong \hat{F}_m$ (Proposition 9.1.9). In particular, Gal(E) is the free profinite group of rank equal to card(C) (Corollary 9.1.10).

PROPOSITION 9.1.1 ([Voe96, Thm. 2.13]):
(a) Let F be a finite Galois extension of $\mathbb{C}(x)$. Let $\mathfrak{p}_1, \ldots, \mathfrak{p}_r$ be the prime divisors of $\mathbb{C}(x)$ which are ramified in F. Then there exist generators $\sigma_1, \ldots, \sigma_r$ of Gal($F/\mathbb{C}(x)$) with $\sigma_1 \cdots \sigma_r = 1$ such that σ_i generates an inertia group over $\mathfrak{p}_i$, $i = 1, \ldots, r$.
(b) If G is a finite group generated by $\sigma_1, \ldots, \sigma_r$ with $\sigma_1 \cdots \sigma_r = 1$, then $\mathbb{C}(x)$ has a finite Galois extension F ramified at most over $\mathfrak{p}_1, \ldots, \mathfrak{p}_r$ such that σ_i generates an inertia group over $\mathfrak{p}_i$, $i = 1, \ldots, r$.

No algebraic proof is known to either parts of Proposition 9.1.1. It would be highly desirable to have one.

Similar transition from topology to complex analysis and then to algebra generalizes Proposition 9.1.1 to Galois extensions of function fields of one variable over $\mathbb{C}$. Following the usual convention of group theory, we set $[x, y] = x^{-1} y^{-1} x y$ for elements x, y of a group G.

PROPOSITION 9.1.2 ([Ser92, Section 6.2]): Let E be a finite extension of $\mathbb{C}(x)$ of genus g and $S = \{\mathfrak{p}_1, \ldots, \mathfrak{p}_r\}$ a set of r prime divisors of $E/\mathbb{C}$.

(a) Let F be a finite Galois extension of E such that $\mathrm{Ram}(F/E) \subseteq S$. Then F has prime divisors $\mathfrak{P}_1, \ldots, \mathfrak{P}_r$ respectively lying over $\mathfrak{p}_1, \ldots, \mathfrak{p}_r$ and there are elements $\sigma_1, \ldots, \sigma_r, \tau_1, \tau_1', \ldots, \tau_g, \tau_g'$ generating $\mathrm{Gal}(F/E)$ such that σ_i generates the decomposition group $D_{\mathfrak{P}_i/\mathfrak{p}_i}$ for $i = 1, \ldots, r$ and

(1) $$\sigma_1 \cdots \sigma_r [\tau_1, \tau_1'] \cdots [\tau_g, \tau_g'] = 1.$$

(b) Let G be a finite group generated by elements $\sigma_1, \ldots, \sigma_r, \tau_1, \tau_1', \ldots, \tau_g, \tau_g'$ satisfying relation (1). Then E has a Galois extension F such that $\mathrm{Gal}(F/E) \cong G$ and F has prime divisors $\mathfrak{P}_1, \ldots, \mathfrak{P}_r$ respectively lying over $\mathfrak{p}_1, \ldots, \mathfrak{p}_r$ such that σ_i generates $D_{\mathfrak{P}_i/\mathfrak{p}_i}$, $i = 1, \ldots, r$.

On the other hand, it is not difficult to replace $\mathbb{C}$ in Propositions 9.1.1 and 9.1.2 by an arbitrary algebraically closed field C of characteristic 0. This depends on the ability to descend from an algebraically closed field to an algebraically closed subfield.

Let E be a function field of one variable over an algebraically closed field K, S a set of prime divisors of E/K, and G a finite group. We denote the set of all Galois extensions F such that $\mathrm{Gal}(F/E) \cong G$ and $\mathrm{Ram}(F/E) \subseteq S$ by $\mathcal{F}(E, S, G)$. If $E = K(x)$ is a field of rational functions, we identify the set of prime divisors of E/K with the set $K \cup \{\infty\}$ and $\mathrm{Ram}(F/E)$ with $\mathrm{Branch}(F/E)$.

LEMMA 9.1.3: Let $K \subseteq L$ be an extension of algebraically closed fields, S a finite subset of $K \cup \{\infty\}$, G a finite group, and x an indeterminate. Suppose $\mathcal{F}(L(x), S, G)$ is a finite set. Then the map $F \mapsto FL$ maps $\mathcal{F}(K(x), S, G)$ bijectively onto $\mathcal{F}(L(x), S, G)$. In particular, $\mathcal{F}(K(x), S, G)$ is a finite set.

Proof: If S is empty, then so is $\mathcal{F}(K(x), S, G)$ and $\mathcal{F}(L(x), S, G)$ (a consequence of the Riemann-Hurwitz formula (Remark 5.8.1(f))). Thus, we may assume that S is nonempty. Applying a Möbius transformation, we may assume that $\infty \in S$.

Since L/K is a regular extension, the map $F \mapsto FL$ maps $\mathcal{F}(K(x), S, G)$ injectively into the set $\mathcal{F}(L(x), S, G)$. The proof that the map is surjective breaks up into two parts.

PART A: Suppose $\mathcal{F}(L(x), S, G)$ consists of only one field F. We denote the set of zeros of a polynomials $h \in L[x]$ by $\mathrm{Zero}(h)$. By [Has80, p. 64], there are polynomials $f_1, \ldots, f_m \in L[X, Y]$ monic in Y and primitive elements $y_1, \ldots, y_m$ of $F/L(x)$ such that $f_i(x, Y)$ is irreducible in $L(x)[Y]$, $f_i(x, y_i) = 0$, and

(2) $$\mathrm{Branch}(F/L(x)) \smallsetminus \{\infty\} = \bigcap_{i=1}^{m} \mathrm{Zero}(\mathrm{discr}(f_i(x, Y))).$$

There exist $u_1, \ldots, u_n \in L$ and polynomials $g_i \in K[U_1, \ldots, U_n, x, Y]$ such that $f_i(x, Y) = g_i(\mathbf{u}, x, Y)$, where $\mathbf{u} = (u_1, \ldots, u_n)$, $F_{\mathbf{u}} = K(\mathbf{u}, x, y_i)$ is a Galois extension of $K(\mathbf{u}, x)$ independent of i with Galois group G. Since L is algebraically closed, we may enlarge the set $\{u_1, \ldots, u_n\}$ if necessary such that it contains $\mathrm{Zero}(\mathrm{discr}(f_i(x, Y)))$ for each i. The same reason implies that the polynomials $f_i(x, Y)$ are absolutely irreducible. Since K is algebraically closed, $\mathbf{u}$ generates an absolutely irreducible variety $U = \mathrm{Spec}(K[\mathbf{u}])$ in $\mathbb{A}_K^n$ defined over K.

By Hilbert [FrJ08, Lemma 13.1.1] and Bertini-Noether [FrJ08, Prop. 9.4.3], U has a nonempty Zariski-open subset U' such that for each $\mathbf{u}' \in U'(K)$ the K-specialization $\mathbf{u} \to \mathbf{u}'$ extends to a $K(x)$-place $'$ of the field $F_{\mathbf{u}}$ with residue field $F_{\mathbf{u}'}$ that has all the properties of the preceding paragraph with $\mathbf{u}'$ replacing $\mathbf{u}$.

Hilbert's Nullstellensatz gives a $\mathbf{u}' \in U'(K)$. Thus, $F_{\mathbf{u}'}$ is a Galois extension of $K(x)$ with Galois group G, $g_i(\mathbf{u}', x, Y)$ is absolutely irreducible (as a polynomial in x, Y), $g_i(\mathbf{u}', x, y_i') = 0$, and $F_{\mathbf{u}'} = K(x, y_i')$. Moreover, $\mathrm{discr}(f_i(x, Y))' = \mathrm{discr}(g_i(\mathbf{u}, x, Y)') = \mathrm{discr}(g_i(\mathbf{u}', x, Y))$ for each i. Hence, by (2),

$$\mathrm{Zero}(\mathrm{discr}(g_i(\mathbf{u}', x, Y))) = \mathrm{Zero}(\mathrm{discr}(f_i(x, Y))')$$
$$= \mathrm{Zero}(\mathrm{discr}(f_i(x, Y)))' \subseteq S' = S.$$

The second equality holds because we assumed that $\mathrm{discr}(f_i(x, Y))$ decomposes into linear factors over $K(\mathbf{u})$.

Again, by [Has80, p. 64],

$$\mathrm{Branch}(F_{\mathbf{u}'}/K(x)) \smallsetminus \{\infty\} \subseteq \bigcap_{i=1}^{m} \mathrm{Zero}(\mathrm{discr}(g_i(\mathbf{u}', x, Y))) \subseteq S.$$

It follows from $\infty \in S$ that $\mathrm{Branch}(F_{\mathbf{u}'}/K(x)) \subseteq S$, so $F_{\mathbf{u}'} \in \mathcal{F}(K(x), S, G)$. By the second paragraph of the proof, $F_{\mathbf{u}'}L \in \mathcal{F}(L(x), S, G) = \{F\}$. Consequently, $F_{\mathbf{u}'}L = F$.

PART B: *The general case.* We list the fields of $\mathcal{F}(L(x), S, G)$ as $F_1, \ldots, F_s$ and set F to be their compositum. Then F is a finite Galois extension of K, say with Galois group H. Moreover, $\mathrm{Branch}(F/L(x)) \subseteq S$ and H is the compositum of all normal subgroups N with $H/N \cong G$. If F' is another field in $\mathcal{F}(L(x), S, H)$, then F' is a compositum of Galois extensions F_i', $i = 1, \ldots, s$, that belong to $\mathcal{F}(L(x), S, G)$. Each of them must be contained in F, so $F' \subseteq F$. Since both fields have the same Galois group over $L(x)$, they coincide. It follows that $\mathcal{F}(L(x), S, H) = \{F\}$.

Part A gives a field $E \in \mathcal{F}(K(x), S, H)$ with $EL = F$. By the definition of H, E is the compositum of s distinct fields $E_1, \ldots, E_s$ with Galois group G. The corresponding composita $E_1L, \ldots, E_sL$ are s distinct fields in $\mathcal{F}(L(x), S, G)$ contained in F. Hence $\{E_1L, \ldots, E_sL\} = \{F_1, \ldots, F_s\}$. Consequently, the map $E \mapsto EL$ from $\mathcal{F}(K(x), S, G)$ to $\mathcal{F}(L(x), S, G)$ is surjective. $\square$

We generalize Lemma 9.1.3 from rational function fields to algebraic function fields.

PROPOSITION 9.1.4: *Let $K \subseteq L$ be an extension of algebraically closed fields, E a function field of one variable over K algebraically independent from L over K, S a finite subset of prime divisors of E/K, and G a finite group. We identify S with a set of prime divisors of EL/L and suppose $\mathcal{F}(EL, S, G)$ is a finite set. Then the map $\lambda \colon \mathcal{F}(E, S, G) \to \mathcal{F}(EL, S, G)$ defined by $\lambda(F) = FL$ is bijective. In particular, $\mathcal{F}(E, S, G)$ is a finite set.*

Proof: The map λ is injective because the fields $\tilde{E}$ and L are linearly disjoint over K. The proof that λ is surjective applies Lemma 9.1.3.

Let x be a separating transcendental element for the extension E/K. Then x is also a separating transcendental element for EL/L. We choose a finite subset T of $K \cup \{\infty\}$ that contains $\mathrm{Ram}(E/K(x))$ and the restriction of S to $K(x)$. Now consider $F' \in \mathcal{F}(EL, S, G)$, let $\hat{F}'$ be the Galois closure of $F'/L(x)$, and set $H = \mathrm{Gal}(\hat{F}'/L(x))$. Then $\hat{F}' \in \mathcal{F}(L(x), T, H)$. By Lemma 9.1.3, there exists $\hat{F} \in \mathcal{F}(K(x), T, H)$ with $\hat{F}L = \hat{F}'$. By linear disjointness, the map $\mathrm{res} \colon \mathrm{Gal}(\hat{F}'/L(x)) \to \mathrm{Gal}(\hat{F}/K(x))$ is an isomorphism. Hence, E has a Galois extension F in $\hat{F}$ satisfying $FL = F'$ and $\mathrm{Gal}(F/E) \cong G$. Finally consider $\mathfrak{p} \in \mathrm{Ram}(F/E)$. Then the unique extension of $\mathfrak{p}$ to a prime divisor of EL/L ramifies in F', so $\mathfrak{p} \in S$. Consequently, $F \in \mathcal{F}(E, S, G)$.   □

PROPOSITION 9.1.5: *Let C be an algebraically closed field of characteristic 0, E a finite extension of $C(x)$ of genus g and $S = \{\mathfrak{p}_1, \ldots, \mathfrak{p}_r\}$ a set of r prime divisors of E/C.*

(a) *Let F be a finite Galois extension of E with $\mathrm{Ram}(F/E) \subseteq S$. Then F has prime divisors $\mathfrak{P}_1, \ldots, \mathfrak{P}_r$ respectively lying over $\mathfrak{p}_1, \ldots, \mathfrak{p}_r$ and there are elements $\sigma_1, \ldots, \sigma_r, \tau_1, \tau_1', \ldots, \tau_g, \tau_g'$ generating $\mathrm{Gal}(F/E)$ such that σ_i generates the decomposition group $D_{\mathfrak{P}_i/\mathfrak{p}_i}$, $i = 1, \ldots, r$, and*

(1) $$\sigma_1 \cdots \sigma_r [\tau_1, \tau_1'] \cdots [\tau_g, \tau_g'] = 1.$$

(b) *Let G be a finite group generated by elements $\sigma_1, \ldots, \sigma_r, \tau_1, \tau_1', \ldots, \tau_g, \tau_g'$ satisfying relation (1). Then E has a Galois extension F such that $\mathrm{Gal}(F/E) \cong G$ and F has prime divisors $\mathfrak{P}_1, \ldots, \mathfrak{P}_r$ respectively lying over $\mathfrak{p}_1, \ldots, \mathfrak{p}_r$ such that σ_i generates $D_{\mathfrak{P}_i/\mathfrak{p}_i}$, $i = 1, \ldots, r$.*

Proof: First we consider the case where $C = \mathbb{C}$. Let E_S be the maximal extension of E that is ramified at most over S. Let Γ be the free profinite group with generators $\sigma_1, \ldots, \sigma_r, \tau_1, \tau_1', \ldots, \tau_g, \tau_g'$ and the unique defining relation (1). Then Γ is finitely generated and, by Lemma 9.1.2, has the same finite quotients as $\mathrm{Gal}(E_S/E)$. Hence, $\Gamma \cong \mathrm{Gal}(E_S/E)$ [FrJ08, Prop. 16.10.7(b)]. It follows from [FrJ08, Lemma 16.10.2] that $\mathcal{F}(E, S, G)$ is finite. Moreover, the cardinality $n(r, g, G)$ of $\mathcal{F}(E, S, G)$ depends only on r, g, and G.

Next we consider the case where $C \subseteq \mathbb{C}$. Without loss we may assume that E is algebraically independent from $\mathbb{C}$ over C and identify S with a set of prime divisors of $E\mathbb{C}/E$ by extending the field of constants from C to $\mathbb{C}$. By the preceding paragraph, $\mathcal{F}(E\mathbb{C}, S, G)$ is finite. Hence, by Proposition 9.1.4, the map $F \mapsto F\mathbb{C}$ maps $\mathcal{F}(E, S, G)$ bijectively onto $\mathcal{F}(E\mathbb{C}, S, G)$. Moreover, $g = \operatorname{genus}(E/C) = \operatorname{genus}(E\mathbb{C}/\mathbb{C})$ [FrJ08, Prop. 3.4.2(b)]. Hence, by the first paragraph, $|\mathcal{F}(E, S, G)| = n(r, g, G)$. By linear disjointness, res: $\operatorname{Gal}(F\mathbb{C}/E\mathbb{C}) \to \operatorname{Gal}(F/E)$ is an isomorphism for each $F \in \mathcal{F}(E, S, G)$. Since res maps the decomposition group over $E\mathbb{C}$ of a prime divisor $\mathfrak{P}$ of $F\mathbb{C}/\mathbb{C}$ isomorphically onto the decomposition group of $\mathfrak{P}|_F$ over E, (a) and (b) of our proposition follow from (a) and (b) of Proposition 9.1.2.

In the general case we find an algebraically closed subfield C_0 of C with a finite transcendence degree over $\mathbb{Q}$, a function field E_0 of one variable over C_0 algebraically independent from C over C_0 with $E_0C = E$, and a set $S_0 = \{\mathfrak{p}_{0,1}, \ldots, \mathfrak{p}_{0,r}\}$ of prime divisors of E_0/C_0 that uniquely extends to S when C_0 extends to C. Without loss we may assume that $C_0 \subseteq \mathbb{C}$. Then, $g = \operatorname{genus}(E/C) = \operatorname{genus}(E_0/C_0) = \operatorname{genus}(E_0\mathbb{C}/\mathbb{C})$. By the preceding paragraph, $|\mathcal{F}(E_0, S_0, G)| = n(r, g, G)$. Moreover, (a) and (b) hold for C_0, E_0, S_0 replacing C, E, S. If $\mathcal{F}(E, S, G)$ had more than $n(r, g, G)$ fields, then we could choose C_0 such that $\mathcal{F}(E_0, S_0, G)$ would also have more that $n(r, g, G)$ fields, in contrast to the previous conclusion. Therefore, $\mathcal{F}(E, S, G)$ is finite. We may therefore apply Proposition 9.1.4 again and conclude that the map $F_0 \mapsto F_0C$ maps the set $\mathcal{F}(E_0, S_0, G)$ bijectively onto the set $\mathcal{F}(E, S, G)$. This map is compatible with restriction of Galois groups and decomposition groups. Therefore, (a) and (b) hold also for C, E, S. $\qquad\square$

Giving a function field E of one variable over a field K and a set S of prime divisors of E/K, we denote (as in the proof of Proposition 9.1.5) the compositum of all finite Galois extensions F of E with $\operatorname{Ram}(F/E) \subseteq S$ by E_S. Thus, E_S is a Galois extension of E. If S' is another set of prime divisors of E/K and $S \subseteq S'$, then $E_S \subseteq E_{S'}$. If S is empty, then E_S is the compositum of all unramified finite Galois extensions of E. In this case we denote E_S also by E_{ur}.

PROPOSITION 9.1.6: *Let C be an algebraically closed field of characteristic 0, E a finite extension of $C(x)$ of genus g, and $S = \{\mathfrak{p}_1, \ldots, \mathfrak{p}_r\}$ a set of r prime divisors of E/C. Then $\operatorname{Gal}(E_S/E)$ is the free profinite group generated by elements $\sigma_1, \ldots, \sigma_r, \tau_1, \tau_1', \ldots, \tau_g, \tau_g'$ satisfying the relation (1) and each σ_i is a generator of the decomposition group of a prime divisor of E_S/C lying over $\mathfrak{p}_i$.*

Proof: We extend the argument of the first paragraph of the proof of Proposition 9.1.5. For each finite Galois extension F of E in E_S we consider the finite set $\mathcal{A}(F/E)$ of all $(2r + 2g)$-tuples

$$(3) \qquad (\mathfrak{P}_1, \ldots, \mathfrak{P}_r, \sigma_1, \ldots, \sigma_r, \tau_1, \tau_1', \ldots, \tau_g, \tau_g')$$

such that $\mathfrak{P}_i$ is a prime divisor of F/C lying over $\mathfrak{p}_i$, σ_i is a generator of the decomposition group $D_{\mathfrak{P}_i/\mathfrak{p}_i}$, $i = 1, \ldots, r$, and $\sigma_1, \ldots, \sigma_r, \tau_1, \tau_1', \ldots, \tau_g, \tau_g'$

are generators of $\mathrm{Gal}(F/E)$ satisfying relation (1). If F' is a finite Galois extension of E in E_S that contains F and $\mathfrak{P}'_i$ is a prime divisor of F'/C lying over $\mathfrak{p}_i$, then $\mathfrak{P}_i = \mathfrak{P}'_i|_F$ is a prime divisor of F/C lying over $\mathfrak{p}_i$ and the epimorphism res: $\mathrm{Gal}(F'/E) \to \mathrm{Gal}(F/E)$ maps $D_{\mathfrak{P}'_i/\mathfrak{p}_i}$ onto $D_{\mathfrak{P}_i/\mathfrak{p}_i}$ [Ser79, Chap. 1, Prop. 22(b)]. Hence res induces a map of $\mathcal{A}(F'/E)$ into $\mathcal{A}(F/E)$. By Proposition 9.1.5(a), each $\mathcal{A}(F/E)$ is nonempty. Therefore, the inverse limit of the sets $\mathcal{A}(F/E)$ is nonempty [FrJ08, Lemma 1.1.3]. Each element of that inverse limit is an $(2r + 2g)$-tuple (3) satisfying relation (1) such that $\mathfrak{P}_i$ is a prime divisor of E_S/C lying over $\mathfrak{p}_i$ and σ_i generates $D_{\mathfrak{P}_i/\mathfrak{p}_i}$.

Now, let Γ be the free profinite group on the generators $\sigma_1, \ldots, \sigma_r$, $\tau_1, \tau'_1, \ldots, \tau_g, \tau'_g$ satisfying relation (1). By Proposition 9.1.5, $\mathrm{Gal}(E_S/E)$ and Γ have the same finite quotients. Consequently, by [FrJ08, Prop. 16.10.7], $\mathrm{Gal}(E_S/E) \cong \Gamma$. $\qquad\square$

COROLLARY 9.1.7: *In the notation of Proposition 9.1.6,*
(a) *If $g = 0$ and $r \geq 2$ or $g \geq 1$ and $r \geq 1$, then $\langle \sigma_i \rangle \cong \hat{\mathbb{Z}}$, $i = 1, \ldots, r$.*
(b) *If $r \geq 1$, then $\mathrm{Gal}(E_S/E) \cong \hat{F}_{r-1+2g}$.*

Proof of (a): In order to prove that $\langle \sigma_i \rangle \cong \hat{\mathbb{Z}}$, it suffices to prove that for each positive integer n the cyclic group C_n of order n is a quotient of $\langle \sigma_i \rangle$.

Let y be a generator of C_n. If $r \geq 2$, we choose $j \neq i$, $1 \leq j \leq r$. Then we map σ_i onto y, σ_j onto y^{-1} and all other generators to 1 to get an epimorphism $\mathrm{Gal}(E_S/E) \to C_n$ that maps $\langle \sigma_i \rangle$ onto C_n.

It remains to consider the case where $r = 1$ and $g \geq 1$. Let D_{2n} be the dihedral group of order $2n$ generated by elements x, y with the defining relations $x^{2n} = 1$, $y^2 = 1$, and $y^{-1}xy = x^{-1}$. Then $[x,y] = x^{-1}y^{-1}xy = x^{-2}$ has order n. Hence, the map $\sigma_1 \mapsto [x,y]^{-1}$, $\tau_1 \mapsto x$, $\tau'_1 \mapsto y$, $\tau_j \mapsto 1$, and $\tau'_j \mapsto 1$ for $j \geq 2$ extends to an epimorphism of $\mathrm{Gal}(E_S/E)$ onto D_{2n} mapping $\langle \sigma_1 \rangle$ onto the cyclic group of order n generated by $[x,y]$.

Proof of (b): By Proposition 9.1.6, $\mathrm{Gal}(E_S/E)$ is the free profinite group generated by the elements $\sigma_2, \ldots, \sigma_r$, $\tau_1, \tau'_1, \ldots, \tau_g, \tau'_g$. The extra generator σ_1 can be expressed in terms of the other generators via (1). $\qquad\square$

Remark 9.1.8: Inverse limit of free profinite groups. One way to construct a free profinite group of arbitrary rank is to start from disjoint sets S, T such that T is finite. For each subset A of S we set $A' = A \cup T$ and consider the free profinite group $\hat{F}_{A'}$ with basis A'. If A, B are finite subsets of S and $A \subseteq B$, then the map $B' \to A'$ that maps each $a \in A'$ onto itself and each $b \in B \smallsetminus A$ onto 1 uniquely extends to an epimorphism $\alpha_{BA} \colon \hat{F}_{B'} \to \hat{F}_{A'}$. The inverse limit F of the groups $\hat{F}_{A'}$ and the maps α_{BA} is isomorphic to the free profinite group $\hat{F}_{S'}$ with basis $S' = S \cup T$. Indeed, for each A let α_{SA} be the limit of all maps α_{BA}, where B ranges over all finite subsets B of S that contain A. For each open normal subgroup N of F there exists a finite subset A of S such that $\mathrm{Ker}(\alpha_{SA}) \leq N$, so $S' \smallsetminus A' \subseteq N$. Thus, S' converges to 1 in the sense of [FrJ08, Section 17.1]. Moreover, if φ_0 is a map of S' into a finite group H that maps the complement of some A' onto 1, then φ_0 decomposes through $\alpha_{SA}\colon S' \to A'$. Since A' is a basis of $\hat{F}_{A'}$, we may extend φ_0 to a continuous homomorphism $\varphi\colon F \to H$.

Using compactness, it is possible to relax the above rigid condition on the maps α_{BA}. Consider a projective limit $G = \varprojlim G_A$ of profinite groups, where A ranges over all finite subsets of S. Assume for each A the group G_A is isomorphic to $\hat{F}_{A'}$, and if $B \supseteq A$, then the associated homomorphism $\rho_{BA}\colon G_B \to G_A$ is surjective. Consider the compact space $(G_A)^{A'}$ (in the product topology) of all functions from A' into G_A. Let Φ_A be a closed subset of $(G_A)^{A'}$. Suppose each $\varphi \in \Phi_A$ satisfies $\langle \varphi(a) \mid a \in A' \rangle = G_A$. Suppose also that if $B \supseteq A$ and $\varphi' \in \Phi_B$, then $\varphi = \rho_{BA} \circ \varphi'|_{A'} \in \Phi_A$ and $\rho_{BA}(\varphi'(b)) = 1$ for each $b \in B \smallsetminus A$. Then φ (resp. φ') uniquely extends to an epimorphism $\varphi_A\colon \hat{F}_{A'} \to G_A$ (resp $\varphi_B\colon \hat{F}_{B'} \to G_B$) such that $\rho_{BA} \circ \varphi_B = \varphi_A \circ \alpha_{BA}$. By [FrJ08, Lemma 17.4.11], φ_A (resp. φ_B) is an isomorphism.

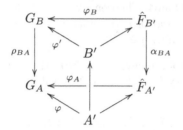

It follows that $\Phi = \varprojlim \Phi_A$ is nonempty [FrJ08, Lemma 1.1.3]. Each $\varphi \in \Phi$ gives an isomorphism of $\hat{F}_{S'}$ onto G. In particular, $\varphi(S')$ is a basis of G and for each A we have $\rho_{SA} \circ \varphi|_A \in \Phi_A$. $\square$

PROPOSITION 9.1.9: *Let C be an algebraically closed field of characteristic 0, E a function field of one variable over C, S an infinite set of cardinality m of prime divisors of E/C. Then $\mathrm{Gal}(E_S/E) \cong \hat{F}_m$.*

Proof: We choose a prime divisor $\mathfrak{p}_1 \in S$ and denote the collection of all finite nonempty subsets of $S \smallsetminus \{\mathfrak{p}_1\}$ by $\mathcal{A}$. We also choose a set T disjoint from S of $2g$ elements, where $g = \mathrm{genus}(E/C)$. For each $A \in \mathcal{A}$ let $G_A = \mathrm{Gal}(E_{\{\mathfrak{p}_1\} \cup A}/E)$ and $A' = A \cup T$. By Corollary 9.1.7(b), $G_A \cong \hat{F}_{A'}$. Let Φ_A be the set of all functions $\varphi\colon A' \to G_A$ such that $G_A = \langle \varphi(a) \mid a \in A' \rangle$ and for each $\mathfrak{p} \in A$, $\varphi(\mathfrak{p})$ generates a decomposition group of a prime divisor of $E_{\{\mathfrak{p}_1\} \cup A}/C$ over $\mathfrak{p}$.

CLAIM: Φ_A *is closed in* $G_A^{A'}$. Indeed, suppose a function $\psi\colon A' \to G_A$ belongs to the closure of Φ_A. Then for each finite Galois extension F of E in $E_{\{\mathfrak{p}_1\} \cup A}$ there exists $\varphi \in \Phi_A$ such that $\psi(a)|_F = \varphi(a)|_F$ for each $a \in A'$. It follows that $\mathrm{Gal}(F/E) = \langle \psi(a)|_F \mid a \in A' \rangle$ and for each $\mathfrak{p} \in A$, $\psi(\mathfrak{p})|_F$ generates a decomposition group of a prime divisor of F/C lying over $\mathfrak{p}$. Taking the limit over all possible F, we find that $\psi \in \Phi_A$, as claimed.

Next note that if $B \in \mathcal{A}$ and $A \subseteq B$, then $E_{\{\mathfrak{p}_1\} \cup A} \subseteq E_{\{\mathfrak{p}_1\} \cup B}$. Let $\rho_{BA} \colon G_B \to G_A$ be the restriction map. Consider $\varphi_B \in \Phi_B$ and $\mathfrak{p} \in B \smallsetminus A$. Then $\varphi_B(\mathfrak{p})$ generates the decomposition group of some prime divisor of $E_{\{\mathfrak{p}_1\} \cup B}/C$ lying over $\mathfrak{p}$. Therefore, $\varphi_B(\mathfrak{p})|_{E_{\{\mathfrak{p}_1\} \cup A}}$ generates the decomposition group of a prime divisor $\mathfrak{P}$ of $E_{\{\mathfrak{p}_1\} \cup A}/C$ lying over $\mathfrak{p}$. Since C is algebraically closed, $D_{\mathfrak{P}/\mathfrak{p}} = I_{\mathfrak{P}/\mathfrak{p}}$. However, $\mathfrak{p}$ is unramified in $E_{\{\mathfrak{p}_1\} \cup A}$, because $\mathfrak{p} \notin A$. Hence, $I_{\mathfrak{P}/\mathfrak{p}} = 1$, so $\rho_{BA}(\varphi_B(\mathfrak{p})) = \varphi_B(\mathfrak{p})|_{E_{\{\mathfrak{p}_1\} \cup A}} = 1$.

Finally observe that ρ_{BA} maps each set of generators of G_B onto a set of generators of G_A. Therefore, $\rho_{BA} \circ \varphi_B|_A \in \Phi_A$.

Consequently, by Remark 9.1.8, $\mathrm{Gal}(E_S/E) \cong \hat{F}_m$. $\square$

If we take S in Proposition 9.1.9 to be the set of all prime divisors of E/C, then $E_S = \tilde{E}$. In this case the group $\mathrm{Gal}(E_S/E)$ becomes the absolute Galois group of E and $\mathrm{card}(S) = \mathrm{card}(C)$.

COROLLARY 9.1.10 ([Dou64, Théorème 2]): *Let C be an algebraically closed field of characteristic 0 and of cardinality m and E a function field of one variable over C. Then $\mathrm{Gal}(E) \cong \hat{F}_m$.*

Since each free profinite group is projective [FrJ08, Cor. 22.4.5], the combination of Corollary 9.1.7(b) and Proposition 9.1.9 gives the following result:

COROLLARY 9.1.11: *Let C be an algebraically closed field of characteristic 0 and S a nonempty set of prime divisors of E/C. Then $\mathrm{Gal}(E_S/E)$ is projective. In particular, $\mathrm{Gal}(E)$ is projective.*

Remark 9.1.12: *Freeness and projectivity.* The projectivity of $\mathrm{Gal}(E_S/E)$ obtained in Corollary 9.1.11 is a much weaker property than the freeness of the group. Yet we generalize it in Theorem 9.5.7 for an arbitrary characteristic by algebraic means and deduce the freeness of $\mathrm{Gal}(E_S/E)$ for infinite sets S (Theorem 9.8.5). If however, $\mathrm{char}(C) > 0$ and S is finite, then $\mathrm{Gal}(E_S/E)$ is not free (Proposition 9.9.4). $\square$

9.2 Fundamental Groups in Positive Characteristic

We continue our survey of the theory of fundamental groups of curves over an algebraically closed fields and move to the case where the characteristic is a prime number p. The results obtained in characteristic 0 can be carried over as long as we "stay away" from p, but are completely different in the general case.

PROPOSITION 9.2.1 ([SGA1, Exposé XIII, Cor. 2.12]): *Let C be an algebraically closed field of characteristic p, E a function field of one variable over C of genus g, and $S = \{\mathfrak{p}_1, \ldots, \mathfrak{p}_r\}$ a finite set of prime divisors of E/C.*
(a) *Let $E_{S,\mathrm{tr}}$ be the compositum of all finite Galois extensions F of E such that $\mathrm{Ram}(F/E) \subseteq S$ and each $\mathfrak{p} \in \mathrm{Ram}(F/E)$ is tamely ramified. Then*

$\mathrm{Gal}(E_{S,\mathrm{tr}}/E)$ *is generated by elements* $\sigma_1,\ldots,\sigma_r,\ \tau_1,\tau_1',\ldots,\tau_g,\tau_g'$ *satisfying the relation*

(1) $$\sigma_1\cdots\sigma_r[\tau_1,\tau_1']\cdots[\tau_g,\tau_g'] = 1.$$

(b) *Let* $E_{S,p'}$ *be the compositum of all finite Galois extensions* F *of* E *of degree not divisible by* p *such that* $\mathrm{Ram}(F/E) \subseteq S$. *Then* $\mathrm{Gal}(E_{S,p'}/E)$ *is the free group generated by elements* $\sigma_1,\ldots,\sigma_r,\ \tau_1,\tau_1',\ldots,\tau_g,\tau_g'$ *with the defining relation* (1) *in the category of profinite groups with order not divisible by* p.

(c) *In both* (a) *and* (b), σ_i *can be chosen to generate a decomposition group of a prime divisor lying over* $\mathfrak{p}_i$, $i = 1,\ldots,r$.

Sketch of proof: One chooses a smooth projective model Γ for E/C. Then one finds a complete discrete valuation ring R, with residue field C, and an algebraically closed quotient field K of characteristic 0 and a projective connected smooth curve Δ over $S = \mathrm{Spec}(R)$ whose special fiber is $\Delta \times_S \mathrm{Spec}(C) \cong \Gamma$. Let Δ_K be the generic fiber of Δ. Then $\mathrm{genus}(\Delta_K) = g$. Let $\mathbf{p}_1,\ldots,\mathbf{p}_r$ be the points of $\Gamma(C)$ corresponding to $\mathfrak{p}_1,\ldots,\mathfrak{p}_r$. By Hensel's lemma, the points $\mathbf{p}_1,\ldots,\mathbf{p}_r$ lift to points $\mathbf{q}_1,\ldots,\mathbf{q}_r$ of $\Delta_K(R)$. Let F be the function field of Δ_K over K and $\mathbf{q}_1,\ldots,\mathbf{q}_r$ the prime divisors corresponding to the points $\mathbf{q}_1,\ldots,\mathbf{q}_r$. Let $T = \{\mathbf{q}_1,\ldots,\mathbf{q}_r\}$. Using knowledge of the behavior of the tamely ramified covers of the curve, one proves that there is a surjective map $\mathrm{Gal}(F_T/F) \to \mathrm{Gal}(E_{S,\mathrm{tr}}/E)$ defining an isomorphism $\mathrm{Gal}(F_{T,p'}/F) \to \mathrm{Gal}(E_{S,p'}/E)$ compatible with decomposition groups. Using Proposition 9.1.6, this gives (a), (b), and (c). $\qquad\square$

Remark 9.2.2: Abhyankar's conjecture. Let C be an algebraically closed field of positive characteristic p, E a function field of one variable of genus g, S a finite nonempty set of r prime divisors of E/C. Proposition 9.2.1 gives us information only on the maximal tamely ramified quotient and the maximal p'-quotient of $\mathrm{Gal}(E_S/E)$. The structure of $\mathrm{Gal}(E_S/E)$ is unknown, even in the case where $E = C(x)$ and S consists of one prime divisor. That is, we do not know the structure of the fundamental group of the affine line in characteristic p. What we do know is the set of finite quotients of $\mathrm{Gal}(E_S/E)$.

Consider a finite Galois extension F of E in E_S. Let $G = \mathrm{Gal}(F/E)$ and denote the normal subgroup of G generated by all p-Sylow subgroups of G by $G(p)$. Let F_0 be the fixed field of $G(p)$ in F. Then F_0 is a finite Galois extension of E in E_S of order not divisible by p. It follows from Proposition 9.2.1(b) that $G/G(p) \cong \mathrm{Gal}(F_0/E)$ is generated by elements $\sigma_1,\ldots,\sigma_r,\tau_1,\tau_1',\ldots,\tau_g,\tau_g'$ satisfying Relation (1). In particular, if E is a field of rational functions over C and $r = 1$, then $g = 0$, so $\sigma_1 = 1$ and $F_0 = E$. This follows also from the Riemann-Hurwitz formula for tamely ramified extensions [FrJ05, Remark 3.6.2(d)]. Therefore, $G = G(p)$ is generated by its p-Sylow subgroups. A finite group having that property is said to be a **quasi-p group**.

This observation led Shreeram Abhyankar in [Abh57] to conjecture that each finite group G such that $G/G(p)$ is generated by $r + 2g$ elements satisfying Relation (1) with $r \geq 1$ appears as a quotient of $\mathrm{Gal}(E_S/E)$. In the case $G(p) = 1$, the conjecture follows from Proposition 9.2.1(b). Jean-Pierre Serre proved Abhyankar's conjecture for solvable groups G [Ser90] using class field theory. Raynaud proved Abhyankar's conjecture for an arbitrary quasi-p group over the affine line [Ray94], and David Harbater settled the general Abhyankar's conjecture [Hrb94a] by reducing it to the case of the affine line proved by Raynaud. Finally, [Pop95] proves that every finite split embedding problem for $\mathrm{Gal}(E_S/E)$ whose kernel is a finite quotient of the fundamental group of the affine line is solvable. If S is nonempty, then $\mathrm{Gal}(E_S/E)$ is a projective group (Corollary 9.5.8). Using Proposition 9.2.1 and Raynaud's result this gives an alternative proof of Abhyankar's conjecture.

Indeed, let G be a finite group such that $G/G(p)$ is generated by $r + 2g$ elements satisfying Relation (1) with $r \geq 1$. By Proposition 9.2.1(b), there is an epimorphism $\varphi \colon \mathrm{Gal}(E_S/E) \to G/G(p)$. Let $\alpha \colon G \to G/G(p)$ be the quotient map. Since $\mathrm{Gal}(E_S/E)$ is projective, there is a homomorphism $\gamma \colon \mathrm{Gal}(E_S/E) \to G$ such that $\alpha \circ \gamma = \varphi$. Denote the fixed field of $\mathrm{Ker}(\gamma)$ by $\hat{E}$. Then there are epimorphisms $\hat{\varphi} \colon \mathrm{Gal}(E_S/E) \to \mathrm{Gal}(\hat{E}/E)$ and $\bar{\varphi} \colon \mathrm{Gal}(\hat{E}/E) \to G/G(p)$ such that $\varphi = \bar{\varphi} \circ \hat{\varphi}$. Moreover, there is an embedding $\bar{\gamma} \colon \mathrm{Gal}(\hat{E}/E) \to G$ such that $\gamma = \bar{\gamma} \circ \hat{\varphi}$ and $\alpha \circ \bar{\gamma} = \bar{\varphi}$. Let $\hat{G} = G \times_{G/G(p)} \mathrm{Gal}(\hat{E}/E)$ be the corresponding fiber product and let $\hat{\alpha} \colon \hat{G} \to \mathrm{Gal}(\hat{E}/E)$ be the projection on the second factor. Then $(\hat{\varphi} \colon \mathrm{Gal}(E_S/E) \to \mathrm{Gal}(\hat{E}/E), \ \hat{\alpha} \colon \hat{G} \to \mathrm{Gal}(\hat{E}/E))$ is a finite split embedding problem for $\mathrm{Gal}(E_S/E)$ whose kernel is isomorphic to $G(p)$, so it is a finite quotient of the fundamental group of the affine line (by Raynaud). It follows from Pop's theorem that the embedding problem is solvable. In particular, G is a quotient of $\mathrm{Gal}(E_S/E)$, as claimed.   □

Remark 9.2.3: Half Riemann existence theorem. One may refer to Proposition 9.2.1 as the **tame Riemann existence theorem**. The best known approximation to Proposition 9.1.1 is the so called **Half Riemann existence theorem**, due to Pop [Pop94]. It applies to an arbitrary Henselian field (K, v). For a positive integer r let $S = \{a_1, b_1, \ldots, a_r, b_r\}$ be a subset of K_s such that $a_i \neq b_i$, $v(a_i - b_i) > v(a_i - b_j)$ for all $i \neq j$, and both $\{a_1, \ldots, a_r\}$ and $\{b_1, \ldots, b_r\}$ are invariant under $\mathrm{Gal}(K)$. Let Π be $\hat{F}_r$ if $\mathrm{char}(K) = \mathrm{char}(\bar{K}_v)$ and the free product of r copies of $\hat{\mathbb{Z}}/\mathbb{Z}_p$ if $\mathrm{char}(K) = 0$ and $\mathrm{char}(\bar{K}_v) = p > 0$. Then the field $K(x)$ of rational functions in x over K has a Galois extension N with $\mathrm{Branch}(N/K(x)) = S$ such that $\mathrm{Gal}(N/K_s(x)) \cong \Pi$ and $\mathrm{Gal}(N/K(x)) = \mathrm{Gal}(K) \ltimes \mathrm{Gal}(N/K_s(x))$. Moreover, one may choose generators $\sigma_1, \ldots, \sigma_r$ for Π such that σ_i generates an inertia group of both a_i and b_i, $i = 1, \ldots, r$. See also [Hrb03, Thm. 4.3.3 and Remark 4.4.4(c)].   □

9.3 Cohomology of Groups

We survey in this section the basic notions and results of the cohomology of
profinite groups needed in this book. Our basic references are [Rib70] and
[Ser79]. In this Chapter we apply a small part of our survey to prove that
$\mathrm{Gal}(E)$ is projective for each extension E of transcendence degree 1 over an
algebraically closed field (Proposition 9.4.6). In Chapter 11 we build on our
survey to prove local-global theorems for Brauer groups. This leads to fields
of transcendence degree 1 over PAC fields with projective absolute Galois
groups.

9.3.1 G-MODULES.

Let G be a profinite group and A a discrete Abelian (additive) group.
We say that A is a G-**module** if G acts continuously on A from the left, that
is there is a continuous map $G \times A \to A$ mapping a pair $(\sigma, a) \in G \times A$ onto
the element σa of A such that

(1a) $(\sigma \tau a) = \sigma(\tau a)$,

(1b) $\sigma(a + b) = \sigma a + \sigma b$, and

(1c) $1a = a$

for all $a, b \in A$ and $\sigma, \tau \in G$. If $\sigma a = a$ for all $\sigma \in G$ and $a \in A$ we say that
A is a **trivial G-module**. Our basic examples occur when $G = \mathrm{Gal}(L/K)$
is a Galois group and A is either the additive group L^+ or the multiplicative
group $L^\times$ of L (where in the latter case we have to switch to a multiplicative
module). We may also take A to be the group of all roots of unity belonging
to L or the group $J(L)$, where J is an Abelian variety defined over K.

For each closed subgroup U of G we write A^U for the fixed module of
A under U. For each $a \in A$ the equality $1a = a$ implies that there exists an
open subgroup U of G such that $\sigma a = a$ for each $\sigma \in U$, i.e. $a \in A^U$. Thus

$$(2) \qquad\qquad\qquad A = \bigcup A^U,$$

where U ranges on all open subgroups (or even open normal subgroups) U
of G.

A map $\varphi \colon A \to B$ between G-modules is a G-**homomorphism** if φ is a
group homomorphism that satisfies $\varphi(\sigma a) = \sigma \varphi(a)$ for all $\sigma \in G$ and $a \in A$.

9.3.2 DEFINITION OF THE COHOMOLOGY GROUPS.

Given a G-module A, we consider for each $q \geq 0$ the group $C^q(G, A)$
of all continuous maps $f \colon G^q \to A$ (called **non-homogeneous q-cochains**)
and the homomorphisms $\partial_{q+1} \colon C^q(G, A) \to C^{q+1}(G, A)$ (called the **non-homogeneous coboundary operators**) defined by $(\partial_1 f)(\sigma) = \sigma f(1) - f(1)$
and for $q \geq 1$ by

$$(3)(\partial_{q+1} f)(\sigma_1, \ldots, \sigma_{q+1}) = \sigma_1 f(\sigma_2, \ldots, \sigma_{q+1})$$

$$+ \sum_{i=1}^{q} (-1)^i f(\sigma_1, \ldots, \sigma_{i-1}, \sigma_i \sigma_{i+1}, \sigma_{i+2}, \ldots, \sigma_{q+1})$$

$$+ (-1)^{q+1} f(\sigma_1, \ldots, \sigma_q).$$

One proves that

(4) $$\partial_{q+1} \circ \partial_q = 0,$$

so that

$$0 \to C^0(G, A) \xrightarrow{\partial_1} C^1(G, A) \xrightarrow{\partial_2} C^2(G, A) \xrightarrow{\partial_3} \cdots$$

is a **complex**. Each element of the group $Z^q(G, A) = \text{Ker}(\partial_{q+1})$ is a q-**cocycle** whereas each element of the group $B^q(G, A) = \text{Im}(\partial_q)$ is a q-**coboundary**. By (4), $B^q(G, A) \leq Z^q(G, A)$. This gives rise to the q-**th cohomology group with coefficients in** A:

$$H^q(G, A) = Z^q(G, A)/B^q(G, A).$$

Note that $C^0(G, A)$ is the set of all functions $f: \{1\} \to A$. Taking $\partial_0 = 0$, we get $B^0(G, A) = 0$ and $H^0(G, A) = Z^0(G, A) = A^G$. A 1-coboundary is a map $f_a: G \to A$ defined by $f_a(\sigma) = \sigma a - a$ for a fixed $a \in A$. A 1-cocycle is a **crossed homomorphism**, namely a map $f: G \to A$ satisfying $f(\sigma\tau) = \sigma f(\tau) + f(\sigma)$. Thus, an element of $H^1(G, A)$ is an equivalence class of crossed homomorphisms modulo coboundaries. If A is a trivial G-module, then each 1-coboundary is 0 and each crossed homomorphism is a homomorphism $G \to A$. Thus, in this case $H^1(G, A) = \text{Hom}(G, A)$.

9.3.3 FUNCTORIALITY OF THE COHOMOLOGY GROUPS.

The cohomology groups are functorial in both variables. Each G-homomorphism $\alpha: A \to B$ of G-modules induces a homomorphism $\alpha: C^q(G, A) \to C^q(G, B)$ of the corresponding cochain groups that commutes with the coboundary operator: $(\alpha f)(\sigma_1, \ldots, \sigma_q) = \alpha(f(\sigma_1, \ldots, \sigma_q))$. It follows that $\alpha(Z^q(G, A)) \leq Z^q(G, B)$ and $\alpha(B^q(G, A)) \leq B^q(G, B)$. Hence, α yields a homomorphism $\alpha: H^q(G, A) \to H^q(G, B)$. Each of the assignments $A \rightsquigarrow C^q(G, A)$, $A \rightsquigarrow Z^q(G, A)$, $A \rightsquigarrow B^q(G, A)$, and $A \rightsquigarrow H^q(G, A)$ is a covariant functor from the category of G-modules to the category of Abelian groups. This means that the composition $\beta \circ \alpha$ of homomorphisms of G-modules is assigned to the composition $\beta \circ \alpha$ of Abelian groups and the identity of G-modules is assigned to the corresponding identity of Abelian groups.

9.3.4 SHORT AND LONG EXACT SEQUENCES.

The most important feature of group cohomology is the theorem about the exact sequences: To each short exact sequence

$$0 \to A \xrightarrow{\alpha} B \xrightarrow{\beta} C \to 0$$

of G-modules there corresponds a long exact sequence

$$0 \to A^G \xrightarrow{\alpha} B^G \xrightarrow{\beta} C^G$$
$$\xrightarrow{\delta} H^1(G, A) \xrightarrow{\alpha} H^1(G, B) \xrightarrow{\beta} H^1(G, C)$$
$$\xrightarrow{\delta} H^2(G, A) \xrightarrow{\alpha} H^2(G, B) \xrightarrow{\beta} H^2(G, C) \xrightarrow{\delta} \cdots,$$

where the **connecting homomorphisms** δ are functorial [Rib70, p. 115, Prop. 4.4].

9.3.5 COMPATIBLE HOMOMORPHISMS.

Generalizing the functoriality of the cohomology to both variables, we consider a G-module A and an H-module B. A pair (φ, β) consisting of a homomorphism of profinite groups $\varphi\colon G \to H$ and a homomorphism $\beta\colon B \to A$ of H and G modules, respectively, is said to be **compatible** if $\sigma(\beta(b)) = \beta(\varphi(\sigma)b)$ for all $\sigma \in G$ and $b \in B$. In this case they define for each $q \geq 0$ a homomorphism $(\varphi, \beta)\colon C^q(H, B) \to C^q(G, A)$ by the formula: $((\varphi, \beta)g)(\sigma_1, \ldots, \sigma_q) = \beta(g(\varphi(\sigma_1), \ldots, \varphi(\sigma_q)))$ for $g \in C^q(H, B)$ and $\sigma_1, \ldots, \sigma_q \in G$. As in Subsection 9.3.3, (φ, β) commutes with the coboundary homomorphisms, so it induces natural homomorphisms

$$(\varphi, \beta)\colon H^q(H, B) \to H^q(G, A).$$

The maps (φ, β) behave functorially in the following sense. If I is a profinite group, C is an I-module, and $\psi\colon H \to I$ and $\gamma\colon C \to B$ are compatible homomorphisms, then $(\psi \circ \varphi, \beta \circ \gamma)$ is a pair of compatible homomorphisms from G to I and C to A and the following triangle is commutative:

(5)
$$
\begin{array}{ccc}
H^q(H, B) & \xleftarrow{\ (\psi, \gamma)\ } & H^q(I, C) \\
 & \searrow{\scriptstyle(\varphi, \beta)} \quad \swarrow{\scriptstyle(\psi \circ \varphi, \beta \circ \gamma)} & \\
 & H^q(G, A) &
\end{array}
$$

9.3.6 INFLATION AND RESTRICTION.

An important example occurs when N is a closed normal subgroup of G. Let A be a G-module and denote the image of an element $\sigma \in G$ in G/N under the quotient map by $\bar{\sigma}$. Then G/N acts on A^N by $\bar{\sigma}a = \sigma a$ and this action is compatible with the inclusion $A^N \to A$. Thus, it induces for each $q \geq 0$ the **inflation homomorphism** $\mathrm{inf}\colon H^q(G/N, A^N) \to H^q(G, A)$. Similarly, the restriction of the action of G on A to N is compatible with the identity map $A \to A$, so it gives rise to the **restriction homomorphisms** $\mathrm{res}\colon H^q(G, A) \to H^q(N, A)$.

LEMMA 9.3.7: *Let G be a profinite group, N a closed normal subgroup, A a G-module, and $q \geq 1$. Suppose $H^i(G, A) = 0$ for $1 \leq i \leq q - 1$. Then the sequence*

(6)
$$0 \to H^q(G/N, A^N) \xrightarrow{\ \mathrm{inf}\ } H^q(G, A) \xrightarrow{\ \mathrm{res}\ } H^q(N, A)$$

is exact. In particular, if $H^1(G, A) = 0$, then the following sequence is exact:

(7)
$$0 \to H^2(G/N, A^N) \xrightarrow{\ \mathrm{inf}\ } H^2(G, A) \xrightarrow{\ \mathrm{res}\ } H^2(N, A).$$

The proof of the lemma for $q = 1$ is carried out by direct verification on cocycles. Then one applies "dimension shifting" to continue the proof for arbitrary q by induction. This is done in [CaF67, p. 100, Prop. 4] for abstract groups and in [Koc70, Sec. 3.7] for profinite groups. Note that the latter source adds two more groups to the sequence (6). The same five terms sequence is also proved to be exact in [Rib70, p. 177, Cor. 5.4] by application of spectral sequences.

9.3.8 CORESTRICTION.

We consider an open subgroup U of a profinite group G and choose a set S of representatives for the left cosets of G modulo U, thus $G = \bigcup_{\sigma \in S} \sigma U$. Then we define for each G-module A a group homomorphism $N_{G/U} \colon A^U \to A^G$ by $N_{G/U}(a) = \sum_{\sigma \in S} \sigma a$. It can be uniquely extended to a natural transformation $\mathrm{cor}_G^U \colon H^q(U, A) \to H^q(G, A)$, called the **corestriction** [Rib70, p. 136]. Composing the corestriction with the restriction gives multiplication with the index of U in G:

$$(8) \qquad \mathrm{cor}_G^U \circ \mathrm{res}_U^G = (G : U)\mathrm{id}.$$

In particular, if G is finite and we apply (8) to an element $x \in H^q(G, A)$ with $q \geq 1$, we find that $\mathrm{res}_1^G(x) \in H^q(1, A) = 0$, hence $|G|x = 0$. In other words, $H^q(G, A)$ is a torsion group and the order of each element of $H^q(G, A)$ divides the order of G.

9.3.9 DIRECT SYSTEMS.

In order to generalize the latter result to profinite groups, we have to be able to take direct limits of cohomology groups. To this end we consider a **direct system** $(A_i, \alpha_{ij})_{i,j \in I}$ of Abelian groups. Thus, I is a partially ordered nonempty set such that for all $i, j \in I$ there exists $k \in I$ with $i, j \leq k$. For all $i, j \in I$ with $i \leq j$ the system has a homomorphism $\alpha_{ij} \colon A_i \to A_j$ such that $\alpha_{jk} \circ \alpha_{ij} = \alpha_{ik}$ if $i \leq j \leq k$. Moreover, $\alpha_{ii} = \mathrm{id}_{A_i}$ for $i \in I$. Let R be the subgroup of $\bigoplus_{i \in I} A_i$ generated by all elements $a_i - a_j$ with $i, j \in I$, $a_i \in A_i$, $a_j \in A_j$ for which there exists $k \geq i, j$ such that $\alpha_{ik}(a_i) = \alpha_{jk}(a_j)$. The factor group $A = \varinjlim A_i = (\bigoplus_{i \in I} A_i)/R$ is called the **direct limit** of the system $(A_i, \alpha_{ij})_{i,j \in I}$. Viewing each A_i as a subgroup of $\bigoplus_{i \in I} A_i$, we may consider the homomorphism $\alpha_i \colon A_i \to \varinjlim A_i$ given by $\alpha_i(a_i) = a_i + R$. The homomorphisms α_i satisfy the compatibility condition $\alpha_j \circ \alpha_{ij} = \alpha_i$ if $i \leq j$. Moreover, given an Abelian group B and homomorphisms $\beta_i \colon A_i \to B$ such that $\beta_j \circ \alpha_{ij} = \beta_i$ whenever $i \leq j$, there is a unique homomorphism $\beta \colon A \to B$ such that $\beta \circ \alpha_i = \beta_i$ for all $i \in I$.

Each $a \in A$ can be written as $a = \sum_{i \in I_0} a_i + R$, where I_0 is a finite subset of I and $a_i \in A_i$ for each $i \in I_0$. We choose $j \in I$ with $i \leq j$ for all $i \in I_0$ and let $a_j = \sum_{i \in I_0} \alpha_{ij}(a_i)$. Then $a_j \in A_j$ and $a = a_j + R = \alpha_j(a_j)$. Consequently, $A = \bigcup_{i \in I} \alpha_i(A_i)$.

If $a_i \in A_i$ and $\alpha_i(a_i) = 0$, then a_i is a sum of elements $a_{ij} - a_{ik}$ in $\bigoplus_{r \in I} A_r$ with $a_{ij} \in A_j$, $a_{ik} \in A_k$, and there exists $l \geq j, k$ with $\alpha_{jl}(a_{ij}) = \alpha_{kl}(a_{ik})$.

Since in a direct sum equality holds if and only if it holds in each coordinate, we may assume that $a_{ij}, a_{ik} \in A_i$ for all j, k. We choose an $m \in I$ greater or equal to i and all of the l's occurring in the above conditions. Then $\alpha_{im}(a_i) = 0$. Of course, if the latter condition holds, then $\alpha_i(a_i) = 0$.

9.3.10 COHOMOLOGY GROUPS AS DIRECTED LIMES.

Now we consider an inverse system $(G_i, \pi_{ji})_{i,j \in I}$ of profinite groups and a directed system $(A_i, \alpha_{ij})_{i,j \in I}$ of Abelian groups such that A_i is a G_i-module and for all $i \leq j$ the pair (π_{ji}, α_{ij}) is compatible. Let $G = \varprojlim G_i$ and $A = \varinjlim A_i$. For each $i \in I$ let $\pi_i \colon G \to G_i$ be the projection on the ith component and $\alpha_i \colon A_i \to A$ the map defined by the embedding of A_i in $\bigoplus_{i \in I} A_i$. Then G is a profinite group, A is an Abelian group, and G acts on A in the following way: Given $\sigma \in G$ and $a \in A$, we choose $i \in I$ and $a_i \in A_i$ with $\alpha_i(a_i) = a$ and set $\sigma_i = \pi_i(\sigma)$. Then we define $\sigma a = \alpha_i(\sigma_i a_i)$. One checks that this definition is good and that the action of G on A is continuous, so that A becomes a G-module. For all $q \geq 0$ and $i \leq j$ the compatibility condition yields a homomorphism $(\pi_{ij}, \alpha_{ij}) \colon H^q(G_i, A_i) \to H^q(G_j, A_j)$. By the commutativity of the triangle (5), this leads to a directed system of cohomological groups $(H^q(G_i, A_i), (\pi_{ij}, \alpha_{ij}))_{i,j \in I}$. By [Rib70, p. 109, Prop. 4.1],

$$(9) \qquad H^q(G, A) = \varinjlim H^q(G_i, A_i).$$

Starting from an arbitrary profinite group G and a G-module A, we present G as an inverse limit $G = \varprojlim G/U$, where U ranges over all open normal subgroups of G, and recall that $A = \bigcup A^U$. Note that if $U' \subseteq U$, then $A^U \leq A^{U'}$. Let $\pi_{U',U} \colon G/U' \to G/U$ be the quotient map and let $\alpha_{U,U'} \colon A^U \to A^{U'}$ be the inclusion map. Then, (9) yields in this case an isomorphism

$$(10) \qquad H^q(G, A) = \varinjlim H^q(G/U, A^U).$$

Given an Abelian group A and a positive integer n we set $A_n = \{a \in A \mid na = 0\}$. For each prime number p we let $A_{p^\infty} = \bigcup_{k=1}^\infty A_{p^k}$ be the p-**primary part** of A. If A is a torsion group, then $A = \bigoplus A_{p^\infty}$. It follows that if $\alpha \colon A \to B$ is a homomorphism of torsion Abelian groups, then $\alpha(A_{p^\infty}) \leq B_{p^\infty}$ for each p. Hence, each exact sequence $A \to B \to C$ of torsion Abelian groups yields an exact sequence $A_{p^\infty} \to B_{p^\infty} \to C_{p^\infty}$ of their p-primary parts.

Since each of the groups G/U is finite, the order of each element of $H^q(G/U, A^U)$ is finite (Subsection 9.3.8). It follows that $H^q(G, A)$ is a torsion Abelian group. As such it has a presentation

$$(11) \qquad H^q(G, A) = \bigoplus_p H^q(G, A)_{p^\infty}.$$

LEMMA 9.3.11: *Let G be a profinite group acting on a vector space V over $\mathbb{Q}$. Then:*
(a) $H^q(G, V) = 0$ *for each $q \geq 1$.*
(b) $H^{q-1}(G, \mathbb{Q}/\mathbb{Z}) \cong H^q(G, \mathbb{Z})$ *for each $q \geq 2$.*

Proof of (a): First we suppose G is finite and consider the restriction map
res: $H^q(G, V) \to H^q(1, V)$ and the corestriction map

$$\text{cor: } H^q(1, V) \to H^q(G, V).$$

Both maps are trivial, so $\alpha = \text{cor} \circ \text{res} \colon H^q(G, V) \to H^q(G, V)$ is also trivial.
By (8), α is multiplication by the order n of G.

Now let $f \colon G^q \to V$ be a q-cocycle. Since V is divisible, there exists a
function $g \colon G^q \to V$ such that $ng = f$. Since division by n is unique, g is
a cocycle. It follows from the preceding paragraph that f is a coboundary.
Consequently, $H^q(G, V) = 0$.

In the general case we use the presentation (10). By the preceding
paragraph, $H^q(G/U, V^U) = 0$ for each U. Consequently, $H^q(G, V) = 0$.

Proof of (b): The short exact sequence of trivial G-modules $0 \to \mathbb{Z} \to \mathbb{Q} \to$
$\mathbb{Q}/\mathbb{Z} \to 0$ induces for each $q \geq 1$ a four terms exact sequence

$$H^{q-1}(G, \mathbb{Q}) \to H^{q-1}(G, \mathbb{Q}/\mathbb{Z}) \to H^q(G, \mathbb{Z}) \to H^q(G, \mathbb{Q}).$$

By (a), the first and the fourth terms of this sequence are 0 for each $q \geq 2$,
so (b) holds. $\square$

9.3.12 INDUCED MODULES.

Let $H \leq G$ be profinite groups. For each H-module A we denote by
$\text{Ind}_H^G(A)$ the Abelian group of all continuous maps $f \colon G \to A$ such that
$f(\eta\sigma) = \eta f(\sigma)$ for all $\eta \in H$ and $\sigma \in G$. The action of G on $\text{Ind}_H^G(A)$
is defined by $(\sigma' f)(\sigma) = f(\sigma\sigma')$. This action is continuous [Rib70, p. 142,
Prop. 7.1], so $\text{Ind}_H^G(A)$ is a G-module. Note that $\text{Ind}_H^G(A)$ is naturally isomorphic to the G-module $\prod_{\sigma \in S} A$, where S is a system of representatives
for the right cosets of G modulo H. Indeed, each continuous map $f \colon S \to A$
uniquely extends to an element $\hat{f}$ of $\text{Ind}_H^G(H)$ by $\hat{f}(\eta\sigma) = \eta f(\sigma)$ for $\eta \in H$
and $\sigma \in S$.

Shapiro's lemma ensures that

(12) $H^q(G, \text{Ind}_H^G(A)) \cong H^q(H, A)$

for each $q \geq 0$ [Rib70, p. 145, Thm. 7.4].

In the special case where $H = 1$, the right hand side of (12) is 0 for each
$q \geq 1$. Hence, $H^q(G, \text{Ind}_1^G(A)) = 0$.

9.3.13 COHOMOLOGICAL TRIVIALITY.

Let G be a finite group and A a G-module. The norm map norm: $A \to A$
is defined by $\text{norm}(a) = \sum_{\sigma \in G} \sigma a$. By [CaF, p. 113, Thm. 9], $H^q(G, A) = 0$
for each $q \geq 1$ if $A^G = \text{norm}(A)$ and $H^1(G, A) = 0$.

9.3.14 COHOMOLOGICAL p-DIMENSION.

Let G be a profinite group, p a prime number, and $n \geq 0$ an integer. We write $n = \mathrm{cd}_p(G)$ if there exists a torsion G-module A such that $H^n(G, A)_{p^\infty} \neq 0$ but $H^q(G, B)_{p^\infty} = 0$ for each $q \geq n + 1$ and every torsion G-module B. In that case $H^q(G, B)_{p^\infty} = 0$ for each $q \geq n + 2$ and every G-module B [Rib70, p. 197, Prop. 1.4]. Finally we note that for the inequality $\mathrm{cd}_p(G) \leq n$ to hold it suffices that $H^{n+1}(G, A) = 0$ for all finite simple p-primary G-modules A [Rib70, p. 200, Prop.1.5]. Here we say that A is a **simple G-module** if the only G-submodules of A are 0 and A itself. In this case $A \cong (\mathbb{Z}/p\mathbb{Z})^r$ for some nonnegative number r.

9.3.15 COHOMOLOGICAL DIMENSION.

The **cohomological dimension**, $\mathrm{cd}(G)$ of a profinite group G is the supremum of $\mathrm{cd}_p(G)$, where p ranges on all prime numbers. Thus, if $n = \mathrm{cd}(G) < \infty$, then there exists a torsion G-module A with $H^n(G, A) \neq 0$ and for all $q \geq n + 1$, all torsion G-modules B, and every prime number p we have $H^q(G, B)_{p^\infty} = 0$. By (11), $H^q(G, B) = 0$. Similarly, the latter equality holds if $q \geq n + 2$ and B is an arbitrary G-module.

9.3.16 GROUP EXTENSIONS.

We consider an exact sequence

$$(13) \qquad\qquad 0 \to A \to E \xrightarrow{\pi} G \to 1$$

of profinite groups, where A is an additive finite Abelian group, choose a continuous section $s\colon G \to E$ of π [FrJ08, Lemma 1.2.7], and define a continuous action of G on A by the formula $(\sigma, a) \mapsto s(\sigma)as(\sigma)^{-1}$. This action does not depend on s. We call (13) an **extension of A by G**. The extension (13) is **equivalent** to another extension $0 \to A \to E \xrightarrow{\pi} G \to 1$ if there exists a homomorphism $E \to E'$ making the diagram

$$
\begin{array}{ccccccccc}
0 & \longrightarrow & A & \longrightarrow & E & \longrightarrow & G & \longrightarrow & 1 \\
  &   & \| &   & \downarrow &   & \| &   &   \\
0 & \longrightarrow & A & \longrightarrow & E' & \longrightarrow & G & \longrightarrow & 1
\end{array}
$$

commutative. Given a profinite group G and a finite G-module A, there is a bijective correspondence between the equivalence classes of extensions of A by G with the given G-action and the elements of $H^2(G, A)$ [Rib70, p. 100, Thm. 3.1]. The class of split extensions corresponds under that correspondence to the 0 element of $H^2(G, A)$ [Rib70, p. 105]. By Subsection 9.3.14 we have for each prime number p that $\mathrm{cd}_p(G) \leq 1$ if and only if $H^2(G, A) = 0$ for all finite simple p-primary G-modules A. Hence, $\mathrm{cd}_p(G) \leq 1$ if and only if each exact sequence $0 \to (\mathbb{Z}/p\mathbb{Z})^r \to E \to G \to 1$ splits. Consequently, by [FrJ08, Cor. 22.4.3], G is projective if and only if $\mathrm{cd}(G) \leq 1$.

By [FrJ08, Prop. 22.10.4] (whose proof does not depend on cohomology), a profinite group G is projective if and only if each of its p-Sylow groups (for all p) is a free pro-p group, alternatively a projective group [FrJ08, Prop. 22.7.6].

9.3.17 GALOIS COHOMOLOGY.

Let L/K be a Galois extension. By the normal basis theorem,

$$H^1(\mathrm{Gal}(L/K), L^+) = 0$$

[Rib70, p. 246, Prop. 1.1]. By the multiplicative form of Hilbert's Theorem 90,

(14) $$H^1(\mathrm{Gal}(L/K), L^\times) = 1$$

[Rib70, p. 246, Prop. 1.2]. If $p = \mathrm{char}(K) > 0$, then $\mathrm{cd}_p(\mathrm{Gal}(K)) \leq 1$ [Rib70, p. 256, Thm. 3.3]. Thus, by Subsection 9.3.16, every p-Sylow subgroup of $\mathrm{Gal}(K)$ is projective. It follows from the last paragraph of Subsection 9.3.16 that $\mathrm{Gal}(K)$ is projective if every l-Sylow subgroup of $\mathrm{Gal}(K)$, for each $l \neq p$, is projective.

9.3.18 BRAUER GROUPS.

Let K be a field. A **central simple K-algebra** is an associative (but not necessarily commutative) K-algebra A whose center is K and with no nontrivial two sided ideals. If A is finitely generated, then by Wedderburn-Artin [Bou58, p. 51, Cor. 2] there exist a division ring D with center K and a positive integer n such that A is isomorphic to the algebra $M_n(D)$ of all $n \times n$ matrices with entries in D. Another finitely generated central simple K-algebra A' is **equivalent** to A if $A' \cong M_{n'}(D)$ for some positive integer n'. We denote the equivalence classes of A by $[A]$. Let $\mathrm{Br}(K)$ be the set of all equivalence classes of finitely generated central simple K-algebras. The operation $([A], [A']) \mapsto [A \otimes_K A']$ makes $\mathrm{Br}(K)$ a group whose unit element is the class of K [Bou58, p. 117]. If L is a field extension of K, then the map $[A] \to [A \otimes_K L]$ is a group homomorphism $\alpha \colon \mathrm{Br}(K) \to \mathrm{Br}(L)$ [Bou58, p. 118, Prop. 6]. The kernel of α consists of all classes $[A]$ such that A **splits** over L, i.e. $A \cong_L M_n(L)$ for some positive integer n. One denotes $\mathrm{Ker}(\alpha)$ by $\mathrm{Br}(L/K)$.

There is an isomorphism $H^2(\mathrm{Gal}(L/K), L^\times) \cong \mathrm{Br}(L/K)$ [Jac96, Thm. 2.5.11] such that if $K \subseteq L \subseteq N$ is a tower of fields and N/K is Galois, then the following diagram is commutative [Lor08, p. 194]

$$
\begin{array}{ccccccc}
0 & \longrightarrow & H^2(\mathrm{Gal}(L/K), L^\times) & \overset{\mathrm{inf}}{\longrightarrow} & H^2(\mathrm{Gal}(N/K), N^\times) & \overset{\mathrm{res}}{\longrightarrow} & H^2(\mathrm{Gal}(N/L), N^\times) \\
& & \downarrow & & \downarrow & & \downarrow \\
0 & \longrightarrow & \mathrm{Br}(L/K) & \longrightarrow & \mathrm{Br}(N/K) & \longrightarrow & \mathrm{Br}(N/L)
\end{array}
$$

where the second arrow in the lower row is the inclusion map and the third one is $[A] \mapsto [A \otimes_K L]$. Since, by (14), $H^1(\mathrm{Gal}(N/K), N^\times) = 1$, Lemma 9.3.7 implies that the upper row is exact. Hence, so is the lower.

If $\mathrm{cd}_p(\mathrm{Gal}(K)) \leq 1$, then $\mathrm{Br}(K)_p = 0$ for each $p \neq \mathrm{char}(K)$ [Rib70, p. 262, Cor.3.7]. If $\mathrm{Br}(L) = 0$ for each finite separable extension L of K, then $\mathrm{cd}(\mathrm{Gal}(K)) \leq 1$ [Rib70, p. 263, Cor. 3.8], so $\mathrm{Gal}(K)$ is projective (Subsection 9.3.16). If $\mathrm{cd}(\mathrm{Gal}(K)) \leq 1$ and K is perfect, then $\mathrm{Br}(K) = 0$ [Rib70, p. 263, Prop. 3.9].

9.4 The Projectivity of $\mathrm{Gal}(C(t))$

Our third main goal in these notes is to prove that for each algebraically closed field C, the group $\mathrm{Gal}(C(x))$ is free. It would then follow that $\mathrm{Gal}(C(x))$ is projective [FrJ08, Lemma 22.3.6]. However, the projectivity of $\mathrm{Gal}(C(x))$ is an essential step in our proof that $\mathrm{Gal}(C(x))$ is free. So, we first prove that $\mathrm{Gal}(C(x))$ is projective. Our proof uses Galois cohomology, but it replaces advanced tools by more basic ones.

Remark 9.4.1: C_i fields. A field K is said to be C_i if every form (i.e. homogeneous polynomial) $f \in K[X_0, \ldots, X_n]$ of positive degree d with $d^i \leq n$ has a nontrivial zero in K^{n+1}. Thus, for K to be C_0 means that every homogeneous polynomial in $K[X_0, X_1]$ has a nontrivial zero in K^2. In other words, K is algebraically closed. The field K is C_1 if and only if each homogeneous polynomial $f \in K[X_0, \ldots, X_n]$ with $\deg(f) \leq n$ has a nontrivial zero in K^{n+1}. For example, every finite field is C_1 (a theorem of Chevalley [FrJ08, Proposition 21.2.4]), every PAC field of characteristic 0 is C_1 [Kol07, Thm. 1], and every perfect PAC field of positive characteristic is C_2 [FrJ08, Thm. 21.3.6]. Moreover, if K is C_i and L is a field extension of K of transcendence degree j, then L is C_{i+j} [FrJ08, Prop. 21.2.12]. In particular, if K is algebraically closed and x is an indeterminate, then every algebraic extension of $K(x)$ is C_1. $\qquad\square$

LEMMA 9.4.2: *Let L/K be a finite Galois extension.*
(a) *If K is C_1, then $\mathrm{norm}_{L/K} L^\times = K^\times$.*
(b) *In the general case, $\mathrm{trace}_{L/K} L = K$.*

Proof of (a): Let $w_1, \ldots, w_d$ be a basis of L/K and set $G = \mathrm{Gal}(L/K)$. Then

$$f(X_1, \ldots, X_d) = \prod_{\sigma \in G} (X_1 w_1^\sigma + \cdots + X_d w_d^\sigma)$$

is a form of degree d with coefficients in K. If $x_1, \ldots, x_d \in K$ and $f(x_1, \ldots, x_d) = 0$, then there exists $\tau \in G$ such that $z = x_1 w_1^\tau + \cdots + x_d w_d^\tau = 0$. Hence, all conjugates of z over K are zero. Thus, $x_1 w_1^\sigma + \cdots + x_d w_d^\sigma = 0$ for all $\sigma \in G$. Since $\det(w_i^\sigma) \neq 0$ [Lan93, p. 266, Cor. 5.4], we have $x_1 = \cdots = x_d = 0$.

Let now $a \in K^\times$. Since K is C_1, there exist $y_0, y_1, \ldots, y_d \in K$, not all 0, such that $f(y_1, \ldots, y_d) = y_0^d a$. By the preceding paragraph, $y_0 \neq 0$. Hence, with $x_i = y_i/y_0$, $i = 1, \ldots, d$, and $b = x_1 w_1 + \cdots + x_d w_d$, we have $\mathrm{norm}_{L/K} b = a$.

Proof of (b): By Artin's theorem about the linear independence of characters [Lan93, p. 283, Thm. 4.1], there exists $x \in L$ with $a = \sum_{\sigma \in G} x^\sigma \neq 0$. Then, $a = \mathrm{trace}_{L/K} x$ and $a \in K$. Consequently, each $b \in K$ can now be written as $b = \mathrm{trace}_{L/K}\left(\frac{b}{a} x\right)$. $\qquad\square$

LEMMA 9.4.3: *Let L/K be a finite cyclic field extension.*
(a) *If K is C_1, then every short exact sequence*

$$(1) \qquad\qquad 1 \longrightarrow L^\times \longrightarrow E \overset{h}{\longrightarrow} \mathrm{Gal}(L/K) \longrightarrow 1$$

   of groups with the usual Galois action of $\mathrm{Gal}(L/K)$ on $L^\times$ splits. Thus, $H^2(\mathrm{Gal}(L/K), L^\times) = 1$.
(b) *In the general case, every short exact sequence*

$$(2) \qquad\qquad 0 \longrightarrow L^+ \longrightarrow E \overset{h}{\longrightarrow} \mathrm{Gal}(L/K) \longrightarrow 1$$

   of groups with the usual Galois action of $\mathrm{Gal}(L/K)$ on L^+ splits. Thus, $H^2(\mathrm{Gal}(L/K), L^+) = 0$.

Proof: Let $n = [L : K]$ and let σ be a generator of $\mathrm{Gal}(L/K)$. We have to find $\varepsilon \in E$ such that $h(\varepsilon) = \sigma$ and $\varepsilon^n = 1$.

By assumption, there exists $\varepsilon \in E$ such that $h(\varepsilon) = \sigma$. For each $y \in L^\times$, we have $y^\varepsilon = y^\sigma$. Also, $\varepsilon^n \in L^\times$. Hence, $(\varepsilon^n)^\sigma = (\varepsilon^n)^\varepsilon = \varepsilon^n$, so $\varepsilon^n \in K^\times$.

By Lemma 9.4.2, there exists $x \in L^\times$ such that $\mathrm{norm}_{L/K} x = \varepsilon^{-n}$ in Case (a) and $\mathrm{trace}_{L/K} x = \varepsilon^{-n}$ in Case (b). For arbitrary elements x, ε of a group G, one proves by induction on n that

$$(x\varepsilon)^n = \varepsilon^n x^{\varepsilon^n} x^{\varepsilon^{n-1}} \cdots x^\varepsilon.$$

In Case (a), this formula gives

$$(3) \quad (x\varepsilon)^n = \varepsilon^n x^{\varepsilon^n} x^{\varepsilon^{n-1}} \cdots x^\varepsilon = \varepsilon^n x^{\sigma^n} x^{\sigma^{n-1}} \cdots x^\sigma = \varepsilon^n \mathrm{norm}_{L/K} x = 1.$$

Therefore, $x\varepsilon$ is the desired element of E.

In Case (b) the operation of L^+ is addition, so we have to replace (3) by

$$(x\varepsilon)^n = \varepsilon^n (x^{\varepsilon^n} + x^{\varepsilon^{n-1}} + \cdots + x^\varepsilon)$$
$$= \varepsilon^n (x^{\sigma^n} + x^{\sigma^{n-1}} + \cdots + x^\sigma) = \varepsilon^n \mathrm{trace}_{L/K} x = 1.$$

Again, $x\varepsilon$ is the desired element of E.

The triviality of the second cohomology groups follows now from Subsection 9.3.16. $\qquad\square$

LEMMA 9.4.4: *Let K be a C_1 field, p a prime number, and E a p-**Sylow extension** of K (i.e. E is the fixed field in K_s of a p-Sylow subgroup of $\mathrm{Gal}(K)$.) Then*
$$H^2(\mathrm{Gal}(E), E_s^\times) = 1.$$

Proof: By Subsection 9.3.10, $H^2(\mathrm{Gal}(E), E_s^\times) = \varinjlim H^2(\mathrm{Gal}(N/E), N^\times)$, where N ranges over all finite Galois extensions of E and the maps involved in the direct limit are inflations. We prove by induction on the degree, that for each finite Galois extension N/L with $E \subseteq L \subseteq N \subseteq K_s$ we have $H^2(\mathrm{Gal}(N/L), N^\times) = 1$.

Indeed, N/L is a p-extension. If this extension is nontrivial, it has a cyclic subextension M/L of degree p. By Remark 9.4.1, L is C_1, hence by Lemma 9.4.3(a), $H^2(\mathrm{Gal}(M/L), M^\times) = 1$. By induction, $H^2(\mathrm{Gal}(N/M), N^\times) = 1$. Finally we use the exactness of the inflation restriction sequence

$$1 \longrightarrow H^2(\mathrm{Gal}(M/L), M^\times) \xrightarrow{\text{inf}} H^2(\mathrm{Gal}(N/L), N^\times) \xrightarrow{\text{res}} H^2(\mathrm{Gal}(N/M), N^\times)$$

(Lemma 9.3.7) to conclude that $H^2(\mathrm{Gal}(N/L), N^\times) = 1$. $\qquad\square$

LEMMA 9.4.5: *Let K be a C_1 field, p a prime number, and E a p-Sylow extension of K. Then, $\mathrm{Gal}(E)$ is projective, hence pro-p free.*

Proof: The statement holds for $p = \mathrm{char}(E)$ by [Rib70, p. 256]. So, we assume that $p \neq \mathrm{char}(E)$.

By Subsection 9.3.16, we have to prove that $H^2(\mathrm{Gal}(E), \mathbb{Z}/p\mathbb{Z}) = 0$. To this end consider the short exact sequence

$$(4) \qquad\qquad 1 \longrightarrow \mu_p \longrightarrow E_s^\times \xrightarrow{p} E_s^\times \longrightarrow 1,$$

where μ_p is the group of roots of unity of order p and the map from $E_s^\times$ to $E_s^\times$ is raising to the pth power. Since $[E(\mu_p) : E]$ divides $p - 1$ and $\mathrm{Gal}(E)$ is a pro-p group, $[E(\mu_p) : E] = 1$, so $\mu_p \subseteq E$ and the action of $\mathrm{Gal}(E)$ on μ_p is trivial. Hence, μ_p is isomorphic to $\mathbb{Z}/p\mathbb{Z}$ as a $\mathrm{Gal}(E)$-module. Now we consider the following segment of the long exact sequence derived from the exact sequence (4) (Subsection 9.3.4):

$$(5) \qquad H^1(\mathrm{Gal}(E), E_s^\times) \longrightarrow H^2(\mathrm{Gal}(E), \mathbb{Z}/p\mathbb{Z}) \longrightarrow H^2(\mathrm{Gal}(E), E_s^\times).$$

The left term of (5) is trivial, by Subsection 9.3.17. The right term of (5) is trivial, by Lemma 9.4.4. Hence, the middle term of (5) is also trivial. $\qquad\square$

PROPOSITION 9.4.6 (Tsen):
(a) *Let E be a C_1 field. Then $\mathrm{Gal}(E)$ is projective.*
(b) *Let E be an extension of transcendence degree 1 over a separably closed field C. Then $\mathrm{Gal}(E)$ is projective.*

Proof of (a): By Lemma 9.4.5, each of the Sylow subgroups of $\mathrm{Gal}(E)$ is projective. It follows from [FrJ08, Prop. 22.10.4] that $\mathrm{Gal}(E)$ is projective. Note that the proof of the latter theorem is carried out without cohomology.

Proof of (b): First we note that $(E\tilde{C})_s/E_s\tilde{C}$ is both a separable extension and a purely inseparable extension, so it is a trivial extension. Thus, $E_s\tilde{C} = (E\tilde{C})_s$. In addition, $E_s \cap E\tilde{C} = E$, hence $\mathrm{Gal}(E) \cong \mathrm{Gal}(E\tilde{C})$. We may therefore assume that C is algebraically closed. By Remark 9.4.1, E is a C_1 field. Hence, by (a), $\mathrm{Gal}(E)$ is projective.   □

Following [FrJ08, Remark 17.4.7], we denote the free profinite group of rank m by $\hat{F}_m$ and rephrase a special case of [FrJ08, Lemma 25.1.8]:

PROPOSITION 9.4.7: *Let m be an infinite cardinal and G a projective group of rank at most m. Suppose every finite split embedding problem for G with a nontrivial kernel has m solutions. Then $G \cong \hat{F}_m$.*

THEOREM 9.4.8: *Let K be a field of characteristic p and cardinality m and let E be a function field of one variable over K. Suppose $\mathrm{Gal}(K)$ is trivial if $p = 0$ or $\mathrm{Gal}(K)$ is a pro-p group if $p > 0$. Then $\mathrm{Gal}(E) \cong \hat{F}_m$.*

Proof: We choose a separating transcendental element x for E/K. Consider a prime number $l \neq p$ and let G_l be an l-Sylow subgroup of $\mathrm{Gal}(K(x))$. Since $\mathrm{Gal}(K_s(x)/K(x)) \cong \mathrm{Gal}(K)$ is trivial if $p = 0$ or a pro-p group if $p > 0$, G_l is an l-Sylow subgroup of $\mathrm{Gal}(K_s(x))$. By Proposition 9.4.6(b), $\mathrm{Gal}(K_s(x))$ is projective. Hence, by Subsection 9.3.16, G_l is projective. It follows from Subsection 9.3.17 that $\mathrm{Gal}(K(x))$ is projective.

By Theorem 5.8.3, K is ample. Hence, by Proposition 8.6.3, every finite split embedding problem for $\mathrm{Gal}(K(x))$ with a nontrivial kernel has m solutions. In particular, $m \geq \mathrm{rank}(\mathrm{Gal}(K(x))) \geq \aleph_0$. By Proposition 9.4.7, $\mathrm{Gal}(K(x)) \cong \hat{F}_m$. It follows from [FrJ08, Prop. 25.4.2] that $\mathrm{Gal}(E) \cong \hat{F}_m$. □

COROLLARY 9.4.9: *Let K be a separably closed field of cardinality m and let E be an algebraic function field of one variable over K. Then $\mathrm{Gal}(E) \cong \hat{F}_m$.*

Remark 9.4.10: An analog of Shafarevich's Conjecture. We denote the extension of a field K generated by all roots of unity by K_{cycl}. As mentioned in Example 5.10.5, Shafarevich's conjecture predicts that $\mathrm{Gal}(K_{\mathrm{cycl}}) \cong \hat{F}_\omega$ for each number field K.

As is the case with several other conjectures (e.g. the Riemann hypothesis), the analog of Shafarevich's conjecture for function fields K of one variable over finite fields is true. In this case $K_{\mathrm{cycl}} = \tilde{\mathbb{F}}_p K$, where $p = \mathrm{char}(K)$. Thus, if we choose a transcendental element x of K over $\mathbb{F}_p$, then K_{cycl} is a finite extension of $\tilde{\mathbb{F}}_p(x)$. Therefore, by Corollary 9.4.9, $\mathrm{Gal}(K_{\mathrm{cycl}}) \cong \hat{F}_\omega$, as claimed.   □

9.5 Projectivity of Fundamental Groups

Let C be an algebraically closed field, E a function field of one variable over C, S a nonempty set of prime divisors of E/C, and E_S the maximal Galois extension of E ramified at most over S. The only known proof of the Riemann existence theorem uses complex analytic methods. It follows, as mentioned in Remark 9.1.12, that the proof of the projectiveness of $\mathrm{Gal}(E_S/E)$ in the case $\mathrm{char}(C) = 0$, stated in Corollary 9.1.11, relies on analytic methods.

The aim of this section is to prove that $\mathrm{Gal}(E_S/E)$ is projective, without any restriction on the characteristic, by algebraic means. This will in particular reproves the projectivity of $\mathrm{Gal}(E_S/E)$ in characteristic 0.

As mentioned in the proof of Proposition 9.4.6, a profinite group G is projective if and only if for each prime number p each p-Sylow subgroup G_p of G is projective [FrJ08, Prop. 22.10.4]. We therefore say that G is p-**projective** if G_p is projective. We say that an embedding problem $(\varphi\colon H \to A,\ \alpha\colon B \to A)$ is **central** if $\mathrm{Ker}(\alpha)$ is contained in the center of B.

LEMMA 9.5.1: *Let p be a prime number.*
(a) *Let G be a profinite group. Suppose for every open subgroup H, each finite nonsplit central embedding problem*

$$
\begin{array}{ccccccccc}
 & & & & & & H & & \\
 & & & & & & \big\downarrow{\varphi} & & \\
0 & \longrightarrow & \mathbb{Z}/p\mathbb{Z} & \longrightarrow & B & \overset{\alpha}{\longrightarrow} & A & \longrightarrow & 1
\end{array}
$$

for which B is a p-group is solvable. Then G is p-projective.
(b) *Let N/E be a Galois extension. Suppose for each finite subextension K of N/E, for each finite p-subextension L/K of N/K, and every nonsplit central exact sequence of p-groups*

$$(1) \qquad\qquad 0 \longrightarrow \mathbb{Z}/p\mathbb{Z} \longrightarrow B \overset{\alpha}{\longrightarrow} \mathrm{Gal}(L/K) \longrightarrow 1$$

there exists a Galois extension $\hat{L}$ of K in N that contains L and there exists an isomorphism $\gamma\colon \mathrm{Gal}(\hat{L}/K) \to B$ such that $\alpha \circ \gamma = \mathrm{res}_L$. Then $\mathrm{Gal}(N/E)$ is p-projective.

Proof: Statement (b) is a reinterpretation of (a) for Galois groups, so we prove (a).

Let G_p be a p-Sylow subgroup of G. In order to prove that G_p is projective, it suffices to prove that each finite embedding problem

$$
(2) \qquad\qquad
\begin{array}{ccccccccc}
 & & & & & & G_p & & \\
 & & & & & & \big\downarrow{\varphi_p} & & \\
1 & \longrightarrow & B_0 & \longrightarrow & B & \overset{\alpha}{\longrightarrow} & A & \longrightarrow & 1
\end{array}
$$

in which B is a p-group and B_0 is a minimal normal subgroup of B is weakly solvable, that is there exists a homomorphism $\gamma\colon G_p \to B$ such that $\alpha \circ \gamma = \varphi_p$ [FrJ08, Lemma 22.3.4 and Lemma 22.4.1]. By elementary group theory, B_0 is isomorphic to $\mathbb{Z}/p\mathbb{Z}$ and lies in the center of B. This means that the short exact sequence in (2) is central.

If the short exact sequence in (2) splits, there exists a homomorphism $\alpha'\colon A \to B$ such that $\alpha \circ \alpha' = \mathrm{id}_A$. Then $\alpha' \circ \varphi_p$ weakly solves (2). Otherwise we choose an open normal subgroup N of G such that $G_p \cap N = \mathrm{Ker}(\varphi_p)$. Let $H = G_p N$. Then H is an open subgroup of G that contains G_p and φ_p extends to a homomorphism $\varphi\colon H \to A$. By assumption, there exists a homomorphism $\gamma\colon H \to B$ such that $\alpha \circ \gamma = \varphi$. The restriction of γ to G_p weakly solves embedding problem (2). Note that since we are now assuming that α does not split, $B_0 \cap \gamma(G_p) \neq 1$, so $B_0 \leq \gamma(G_p)$. Therefore, $\gamma|_{G_p}$ is even surjective. $\square$

LEMMA 9.5.2: *Let A be a finite group, p a prime number, and let*

$$0 \longrightarrow \mathbb{Z}/p\mathbb{Z} \longrightarrow E_i \xrightarrow{\varepsilon_i} A \longrightarrow 1$$

$i = 1, 2$, *be central group extensions. Then there exists an isomorphism $\varphi\colon E_1 \to E_2$ such that $\varepsilon_2 \circ \varphi = \varepsilon_1$ if and only if the two group extensions*

$$0 \longrightarrow \mathbb{Z}/p\mathbb{Z} \longrightarrow E_1 \times_A E_2 \xrightarrow{\pi_i} E_i \longrightarrow 1,$$

where $\pi_i\colon E_1 \times_A E_2 \to E_i$ is the projection onto E_i, split.

Proof: Suppose there exists an isomorphism $\varphi\colon E_1 \to E_2$ such that $\varepsilon_2 \circ \varphi = \varepsilon_1$. Then φ induces a homomorphism $\varphi'\colon E_1 \to E_1 \times_A E_2$ such that $\pi_1 \circ \varphi' = \mathrm{id}_{E_1}$ [FrJ08, Prop. 22.2.1]. Applying the same argument to φ^{-1} yields the splitting of π_2.

Conversely, suppose $\pi_1\colon E_1 \times_A E_2 \to E_1$ has a group theoretic section $\pi_1'\colon E_1 \to E_1 \times_A E_2$, that is $\pi_1 \circ \pi_1' = \mathrm{id}_{E_1}$. Let $\psi_2 = \pi_2 \circ \pi_1'$. Then, $\varepsilon_2 \circ \psi_2 = \varepsilon_2 \circ \pi_2 \circ \pi_1' = \varepsilon_1 \circ \pi_1 \circ \pi_1' = \varepsilon_1$, so $\mathrm{Ker}(\psi_2) \leq \mathrm{Ker}(\varepsilon_1)$. If $\mathrm{Ker}(\psi_2) = 1$, then $\psi_2\colon E_1 \to E_2$ is the desired isomorphism φ. Otherwise, since $\mathrm{Ker}(\varepsilon_1) = \mathbb{Z}/p\mathbb{Z}$, we have $\mathrm{Ker}(\psi_2) = \mathrm{Ker}(\varepsilon_1)$. Therefore, ψ_2 induces a monomorphism $\psi_2'\colon A \to E_2$ such that $\varepsilon_2 \circ \psi_2' = \mathrm{id}_A$. It follows that $E_2 = \mathbb{Z}/p\mathbb{Z} \times \psi_2'(A)$.

Arguing with π_2, we are reduced to the case where the latter consequence of the preceding paragraph holds and in addition $E_1 = \mathbb{Z}/p\mathbb{Z} \times \psi_1'(A)$, where $\psi_1'\colon A \to E_1$ is a group theoretic section of ε_1. Now we define a map $\varphi\colon E_1 \to E_2$ whose restriction to $\mathbb{Z}/p\mathbb{Z}$ is the identity map and $\varphi(\psi_1'(a)) = \psi_2'(a)$ for each $a \in A$. Then φ is an isomorphism such that $\varepsilon_2 \circ \varphi = \varepsilon_1$, as desired. $\square$

LEMMA 9.5.3: *Let L/K be a Galois extension, $p \neq \mathrm{char}(K)$ a prime number, and (1) a nonsplit central exact sequence of p-groups. Suppose K contains a root ζ of 1 of order p and let $L(x^{1/p})$ be a solution field of (1) with $x \in L^\times$. Then the set of solution fields of (1) coincides with the set of fields $L((ax)^{1/p})$, $a \in K^\times$.*

Proof: Set $x_1 = x$, $N_1 = L(x_1^{1/p})$, $E_1 = \mathrm{Gal}(N_1/K)$, and $A = \mathrm{Gal}(L/K)$. Let $\varepsilon_1\colon E_1 \to A$ be the restriction map. By assumption, N_1 is a solution field of (1). Hence, $0 \to \mathbb{Z}/p\mathbb{Z} \to E_1 \xrightarrow{\varepsilon_1} A \to 1$ is a nonsplit central extension (because (1) is).

Now consider an $a \in K^\times$. Set $x_2 = ax_1$, $N_2 = L(x_2^{1/p})$, and $N = N_1 N_2$. Then $N = N_1 K(a^{1/p})$ is a Galois extension of K. Moreover, if $\sigma \in \mathrm{Gal}(N/L)$, then $\sigma|_{N_1}$ is in the center of $\mathrm{Gal}(N_1/K)$ (by assumption) and $\sigma|_{L(a^{1/p})}$ is in $\mathrm{Gal}(L(a^{1/p})/L)$, hence is also in the center of $\mathrm{Gal}(L(a^{1/p})/K)$ (because $\mathrm{Gal}(L(a^{1/p})/K) = \mathrm{Gal}(L(a^{1/p})/L) \times \mathrm{Gal}(L(a^{1/p})/K(a^{1/p}))$ and $\mathrm{Gal}(L(a^{1/p})/L)$ is cyclic). Thus, $\mathrm{Gal}(N/L)$ is contained in the center of $\mathrm{Gal}(N/K)$. It follows that N_2 (that lies between L and N) is a Galois extension of K and $\mathrm{Gal}(N_2/L)$ is contained in the center of $\mathrm{Gal}(N_2/K)$.

Assuming $N_1 \neq N_2$, we set $E_2 = \mathrm{Gal}(N_2/K)$ and let $\varepsilon_2 : E_2 \to A$ be the restriction map. Then $\mathrm{Gal}(N_2/L) \cong \mathbb{Z}/p\mathbb{Z}$ (otherwise, $a^{1/p} \in N_1$, so the ε_1 splits)   and

$$0 \to \mathbb{Z}/p\mathbb{Z} \to E_2 \xrightarrow{\varepsilon_2} A \to 1$$ is a central exact sequence. Moreover,

$$\mathrm{Gal}(N/K) \cong E_1 \times_A E_2$$

[FrJ08, Example 22.2.7(a)] is a split extension of both E_1 and E_2. Hence, by Lemma 9.5.2, there exists an isomorphism $\varphi : E_1 \to E_2$ that commutes with restriction to L. Therefore, N_2 is also a solution field of (1).

Conversely, suppose $N_2 = L(x_2^{1/p})$ with $x_2 \in L^\times$ is a solution field of embedding problem (1) and $N_2 \neq N_1$ and let E_2 and ε_2 be as above. Then there exists an isomorphism $\varphi : E_1 \to E_2$ such that $\varepsilon_2 \circ \varphi = \varepsilon_1$. Then, with $N = N_1 N_2$, $\mathrm{Gal}(N/K) \cong E_1 \times_A E_2$. By Lemma 9.5.2, the group extension $0 \to \mathbb{Z}/p\mathbb{Z} \to \mathrm{Gal}(N/K) \to \mathrm{Gal}(N_1/K) \to 1$ splits, which implies that $N = N_1(a^{1/p})$ with $a \in K^\times$. But $N = N_1(x_2^{1/p})$, so by Kummer theory, $x_2 a^{-1} \in (N_1^\times)^p$ (replacing a by a power of a if necessary). Hence, $L((x_2 a^{-1})^{1/p}) \subseteq N_1$. If equality holds, then by Kummer theory, $x_2 a^{-1} x_1^{-1} \in (L^\times)^p$ (replacing x_1 by some power of itself if necessary), therefore $N_2 = L(x_2^{1/p}) = L((ax)^{1/p})$, as claimed.

Otherwise, $x_2 a^{-1} \in (L^\times)^p$, so $N_2 = L(x_2^{1/p}) = L(a^{1/p})$. Hence, the short exact sequence $1 \to \mathrm{Gal}(N_2/L) \to \mathrm{Gal}(N_2/K) \to \mathrm{Gal}(L/K) \to 1$ splits. Therefore, also the short exact sequence $1 \to \mathrm{Gal}(N_1/L) \to \mathrm{Gal}(N_1/K) \to \mathrm{Gal}(L/K) \to 1$ splits (because both N_1 and N_2 are solution fields of (1)). This contradicts the assumption that embedding problem (1) does not split. $\square$

LEMMA 9.5.4: *Let K be a function field of one variable over an algebraically closed field C, S a finite nonempty set of prime divisors of K/C, $p \neq \mathrm{char}(K)$ a prime number, and L/K a finite Galois subextension of K_S/K. Suppose (1) is a nonsplit central p-embedding problem which is solvable in K_s. Then (1) has a solution field $\tilde{L}$ in K_S.*

Proof: Let $L(x^{1/p})$ be a solution field of (1) in K_s. By Lemma 9.5.3, it suffices to find $a \in K^\times$ such that $L((ax)^{1/p}) \subseteq K_S$.

We extend each $\sigma \in \mathrm{Gal}(L/K)$ to an element σ of $\mathrm{Gal}(L(x^{1/p})/K)$. Then $L((x^{1/p})^\sigma) = L(x^{1/p})$. Hence, $(x^{1/p})^\sigma = x^{i/p} u$ for some $0 \leq i \leq p-1$ and $u \in$

$L^\times$ (by Kummer theory). Now we consider the element $\tau \in \mathrm{Gal}(L(x^{1/p})/L)$ defined by $(x^{1/p})^\tau = \zeta x^{1/p}$, where ζ is a root of unity of order p. Then $(x^{1/p})^{\tau\sigma} = (\zeta x^{1/p})^\sigma = \zeta x^{i/p} u$ and $(x^{1/p})^{\sigma\tau} = (x^{i/p} u)^\tau = \zeta^i x^{i/p} u$. Since (1) is central, $\tau\sigma = \sigma\tau$, so $i = 1$. It follows that $x^\sigma = xu^p$, so $\mathrm{div}(x^\sigma) \equiv \mathrm{div}(x) \bmod p\mathrm{Div}(L/C)$. Hence, $v_{\mathfrak{P}}(x^\sigma) \equiv v_{\mathfrak{P}}(x) \bmod p$ for each prime divisor $\mathfrak{P}$ of L/C. Since $v_{\mathfrak{P}}(x^\sigma) = v_{\mathfrak{P}^{\sigma^{-1}}}(x)$, this implies that $v_{\mathfrak{P}^\sigma}(x) \equiv v_{\mathfrak{P}}(x) \bmod p$ for all $\mathfrak{P}$ and σ. Since the set of prime divisors of L/C lying over each prime divisor $\mathfrak{p}$ of K/C form a conjugacy class under the action of $\mathrm{Gal}(L/K)$, we may denote the common residue modulo p of $v_{\mathfrak{P}}(x)$ for all $\mathfrak{P}$ dividing $\mathfrak{p}$ by $n_{\mathfrak{p}}$ and write $\mathrm{div}(x) \equiv \sum_{\mathfrak{p}} n_{\mathfrak{p}} \sum_{\mathfrak{P}|\mathfrak{p}} \mathfrak{P} \bmod p\mathrm{Div}(L/C)$, where $\mathfrak{p}$ ranges over the prime divisors of K/C.

If $\mathfrak{p} \notin S$, then $\mathfrak{p}$ is unramified in L, so $\mathfrak{p} = \sum_{\mathfrak{P}|\mathfrak{p}} \mathfrak{P}$. Hence,

$$\mathrm{div}(x) = \sum_{\mathfrak{p}\notin S} n_{\mathfrak{p}} \sum_{\mathfrak{P}|\mathfrak{p}} \mathfrak{P} + \sum_{\mathfrak{p}\in S} n_{\mathfrak{p}} \sum_{\mathfrak{P}|\mathfrak{p}} \mathfrak{P} \bmod p\mathrm{Div}(L/C)$$

$$\equiv \mathfrak{a} + \mathfrak{B} \bmod p\mathrm{Div}(L/C),$$

where $\mathfrak{a} \in \mathrm{Div}(K/C)$ and $\mathfrak{B}$ is a divisor of L/C that involves only primes over S.

We choose $\mathfrak{o} \in S$. By Subsection 6.3.2, there exists $a \in K^\times$ with $\mathrm{div}(a) + \mathfrak{a} - \deg(\mathfrak{a})\mathfrak{o} \equiv 0 \bmod p\mathrm{Div}(K/C)$. Therefore, $\mathrm{div}(ax) \equiv \deg(\mathfrak{a})\mathfrak{o} + \mathfrak{B} \bmod p\mathrm{Div}(L/C)$. This implies that $v_{\mathfrak{P}}(ax) \equiv 0 \bmod p$ for each $\mathfrak{P}$ which does not lie over S. Such $\mathfrak{P}$ is unramified in $L((ax)^{1/p})$ [FrJ08, Example 2.3.8]. Consequently, $L((ax)^{1/p}) \subseteq K_S$.   □

In order to prove an analog of Lemma 9.5.3 also for $p = \mathrm{char}(C) > 0$, we have to replace Kummer theory in the above arguments by Artin-Schreier theory. To that end we consider till the end of the proof of Lemma 9.5.6 only fields of characteristic p. Let $\wp$ be the additive operator defined on fields of characteristic p by $\wp(x) = x^p - x$. Recall that if L/K is a cyclic extension of degree p, then $L = K(x)$, where $\wp(x) \in K \smallsetminus \wp(K)$ [Lan93, p. 290, Thm. 6.4]. For each subgroup A of the additive group of K we have $[K(\wp^{-1}(A)) : K] = [A + \wp(K) : \wp(K)]$ [Lan93, p. 296, Thm. 8.3]. In particular, let $x, y, z \in K_s$ with $\wp(x), \wp(y), \wp(z) \in K$. Then

(3a) $K(x) = K$ if and only if $\wp(x) \in \wp(K)$.

(3b) If $K(x) = K(y)$, then there exist $k, l \in \mathbb{Z}$ not both divisible by p such that $\wp(kx) + \wp(ly) \equiv 0 \bmod \wp(K)$. Conversely, if neither of k, l is divisible by p and $\wp(kx) + \wp(ly) \equiv 0 \bmod \wp(K)$, then $K(x) = K(y)$.

(3c) If $x \notin K$ and $K(x) = K(y)$, then there exists $k \in \mathbb{Z}$ such that $p \nmid k$ and $\wp(y) \equiv \wp(kx) \bmod \wp(K)$.

(3d) If $x_i \in K_s$, $a_i = \wp(x_i) \in K$ for $i = 1, \ldots, n$, and $a_1, \ldots, a_n$ are linearly independent over $\mathbb{F}_p$ modulo $\wp(K)$, then the fields $K(x_1), \ldots, K(x_n)$ are linearly disjoint cyclic extensions of K of degree p.

We use the rules (3) in the proof of the following additive analog of Lemma 9.5.3.

LEMMA 9.5.5: *Let L/K be a finite Galois extension of fields of positive characteristic p and let (1) be a nonsplit central exact sequence. Suppose $L(x)$ is a solution field of (1) and $\wp(x) \in L$. Then $\hat{L}$ is a solution field of (1) if and only if $\hat{L} = L(y)$ with $\wp(y) \in L$ and $\wp(y) \equiv \wp(x) \bmod K + \wp(L)$.*

Proof: First suppose y is an element of K_s such that $\wp(y) \in L$ and $\wp(y) \equiv \wp(x) + a \bmod \wp(L)$ with $a \in K$. Set $A = \mathrm{Gal}(L/K)$ and $N = L(x, y)$ and assume $L(x) \neq L(y)$. We choose $z \in K_s$ such that $\wp(z) = a$. Then $K(z)/K$ is a cyclic extension of degree 1 or p and $\wp(y) \equiv \wp(x + z) \bmod \wp(L)$. Hence, $L(x, y) = N = L(x, x + z) = L(x)K(z)$ (by (3b)). Therefore, the extension

(4) $1 \longrightarrow \mathrm{Gal}(N/L(x)) \longrightarrow \mathrm{Gal}(N/K) \longrightarrow \mathrm{Gal}(L(x)/K) \longrightarrow 1$

splits. Next note that $\mathrm{Gal}(L(z)/K) = \mathrm{Gal}(L(z)/L) \times \mathrm{Gal}(L(z)/K(z))$ and $\mathrm{Gal}(L(z)/L)$ is cyclic. Hence, $\mathrm{Gal}(L(z)/L)$ is contained in the center of $\mathrm{Gal}(L(z)/K)$. In addition, by assumption, $\mathrm{Gal}(L(x)/L)$ is contained in the center of $\mathrm{Gal}(L(x)/K)$. Hence, $\mathrm{Gal}(N/L)$ is contained in the center of $\mathrm{Gal}(N/K)$. It follows that $L(y)/K$ is Galois and $1 \to \mathrm{Gal}(L(y)/L) \to \mathrm{Gal}(L(y)/K) \to \mathrm{Gal}(L/K) \to 1$ is a central exact sequence. Moreover, the relation $\wp(y - z) \equiv \wp(x) \bmod \wp(L)$ implies that $L(y)K(z) = N$, so the extension

(5) $1 \longrightarrow \mathrm{Gal}(N/L(y)) \longrightarrow \mathrm{Gal}(N/K) \longrightarrow \mathrm{Gal}(L(y)/K) \longrightarrow 1$

splits. Then $\mathrm{Gal}(N/K) \cong \mathrm{Gal}(L(x)/K) \times_A \mathrm{Gal}(L(y)/K)$ and the restriction maps on $L(x)$ and $L(y)$ correspond to the projections on the groups $\mathrm{Gal}(L(x)/K)$ and $\mathrm{Gal}(L(y)/K)$. By Lemma 9.5.2 there exists an isomorphism $\varphi\colon \mathrm{Gal}(L(y)/K) \to \mathrm{Gal}(L(x)/K)$ that commutes with the restriction to L. It follows that $L(y)$ is a solution field of (1).

Conversely, suppose $\hat{L}$ is a solution field of (1). In particular, $\hat{L}$ is a cyclic extension of degree p of L. Hence $\hat{L} = L(y_0)$ with $\wp(y_0) \in L$ and there exists an isomorphism $\varphi\colon \mathrm{Gal}(L(y_0)/K) \to \mathrm{Gal}(L(x)/K)$ that commutes with the restriction to L. Hence, with y_0 replacing y, both extensions (4) and (5) split (Lemma 9.5.2). This implies that $N = L(x, z)$ with $\wp(z) \in K$. If $L(y_0) = L(x)$, then $\wp(ky_0) \equiv \wp(x) \bmod \wp(L)$ for some $k \in \mathbb{Z}$ with $p \nmid k$ (by (3c)).

If $L(y_0) \neq L(x)$, then there exist $k, l \in \mathbb{Z}$ with $p \nmid k$ such that $\wp(ky_0) + \wp(lz) \equiv \wp(x) \bmod \wp(L)$ (by (3d)), so $\wp(ky_0) \equiv \wp(x) \bmod K + \wp(L)$. In both cases $y = ky_0$ satisfies the requirements of the lemma. $\square$

LEMMA 9.5.6: *Let K be a function field of one variable over an algebraically closed field C of characteristic $p > 0$. Let S be a finite nonempty set of prime divisors of K/C. Let L/K be a finite Galois subextension of K_S/K. Suppose the central nonsplit embedding problem (1) has a solution. Then (1) has a solution field $\hat{L}$ in K_S.*

Proof: By assumption there exists $u \in L \smallsetminus \wp(L)$ and there exists $x \in K_s$ such that $\wp(x) = u$ and $L(x)$ solves (1). If $\mathfrak{P}$ is a prime divisor of L/C that does not lie over S, then $\mathfrak{P}$ is unramified over K. Moreover, the residue field of L at $\mathfrak{P}$ is C (because C is algebraically closed), hence equals to the residue field of K at $\mathfrak{P}|_K$. Therefore, K is $\mathfrak{P}$-dense in L. Since S is nonempty, the strong approximation theorem [FrJ08, Prop. 3.3.1] gives an $a \in K$ such that

$$(6) \qquad \begin{aligned} v_{\mathfrak{P}}(a - u) &\geq 0 \text{ if } \mathfrak{P}|_K \notin S \wedge v_{\mathfrak{P}}(u) < 0 \\ v_{\mathfrak{P}}(a) &\geq 0 \text{ if } \mathfrak{P}|_K \notin S \wedge v_{\mathfrak{P}}(u) \geq 0. \end{aligned}$$

We choose $y \in K_s$ such that $\wp(y) = u - a$. By Lemma 9.5.5, $L(y)$ is a solution field of (1). By (6), $v_{\mathfrak{P}}(u - a) \geq 0$ for each $\mathfrak{P}$ that does not lie over S, hence by [FrJ08, Example 2.3.9], each such $\mathfrak{P}$ is unramified in $L(y)$. Consequently, $L(y) \subseteq K_S$, as desired. □

We combine Lemmas 9.5.4 and 9.5.6 with Lemma 9.5.1(b):

THEOREM 9.5.7: *Let E be a function field of one variable over an algebraically closed field C and S a finite nonempty set of prime divisors of E/C. Then $\mathrm{Gal}(E_S/E)$ is projective.*

Proof: Consider a finite extension K of E in E_S, a prime number p, a finite p-extension L of K in E_S, and a central nonsplit embedding problem (1). By Lemma 9.5.1(b) it suffices to solve (1) in E_S. Let S' be the set of prime divisors of K/C that lie over S. Then $K_{S'} = E_S$. Hence, without loss, we may assume that $K = E$ and $S' = S$. Therefore, by Lemmas 9.5.4 and 9.5.6, it suffices to solve embedding problem (1) in the separable closure K_s of K.

By Proposition 9.4.6(b), $\mathrm{Gal}(K)$ is projective. Hence, there exists a homomorphism $\gamma \colon \mathrm{Gal}(K) \to B$ with $\alpha \circ \gamma = \mathrm{res}_L$. In particular,

$$\alpha(\gamma(\mathrm{Gal}(K))) = \mathrm{Gal}(L/K).$$

If $\mathbb{Z}/p\mathbb{Z} \cap \gamma(\mathrm{Gal}(K))$ is trivial, then α has a group theoretic section, in contrast to our assumption. Therefore, $\mathbb{Z}/p\mathbb{Z} \subseteq \gamma(\mathrm{Gal}(K))$, so γ is surjective. The fixed field of $\mathrm{Ker}(\gamma)$ in K_s is the desired field $L(x)$. □

COROLLARY 9.5.8: *Let E be a function field of one variable over an algebraically closed field C and S a nonempty set of prime divisors of E/C. Then $\mathrm{Gal}(E_S/E)$ is projective.*

Proof: Every finite embedding problem for $\mathrm{Gal}(E_S/E)$ is equivalent to an embedding problem of the form

$$(7) \qquad (\mathrm{res} \colon \mathrm{Gal}(E_S/E) \to \mathrm{Gal}(F/E),\ \alpha \colon B \to \mathrm{Gal}(F/E)),$$

where F is a finite Galois extension of E in E_S, B is a finite group, and α is an epimorphism. The case $F = E$ being trivial, we may assume that F is a proper extension of E. Then the set T of all prime divisors of E/C ramified in F is finite and we have $F \subseteq E_T \subseteq E_S$. Since S is nonempty, we may extend F in E_S, if necessary, to assume that T is nonempty. By Theorem 9.5.7, there is a homomorphism $\gamma \colon \mathrm{Gal}(E_T/E) \to B$ such that $\alpha \circ \gamma = \mathrm{res}_{E_T/F}$. It follows that the homomorphism $\gamma' = \gamma \circ \mathrm{res}_{E_S/E_T}$ weakly solves embedding problem (7). Consequently, $\mathrm{Gal}(E_S/E)$ is a projective group. □

9.6 Maximal Unramified Extensions

Let C be an algebraically closed field, E a function field of one variable over C, and S a set of prime divisors of E/C. Theorem 9.5.7 states that $\mathrm{Gal}(E_S/E)$ is projective if S is nonempty. In this section we consider the case when S is empty and redenote E_S by E_{ur}. Thus E_{ur} is the maximal unramified extension of E. In this case Theorem 9.5.7 is false, that is $\mathrm{Gal}(E_{\mathrm{ur}}/E)$ is not projective. We prove it in two ways. The first method uses Proposition 9.2.1, hence the Riemann existence theorem. The second method is algebraic and involves the Jacobian of E.

PROPOSITION 9.6.1: *Let E be a function field of one variable over an algebraically closed field C of positive genus g. Then $\mathrm{Gal}(E_{\mathrm{ur}}/E)$ is not projective.*

First proof: Let $p = \mathrm{char}(C)$ and choose a prime number $l \neq p$. We denote the compositum of all finite unramified Galois extensions of E of degree not divisible by p by E'_{ur} and of an l-power degree by $E_{\mathrm{ur}}^{(l)}$. Then $E \subseteq E_{\mathrm{ur}}^{(l)} \subseteq E'_{\mathrm{ur}} \subseteq E_{\mathrm{ur}}$. Assume $\mathrm{Gal}(E_{\mathrm{ur}}/E)$ is projective. Then $\mathrm{Gal}(E_{\mathrm{ur}}^{(l)}/E)$, being the maximal pro-$l$ quotient of $\mathrm{Gal}(E_{\mathrm{ur}}/E)$, is also projective [FrJ08, Prop. 22.4.8], hence pro-l free [FrJ08, Prop. 22.7.6]. On the other hand, by Proposition 9.2.1(b), $\mathrm{Gal}(E'_{\mathrm{ur}}/E)$ is the free group generated by elements $\tau_1, \tau_1', \ldots, \tau_g, \tau_g'$ with the defining relation

$$(1) \qquad\qquad [\tau_1, \tau_1'] \cdots [\tau_g, \tau_g'] = 1$$

in the category of profinite groups with order not divisible by p. Since $\mathrm{Gal}(E_{\mathrm{ur}}^{(l)}/E)$ is also the maximal pro-l quotient of $\mathrm{Gal}(E'_{\mathrm{ur}}/E)$, it is the free pro-$l$ group generated by elements $\tau_1, \tau_1', \ldots, \tau_g, \tau_g'$ with the defining relation (1). Now choose a basis $t_1, t_1', \ldots, t_g, t_g'$ for the $\mathbb{F}_l$-vector space $\mathbb{F}_l^{2g}$. The map $\tau_i \mapsto t_i$ and $\tau_i' \mapsto t_i'$ for $i = 1, \ldots, g$ extends to an epimorphism of $\mathrm{Gal}(E_{\mathrm{ur}}^{(l)}/E)$ onto $\mathbb{F}_l^{2g}$. Since the rank of the latter group is $2g$ and that of the former one is at most $2g$, we deduce that $\mathrm{rank}(E_{\mathrm{ur}}^{(l)}/E) = 2g$. It follows from [FrJ08, Lemma 17.4.6(b)] that $\tau_1, \tau_1', \ldots, \tau_g, \tau_g'$, viewed as generators of $\mathrm{Gal}(E_{\mathrm{ur}}^{(l)}/E)$ form a basis of that group. Thus, every map of the basis into an l-group A extends to a homomorphism of $\mathrm{Gal}(E_{\mathrm{ur}}^{(l)}/E)$ into A. In particular, this is the case if we choose A to be noncommutative and a_1, a_1' elements of A with $[a_1, a_1'] \neq 1$. Then the map $\tau_1 \mapsto a_1$, $\tau_1' \mapsto a_1'$, $\tau_i \mapsto 1$, and $\tau_i' \mapsto 1$ for $i \geq 2$ extends to a homomorphism into A. It follows from (1) that $[a_1, a_1'] = 1$. This contradiction proves that $\mathrm{Gal}(E_{\mathrm{ur}}/E)$ is not projective. $\qquad\square$

The second proof of Proposition 9.6.1 depends on the following piece of information.

LEMMA 9.6.2: *Let E be a function field of one variable of genus g over an algebraically closed field C. Let $l \neq \mathrm{char}(C)$ be a prime number and A the* subgroup of $E^\times/(E^\times)^l$ consisting of all cosets $x(E^\times)^l$ such that $l|v_\mathfrak{p}(x)$ for all prime divisors $\mathfrak{p}$ of E/C. Then, $A \cong (\mathbb{Z}/l\mathbb{Z})^{2g}$.

Proof: We distinguish between two cases.

CASE A: $g = 0$. Then $E = C(t)$ is the field of rational functions over C in an indeterminate t [FrJ08, Example 3.2.4]. In this case, each finite prime divisor of E/C has a prime element of the form $t - a$ with some $a \in C$. Thus, if $x(E^\times)^l \in A$, then $x = c\prod_{a \in C}(t - a)^{lk(a)}$ with $c \in C^\times$ and with $k(a) \in \mathbb{Z}$ such that $k(a) = 0$ for all but finitely many a's. Moreover, since C is algebraically closed, c is an l-power in C. Hence, $x(E^\times)^l$ is the unit element of $E^\times/(E^\times)^l$. Consequently, A is trivial.

CASE B: $g \geq 1$. We consider the group $\mathrm{Div}_0(E/C)$ of divisors of E/C of degree 0, its subgroup $\mathrm{div}(E^\times)$ of principal divisors, and the Jacobian variety J of E/C (which exists since $\mathrm{genus}(E/C) > 0$). For each $x(E^\times)^l \in A$ there exists a divisor $\mathfrak{a}$ of E/C such that $\mathrm{div}(x) = l\mathfrak{a}$. It satisfies $0 = l\deg(\mathfrak{a})$, so $\deg(\mathfrak{a}) = 0$. We map $x(E^\times)^l$ onto $\mathfrak{a} + \mathrm{div}(E^\times)$. If $y \in E^\times$, then $\mathrm{div}(xy^l) = l(\mathfrak{a} + \mathrm{div}(y))$, so our map defines a homomorphism $\alpha\colon A \to \mathrm{Div}_0(E/C)/\mathrm{div}(E^\times)$. If $x(E^\times)^l \in \mathrm{Ker}(\alpha)$, then $\mathfrak{a} = \mathrm{div}(z)$ for some $z \in E^\times$, so $\mathrm{div}(xz^{-l}) = 0$. Hence, $xz^{-l} \in C^\times$ [FrJ08, Sec. 3.1]. Since C is algebraically closed, there exists $c \in C^\times$ such that $x = (cz)^l$. It follows that α is injective. Note that since $l\mathfrak{a} = \mathrm{div}(x)$, we have $l(\mathfrak{a} + \mathrm{div}(E^\times)) = 0$. Thus, the image of α lies in the subgroup $\mathcal{D}$ of $\mathrm{Div}_0(E/C)/\mathrm{div}(E^\times)$ of all elements annihilated by l. Conversely, if $\mathfrak{a} + \mathrm{div}(E^\times) \in \mathcal{D}$, then there exists $x \in E^\times$ such that $l\mathfrak{a} = \mathrm{div}(x)$, so $\alpha(x(E^\times)^l) = \mathfrak{a} + \mathrm{div}(E^\times)$. It follows that $\mathrm{Im}(\alpha) = \mathcal{D}$. Hence $A \cong \mathcal{D}$.

As mentioned in Subsection 6.3.2, there is an isomorphism

$$\mathrm{Div}_0(E/C)/\mathrm{div}(E^\times) \cong J(C).$$

Hence, $\mathcal{D} \cong J(C)_l$. By Subsection 6.3.1, $J(C)_l \cong (\mathbb{Z}/l\mathbb{Z})^{2g}$. Consequently, $A \cong (\mathbb{Z}/l\mathbb{Z})^{2g}$. $\qquad\square$

Next we apply Kummer theory.

LEMMA 9.6.3: Let E be a function field of one variable over an algebraically closed field C and let $l \neq \mathrm{char}(C)$ be a prime number. Denote the maximal unramified pro-l extension of E by $E_{\mathrm{ur}}^{(l)}$ and set $g = \mathrm{genus}(E/C)$. Then $\mathrm{rank}(\mathrm{Gal}(E_{\mathrm{ur}}^{(l)}/E)) = 2g$.

Proof: Denote the compositum of all cyclic unramified extensions of E of degree l by F. By [FrJ08, Lemma 22.7.4], $\mathrm{Gal}(E_{\mathrm{ur}}^{(l)}/F)$ is the Frattini subgroup of $\mathrm{Gal}(E_{\mathrm{ur}}^{(l)}/E)$ and $\mathrm{Gal}(F/E) \cong (\mathbb{Z}/l\mathbb{Z})^r$, where $r = \mathrm{rank}(\mathrm{Gal}(E_{\mathrm{ur}}^{(l)}/E))$. On the other hand, since E contains a root of unity of order l, each cyclic extension of E of degree l has the form $E(x^{1/l})$ with $x \in E^\times$. That extension is unramified over E if and only if $l|v_\mathfrak{p}(x)$ for each prime divisor $\mathfrak{p}$

of E/C [FrJ08, Example 2.3.8]. Thus, by Kummer Theory [Lan93, p. 295, Thm. 8.2], $\mathrm{Gal}(F/E) \cong A$, where A is as in Lemma 9.6.2, hence by that lemma $\mathrm{Gal}(F/E) \cong (\mathbb{Z}/l\mathbb{Z})^{2g}$. Combining that with the opening statement of the proof, we conclude that $\mathrm{rank}(E_{\mathrm{ur}}^{(l)}/E) = 2g$. $\qquad\square$

PROPOSITION 9.6.1: *Let E be a function field of one variable over an algebraically closed field C of positive genus g. Then $\mathrm{Gal}(E_{\mathrm{ur}}/E)$ is not projective.*

Second proof: Let $p = \mathrm{char}(C)$ and choose a prime number $l \neq p$. We denote the compositum of all finite unramified Galois extensions of E of degree not divisible by p by E'_{ur} and of an l-power degree by $E_{\mathrm{ur}}^{(l)}$. Then $E \subseteq E_{\mathrm{ur}}^{(l)} \subseteq E'_{\mathrm{ur}} \subseteq E_{\mathrm{ur}}$. Assume that $\mathrm{Gal}(E_{\mathrm{ur}}/E)$ is projective. Then $G = \mathrm{Gal}(E_{\mathrm{ur}}^{(l)}/E)$, being the maximal pro-$l$ quotient of $\mathrm{Gal}(E_{\mathrm{ur}}/E)$, is also projective [FrJ08, Prop. 22.4.8], hence pro-l free [FrJ08, Prop. 22.7.6]. By Lemma 9.6.3, $\mathrm{rank}(G) = 2g$.

Now we choose a proper finite extension F of E in $E_{\mathrm{ur}}^{(l)}$ and set $h = \mathrm{genus}(F/C)$. Since F is unramified over E, Riemann-Hurwitz genus formula simplifies to $2h - 2 = [F : E](2g - 2)$ (Remark 5.8.1(f)). hence

(2) $$h - 1 = [F : E](g - 1).$$

On the other hand, $H = \mathrm{Gal}(E_{\mathrm{ur}}^{(l)}/F)$ is an open subgroup of G of index $[F : E]$. Hence, by Nielsen-Schreier [FrJ08, Prop. 17.5.7], $\mathrm{rank}(H) - 1 = [F : E](\mathrm{rank}(G) - 1)$. Note that $F_{\mathrm{ur}}^{(l)} = E_{\mathrm{ur}}^{(l)}$, so by Lemma 9.6.3, $\mathrm{rank}(H) = 2h$. Hence,

(3) $$2h - 1 = [F : E](2g - 1)$$

Substituting the value of h from (2) in (3) leads to $[F : E] = 1$. This contradiction to our assumption proves that $\mathrm{Gal}(E_{\mathrm{ur}}/E)$ is not projective. $\square$

9.7 Embedding Problems with Given Branching

We fix for the whole section a rational function field $E = C(x)$ over an algebraically closed field C. Assume that C is complete with respect to an ultrametric absolute value. We show in this section how to solve finite split embedding problems with an extra information on the branch points of the solution fields. This prepares the way in the next section to prove for general C that $\mathrm{Gal}(E_S/E)$ is free of rank m, if $|S| = m = \mathrm{card}(C)$.

LEMMA 9.7.1: *For each integer $n > 1$ there exists a cyclic extension F/E of degree n such that $\mathrm{Branch}(F/E) = \{1, \infty\}$.*

Proof: The lemma follows from Lemma 4.2.5 by applying a suitable Möbius transformation. Nevertheless, we supply a direct proof to the special case at hand.

If $\mathrm{char}(C) \nmid n$, let $F = E(y)$, where $y^n = x - 1$. If $n = p = \mathrm{char}(C) > 0$, let $F = E(y)$, where $y^p - y = \frac{x^2}{x-1}$. Then $\mathrm{Branch}(F/E) = \{1, \infty\}$ [FrJ08, Examples 2.3.8 and 2.3.9]. In each case F/E is a cyclic extension of degree n.

The rest of the proof reduces the general case to these two cases.

PART A: *Without loss of generality n is a prime power.* Indeed, if $n = \prod_{i=1}^{m} p_i^{r_i}$, where $p_1, \ldots, p_m$ are distinct primes, and for each $1 \leq i \leq m$ there is a cyclic extension F_i/E of degree $p_i^{r_i}$, ramified at $\{1, \infty\}$ and unramified elsewhere, then the compositum $F = \prod_{i=1}^{m} F_i$ has the required properties.

PART B: *Without loss of generality n is prime.* Indeed, assume that n is a power of a prime p and there is a cyclic extension F_1/E of degree p, whose branch points are $1, \infty$. Let $S = \{1, \infty\}$. By Theorem 9.5.7, the embedding problem

$$(\alpha\colon \mathbb{Z}/n\mathbb{Z} \to \mathbb{Z}/p\mathbb{Z} = \mathrm{Gal}(F_1/E),\ \mathrm{res}\colon \mathrm{Gal}(E_S/E) \to \mathrm{Gal}(F_1/E))$$

for $\mathrm{Gal}(E_S/E)$ has a weak solution, say, $\psi\colon \mathrm{Gal}(E_S/E) \to \mathbb{Z}/n\mathbb{Z}$. But ψ is surjective, because $\alpha(\psi(\mathrm{Gal}(E_S/E))) = \mathbb{Z}/p\mathbb{Z}$ and $\mathbb{Z}/n\mathbb{Z}$ is the only subgroup H of $\mathbb{Z}/n\mathbb{Z}$ with $\alpha(H) = \mathbb{Z}/p\mathbb{Z}$. The fixed field F of $\mathrm{Ker}(\psi)$ has the required properties. $\qquad\square$

LEMMA 9.7.2: *Suppose C is complete with respect to an ultrametric absolute value $|\ |$. Let $c \in C$, $r \in C^\times$, and set $w = \frac{r}{x-c}$. Let $n > 1$ be an integer. Then there exists $0 < \varepsilon < |r|$ such that for all distinct $b_1, b_2 \in C$ with $|b_1 - c|, |b_2 - c| \leq \varepsilon$ there is a cyclic extension F/E of degree n, with $\mathrm{Branch}(F/E) = \{b_1, b_2\}$ and $F \subseteq \mathrm{Quot}(C\{w\})$.*

Proof: Lemma 9.7.1 gives a cyclic extension F_1/E of degree n with $\mathrm{Branch}(F_1/E) = \{1, \infty\}$. Since F_1/E is unramified at 0 and C is algebraically closed, we have $F_1 \subset C((x))$. Let y be a primitive element of F_1/E integral over $C[x]$. By Proposition 2.4.5, y converges at some point $b \in C$. Thus, if we write $y = \sum_{n=0}^{\infty} a_n x^n$, then the series $\sum_{n=0}^{\infty} a_n b^n$ converges. Set $\varepsilon = \min(1, |rb|, \frac{|r|}{2})$. Then for each $a \in C^\times$ with $|a| \leq |rb|$ we have $\mu_a(y) = \sum_{n=0}^{\infty} a_n a^n x^n = \sum_{n=0}^{\infty} a_n \left(\frac{a}{r}\right)^n (rx)^n$, so the latter series in rx converges. This means that $\mu_a(y) \in C\{rx\}$ and $\mu_a(F_1) \subseteq \mathrm{Quot}(C\{rx\})$.

Let $b_1, b_2 \in C$ such that $|b_1 - c|, |b_2 - c| \leq \varepsilon$. Set $a = b_2 - b_1$ and $F_2 = \mu_a(F_1)$. Then $|a| \leq \varepsilon \leq |rb|$, so $F_2 \subseteq \mathrm{Quot}(C\{rx\})$. By Remark 4.1.4,

$$\mathrm{Branch}(F_2/E) = (\mu_a')^{-1}(\mathrm{Branch}(F_1/E))$$

$$= \frac{1}{a}\{1, \infty\} = \{\frac{1}{b_2 - b_1}, \infty\}.$$

Let θ be the C-automorphism of E given by $\theta(x) = \frac{1}{x-c}$, so that $\theta(rx) = w$. Extend θ to an isomorphism of fields $\theta\colon F_2 \to F_3$. Then $F_3 \subseteq \mathrm{Quot}(C\{w\})$

and by Remark 4.1.4,

$$\text{Branch}(F_3/E) = (\theta')^{-1}(\text{Branch}(F_2/E))$$
$$= (\theta')^{-1}\{\frac{1}{b_2 - b_1}, \infty\} = \{c + b_2 - b_1, c\}.$$

Let $d = c - b_1$. Then $|d| \leq \varepsilon \leq 1$. Let λ be the automorphism of $C\{w\}$ that maps $f = \sum_{n=0}^{\infty} a_n w^n$ onto

$$\lambda(f) = \sum_{n=0}^{\infty} a_n(w + d)^n = \sum_{n=0}^{\infty} a_n \sum_{k=0}^{n} \binom{n}{k} d^{n-k} w^k$$
$$= \sum_{k=0}^{\infty} \Big(\sum_{n=k}^{\infty} \binom{n}{k} a_n d^{n-k} \Big) w^k.$$

Then $\binom{n}{k} a_n d^{n-k} \to 0$ as $n \to \infty$, so the series $\sum_{n=k}^{\infty} \binom{n}{k} a_n d^{n-k}$ converges in C, hence λ is well defined. Moreover,

$$\Big| \sum_{n=k}^{\infty} \binom{n}{k} a_n d^{n-k} \Big| \leq \max_{n \geq k} |a_n|.$$

We extend λ to an automorphism of $\text{Quot}(C\{w\})$. The restriction of λ to E is the map $w \mapsto w + d$. Let $F = \lambda(F_3)$. Then $F \subseteq \text{Quot}(C\{w\})$ and

$$\text{Branch}(F/E) = (\lambda')^{-1}(\text{Branch}(F_3/E)) = \{c+b_2-b_1-d, c-d\} = \{b_2, b_1\}. \quad \square$$

Remark 9.7.3: A **disk** in $C \cup \{\infty\}$ is a set of the form

$$D = \theta(\{a \in C \mid |a| \leq \varepsilon\})$$

where $\varepsilon > 0$ and θ is a Möbius transformation over C. Thus, each set of the form $D = \{a \in C \mid |a - c| \leq \varepsilon\}$ or $D = \{a \in C \mid |a| \geq \varepsilon\} \cup \{\infty\}$, where $c \in C$, is a disk. (In fact, each disk is of this form; but we shall not use this fact.) Note that the cardinality of a disk is the same as the cardinality of C.
$\square$

LEMMA 9.7.4: *Assume C is complete with respect to an ultrametric absolute value. Let F_1/E be a finite Galois extension with group G_1 and*

(1) $$\alpha: G = G_1 \ltimes H \to G_1 = \text{Gal}(F_1/E)$$

a finite split embedding problem for $\text{Gal}(E)$ with a nontrivial kernel H. Consider a finite set J that does not contain 1 and let $\{G_i\}_{i \in J}$ be a finite family of nontrivial cyclic subgroups of G that generate H. Then there exists a family of pairwise disjoint disks $\{D_i\}_{i \in J}$ in C such that for every $B \subset \bigcup_{i \in J} D_i$ with $\text{card}(B \cap D_i) = 2$ for each $i \in J$, there exists a solution field F of (1) with $\text{Branch}(F^{G_1}/E) = B$.

Proof: Let $I = J \cup \{1\}$. Then $G = \langle G_i \mid i \in I \rangle$. We choose distinct elements $c_i \in C$, $i \in I$, and an element $r \in C^{\times}$ with $|r| \leq |c_i - c_j|$ for all distinct $i, j \in I$. Then let $w_i = \frac{r}{x - c_i}$, $P_i = \text{Quot}(C\{w_j \mid j \neq i\})$ and $P_i' = \text{Quot}(C\{w_i\})$.

CLAIM: *We may assume that $F_1 \subseteq P_1'$.* Indeed, since C is algebraically closed, every prime divisor of F_1/C is of degree 1. In particular, F_1/C has an unramified prime divisor of degree 1. By Lemma 4.3.7, there is a C-automorphism of E that extends to an embedding $\theta\colon F_1 \to P_1'$. Let $F_1' = \theta(F_1)$ and extend θ to an automorphism of E_s. Then θ defines isomorphisms $\theta_*\colon \mathrm{Gal}(F_1/E) \to \mathrm{Gal}(F_1'/E)$ and $\theta_*\colon \mathrm{Gal}(E) \to \mathrm{Gal}(E)$ such that the following diagram commutes

$$
\begin{array}{ccc}
\mathrm{Gal}(E) & \xrightarrow{\;\theta_*\;} & \mathrm{Gal}(E) \\
\big\downarrow{\scriptstyle\mathrm{res}} & & \big\downarrow{\scriptstyle\mathrm{res}} \\
\end{array}
$$

$$
G \xrightarrow{\;\alpha\;} \mathrm{Gal}(F_1/E) \xrightarrow{\;\theta_*\;} \mathrm{Gal}(F_1'/E).
$$

Suppose that there is a family of disjoint disks $\{D_i'\}_{i\in J}$ in C such that for every $B' \subset \bigcup_{i\in J} D_i'$ with $\mathrm{card}(B' \cap D_i') = 2$, for each $i \in J$, the embedding problem

$$
(\theta_* \circ \alpha\colon G \to \mathrm{Gal}(F_1'/E),\ \mathrm{res}\colon \mathrm{Gal}(E) \to \mathrm{Gal}(F_1'/E))
$$

has a solution field F' with $\mathrm{Branch}(F'^{G_1}/E) = B'$. Let θ' be the permutation of $C \cup \{\infty\}$ induced by θ as in Remark 4.1.4. Then the disks $D_i = \theta'(D_i)$, for $i \in J$, have the required property.

Indeed, if $B \subset \bigcup_{i\in J} D_i$ and $\mathrm{card}(B \cap D_i) = 2$, for each $i \in J$, we put $B' = (\theta')^{-1}(B)$, let F' be as above, and extend θ to an automorphism of E_s. Then $F = \theta^{-1}(F')$ solves (1) and $\theta(F^{G_1}) = F'^{G_1}$. By Remark 4.1.4,

$$
\theta'(\mathrm{Branch}(F'^{G_1}/E)) = \mathrm{Branch}(F^{G_1}/E).
$$

Hence, $B = \mathrm{Branch}(F^{G_1}/E)$, as desired.

Thus, replacing F_1 by F_1' we may assume that $F_1 \subseteq P_1'$.

By Lemma 9.7.2, there is an $0 < \varepsilon < |r|$ such that the (necessarily disjoint) disks $D_i = \{a \in C \mid |a - c_i| \leq \varepsilon\}$, for $i \in J$, have the following property: For every $B \subset \bigcup_{i\in J} D_i$ with $\mathrm{card}(B \cap D_i) = 2$, for each $i \in J$, there exist Galois extensions F_i/E with the cyclic Galois group G_i and $\mathrm{Branch}(F_i/E) = B \cap D_i$ and $F_i \subseteq \mathrm{Quot}(C\{w_i\}_{i\in I})$, for each $i \in J$. Let $P = \mathrm{Quot}(C\{w_i\}_{i\in I})$.

By Proposition 3.4.5, $\mathcal{E} = (E, F_i, P_i, Q; G_i, G)_{i\in I}$ is patching data. Its compound F is, by Lemma 1.3.1(c), a Galois extension of E that solves (1). By Lemma 7.2.3(c),

$$
\mathrm{Branch}(F^{G_1}/E) = \bigcup_{i\in J} \mathrm{Branch}(F_i/E) = \bigcup_{i\in J} B \cap D_i = B. \qquad \square
$$

9.8 Descent

We wish to apply Lemma 9.7.4 to a sufficiently large complete extension of a given algebraically closed field. Thus we consider the following situation. Let $C_1 \subseteq C_2$ be two algebraically closed fields and x an intermediate. We set $E_1 = C_1(x)$, $E_2 = C_2(x)$, and let

$$(1) \qquad \rho\colon G = G_1 \ltimes H \to G_1 = \mathrm{Gal}(F_1/E_1)$$

be a finite split embedding problem for $\mathrm{Gal}(E_1)$ with a nontrivial kernel H. Let $F_2 = F_1 E_2$. Then the restriction map $\mathrm{Gal}(F_2/E_2) \to \mathrm{Gal}(F_1/E_1)$ is an isomorphism. We identify $\mathrm{Gal}(F_2/E_2)$ with $G_1 = \mathrm{Gal}(F_1/E_1)$ via this map. Then (1) induces a finite split embedding problem

$$(2) \qquad \rho\colon G = G_1 \ltimes H \to G_1 = \mathrm{Gal}(F_2/E_2)$$

for $\mathrm{Gal}(E_2)$ with a nontrivial kernel.

Before dealing with embedding problems let us notice a simple fact:

Remark 9.8.1: Let A be an infinite subset of a field K. Then every nonempty Zariski K-open subset of $\mathbb{A}^n$ meets A^n. Indeed, the only polynomial in n variables over K that vanishes on A^n is 0. $\qquad\square$

LEMMA 9.8.2: *Let A be an infinite subset of C_1. Suppose (2) has a solution field L_2 such that $\infty \notin \mathrm{Branch}(L_2^{G_1}/E_2)$ and the elements of $\mathrm{Branch}(L_2^{G_1}/E_2)$ are algebraically independent over C_1. Then (1) has a solution field L_1 with $\mathrm{Branch}(L_1^{G_1}/E_1) \subseteq A$.*

Proof: There is an irreducible monic polynomial $h \in C_2[x, Z]$ such that $L_2 = E_2(z)$, with $h(x, z) = 0$. Furthermore, there are irreducible polynomials $f_1, \ldots, f_r \in C_2[x, Z]$ such that a root z_j of f_j is a primitive element of $L_2^{G_1}/E_2$ (hence also of L_2/F_2), and

$$(3) \qquad \mathrm{Branch}(L_2^{G_1}/E_2) = \bigcap_{j=1}^{r} \mathrm{Zero}(\mathrm{discr}(f_j))$$

[Has80, p. 64].

We set $\mathrm{Branch}(L_2^{G_1}/E_2) = \{u_1, \ldots, u_k\}$ and choose $u_{k+1}, \ldots, u_l \in C_2$ such that $h, f_1, \ldots, f_r \in C_1[\mathbf{u}][x, Z]$. We also set $E_{\mathbf{u}} = C_1(\mathbf{u}, x)$, $F_{\mathbf{u}} = F_1(\mathbf{u})$ and add more elements of C_2 to $\{u_1, \ldots, u_l\}$, if necessary, such that $L_{\mathbf{u}} = E_{\mathbf{u}}(z)$ is a Galois extension of $E_{\mathbf{u}}$ that solves the embedding problem $G \to \mathrm{Gal}(F_{\mathbf{u}}/E_{\mathbf{u}})$ induced from (1), and $L_{\mathbf{u}}^{G_1} = E_{\mathbf{u}}(z_j)$, $j = 1, \ldots, r$.

Let $U = \mathrm{Spec}(C_1[\mathbf{u}])$ be the irreducible variety that $\mathbf{u}$ generates over C_1. For each $\mathbf{u}' \in U(C_1)$ the C_1-specialization $\mathbf{u} \to \mathbf{u}'$ first extends to an F_1-place $'\colon F_{\mathbf{u}} \to F_1 \cup \{\infty\}$, and then to a place $'\colon L_{\mathbf{u}} \to \tilde{E}_1 \cup \{\infty\}$. Let $B = \{u_1', \ldots, u_k'\} \subset C_1$ be the image of $\mathrm{Branch}(L_2^{G_1}/E_2) = \{u_1, \ldots, u_k\}$.

The variety U has a nonempty Zariski-open subset U' such that for all $\mathbf{u}' \in U'$ the following statements hold:

(4a) $h', f'_1, \ldots, f'_r \in C_1[x, Z]$ are irreducible over $C_1(x)$ [FrJ08, Prop. 9.4.3];

(4b) $L_1 = E_1(z')$ is Galois over E_1 and L_1 solves embedding problem (1) [FrJ08, Lemma 13.1.1];

(4c) the respective roots $z'_1, \ldots, z'_r$ of $f'_1, \ldots, f'_r$ are primitive elements for $L_1^{G_1}/E_1$.

From (3), $B = \bigcap_{j=1}^r \mathrm{Zero}(\mathrm{discr}(f'_j))$. Since $L_1^{G_1}/E_1$ is unramified at each point outside $\mathrm{Zero}(\mathrm{discr}(f'_j))$, $j = 1, \ldots, r$ (by [Has80, p. 64]),

(4d) $\mathrm{Branch}(L_1^{G_1}/E_1) \subseteq B$.

By assumption, $u_1, \ldots, u_k$ are algebraically independent over C_1. Therefore, the projection on the first k coordinates $\mathrm{pr}: U \to \mathbb{A}^k$ is a dominant map, hence $\mathrm{pr}(U')$ contains a Zariski-open subset of $\mathbb{A}^k$ [Lan58, p. 88, Prop. 4]. By Remark 9.8.1, we may choose $\mathbf{u}' \in U'(C_1) \cap \mathrm{pr}^{-1}(\mathbb{A}^k)$, so $B = \{u'_1, \ldots, u'_k\} \subset A$. Consequently, $\mathrm{Branch}(L_1^{G_1}/E_1) \subseteq A$. $\square$

To achieve the algebraic independence in Lemma 9.8.2 we use:

LEMMA 9.8.3: *Let $C_1 \subset C_2$ be two algebraically closed fields such that $\mathrm{card}(C_1) < \mathrm{card}(C_2)$. Let $\{D_j\}_{j \in J}$ be a finite collection of pairwise disjoint subsets of C_2 of cardinality $\mathrm{card}(C_2)$. Then there exists a set $B \subseteq \bigcup_{j \in J} D_j$ such that $\mathrm{card}(B \cap D_j) = 2$ for each $j \in J$ and the elements of B are algebraically independent over C_1.*

Proof: Write J as $\{1, \ldots, k\}$, and suppose, by induction, that we have already found $b_j, b'_j \in D_j$, for $j = 1, \ldots, k-1$, such that $b_1, b'_1, \ldots, b_{k-1}, b'_{k-1}$ are algebraically independent over C_1. The cardinality of the algebraic closure C'_1 of $C_1(b_1, b'_1, \ldots, b_{k-1}, b'_{k-1})$ in C_2 is $\mathrm{card}(C_1) < \mathrm{card}(C_2) = \mathrm{card}(D_k)$, so there exist $b_k, b'_k \in D_k$ algebraically independent over C'_1. Thus, $b_1, b'_1, \ldots, b_k, b'_k$ are algebraically independent over C_1. $\square$

LEMMA 9.8.4: *Let G be a projective group of rank m. Set $m' = 1$ if $m = \aleph_0$ and $m' = m$ if $m > \aleph_0$. Suppose every finite split embedding problem for G with a nontrivial kernel has m' solutions. Then $G \cong \hat{F}_m$.*

Proof: The case where $m > \aleph_0$ is settled in [FrJ08, Lemma 21.5.8]. Consider the case where $m = \aleph_0$. By Iwasawa, it suffices to prove that every finite embedding problem

(5) $(\varphi: G \to A, \ \alpha: B \to A)$

is solvable [FrJ08, Cor. 24.8.3]. Indeed, since G is projective, there exists a homomorphism $\gamma: G \to B$ with $\alpha \circ \gamma = \varphi$. Then $\mathrm{Ker}(\gamma)$ is an open normal subgroup of G, so $\hat{A} = G/\mathrm{Ker}(\gamma)$ is a finite group. Let $\hat{\varphi}: G \to \hat{A}$ be the quotient map and $\bar{\varphi}: \hat{A} \to A$ and $\hat{\gamma}: \hat{A} \to B$ the homomorphisms induced by φ and γ, respectively. In particular $\alpha \circ \hat{\gamma} = \bar{\varphi}$. Next consider the fiber product $\hat{B} = B \times_A \hat{A}$ with the corresponding projections $\beta: \hat{B} \to B$ and $\hat{\alpha}: \hat{B} \to \hat{A}$. The defining property of the fiber product gives a homomorphism $\hat{\alpha}': \hat{A} \to \hat{B}$ such that $\hat{\alpha} \circ \hat{\alpha}' = \mathrm{id}_{\hat{A}}$. In other words, $\hat{\alpha}$ splits. By assumption there exists an epimorphism $\delta\colon G \to \hat{B}$ such that $\hat{\alpha} \circ \delta = \hat{\varphi}$. Thus, $\beta \circ \delta$ solves embedding problem (5). Consequently, $G \cong \hat{F}_\omega$. $\qquad\square$

The preceding lemmas yield the main result of this chapter:

THEOREM 9.8.5: *Let C be an algebraically closed field of cardinality m, $E = C(x)$ the field of rational functions over C, and S a subset of $C \cup \{\infty\}$ of cardinality m. Then $\mathrm{Gal}(E_S/E)$ is isomorphic to the free profinite group of rank m.*

Proof: Put $C_1 = C$ and $E_1 = E$. By Corollary 9.5.8, $\mathrm{Gal}(E_S/E)$ is projective. Therefore, by Lemma 9.8.4, it suffices to show that every finite split embedding problem (1) for $\mathrm{Gal}(E_S/E)$ with a nontrivial kernel has m' solution fields, where $m' = 1$ if $m = \aleph_0$, and $m' = m$ otherwise.

Let $\beta < m$ be an ordinal number. Suppose, by transfinite induction, that $\{N_\alpha\}_{\alpha<\beta}$ is a family of distinct solution fields of (1). For each α, the set $\mathrm{Branch}(N_\alpha/E)$ is finite. Hence, $A = S \smallsetminus \bigcup_{\alpha<\beta} \mathrm{Branch}(N_\alpha/E)$ is infinite.

We choose an algebraically closed field C_2 that contains C and is complete with respect to a nontrivial ultrametric absolute value such that $\mathrm{card}(C) < \mathrm{card}(C_2)$. For instance, choose a field C' that contains C such that $\mathrm{card}(C) < \mathrm{card}(C')$, and let C_2 be the completion of the algebraic closure of $C'((t))$. We consider the induced embedding problem (2).

Let $\{G_j \mid j \in J\}$ be a nonempty set of nontrivial cyclic groups that generate H with $1 \notin J$. By Lemma 9.7.4, there exists a family of disks $\{D_j\}_{j\in J}$ in C_2 such that for every $B \subset \bigcup_{j\in J} D_j$ with $\mathrm{card}(B \cap D_j) = 2$ for each $j \in J$ there exists a solution field L_2 to (2) with $\mathrm{Branch}(L_2^{G_1}/C_2(x)) = B$. We choose such a set B. By Remark 9.7.3, $\mathrm{card}(D_j) = \mathrm{card}(C_2)$. By Lemma 9.8.3, we may assume that the elements of B are algebraically independent over C. Therefore, by Lemma 9.8.2, (1) has a solution field $N = N_\beta$ such that $\mathrm{Branch}(N^{G_1}/E) \subseteq A$.

Since $N = F_1 N^{G_1}$, we have

$$\mathrm{Branch}(N/E) = \mathrm{Branch}(F_1/E) \cup \mathrm{Branch}(N^{G_1}/E)$$

(Remark 4.1.1). Furthermore,

$$\mathrm{Branch}(F_1/E), \mathrm{Branch}(N^{G_1}/E) \subseteq S,$$

so $\mathrm{Branch}(N/E) \subseteq S$. Since $\mathrm{Branch}(N^{G_1}/E) \subseteq A$, we have

$$\mathrm{Branch}(N^{G_1}/E) \cap \mathrm{Branch}(N_\alpha/E) = \emptyset$$

for each $\alpha < \beta$. In addition, $[N^{G_1} : E] = |H| > 1$, so by the Riemann-Hurwitz genus formula (Remark 5.8.1(f)), $\mathrm{Branch}(N^{G_1}/E) \neq \emptyset$. Since

$$\mathrm{Branch}(N^{G_1}/E) \subset \mathrm{Branch}(N/E),$$

it follows that $\mathrm{Branch}(N/E) \neq \mathrm{Branch}(N_\alpha/E)$ for each $\alpha < \beta$. Consequently $N \neq N_\alpha$ for each $\alpha < \beta$. $\qquad\square$

Remark 9.8.6: Fundamental groups. In the special case of Theorem 9.8.5, where S is the set of all prime divisors, $C(x)_S = C(x)_s$. Thus, $\mathrm{Gal}(C(x)) \cong \hat{F}_m$. If F is a finite extension of $C(x)$, then $\mathrm{Gal}(F)$ is isomorphic to an open subgroup of $\mathrm{Gal}(C(x))$, so $\mathrm{Gal}(F) \cong \hat{F}_m$ [FrJ08, Prop. 25.2.2]. It follows that Theorem 9.8.5 is essentially a generalization of Corollary 9.4.9.

One may try to generalize the latter observation to a proper finite extension F of $C(x)$ and a set S of prime divisors of F/C of cardinality $\mathrm{card}(C)$. Let T be the set of all prime divisors of $C(x)/C$ that lie under S and those that ramify in F. Then $F_S \subseteq C(x)_T$, so $\mathrm{Gal}(C(x)_T/F)$ is an open subgroup of $\mathrm{Gal}(C(x)_T/C(x))$. By Theorem 9.8.5 and [FrJ08, Prop. 25.2.2], $\mathrm{Gal}(C(x)_T/F) \cong \hat{F}_m$. However, $\mathrm{Gal}(F_S/F)$ might be a proper quotient of $\mathrm{Gal}(C(x)_T/F)$, so our methods fail to prove that the latter group is also isomorphic to $\hat{F}_m$.

Nevertheless, using formal patching, Harbater proved that $\mathrm{Gal}(F_S/F) \cong \hat{F}_m$ if the complement of S is finite [Hrb95, Thm. 4.4]. Using rigid analytic patching, Pop proved the latter isomorphism under the weaker condition that $\mathrm{card}(S) = m$ [Pop95, p. 556, Thm. A]. Note that both Harbater and Pop consider a smooth projective model X for F/C, reinterpret S as a subset of $X(C)$, call $\mathrm{Gal}(F_S/F)$ the **fundamental group of** $X \smallsetminus S$, and denote it by $\Pi(X \smallsetminus S)$.   □

9.9 Fundamental Groups with S Finite

We consider again a function field E of one variable of genus g over an algebraically closed field C of characteristic p. Let S be a finite nonempty set of prime divisors of E/C. As before we denote the maximal Galois extension of E ramified at most over S by E_S. By Corollary 9.1.7, $\mathrm{Gal}(E_S/E)$ is a free profinite group if $p = 0$. We prove in this section that this is false if $p > 0$.

For each prime number l we denote the maximal pro-l extension of E which is ramified at most over S by $E_S^{(l)}$.

LEMMA 9.9.1: *Let E be a function field of one variable over an algebraically closed field C, S a finite nonempty set of prime divisors of E/C, and l a prime number. Then $\mathrm{Gal}(E_S^{(l)}/E)$ is a free pro-l group.*

Proof: The group $\mathrm{Gal}(E_S^{(l)}/E)$ is the maximal pro-l quotient of $\mathrm{Gal}(E_S/E)$. By Theorem 9.5.7, the latter group is projective. Hence, by [FrJ08, Prop. 22.4.8], so is the former. Alternatively, one may repeat the proof of Theorem 9.5.7.   □

LEMMA 9.9.2: *Let C be an infinite field of positive characteristic p and cardinality m. Let E be a function field of one variable over C and S a nonempty set of prime divisors of E/C. Then $\mathrm{rank}(\mathrm{Gal}(E_S^{(p)}/E)) = m$.*

Proof: Since $\mathrm{card}(E) = m$, the field E has at most m finite extensions in $E_S^{(p)}$, hence $\mathrm{rank}(\mathrm{Gal}(E_S^{(p)}/E)) \le m$ [FrJ08, Prop. 17.1.2]. Thus, it suffices to prove that $\mathrm{rank}(\mathrm{Gal}(E_S^{(p)}/E)) \geq m$. The rest of the proof breaks up into two parts.

PART A: *Assume $E = C(x)$ with a transcendental element x over C.* Since $\mathrm{Gal}(C(x)_S^{(p)}/C(x))$ is a pro-p group, it suffices to construct m linearly disjoint cyclic extensions in $C(x)_S$ of degree p [FrJ08, Lemma 22.7.1].

We apply a Möbius transformation on $C(x)$, if necessary, to assume that the pole $\mathfrak{p}_{x,\infty}$ of x belongs to S. Then $C(x)_{\{\mathfrak{p}_{x,\infty}\}}^{(p)} \subseteq C(x)_S^{(p)}$. Therefore, we may further assume that $S = \{\mathfrak{p}_{x,\infty}\}$. Since C is infinite, the dimension of C as a vector space over $\mathbb{F}_p$ is m. Let B be a basis of C over $\mathbb{F}_p$. For each $b \in B$ let y_b be an element of $C(x)_s$ such that $y_b^p - y_b = bx$. Then $C(x, y_b)$ is a cyclic extension of degree p. Moreover, since $\mathfrak{p}_{x,\infty}$ is the only pole of bx, no prime divisor of $C(x)/C$ but $\mathfrak{p}_{x,\infty}$ is ramified in $C(x, y_b)$. This means that $C(x, y_b) \subseteq C(x)_S^{(p)}$. To conclude the proof we have now to prove that the elements of Bx are linearly independent over $\mathbb{F}_p$ modulo $\wp(C(x))$ (Statement (3d) of Section 9.5).

To that end consider distinct elements $b_1, \ldots, b_n$ of B and arbitrary elements $\beta_1, \ldots, \beta_n \in \mathbb{F}_p$. Assume there exists $u \in C(x)$ with

$$(1) \qquad \sum_{i=1}^{n} \beta_i b_i x = u^p - u.$$

Then u is integral over $C[x]$, and because $C[x]$ is integrally closed, $u \in C[x]$. If $\deg(u) \geq 1$, then the degree of the right hand side of (1) is greater than 1 while the degree of the left hand side of (1) is 1. If $\deg(u) = 0$, then $\sum_{i=1}^{n} \beta_i b_i = 0$. Consequently, $\beta_i = 0$ for each i, as contended.

PART B: *The general case.* We choose a transcendental element x of E over C and denote the set of prime divisors of $C(x)/C$ lying under S by T. By Part A, $\mathrm{rank}(\mathrm{Gal}(C(x)_T^{(p)}/C(x))) = m$. Since E is a finite extension of $C(x)$, so is $E_0 = C(x)_T^{(p)} \cap E$, hence $\mathrm{Gal}(C(x)_T^{(p)}/E_0)$ is an open subgroup of $\mathrm{Gal}(C(x)_T^{(p)}/C(x))$, hence $\mathrm{rank}(\mathrm{Gal}(C(x)_T^{(p)}/E_0)) = m$ [FrJ08, Cor. 17.1.5]. Now observe that $C(x)_T^{(p)} \subseteq E_S^{(p)}$, so $\mathrm{Gal}(C(x)_T^{(p)}/E_0)$ is a quotient of the group $\mathrm{Gal}(E_S^{(p)}/E)$. Therefore, by [FrJ08, Cor. 17.1.4], $\mathrm{rank}(\mathrm{Gal}(E_S^{(p)}/E)) \geq m$, as contended. $\qquad\square$

LEMMA 9.9.3: *Let C be an algebraically closed field, E a function field of one variable over C, S a finite set of prime divisors of E/C, and l a prime number that does not divide $\mathrm{char}(C)$. Then $\mathrm{rank}(\mathrm{Gal}(E_S^{(l)}/E)) < \infty$.*

Proof: Let E' be the compositum of all cyclic extensions of E of degree l in $E_S^{(l)}$. Since $\mathrm{Gal}(E_S^{(l)}/E')$ is a pro-l group and $\mathrm{Gal}(E_S^{(l)}/E')$ is the Frattini subgroup of $\mathrm{Gal}(E_S^{(l)}/E)$, the rank of $\mathrm{Gal}(E_S^{(l)}/E)$ is equal to that of

$\mathrm{Gal}(E'/E)$ [FrJ08, Lemma 22.7.4]. Thus, it suffices to prove that $\mathrm{Gal}(E'/E)$ is finite.

Let S' be the complement of S in the set of all prime divisors of E/C. Denote the subgroup of $E^\times$ consisting of all elements x satisfying $l|v_\mathfrak{p}(x)$ for all $\mathfrak{p} \in S'$ by B'. Then $E' = E(x^{1/l} \mid x \in B')$ [FrJ08, Example 2.3.8]. Let B be the subgroup of $E^\times/(E^\times)^l$ consisting of all $x(E^\times)^l$ with $x \in B'$. By Kummer theory [Lan93, p. 294, Thm. 8.1], $B \cong \mathrm{Gal}(E'/E)$. Thus, we have to prove that B is finite.

To that end consider the map $\nu \colon B \to (\mathbb{Z}/l\mathbb{Z})^S$ defined by

$$\nu(x(E^\times)^l) = (v_\mathfrak{p}(x) + l\mathbb{Z})_{\mathfrak{p} \in S}.$$

Then $\mathrm{Ker}(\nu)$ consists of all left classes $x(E^\times)^l$ such that $l|v_\mathfrak{p}(x)$ for all prime divisors of E/C. By Lemma 9.6.2, $\mathrm{Ker}(\nu)$ is finite. Since $(\mathbb{Z}/l\mathbb{Z})^S$ is finite, it follows that B is also finite. $\square$

PROPOSITION 9.9.4: *Let C be an algebraically closed field of positive characteristic, E a function field of one variable over C, and S a set of prime divisors of E/C with $\mathrm{card}(S) < \mathrm{card}(C)$. Suppose that S is nonempty or E is not rational. Then $\mathrm{Gal}(E_S/E)$ is not a free profinite group.*

Proof: If S is empty, then $E_S = E_{\mathrm{ur}}$. By assumption, $\mathrm{genus}(E) > 0$, so $\mathrm{Gal}(E_S/E)$ is not projective (Proposition 9.6.1). It follows that $\mathrm{Gal}(E_S/E)$ is not free [FrJ08, Cor. 22.4.5].

Assume S is nonempty and $\mathrm{Gal}(E_S/E)$ is a free profinite group of rank m. Then, for each prime number l, the maximal pro-l quotient $\mathrm{Gal}(E_S^{(l)}/E)$ of $\mathrm{Gal}(E_S/E)$ is a free pro-l group of rank m [FrJ08, Lemma 17.4.10]. Applying Lemma 9.9.2 for the case $l = \mathrm{char}(C)$, we conclude that

$$m = \mathrm{rank}(\mathrm{Gal}(E_S^{(l)}/E)) = \mathrm{card}(C)$$

is infinite.

On the other hand consider the case $l \neq \mathrm{char}(C)$. If S is finite, then $\mathrm{rank}(\mathrm{Gal}(E_S^{(l)}/E)) < \infty$ (Lemma 9.9.3). This contradicts the conclusion of the preceding paragraph.

If S is infinite, then $\aleph_0 \leq \mathrm{card}(S) < m$. Let $\mathcal{A}$ be the collection of all finite subsets of S. Then $\mathrm{card}(\mathcal{A}) = \mathrm{card}(S)$ and $E_S^{(l)} = \bigcup_{A \in \mathcal{A}} E_A^{(l)}$. Since m is infinite, $\mathrm{rank}(\mathrm{Gal}(E_S^{(l)}/E))$ is equal to the cardinality of the set of all finite extensions of E in $E_S^{(l)}$ [FrJ08, Prop. 17.1.2]. Each of these extensions is contained in $E_A^{(l)}$ for some $A \in \mathcal{A}$. For each $A \in \mathcal{A}$, E has at most countably many finite extensions in $E_A^{(l)}$ (because $\mathrm{rank}(\mathrm{Gal}(E_A^{(l)}/E)) < \infty$). It follows that $\mathrm{rank}(\mathrm{Gal}(E_S/E)) \leq \mathrm{card}(\mathcal{A})\aleph_0 = \mathrm{card}(S)\aleph_0 < m$. Again, this is a contradiction to the conclusion of the second paragraph of the proof. We conclude from this contradiction that $\mathrm{Gal}(E_S/E)$ is not a free profinite group. $\square$

Remark 9.9.5: Non-isomorphic fundamental groups. Let E be a function field of one variable of genus $g > 0$ over an algebraically closed field of characteristic p and let S be a finite set of prime divisors of r elements. By Proposition 9.1.6, $\mathrm{Gal}(E_S/E)$ is uniquely determined up to an isomorphism by r and g if $p = 0$. This is not the case if $p > 0$.

Indeed, for $p \neq 0, 2$ consider elements a and a' in $\tilde{\mathbb{F}}_p$ such that the elliptic curve Γ defined over $\tilde{\mathbb{F}}_p$ with j-invariant a is ordinary and the elliptic curve Γ' defined over $\tilde{\mathbb{F}}_p$ with j-invariant a' is supersingular. Let F (resp. F') be the function field of Γ (resp. Γ'). Then $\mathrm{Gal}(F_{\mathrm{ur}}/F) \not\cong \mathrm{Gal}(F'_{\mathrm{ur}}/F')$ although genus$(F/\tilde{\mathbb{F}}_p) = 1 = $ genus$(F'/\tilde{\mathbb{F}}_p)$. Indeed, F has a unique unramified $\mathbb{Z}/p\mathbb{Z}$-extension while F' has none [Hrb77, p. 338, Exercise 4.8].

Similarly, let $E = \tilde{\mathbb{F}}_p(x)$, $S = \{0, 1, \infty, a\}$ and $S' = \{0, 1, \infty, a'\}$ with distinct $a, a' \in \tilde{\mathbb{F}}_p$. By [Hrb94b, Thm. 1.8], $\mathrm{Gal}(E_S/E) \not\cong \mathrm{Gal}(E_{S'}/E)$ although they have the same invariants, $r = 4$ and $g = 0$. $\qquad\square$

Notes

The proof of Proposition 9.1.1 and its generalization, Proposition 9.1.2, uses non-algebraic tools such as algebraic topology and the theory of Riemann surfaces, so it goes beyond the scope of this book. For a complete detailed proof of Proposition 9.1.1 we refer the reader to Helmut Völklein's book [Voe96]. See also [MaM99, Chap. 1, Thms. 1.3 and 1.4]. A proof of Proposition 9.1.2 can be found in [Dou79], [Matz87, p. 30, Satz 1], and [Ser92, Section 6.2]. Proposition 9.1.5 is reduced to Proposition 9.1.2 via Proposition 9.1.4. This reduction goes also under the name of "Grothendieck specialization theorem". Standard projective limit argument allows us to deduce Proposition 9.1.6 from Proposition 9.1.5. This transition appears also in [Matz87, p. 37, Satz 3]. Corollary 9.1.10 is due to Douady. It is a special case of Proposition 9.1.9. The proof of the latter theorem applies Proposition 9.1.6 and a projective limit argument over all finite subsets of S. Although we do not include information about the decomposition groups in Proposition 9.1.9 that information enters into the limit argument in an essential way. We have borrowed that ingredient of the proof from the proof of [Rib70, p. 70, Thm. 8.1].

A survey of Abhyankar's conjecture, Raynaud's proof of the conjecture for the affine line, and Pop's reduction to Raynaud's result appears in [MaM99, Sections 5.2, 5.3, and 5.4].

Section 9.3 surveys the cohomology of groups and Galois cohomology to the extent needed in the book. Our main source is [Rib70].

Section 9.4 reproduces [Jar99, Sec. 1], which by itself puts together well known arguments. The standard proof of Lemma 9.4.4 uses a special case of cohomological triviality: Let G be a finite group and let A be a G-module. If $\hat{H}^0(G, A) = A^G/NA = 0$ (where $Na = \Sigma_{\sigma \in G}\sigma a$) and $H^1(G, A) = 0$, then $H^2(G, A) = 0$ [CaF67, p. 113, Thm. 9]. In our case, $G = \mathrm{Gal}(N/K)$, $A = N^\times$ and $A^G/NA = K^\times/\mathrm{norm}_{N/K}N^\times = 1$. Also, $H^1(G, N^\times) = 1$, by Hilbert's theorem 90. So, indeed, $H^2(G, A) = 1$. Replacing cohomological triviality in the proof of Lemma 9.4.4 by the more elementary argument is due to Sigrid Böge (private communication).

Proposition 9.4.6 is usually referred to as **Tsen's theorem**, because Tsen proved the essential ingredient of its proof, namely that the field of ratioanl functions over an algebraically closed field is C_1 [Tse33].

The proof that $\mathrm{Gal}(E_S/E)$ is projective in Section 9.5 is based on tips of Heinrich Matzat.

Lemma 9.5.2 is a rewrite of [Son94, Lemma 2.6]. Lemma 9.5.3 is due to Shafarevich [Sha89, p. 109]. See also [Son94, Prop. 2.5]. The proof of Lemma 9.5.4 is a modification of the proof of [Son94, Prop. 3.2]. Theorem 9.5.7 is proved by Serre [Ser90, Prop. 1], using étale cohomology.

The second proof of Proposition 9.6.1 that do not use Riemann's existence theorem arose from discussions with Gerhard Frey. The same goes for the proof of Lemma 9.6.2.

Harbater proves Theorem 9.8.5 in the case where $C \cup \{\infty\} \smallsetminus S$ is finite by formal patching [Har95, Thm. 4.1]. Pop proves Theorem 9.8.5 in its full strength by rigid methods [Pop05, p. 556, Cor.]. We follow [HaJ00a]. Corollary 9.4.9 is a special case of Theorem 9.8.5 and Theorem 9.4.8 is a slight generalization of Corollary 9.4.9.

Shafarevich discussed his conjecture on the freeness of $\mathrm{Gal}(\mathbb{Q}_{\mathrm{ab}})$ during a talk in Oberwolfach in 1964. Latter it appeared in [Bey80].

A Galois theoretic version of Lemma 9.8.4 appears in [Matz87, p. 231, Lemma 1].

Chapter 10.
Semi-Free Profinite Groups

We have already pointed out that a profinite group G of an infinite rank m is free of rank m if (and only if) G is projective and every finite split embedding problem for G with a nontrivial kernel has m solutions (Proposition 9.4.7). Dropping the condition on G to be projective leads to the notion of a "quasi-free profinite group" (Section 10.6).

A somewhat stronger condition is that of a "semi-free profinite group". We say that G is **semi-free** if every finite split embedding problem for G with a nontrivial kernel has m independent solutions (Definition 10.1.5). The advantage of the latter notion on the former one is that the known conditions on a closed subgroup of a free profinite group of rank m to be free of rank m go over to semi-free groups. Indeed, even the method of proof that applies twisted wreath products goes over from free profinite groups to semi-free profinite groups (Section 10.3). Thus, every open subgroup of a semi-free group G is semi-free (Lemma 10.4.1), every normal closed subgroup N of G with G/N Abelian is semi-free, every proper open subgroup of a closed normal subgroup of G is semi-free, and in general every closed subgroup M of G that is "contained in a diamond" is semi-free (Theorem 10.5.3).

An application of the diamond theorem to function fields of one variable over PAC fields appears in the next chapter.

10.1 Independent Subgroups

We introduce a group theoretic counterpart of "linear disjointness of fields".

Definition 10.1.1: Independent subgroups. Let F be a profinite group.
(a) Open subgroups $M_1, \ldots, M_n$ of F are F**-independent**, if

$$\left(F : \bigcap_{i=1}^{n} M_i\right) = \prod_{i=1}^{n} (F : M_i).$$

(b) A family $\mathcal{M}$ of open subgroups of F is F**-independent** if $M_1, \ldots, M_n$ are F-independent, for all distinct $M_1, \ldots, M_n \in \mathcal{M}$. □

Remark 10.1.2:
(a) Let F be a profinite group, μ the normalized Haar measure of F, and $M_1, \ldots, M_n$ open subgroups of F. Then $M_1, \ldots, M_n$ are F-independent if and only if $M_1, \ldots, M_n$ are μ-independent, that is $\mu(\bigcap_{i=1}^{n} M_i) = \prod_{i=1}^{n} \mu(M_i)$ [FrJ08, Lemma 18.3.7].
(b) Suppose in the notation of (a) that $M_1, \ldots, M_n$ are normal. Then $M_1, \ldots, M_n$ are F-independent if and only if $F/\bigcap_{i=1}^{n} M_i \cong \prod_{i=1}^{n} F/M_i$.

M. Jarden, *Algebraic Patching*, Springer Monographs in Mathematics,
DOI 10.1007/978-3-642-15128-6_10, © Springer-Verlag Berlin Heidelberg 2011

(c) Let $L_1, \ldots, L_n$ be finite separable extensions of a field K. Then $\mathrm{Gal}(L_1), \ldots, \mathrm{Gal}(L_n)$ are $\mathrm{Gal}(K)$-independent if and only if $L_1, \ldots, L_n$ are linearly disjoint over K [FrJ08, Lemma 18.5.1].   □

The following basic rules for F-independency of open subgroups of a profinite group F can be deduced from the corresponding properties of linear disjointness [FrJ08, Section 2.5], using a realization of F as a Galois group of a Galois extension [FrJ08, Cor. 1.3.4]. We give here direct proofs.

LEMMA 10.1.3: *Let $M_1, \ldots, M_n$ be open subgroups of a profinite group F. Then:*

(a) $(F : \bigcap_{i=1}^n M_i) \le \prod_{i=1}^n (F : M_i)$.

(b) *M_1, M_2 are F-independent if and only if $(M_1 : M_1 \cap M_2) = (F : M_2)$.*

(c) *Suppose $M_1 \triangleleft F$. Then M_1, M_2 are F-independent if and only if $F = M_1 M_2$.*

(d) *Let $M_1 \le N_1 \le F$. Then M_1, M_2 are F-independent if and only if N_1, M_2 are F-independent and $M_1, N_1 \cap M_2$ are N_1-independent (the* **tower property***).*

(e) *$M_1, \ldots, M_n$ are F-independent if and only if $M_1, \ldots, M_{n-1}$ are F-independent and $\bigcap_{i=1}^{n-1} M_i, M_n$ are F-independent.*

(f) *Suppose $M_i \le N_i \le F$ for $i = 1, \ldots, n$. If $M_1, \ldots, M_n$ are F-independent, then so are $N_1, \ldots, N_n$.*

Proof of (a): The map $F/\bigcap_{i=1}^n M_i \to \prod_{i=1}^n F/M_i$ of quotient spaces defined by $f \bigcap_{i=1}^n M_i \mapsto (fM_1, \ldots, fM_n)$ is injective, hence (a) is true.

Proof of (b): The statement follows from the identity $(F : M_1)(M_1 : M_1 \cap M_2) = (F : M_1 \cap M_2)$.

Proof of (c): The assumption $M_1 \triangleleft F$ implies that $(M_2 : M_1 \cap M_2) = (M_1 M_2 : M_1) \le (F : M_1)$. Now we apply (b) with the indices 1 and 2 exchanged to conclude (c).

Proof of (d): First assume that N_1, M_2 are F-independent and $M_1, N_1 \cap M_2$ are N_1-independent. Then, by (b),

(1)
$$(N_1 : N_1 \cap M_2) = (F : M_2) \text{ and } (N_1 : M_1 \cap M_2) = (N_1 : M_1)(N_1 : N_1 \cap M_2).$$

Hence,

$$
\begin{aligned}
(F : M_1 \cap M_2) &= (F : N_1)(N_1 : M_1 \cap M_2) \\
&= (F : N_1)(N_1 : M_1)(N_1 : N_1 \cap M_2) = (F : M_1)(F : M_2),
\end{aligned}
$$

so M_1, M_2 are F-independent.

Conversely, suppose $(F : M_1 \cap M_2) = (F : M_1)(F : M_2)$. Then, by (a),

$$
\begin{aligned}
(F : M_1)(F : M_2) &= (F : M_1 \cap M_2) = (F : N_1)(N_1 : M_1 \cap M_2) \\
&\le (F : N_1)(N_1 : M_1)(N_1 : N_1 \cap M_2) \le (F : M_1)(F : M_2).
\end{aligned}
$$

Hence, $(N_1 : M_1 \cap M_2) = (N_1 : M_1)(N_1 : N_1 \cap M_2)$, so by definition, $M_1, N_1 \cap M_2$ are N_1-independent. Also, $(N_1 : N_1 \cap M_2) = (F : M_2)$, so by (b), N_1, M_2 are F-independent.

Proof of (e):   First we suppose $M_1, \ldots, M_{n-1}$ are F-independent and $\bigcap_{i=1}^{n-1} M_i, M_n$ are F-independent. Then, by (b),

$$
\begin{aligned}
&(F : M_1 \cap \cdots \cap M_n) \\
&= (F : M_1 \cap \cdots \cap M_{n-1})(M_1 \cap \cdots \cap M_{n-1} : M_1 \cap \cdots \cap M_n) \\
&= (F : M_1) \cdots (F : M_{n-1})(F : M_n),
\end{aligned}
$$

so $M_1, \ldots, M_n$ are F-independent.

Conversely, suppose $M_1, \ldots, M_n$ are F-independent. Then, by (a),

$$
\prod_{i=1}^{n}(F : M_i) = (F : \bigcap_{i=1}^{n} M_i) \le (F : \bigcap_{i=1}^{n-1} M_i)(F : M_n) \le \prod_{i=1}^{n-1}(F : M_i) \cdot (F : M_n),
$$

hence $(F : \bigcap_{i=1}^{n} M_i) = (F : \bigcap_{i=1}^{n-1} M_i)(F : M_n)$, as desired.

Proof of (f):   By definition, $M_1, \ldots, M_{n-1}$ are F-independent. Hence, by an induction hypothesis, $N_1, \ldots, N_{n-1}$ are F-independent. By the tower property (d), $N_n, \bigcap_{i=1}^{n-1} M_i$ are F-independent. Hence, by (d) again, $N_n, \bigcap_{i=1}^{n-1} N_i$ are F-independent. It follows from (e) that $N_1, \ldots, N_n$ are F-independent. $\square$

LEMMA 10.1.4: *Let* $\mathcal{M} = (M_\alpha \mid \alpha < \lambda)$ *be a transfinite sequence of open normal subgroups of a profinite group* F. *Suppose* $M_\kappa \bigcap_{\alpha < \kappa} M_\alpha = F$ *for each* $\kappa < \lambda$. *Then* $\mathcal{M}$ *is* F-*independent.*

Proof: Let $\alpha_1 < \cdots < \alpha_n$ be ordinal numbers smaller than λ. By assumption, $\bigcap_{\alpha < \alpha_n} M_\alpha, M_{\alpha_n}$ are F-independent. Hence, by Lemma 10.1.3(d), $\bigcap_{i=1}^{n-1} M_{\alpha_i}, M_{\alpha_n}$ are F-independent. By an induction hypothesis on n,

$$
M_{\alpha_1}, \ldots, M_{\alpha_{n-1}}
$$

are F-independent.   Hence, by Lemma 10.1.3(e), $M_{\alpha_1}, \ldots, M_{\alpha_n}$ are F-independent. Consequently, $\mathcal{M}$ is F-independent. $\square$

Definition 10.1.5: Semi-free profinite group. Solutions of an embedding problem

$$
\text{(2)} \qquad\qquad (\varphi \colon F \to A, \; \alpha \colon B \to A)
$$

of a profinite group F are **independent** if their kernels are $\mathrm{Ker}(\varphi)$-independent.

Note that if the kernel of (2) is trivial, then α is an isomorphism and $\psi = \alpha^{-1} \circ \varphi$ is a solution. In this case, if for each i in a set I we set

$\psi_i = \psi$, then $\mathrm{Ker}(\psi_i) = \mathrm{Ker}(\varphi)$, so $\{\mathrm{Ker}(\psi_i) \mid i \in I\}$ are $\mathrm{Ker}(\psi)$-independent. Therefore, $\{\psi_i \mid i \in I\}$ is a set of independent solutions of (2).

A profinite group F of infinite rank m is **semi-free** if every finite split embedding problem for F with a nontrivial kernel has m independent solutions. $\qquad\square$

Remark 10.1.6: Let M be a profinite group and m an infinite cardinal number. Suppose each finite split embedding problem with a nontrivial kernel of M has m independent solutions. In particular, if G is a nontrivial finite group, then the embedding problem $(M \to 1, G \to 1)$ has m independent solutions. Thus, M has m independent open normal subgroups M_i with $M/M_i \cong G$. In particular $\mathrm{rank}(M) \geq m$ [FrJ08, Lemma 17.1.2]. It follows that if in addition, $\mathrm{rank}(M) \leq m$, then $\mathrm{rank}(M) = m$ and M is semi-free. $\square$

LEMMA 10.1.7: *Let F be a profinite group and $\mathcal{M}$ an infinite family of pairwise F-independent normal open subgroups of F. Then $\mathcal{M}$ contains an F-independent subfamily $\mathcal{M}_0$ of cardinality $\mathrm{card}(\mathcal{M})$.*

Proof: By Zorn's lemma, $\mathcal{M}$ has a maximal F-independent subfamily $\mathcal{M}_0$. We prove that $\mathrm{card}(\mathcal{M}_0) = \mathrm{card}(\mathcal{M})$.

Otherwise, $\mathrm{card}(\mathcal{M}_0) < \mathrm{card}(\mathcal{M})$. Let $\mathcal{M}_1$ be the family of all finite intersections of elements of $\mathcal{M}_0$. If $\mathcal{M}_0$ is finite, then so is $\mathcal{M}_1$. If $\mathcal{M}_0$ is infinite, then $\mathrm{card}(\mathcal{M}_1) = \mathrm{card}(\mathcal{M}_0)$. In both cases, $\mathrm{card}(\mathcal{M}_1) < \mathrm{card}(\mathcal{M})$.

Next we denote the family of all subgroups of F that contain a group belonging to $\mathcal{M}_1$ by $\mathcal{M}_2$. Again, if $\mathcal{M}_1$ is finite, then so is $\mathcal{M}_2$. If $\mathcal{M}_1$ is infinite, then $\mathrm{card}(\mathcal{M}_2) = \mathrm{card}(\mathcal{M}_1)$. In both cases, $\mathrm{card}(\mathcal{M}_2) < \mathrm{card}(\mathcal{M})$.

By Lemma 10.1.3(d), each open proper subgroup of F contains at most one group $M \in \mathcal{M}$. Hence, there exists $M \in \mathcal{M}$ not contained in any proper subgroup of F that belongs to $\mathcal{M}_2$. We claim that the family $\mathcal{M}_0 \cup \{M\}$ is F-independent.

To prove the claim we consider $M_1, \ldots, M_n \in \mathcal{M}_0$. Then, $M_1, \ldots, M_n$ are F-independent. Moreover, $M \bigcap_{i=1}^n M_i \in \mathcal{M}_2$. Hence, by the choice of M we have $M \bigcap_{i=1}^n M_i = F$. By Lemma 10.1.3(c), $M, \bigcap_{i=1}^n M_i$ are F-independent. Therefore, by Lemma 10.1.3(e), $M_1, \ldots, M_n, M$ are F-independent. This completes the proof of the claim of the preceding paragraph and gives the desired contradiction to the maximality of $\mathcal{M}_0$. $\qquad\square$

Taking into account Remark 10.1.6, Lemma 10.1.7 yields the following result:

COROLLARY 10.1.8: *Let m be an infinite cardinal number and F a profinite group of rank at most m. Then F is semi-free of rank m if and only if every finite split embedding problem with a nontrivial kernel has m pairwise independent solutions.*

Definition 10.1.9: *Weight.* Let M be a closed subgroup of a profinite group F. We define the **weight** of the quotient space F/M as 1 if M is open and as the cardinality of the set of all open subgroups of F that contain M if $(F : M) = \infty$.

Let m be an infinite cardinal number. If $L = \bigcap_{i \in I} L_i$, L_i is a closed subgroup of F, weight$(F/L_i) < m$ for each $i \in I$, and card$(I) < m$, then weight$(F/L) < m$ [FrJ08, Lemma 25.2.1(b)].

If N is a closed subgroup of a closed subgroup M of F, and weight(F/M), weight$(M/N) < m$, then weight$(F/N) < m$ [FrJ08, Lemma 25.2.1(d)]. $\square$

Definition 10.1.10: Small quotient spaces. Let M be a closed subgroup of a profinite group F. We denote the set of all open subgroups of F that contain M by Open(F/M). We say that the quotient space F/M is **small** if for each positive integer n, the set Open(F/M) has only finitely many groups of index at most n. In particular if $M \lhd F$, then the quotient group F/M is small in the sense of [FrJ08, p. 329].

Note that if M' is a closed subgroup of M and F/M' is small, then so is F/M. In particular, if $M' \lhd F$ and F/M' is finitely generated, then F/M' is a small group [FrJ08, Lemma 16.10.2], so F/M is also a small quotient space.

Let $\hat{M} = \bigcap_{\sigma \in F} M^\sigma$ be the **normal core** of M in F. If $F/\hat{M}$ is finitely generated, then by the preceding paragraph, F/M is small. If in addition N is an open normal subgroup of M, then there is an open normal subgroup L of F such that $L \cap \hat{M} \le N \cap \hat{M}$. Since $F/(L \cap \hat{M})$ embeds into $F/L \times F/\hat{M}$ and the latter group is finitely generated, $F/(L \cap \hat{M})$ is small, so F/N is small. $\square$

LEMMA 10.1.11: *Let F be a profinite group, M a closed subgroup of F, and $\mathcal{E}$ an F-independent infinite family of open normal subgroups of F. If*
(a) *weight$(F/M) < $ card$(\mathcal{E})$, or*
(b) *F/M is small and there exists a positive integer n such that $(F : E) \le n$ for each $E \in \mathcal{E}$,*

then there exists $E \in \mathcal{E}$ such that $EM = F$.

Proof: Every proper open subgroup F_0 of F contains at most one group of $\mathcal{E}$. In Case (a), card$($Open$(F/M)) < $ card$(\mathcal{E})$. Hence, there exists $E \in \mathcal{E}$ which is contained in no proper open subgroup of F that contains M. Therefore, $EM = F$.

In Case (b), $(F : EM) \le (F : E) \le n$, so only finitely many $E \in \mathcal{E}$ satisfy $EM < F$. Since $\mathcal{E}$ is infinite, there exists $E \in \mathcal{E}$ with $EM = F$. $\square$

LEMMA 10.1.12: *Let F be a profinite group, M an open normal subgroup, and $\mathcal{E}$ an infinite F-independent family of open normal subgroups of F. Then, for each n there exist $E_1, \ldots, E_n \in \mathcal{E}$ such that $E_1, \ldots, E_n, M$ are F-independent.*

Proof: We assume inductively that $E_1, \ldots, E_{n-1} \in \mathcal{E}$ and $E_1, \ldots, E_{n-1}, M$ are F-independent. Then $M' = E_1 \cap \cdots E_{n-1} \cap M$ is an open subgroup of F. By Lemma 10.1.11, there exists $E_n \in \mathcal{E}$ with $E_n M' = F$. Hence, by Lemma 10.1.3, $E_1, \ldots, E_n, M$ are F-independent. This concludes the induction. $\square$

We conclude this section with two results that line the notion of a "semi-free profinite group" to that of a "free profinite group".

PROPOSITION 10.1.13: *Let F be a free profinite group of infinite rank m. Then every finite embedding problem (2) for F with a nontrivial kernel has m independent solutions. In particular, F is semi-free.*

Proof: Let $\lambda < m$ be an ordinal number and assume $\{\gamma_\kappa \mid \kappa < \lambda\}$ are independent solutions of (2). We prove the existence of a solution γ_λ such that the set of solutions $\{\gamma_\kappa \mid \kappa \leq \lambda\}$ of (2) is independent. Applying transfinite induction and Lemma 10.1.4, this will give m independent solutions of (2).

Let $E = \mathrm{Ker}(\varphi)$. Since $\alpha \circ \gamma_\kappa = \varphi$, we have $\mathrm{Ker}(\gamma_\kappa) \leq E$, so $M = \bigcap_{\kappa < \lambda} \mathrm{Ker}(\gamma_\kappa) \leq E$. Since $\mathrm{card}(\lambda) < m$ and the $\mathrm{Ker}(\gamma_\kappa)$'s are open in F, $\mathrm{weight}(F/M) < m$ (Definition 10.1.9). Set $\hat{A} = F/M$ and let $\hat{\varphi} \colon F \to \hat{A}$ be the quotient map. Then there exists an epimorphism $\bar{\varphi} \colon \hat{A} \to A$ such that $\bar{\varphi} \circ \hat{\varphi} = \varphi$. This gives rise to a commutative diagram

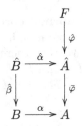

in which the square is cartesian. By [FrJ08, Lemma 25.1.3], every finite embedding problem for F has m solutions. Hence, by [FrJ08, Lemma 25.1.5], there exists an epimorphism $\hat{\gamma} \colon F \to \hat{B}$ such that $\hat{\alpha} \circ \hat{\gamma} = \hat{\varphi}$. Thus, $\gamma = \hat{\beta} \circ \hat{\gamma}$ satisfies $\alpha \circ \gamma = \varphi$, that is γ is a solution of (2).

Let $D = \mathrm{Ker}(\gamma)$ and $N = \mathrm{Ker}(\hat{\gamma})$. Then $D \cap M = N$, because if $x \in D \cap M$, then $\hat{\beta}(\hat{\gamma}(x)) = \gamma(x) = 1$ and $\hat{\alpha}(\hat{\gamma}(x)) = \hat{\varphi}(x) = 1$, hence $\hat{\gamma}(x) = 1$, so $x \in N$.

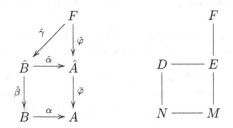

Then $E/D \cong \mathrm{Ker}(\alpha)$ and $M/N \cong \mathrm{Ker}(\hat{\alpha})$. By [FrJ08, Lemma 22.2.5], $\mathrm{Ker}(\alpha) \cong \mathrm{Ker}(\hat{\alpha})$. Hence, $(E : D) = (M : N)$. Therefore, by Lemma 10.1.3(b), D, M are E-independent. Consequently, γ is independent of the set of solutions $\{\gamma_\kappa \mid \kappa < \lambda\}$.   □

The converse of Proposition 10.1.13 is a reformulation of Proposition 9.4.7.

PROPOSITION 10.1.14: *Let F be a projective semi-free profinite group of infinite rank m. Then F is a free profinite group.*

10.2 Fiber Products

Fiber products of pairs of profinite groups are introduced in [FrJ08, Sec. 13.7]. Here we consider fiber products of finitely many profinite groups and use them in the next section to construct wreath products.

Definition 10.2.1: Fiber products. For each $1 \le i \le n$ let $\alpha_i \colon H_i \to G$ be an epimorphism of profinite groups. The **fiber product** $\prod_G H_i = H_1 \times_G \cdots \times_G H_n$ with respect to the α_i's is the group

$$\prod_G H_i = \{(h_1, \ldots, h_n) \in \prod_{i=1}^{n} H_i \mid \alpha_1(h_1) = \cdots = \alpha_n(h_n)\}.$$

For each $g \in G$ we can choose $h_i \in H_i$ with $\alpha_i(h_i) = g$. Hence, the projection $\mathrm{pr}_j \colon \prod_G H_i \to H_j$ on the jth coordinate is surjective. It follows that the homomorphism $\alpha^{(n)} = \alpha_j \circ \mathrm{pr}_j \colon \prod_G H_i \to G$ is independent of j and is surjective.

If all of the H_i's are the same group H and all of the α_i's are the same map α, we also write H_G^n for $\prod_G H_i$.

Note that the fiber product is associative in the following sense: There is a natural isomorphism $\prod_G H_i \cong (H_1 \times_G \cdots \times_G H_m) \times_G (H_{m+1} \times_G \cdots \times_G H_n)$ for each $1 \le m \le n$. $\qquad\qquad\square$

Example 10.2.2: Let $G, A_1, \ldots, A_n$ be profinite groups. Suppose G acts on each A_i. Let $\alpha_i \colon G \ltimes A_i \to G$ be the projection on the first coordinate, $i = 1, \ldots, n$. Then $\prod_G (G \ltimes A_i)$ consists of all n-tuples $(\sigma a_1, \ldots, \sigma a_n)$ with $\sigma \in G$ and $a_i \in A_i$, $i = 1, \ldots, n$. For each j, the image of such an n-tuple under $\alpha_j \circ \mathrm{pr}_j$ is σ. On the other hand, G acts on $\prod A_i$ componentwise and the projection $G \ltimes \prod A_i \to G$ on the first coordinate maps an element $\sigma(a_1, \ldots, a_n)$ of $G \ltimes \prod A_i$ onto σ. We may therefore identify $G \ltimes \prod A_i$ with $\prod_G (G \ltimes A_i)$ and the projection on the first coordinate with $\alpha^{(n)}$.

In particular, if all of the A_i's are the same group A, we have $(G \ltimes A)_G^n = G \ltimes A^n$ and $\alpha^{(n)} \colon (G \ltimes A)_G^n \to G$ is the projection on the first coordinate. $\square$

A key property of fiber products, in our setting, is that weak solutions ψ_i of embedding problems $(\varphi \colon F \to G, \alpha_i \colon H_i \to G)$, $i = 1, \ldots, n$, give rise to a canonical weak solution, $\psi = \prod_{i=1}^{n} \psi_i$, of the embedding problem $(\varphi \colon F \to G, \alpha^{(n)} \colon \prod_G H_i \to G)$. The homomorphism $\psi \colon F \to \prod_G H_i$ is defined by $\psi(x) = (\psi_1(x), \ldots, \psi_n(x))$. In particular, taking $F = G$ and $\varphi = \mathrm{id}$, we find that if all of the α_i's split, so does $\alpha^{(n)}$.

In the case where all of the H_i's are the same group H and all of the α_i's are the same map α, we prove that independency of solutions $\psi_1, \ldots, \psi_n$ of embedding problems is equivalent to a solution of the corresponding embedding problem onto H_G^n.

LEMMA 10.2.3: Let $\mathcal{E} = (\varphi \colon F \to G, \ \alpha \colon H \to G)$ be a finite embedding problem of a profinite group F and $\psi_1, \ldots, \psi_n$ weak solutions of $\mathcal{E}$. Then $\psi_1, \ldots, \psi_n$ are independent solutions if and only if the weak solution $\psi \colon F \to H_G^n$ of the embedding problem $\mathcal{E}_n = (\varphi \colon F \to G, \ \alpha^{(n)} \colon H_G^n \to G)$ defined by $\psi(x) = (\psi_1(x), \ldots, \psi_n(x))$ is a solution of that problem, that is $\psi(F) = H_G^n$.

Proof: Let $E = \mathrm{Ker}(\varphi)$, $M_i = \mathrm{Ker}(\psi_i)$, and $M = \bigcap_{i=1}^n M_i$. Then $(F : M_i) \leq |H|$, $M = \mathrm{Ker}(\psi)$, and $(F : M) \leq |H_G^n|$. By definition,

$$H_G^n = \{(h_1, \ldots, h_n) \in H^n \mid \alpha(h_1) = \cdots = \alpha(h_n)\} = \bigcup_{g \in G} (\alpha^{-1}(g))^n,$$

so $|H_G^n| = |G|\left(\frac{|H|}{|G|}\right)^n$.

Now we assume that $\psi_1, \ldots, \psi_n$ are independent solutions of $\mathcal{E}$ and prove that ψ is a solution of $\mathcal{E}_n$. Since $\alpha \circ \mathrm{pr}_j \circ \psi = \alpha \circ \psi_j = \varphi$, it suffices to prove that $\psi(F) = H_G^n$. Indeed, $\psi_i(F) = H$ and $(E : M) = \prod_{i=1}^n (E : M_i)$, hence

$$|\psi(F)| = (F : M) = (F : E)(E : M) = (F : E) \prod_{i=1}^n (E : M_i)$$

$$= (F : E) \prod_{i=1}^n \frac{(F : M_i)}{(F : E)} = |G|\frac{|H|^n}{|G|^n} = |H_G^n|.$$

Consequently, $\psi(F) = H_G^n$.

Conversely, suppose ψ is a solution of $\mathcal{E}_n$. Then by Lemma 10.1.3(a),

$$|H_G^n| = |\psi(F)| = (F : E)(E : M) \leq (F : E) \prod_{i=1}^n (E : M_i)$$

$$= (F : E) \prod_{i=1}^n \frac{(F : M_i)}{(F : E)} \leq |G|\frac{|H|^n}{|G|^n} = |H_G^n|,$$

hence $(E : M) = \prod_{i=1}^n (E : M_i)$ and $(F : M_i) = |H|$ for each i. This means that $\psi_1, \ldots, \psi_n$ are independent solutions of $\mathcal{E}$. $\square$

PROPOSITION 10.2.4: If every finite split embedding problem of a profinite group F is solvable, then every finite split embedding problem with a nontrivial kernel has $\aleph_0$ independent solutions. In particular, if $\mathrm{rank}(F) = \aleph_0$, then F is semi-free.

Proof: We consider a finite split embedding problem

$$\mathcal{E} = (\varphi \colon F \to G, \ \alpha \colon H \to G).$$

By induction we suppose $\psi_1, \ldots, \psi_n$ are independent solutions of $\mathcal{E}$. By Lemma 10.2.3, the map $\psi\colon F \to H_G^n$ defined by $\psi(x) = (\psi_1(x), \ldots, \psi_n(x))$ is a solution of the embedding problem $\mathcal{E}_n = (\varphi\colon F \to G,\ \pi\colon H_G^n \to G)$, where $\pi = \alpha \circ \mathrm{pr}_j$ is independent of j.

Let $\pi'\colon H_G^{n+1} \to H$ be the projection on the $(n+1)$th coordinate and $\alpha'\colon H_G^{n+1} \to H_G^n$ the projection on the first n-coordinates. Then we observe that the rectangle in diagram (1) is cartesian (Definition 10.2.1). Since α split, so does α' (comments preceding Lemma 10.2.3). By assumption, there exists an epimorphism $\psi'\colon F \to H_G^{n+1}$ such that $\alpha' \circ \psi' = \psi$.

(1)

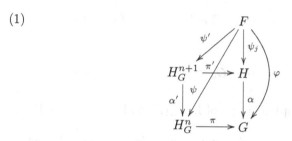

Thus, $\alpha \circ \pi' \circ \psi' = \pi \circ \alpha' \circ \psi' = \pi \circ \psi = \varphi$. It follows that ψ' is a solution of the embedding problem $\mathcal{E}_{n+1} = (\varphi\colon F \to G,\ \alpha \circ \mathrm{pr}_i\colon H_G^{n+1} \to G)$ for each $1 \le i \le n+1$. By Lemma 10.2.3, $\mathcal{E}_{n+1}$ has independent solutions $\psi_1', \ldots, \psi_{n+1}'$ such that $\psi'(x) = (\psi_1'(x), \ldots, \psi_{n+1}'(x))$ for each $x \in F$. Since $\alpha' \circ \psi' = \psi$, we have $\psi_i'(x) = \psi_i(x)$ for $i = 1, \ldots, n$ and every $x \in F$. We set $\psi_{n+1} = \psi_{n+1}'$ to conclude that $\psi_1, \ldots, \psi_n, \psi_{n+1}$ are independent solutions of $\mathcal{E}$, as desired. $\qquad\square$

10.3 Twisted Wreath Products

Following the works [Har99a] and [Har99b], twisted wreath products have been introduced in [FrJ08, Section 13.7] in order to prove the diamond theorem for Hilbertian fields [FrJ08, Thm. 13.8.3] and the diamond theorem for free profinite groups [FrJ08, Thm. 25.4.3]. Here we consider those products once more in order to generalize the latter theorem to a diamond theorem for semi-free profinite groups.

Definition 10.3.1: Twisted wreath product. Let A and G be finite groups and G_0 a subgroup of G. Suppose G_0 acts on A from the right and let $\mathrm{Ind}_{G_0}^G(A)$ be the set of all functions $f\colon G \to A$ such that $f(\sigma\tau) = f(\sigma)^\tau$ for all $\sigma \in G$ and $\tau \in G_0$. We make $\mathrm{Ind}_{G_0}^G(A)$ a group by the rule $(fg)(\sigma) = f(\sigma)g(\sigma)$ and let G acts on $\mathrm{Ind}_{G_0}^G(A)$ by $f^\sigma(\rho) = f(\sigma\rho)$ for all $\sigma, \rho \in G$. This gives rise to the semidirect product $G \ltimes \mathrm{Ind}_{G_0}^G(A)$ that we call the **twisted wreath product** of G and A over G_0 and denote by $G\mathrm{wr}_{G_0}A$. In particular, the map $\sigma f \mapsto \sigma$ for $\sigma \in G$ and $f \in \mathrm{Ind}_{G_0}^G(A)$ is a split epimorphism onto G with kernel $\mathrm{Ind}_{G_0}^G(A)$.

The map $\pi_0\colon \operatorname{Ind}_{G_0}^G(A) \to A$ defined by $\pi_0(f) = f(1)$ is an epimorphism. It commutes with the action of G_0. Indeed, for $f \in \operatorname{Ind}_{G_0}^G(A)$ and $\tau \in G_0$ we have $\pi_0(f^\tau) = f^\tau(1) = f(\tau) = f(1)^\tau = \pi_0(f)^\tau$. Thus, π_0 extend to an epimorphism $\pi\colon G_0 \ltimes \operatorname{Ind}_{G_0}^G(A) \to G_0 \ltimes A$ defined by $\pi(\tau f) = \tau f(1)$ giving rise to the following commutative diagram of short exact sequences:

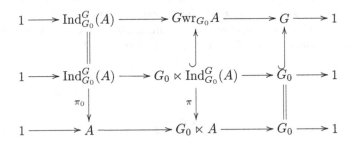

We call π_0 and π the **Shapiro maps** of $\operatorname{Ind}_{G_0}^G(A)$ and $G_0 \ltimes \operatorname{Ind}_{G_0}^G(A)$, respectively.

If B is a normal subgroup of A invariant under G_0, then the action of G_0 on A induces an action on A/B and the quotient map $A \to A/B$ gives rise to epimorphisms $G_0 \ltimes A \to G_0 \ltimes A/B$, $\operatorname{Ind}_{G_0}^G(A) \to \operatorname{Ind}_{G_0}^G(A/B)$, $G_0 \ltimes \operatorname{Ind}_{G_0}^G(A) \to G_0 \ltimes \operatorname{Ind}_{G_0}^G(A/B)$, and $\operatorname{Gwr}_{G_0} A \to \operatorname{Gwr}_{G_0} A/B$. The second and the third maps commute with π. $\qquad\square$

Remark 10.3.2: *Distributive law for twisted wreath products.* Let G be a finite group, G_0 a subgroup, and $A_1, \ldots, A_n$ finite groups. Suppose G_0 acts on each A_i. Then $\operatorname{Ind}_{G_0}^G(\prod A_i) = \prod \operatorname{Ind}_{G_0}^G A_i$ and $\prod_G (\operatorname{Gwr}_{G_0} A_i) \cong \operatorname{Gwr}_{G_0} \prod A_i$, where G_0 acts on $\prod A_i$ componentwise.

Indeed, each element of $\operatorname{Ind}_{G_0}^G(\prod A_i)$ can be identified as an n-tuple $(f_1, \ldots, f_n)$ with $f_i \in \operatorname{Ind}_{G_0}^G A_i$. A combination of this observation with Example 10.2.2 gives $\prod_G (\operatorname{Gwr}_{G_0} A_i) = \operatorname{Gwr}_{G_0} \prod A_i$. Explicitly, each element of $\prod_G (\operatorname{Gwr}_{G_0} A_i)$ has the form $(\sigma f_1, \ldots, \sigma f_n)$, where $\sigma \in G$ and $f_i \in \operatorname{Ind}_{G_0}^G(A_i)$, $i = 1, \ldots, n$. We identify that element with the element $\sigma(f_1, \ldots, f_n)$ of $\operatorname{Gwr}_{G_0} \prod A_i$. $\qquad\square$

The next technical lemma induces finite split embedding problems for a closed subgroup M of a profinite group F to finite split embedding problems for F. Under an additional assumption, independent solutions of the problems of F yield independent solutions of the problems of M.

LEMMA 10.3.3: *Let F be a profinite group, M a closed subgroup and*

$$\mathcal{E}_1(A) = (\mu\colon M \to G_1,\ \alpha_1\colon G_1 \ltimes A \to G_1)$$

a finite split embedding problem for M. Let F_0 be an open subgroup of F, D and L open normal subgroups of F, and N a closed normal subgroup of

F as in the following diagram:

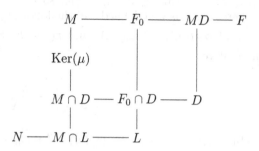

Set $G = F/L$, $G_0 = F_0/L$, and let $\varphi\colon F \to G$ and $\varphi_0\colon F_0 \to G_0$ be the quotient maps.

(a) There exists an epimorphism $\bar{\varphi}_1\colon G_0 = F_0/L \to G_1 = M/\mathrm{Ker}(\mu)$ such that $\mu = \bar{\varphi}_1 \circ \varphi|_M$ (after identifying $M/M \cap L$ with ML/L).

(b) Let $\rho\colon G_0 \ltimes A \to G_1 \ltimes A$ be the extension of $\bar{\varphi}_1$ by id_A. Consider the finite split embedding problem

$$\mathcal{E}(A) = (\varphi\colon F \to G,\ \beta\colon \mathrm{Gwr}_{G_0} A \to G),$$

where G_0 acts on A via $\bar{\varphi}_1$ and β is the projection on the first coordinate. Let $\pi\colon G_0 \ltimes \mathrm{Ind}_{G_0}^{G}(A) \to G_0 \ltimes A$ be the Shapiro map.

For a positive integer n we assume that

(*) no finite split embedding problem $\mathcal{E}(\bar{A}) = (\bar{\varphi}\colon F/N \to G,\ \bar{\beta}\colon \mathrm{Gwr}_{G_0} \bar{A} \to G)$, where $\bar{A}$ is a nontrivial quotient of A^n and $\bar{\varphi}\colon F/N \to F/L = G$ is the quotient map, has a solution.

Finally, let $\psi_1, \ldots, \psi_n$ be independent solutions of $\mathcal{E}(A)$. Then $\nu_i = \rho \circ \pi \circ \psi_i|_M$, $i = 1, \ldots, n$, are independent solutions of $\mathcal{E}_1(A)$.

Proof: Since $M \cap D \le \mathrm{Ker}(\mu)$, the map $\mu\colon M \to G_1$ extends to an epimorphism $\mu'\colon MD \to G_1$ by $\mu'(md) = \mu(m)$. In particular, μ' is trivial on L, so $\varphi_1 = \mu'|_{F_0}$ decomposes as $\varphi_1 = \bar{\varphi}_1 \circ \varphi_0$, where $\bar{\varphi}_1$ is an epimorphism from G_0 onto G_1.

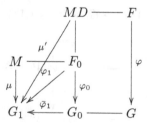

Since $\varphi_0 = \varphi|_{F_0}$, we have that $\mu = \bar{\varphi}_1 \circ \varphi|_M$. This proves (a).

The proof of (b) breaks up into three parts.

PART A: *The maps ν_i are weak solutions of $\mathcal{E}_1(A)$.* For each $1 \le i \le n$ we note that $\beta(\psi_i(M)) = \varphi(M) \le \varphi(F_0) = G_0$, so $\psi_i(M) \le \beta^{-1}(G_0) = G_0 \ltimes \mathrm{Ind}_{G_0}^G(A)$. Therefore, ν_i is well defined and the following diagram where α is the restriction of β to $G_0 \ltimes \mathrm{Ind}_{G_0}^G(A)$ and π_0 is the Shapiro map of $\mathrm{Ind}_{G_0}^G(A)$ is commutative:

(1)

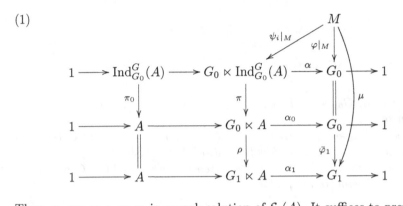

Thus, $\alpha_1 \circ \nu_i = \mu$, so ν_i is a weak solution of $\mathcal{E}_1(A)$. It suffices to prove that ν_i is surjective and $\nu_1, \ldots, \nu_n$ are independent.

PART B: *The map ν.* We use Remark 10.3.2 to identify $\mathrm{Ind}_{G_0}^G(A^n)$ with $\mathrm{Ind}_{G_0}^G(A)^n$, and $\mathrm{Gwr}_{G_0} A^n$ with $(\mathrm{Gwr}_{G_0} A)_G^n$, and denote the corresponding Shapiro maps by $\pi_0^{(n)}$ and $\pi^{(n)}$. By Lemma 10.2.3, $\psi_1, \ldots, \psi_n$ define a solution ψ of the embedding problem

$$\mathcal{E}(A^n) = (\varphi \colon F \to G, \ \beta^{(n)} \colon \mathrm{Gwr}_{G_0} A^n \to G),$$

where $\beta^{(n)} = \mathrm{pr}_j \circ \beta$ for each $1 \le j \le n$. Replacing A in (1) by A^n, we get a commutative diagram

(2)

where $\alpha^{(n)}$, $\alpha_0^{(n)}$, and $\alpha_1^{(n)}$ are the projections on the first factor, and $\rho^{(n)}$ is the extension of $\bar{\varphi}_1$ by id_{A^n}. Identifying $G_1 \ltimes A^n$ with $(G_1 \ltimes A)_G^n$ (Example

10.2.2), we first observe that the combined map $\nu = \rho^{(n)} \circ \pi^{(n)} \circ \psi|_M$ satisfies $\nu = \nu_1 \times \cdots \times \nu_n$.

Indeed, by Remark 10.3.2, for each $x \in M$ there are $\sigma_0 \in G_0$ and $f_1, \ldots, f_n \in \mathrm{Ind}_{G_0}^G(A)$ such that

$$(\psi_1(x), \ldots, \psi_n(x)) = \psi(x) = \sigma_0(f_1, \ldots, f_n) = (\sigma_0 f_1, \ldots, \sigma_0 f_n).$$

Hence,

$$
\begin{aligned}
\nu(x) &= \rho^{(n)}(\pi^{(n)}(\psi(x))) = \rho^{(n)}(\pi^{(n)}(\sigma_0(f_1, \ldots, f_n))) \\
&= \rho^{(n)}(\sigma_0(f_1(1), \ldots, f_n(1))) \\
&= \bar{\varphi}_1(\sigma_0)(f_1(1), \ldots, f_n(1)) = (\rho(\sigma_0 f_1(1)), \ldots, \rho(\sigma_0 f_n(1))) \\
&= (\rho(\pi(\sigma_0 f_1)), \ldots, \rho(\pi(\sigma_0 f_n))) = (\rho(\pi(\psi_1(x))), \ldots, \rho(\pi(\psi_n(x)))) \\
&= (\nu_1(x), \ldots, \nu_n(x)),
\end{aligned}
$$

as claimed.

In order to prove that $\nu_1, \ldots, \nu_n$ are surjective and independent, it suffices to prove that ν is surjective (Lemma 10.2.3). Since $\alpha_1^{(n)}(\nu(M)) = \mu(M) = G_1$, it suffices to prove that $A^n \le \nu(M)$. This will follow once we prove that $\pi^{(n)}(\psi(N)) = A^n$.

PART C: *A proof that $\pi^{(n)}(\psi(N)) = A^n$.* Since $N \le L = \mathrm{Ker}(\varphi)$, we have $\varphi(N) = 1$. In addition, $N \lhd F$, so $\psi(N) \le \mathrm{Ind}_{G_0}^G(A^n)$ and $\psi(N) \lhd \mathrm{Gwr}_{G_0} A^n$. Hence, $\psi(N)$ is a normal G-invariant subgroup of $\mathrm{Ind}_{G_0}^G(A^n)$. Thus, $B = \pi^{(n)}(\psi(N))$ is a normal G_0-invariant subgroup of A^n. Therefore, G_0 acts on A^n/B. This gives a commutative diagram of two short exact sequences

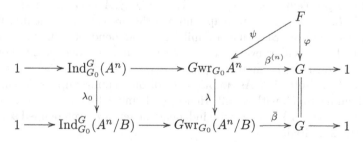

in which λ_0 and λ are defined by the quotient map $A^n \to A^n/B$.

Now $\psi(N) \le (\pi^{(n)})^{-1}(B) = \{f \in \mathrm{Ind}_{G_0}^G(A^n) \mid f(1) \in B\}$ and, as mentioned above, $\psi(N)$ is a G-invariant subgroup of $\mathrm{Ind}_{G_0}^G(A^n)$. Hence,

$$
\begin{aligned}
\psi(N) &\le \bigcap_{\sigma \in G} \{f \in \mathrm{Ind}_{G_0}^G(A^n) \mid f(1) \in B\}^\sigma \\
&= \bigcap_{\sigma \in G} \{f \in \mathrm{Ind}_{G_0}^G(A^n) \mid f(\sigma) \in B\} = \mathrm{Ker}(\lambda).
\end{aligned}
$$

It follows that $\lambda \circ \psi$ defines an epimorphism $\bar{\psi}: F/N \rightarrow \mathrm{Gwr}_{G_0}(A^n/B)$ that solves embedding problem $\mathcal{E}(\bar{A})$ with $\bar{A} = A^n/B$. By assumption, this cannot happen, unless $B = A^n$, as claimed. $\square$

As a corollary to Lemma 10.3.3 we prove a sufficient condition for a closed subgroup of a semi-free profinite group to be semi-free.

PROPOSITION 10.3.4: *Let F be a semi-free profinite group of infinite rank m and let M be a closed subgroup of F. Suppose for every open normal subgroup D of F and for every finite group A there exists*

(a) *an open subgroup F_0, an open normal subgroup L, and a closed normal subgroup N of F such that*

(b) $M \leq F_0$, $L \leq F_0 \cap D$, and $N \leq M \cap L$; and

(c) *no finite split embedding problem*

$$(\varphi: F/N \rightarrow F/L, \ \alpha: F/\mathrm{Lwr}_{F_0/L}\bar{A} \rightarrow F/L),$$

where $\bar{A}$ is a nontrivial quotient of A^2, φ is the quotient map, and α is the projection on the first factor, is solvable.

Then M is semi-free of rank m.

Proof: By [FrJ08, Cor. 17.1.4], $\mathrm{rank}(M) \leq \mathrm{rank}(F) = m$. Thus, by Corollary 10.1.8, it suffices to prove that every finite split embedding problem $\mathcal{E}_1(A) = (\mu: M \rightarrow G_1, \alpha_1: G_1 \ltimes A \rightarrow G_1)$ has m pairwise independent solutions.

To this end we choose a proper open normal subgroup D of F with $M \cap D \leq \mathrm{Ker}(\mu)$. By assumption there exist subgroups F_0, L, and N as in (a) such that (b) and (c) hold. As in Lemma 10.3.3 we consider the finite split embedding problem $\mathcal{E}(A) = (\varphi: F \rightarrow G, \alpha: \mathrm{Gwr}_{G_0}A \rightarrow G)$, where $G = F/L$, $G_0 = F_0/L$, φ is the quotient map, and α is the projection on the first factor.

By assumption, $\mathcal{E}(A)$ has a family Ψ of independent solutions of cardinality m. In particular, every pair of solutions in Ψ is independent. For each $\psi \in \Psi$ we consider the map $\nu = \rho \circ \pi \circ \psi|_M: M \rightarrow G_1 \ltimes A$, where π and ρ are as in (1). Note that Assumption (c) of our lemma implies Assumption (*) of Lemma 10.3.3 with $n = 2$. Hence, by Lemma 10.3.3 in the case $n = 2$, $\{\rho \circ \pi \circ \psi|_M \mid \psi \in \Psi\}$ is a pairwise independent set of m well defined solutions of $\mathcal{E}_1(A)$, as desired. $\square$

10.4 Closed Subgroups of Semi-free Profinite Groups

We prove in this section that the property of a profinite group F to be semi-free is inherited to each closed subgroup M which does not lie too deep in F. By that we mean that either the cardinality of all open subgroups of F that contain M is less than $\mathrm{rank}(F)$ or for each positive integer n there are only finitely many open subgroups that contain M. In particular, each open subgroup of F is semi-free. The proof of the latter statement uses the machinery of twisted wreath products developed in the preceding section. The proof of the two major statements is then reduced to the statement about open subgroups.

LEMMA 10.4.1: *Let M be an open subgroup of a semi-free profinite group F of rank m. Then M is semi-free of rank m.*

Proof: Let D be an open normal subgroup of F. We choose an open normal subgroup L of F with $L \leq M \cap D$ and set $F_0 = M$ and $N = L$. Then for each nontrivial finite group $\bar{A}$ on which F_0/L acts, the finite split embedding problem $(F/N \to F/L, F/\mathrm{Lwr}_{F_0/L}\bar{A} \to F/L)$ has no solution because $|F/\mathrm{Lwr}_{F_0/L}\bar{A}| > |F/N|$. It follows from Proposition 10.3.4 that M is semi-free of rank m.   □

LEMMA 10.4.2: *Let M be a closed subgroup of a semi-free profinite group F of infinite rank m. Suppose weight$(F/M) < m$. Then M is semi-free of rank m.*

Proof: Let $\mathcal{E}_M = (\mu\colon M \to G, \alpha\colon H \to G)$ be a finite split embedding problem with a nontrivial kernel. We use transfinite induction to construct for each $\lambda < m$ a solution ψ_λ of $\mathcal{E}_M$ such that the set $\{\psi_\lambda \mid \lambda < m\}$ of solutions is independent.

Let $M_1 = \mathrm{Ker}(\mu)$ and consider an ordinal number $\lambda < m$. Inductively suppose we have constructed an independent family $\{\psi_\kappa \mid \kappa < \lambda\}$ of solutions of $\mathcal{E}_M$. Then $\{\mathrm{Ker}(\psi_\kappa) \mid \kappa < \lambda\}$ is an M_1-independent family of open subgroups of M_1. By Definition 10.1.9, $N = \bigcap_{\kappa < \lambda} \mathrm{Ker}(\psi_\kappa)$ is a closed normal subgroup of M and weight$(M/N) < m$, hence weight$(F/N) < m$.

By [FrJ08, Lemma 1.2.5(c)], μ extends to an epimorphism $\varphi\colon E \to G$ for some open subgroup E of F containing M. In particular, $E_1 = \mathrm{Ker}(\varphi)$ satisfies $E_1 \cap M = M_1$. By Lemma 10.4.1, E is semi-free of rank m. Hence, the finite split embedding problem $\mathcal{E}_E = (\varphi\colon E \to G, \alpha\colon H \to G)$ has an independent family Ψ of solutions of cardinality m. Thus, $\{\mathrm{Ker}(\psi) \mid \psi \in \Psi\}$ is an E_1-independent family of open normal subgroups of E_1. Since E_1 is open in F, we have weight$(E_1/N) = \mathrm{weight}(E/N) < m$. Hence by Lemma 10.1.11, there exists $\psi \in \Psi$ such that $\mathrm{Ker}(\psi)N = E_1$. Let $\psi_\lambda = \psi|_M$. By Lemma 10.1.3(d), $\mathrm{Ker}(\psi_\lambda)N = M_1$. Hence, ψ_λ is a solution of $\mathcal{E}_M$ independent of $\{\psi_\kappa \mid \kappa < \lambda\}$. This concludes the transfinite induction and proves, by Lemma 10.1.4, that $\{\psi_\lambda \mid \lambda < m\}$ is an independent set of solutions of $\mathcal{E}_M$.   □

LEMMA 10.4.3: *Let M be a closed subgroup of a semi-free profinite group F of an infinite rank m. Suppose F/M is small. Then M is semi-free of rank m.*

Proof: Since weight$(F/M) \leq \aleph_0$, the case where $m > \aleph_0$ is a special case of Lemma 10.4.2. Thus, we have only to prove the lemma under the additional assumption that $m = \aleph_0$.

By Proposition 10.2.4, it suffices to prove that every finite split embedding problem with a nontrivial kernel for M is solvable. Let $\mathcal{E}_M = (\mu\colon M \to G, \alpha\colon H \to G)$ be such an embedding problem.

Let $M_1 = \mathrm{Ker}(\mu)$. By [FrJ08, Lemma 1.2.5(c)], μ extends to an epimorphism $\varphi\colon E \to G$ for some open subgroup E of F containing M. In particular, $E_1 = \mathrm{Ker}(\varphi)$ is an open normal subgroup of E with $E_1 \cap M = M_1$ and $E_1 M = E$.

By Lemma 10.4.1, E is semi-free of rank $\aleph_0$. Hence, the finite split embedding problem $\mathcal{E}_E = (\varphi\colon E \to G,\ \alpha\colon H \to G)$ has an infinite independent family Ψ of solutions. Thus, $\mathcal{K} = \{\mathrm{Ker}(\psi)\mid \psi \in \Psi\}$ is an E_1-independent family of open subgroups of E_1 and each of them is normal in E. In particular, by Lemma 10.1.3, each proper open subgroup of E_1 contains at most one group belonging to $\mathcal{K}$.

Let $K \in \mathcal{K}$. Then $(KM : KM_1) = (M : M_1) = (E : E_1) = (KM : E_1 \cap KM)$, hence $KM_1 = E_1 \cap KM$. In addition, $(F : KM) \le (F : E)(E : K) = (F : E)|H|$. Since F/M is small, F has only finitely many open subgroups of the form KM with $K \in \mathcal{K}$.

Now assume the set $\mathcal{K}_0 = \{K \in \mathcal{K}\mid KM_1 < E_1\}$ is infinite. Then, by the preceding paragraph, there exist distinct $K_1, K_2 \in \mathcal{K}_0$ with $K_1 M = K_2 M$. By the preceding paragraph, $K_1 M_1 = E_1 \cap K_1 M = E_1 \cap K_2 M = K_2 M_1$. This contradicts the independency of K_1, K_2 in E_1. We conclude from that contradiction, that $\mathcal{K}_0$ is finite and choose $K \in \mathcal{K} \smallsetminus \mathcal{K}_0$. Let $\psi \in \Psi$ with $\mathrm{Ker}(\psi) = K$. Then $\psi|_M$ is a solution of $\mathcal{E}_M$. $\qquad\square$

10.5 The Diamond Theorem

The diamond theorem for Hilbertian fields gives a convenient condition on a separable algebraic extension M of a Hilbertian field K to be Hilbertian. The condition requires the existence of Galois extensions M_1 and M_2 of K such that $M \not\subseteq M_1$, $M \not\subseteq M_2$, and $M \subseteq M_1 M_2$ [FrJ08, Thm. 13.8.2]. In that case we say that M is **contained in a K-diamond**. An analog of that theorem holds for free profinite groups: Let F be a free profinite group of infinite rank m and M a closed subgroup of F that is contained in an F-diamond (Definition 10.5.2). Then M is free of rank m. The proofs of both theorems are similar, both utilize twists wreath products. It turns out that the same method applies also to semi-free groups.

We start with a technical lemma that emphasizes the incommutativity of twisted wreath products.

LEMMA 10.5.1 ([FrJ08, Lemma 13.7.4]): *Let* $\alpha\colon \mathrm{Gwr}_{G_0} A \to G$ *be a twisted wreath product of finite groups,* $H_1 \lhd \mathrm{Gwr}_{G_0} A$, *and* $h_2 \in \mathrm{Gwr}_{G_0} A$. *Put* $I = \mathrm{Ind}_{G_0}^G(A) = \mathrm{Ker}(\alpha)$ *and* $G_1 = \alpha(H_1)$. *Suppose* $A \ne 1$.
(a) *Suppose* $\alpha(h_2) \notin G_0$ *and* $(G_1 G_0 : G_0) > 2$. *Then there is an* $h_1 \in H_1 \cap I$ *with* $h_1 h_2 \ne h_2 h_1$.
(b) *Suppose* $G_1 \not\le G_0$ *and* $\alpha(h_2) \notin G_1 G_0$. *Then there is an* $h_1 \in H_1 \cap I$ *with* $h_1^{h_2} \notin \langle h_1 \rangle^{h'}$ *for all* $h' \in \alpha^{-1}(G_1 G_0)$. *In particular,* $h_1 h_2 \ne h_2 h_1$.

Proof: Put $\sigma_2 = \alpha(h_2)$. Consider $\sigma_1 \in G_1$ and $g \in I$. By definition, there are $f_1, f_2 \in I$ with $\sigma_1 f_1 \in H_1$ and $h_2 = \sigma_2 f_2$. Put $h_1 = g^{\sigma_1 f_1} g^{-1}$. Then

$h_1 = [\sigma_1 f_1, g^{-1}] \in [H_1, I] \leq H_1 \cap I$. For each $\tau \in G$

$$h_1(\tau) = ((g^{\sigma_1})^{f_1})(\tau)g(\tau)^{-1} = g(\sigma_1\tau)^{f_1(\tau)}g(\tau)^{-1}.$$

Hence, for all $\tau \in G$ and $f' \in I$ we have:

(1a) $h_1^{h_2}(1) = h_1^{\sigma_2 f_2}(1) = h_1(\sigma_2)^{f_2(1)} = g(\sigma_1\sigma_2)^{f_1(\sigma_2)f_2(1)}g(\sigma_2)^{-f_2(1)}$,

(1b) $h_1^{\tau f'}(1) = h_1(\tau)^{f'(1)} = g(\sigma_1\tau)^{f_1(\tau)f'(1)}g(\tau)^{-f'(1)}$, and

(1c) $h_1(1) = g(\sigma_1)^{f_1(1)}g(1)^{-1}$.

We apply (1) in the proofs of (a) and (b) to special elements σ_1 and g. Choose $a \in A$, $a \neq 1$.

Proof of (a): Since $(G_1 G_0 : G_0) > 2$, there is a $\sigma_1 \in G_1$ with distinct cosets $\sigma_1^{-1}G_0, \sigma_2 G_0, G_0$. Thus, none of the cosets $\sigma_1 G_0, \sigma_2 G_0, \sigma_1\sigma_2 G_0$ is G_0. Therefore, by definition of I, there is a $g \in I$ with $g(\sigma_1) = g(\sigma_2) = g(\sigma_1\sigma_2) = 1$ and $g(1) = a$. By (1a), $h_1^{h_2}(1) = 1$. By (1c), $h_1(1) \neq 1$. Consequently, $h_1^{h_2} \neq h_1$, as desired.

Proof of (b): Since $H_1 \lhd \mathrm{Gwr}_{G_0}A$, we have $G_1 \lhd G$, so $G_1^{\sigma_2} = G_1 \not\leq G_0$, hence $G_1 \not\leq G_0^{\sigma_2^{-1}}$. Hence, $G_1 \cap G_0$ and $G_1 \cap G_0^{\sigma_2^{-1}}$ are proper subgroups of G_1. Their union is a proper subset of G_1. Thus, there is an element $\sigma_1 \in G_1 \smallsetminus (G_0 \cup G_0^{\sigma_2^{-1}})$. It follows that $\sigma_2 \notin \sigma_1\sigma_2 G_0$. By assumption, $\sigma_2 \notin G_1 G_0$. Therefore, there is a $g \in I$ with $g(G_1 G_0) = 1$, $g(\sigma_1\sigma_2) = 1$, and $g(\sigma_2) = a^{-1}$.

 Consider $\tau \in G_1 G_0$ and $f' \in I$. By (1a), $h_1^{h_2}(1) = a^{f_2(1)} \neq 1$. By (1b), $h_1^{\tau f'}(1) = 1$. Hence, $(h_1^k)^{\tau f'}(1) = 1$ for all integers k. It follows that $h_1^{h_2} \notin \langle h_1 \rangle^{h'}$ for all $h' \in \alpha^{-1}(G_1 G_0)$. $\square$

Definition 10.5.2: A closed subgroup M of a profinite group F is said to be **contained in an F-diamond** if F has closed normal subgroups M_1, M_2 such that $M_1 \cap M_2 \leq M$, $M_1 \not\leq M$, and $M_2 \not\leq M$. The following diagram of profinite groups reveals why we have chosen that name:

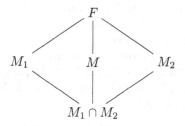

THEOREM 10.5.3 (The diamond theorem for semi-free profinite groups): *Let F be a semi-free profinite group of infinite rank m. If a closed subgroup M of F is contained in an F-diamond, then M is semi-free of rank m.*

Proof: Let M_1, M_2 be as in Definition 10.5.2. We use Lemma 10.4.1 to assume $(F : M) = \infty$. Then we first prove the theorem under an additional assumption:

(2) Either $M_1 M_2 = F$ or $(M_1 M : M) > 2$.

The proof of the theorem in this case utilizes Proposition 10.3.4. It has two parts.

PART A: *Construction of L, F_0, and N.* We consider an open normal subgroup D of F, choose another open normal subgroup L of F in D, and set $F_0 = ML$. Let $G = F/L$ and $\varphi \colon F \to G$ the quotient map, $G_0 = \varphi(M) = F_0/L$, $G_1 = \varphi(M_1)$, and $G_2 = \varphi(M_2)$. Then

(3a) $G_1, G_2 \triangleleft G$.

Moreover, choosing L sufficiently small, the following holds:

(3b) $G_1, G_2 \not\leq G_0$ (use $M_1, M_2 \not\leq M$).

(3c) $(G : G_0) > 2$ (use $(F : M) = \infty$).

(3d) $G_1 G_2 = G$ or $(G_1 G_0 : G_0) > 2$ (use (2)).

This implies:

(4) $G_2 \not\leq G_1 G_0$ or $(G_1 G_0 : G_0) > 2$.

Indeed, suppose both $G_2 \leq G_1 G_0$ and $G_1 G_2 = G$. Then $G = G_1 G_0$, so by (3c), $(G_1 G_0 : G_0) > 2$.

Now let $N = L \cap M_1 \cap M_2$. Then $N \leq M$.

PART B: *An embedding problem.* Suppose G_0 acts on a nontrivial finite group $\bar{A}$ and set $H = G\mathrm{wr}_{G_0}\bar{A}$. Consider the embedding problem

(5) $(\varphi \colon F \to G, \ \alpha \colon H \to G)$

where α is the quotient map. We have to prove that (5) has no solution that factors through F/N.

Assume $\psi \colon F \to H$ is an epimorphism with $\alpha \circ \psi = \varphi$ and $\psi(N) = 1$. For $i = 1, 2$ put $H_i = \psi(M_i)$. Then $H_i \triangleleft H$ and $\alpha(H_i) = \varphi(M_i) = G_i$.

We use (4) to find $h_1 \in H_1$ and $h_2 \in H_2$ with $\alpha(h_1) = 1$ and $[h_1, h_2] \neq 1$. First suppose $G_2 \not\leq G_1 G_0$. Then there is an $h_2 \in H_2$ with $\alpha(h_2) \notin G_1 G_0$. By (3b), $G_1 \not\leq G_0$, so Lemma 10.5.1(b) provides the required $h_1 \in H_1$. Now suppose $(G_1 G_0 : G_0) > 2$. We use (3b) to find $h_2 \in H_2$ with $\alpha(h_2) \notin G_0$. Lemma 10.5.1(a) gives the required $h_1 \in H_1$.

Having chosen h_i, we choose $f_i \in M_i$ with $\psi(f_i) = h_i$. Then $\varphi(f_1) = \alpha(h_1) = 1$, so $f_1 \in L$. Then $[f_1, f_2] \in [L, M_2] \cap [M_1, M_2] \leq L \cap (M_1 \cap M_2) = N$. Therefore, $[h_1, h_2] = [\psi(f_1), \psi(f_2)] \in \psi(N) = 1$. This contradiction proves that ψ as above does not exist.

CONCLUSION OF THE PROOF: In the general case we use $M_1 \not\leq M$ to conclude that $(M_1 M : M) \geq 2$. The case $(M_1 M : M) > 2$ is covered by the special case proved above. Suppose $(M_1 M : M) = 2$. Choose an open subgroup K_2 of F containing M but not $M_1 M$. Then, $K_2 \cap M_1 M = M$. Put $K = K_2 M_1 M$. Then $(K : K_2) = (M_1 M : M) = 2$, hence $K_2 \triangleleft K$. Observe: $M_1 K_2 = K$ and $K_2 \cap M_1 \leq K_2 \cap M_1 M = M \leq K$. Furthermore, $K_2 \not\leq M$, because $(K_2 : M) = \infty$.

By Lemma 10.4.1, K is semi-free of rank m, so the first alternative of (2) applies with K replacing F and K_2 replacing M_2. Consequently, M is semi-free of rank m. □

THEOREM 10.5.4 (Bary-Soroker, Haran, and Harbater): *Let F be a semi-free profinite group of an infinite rank m and let M be a closed subgroup. Then M is semi-free of rank m in each of the following cases:*
(a) *M is an open subgroup of a closed subgroup M_0 of F and M_0 contains a closed normal subgroup N of F that is not contained in M (an analog of a theorem of Weissauer).*
(b) *M is a proper subgroup of finite index of a closed normal subgroup of F.*
(c) *$M \lhd F$ and F/M is Abelian (an analog of a theorem of Kuyk).*
(d) *$M \lhd F$, F/M is pronilpotent of order divisible by at least two prime numbers.*

Proof of (a): The case where M is open is taken care of by Lemma 10.4.1. Thus, we may assume that $(F : M) = \infty$ and choose an open normal subgroup M_1 of F with $M_1 \cap M_0 \leq M$. In particular, $M_1 \not\leq M$, $N \not\leq M$, and $M_1 \cap N \leq M$. It follows from the Diamond theorem 10.5.3 that M is semi-free of rank m.

Proof of (b): Take $N = M_0$ in (a).

Proof of (c): If $M = F$, there is nothing to prove. Otherwise, we choose $\sigma \in F \smallsetminus M$. Then $\langle M, \sigma \rangle / M$ is a nontrivial Abelian group. Hence, $\langle M, \sigma \rangle$ has a proper open subgroup L that contains M. By (a), L is semi-free of rank m. In addition, L/M is a procyclic group. Hence, by Proposition 10.4.3, M is semi-free of rank m.

Proof of (d): Since each Sylow subgroup of a pronilpotent group is normal, F has closed normal subgroups P_1, P_2 that properly contain M such that P_1/M is the p_1-Sylow subgroup of F/M, P_2/M is the p_2-Sylow subgroup of F/M, and $p_1 \neq p_2$. In particular, $P_1 \cap P_2 = M$. By the Diamond theorem 10.5.3, M is semi-free of rank m. □

Remark 10.5.5: One may add two more cases to the list of Theorem 10.5.4:
(e) M is a **sparse subgroup** of F. That is, for each positive integer n there exists an open subgroup K of F that contains M such that for each proper open subgroup L of K that contains M we have $[K : L] \geq n$.
(f) We have $(F : M) = \prod p^{\alpha(p)}$, where $\alpha(p) < \infty$ for each prime number p.

The reader may try to settle these cases by himself, applying Lemma 10.4.1 or consult [BHH10 Section 4]. □

Applying Proposition 10.1.14, the results we have proved so far about closed subgroups of semi-free profinite groups yield the corresponding results about closed subgroups of free profinite groups. This gives a new proof for results proved in [FrJ08, Section 25.4].

THEOREM 10.5.6: *Let F be a free profinite group of infinite rank m and let M be a closed subgroup of F. Then each of the following conditions on M suffices for M to be free of rank m:*
(a) *M is open in F.*
(b) *weight$(F/M) < m$.*
(c) *F/M is small.*
(d) *M is contained in an F-diamond.*
(e) *M is an open subgroup of a closed subgroup M_0 of F and M_0 contains a closed normal subgroup of F that is not contained in M.*
(f) *M is a proper subgroup of finite index of a closed normal subgroup of F.*
(g) *$M \lhd F$ and F/M is Abelian.*
(h) *$M \lhd F$ and F/M is pronilpotent of order divisible by at least two prime numbers.*
(i) *M is a sparse subgroup of F.*
(j) *$(F : M) = \prod p^{\alpha(p)}$, where $\alpha(p) < \infty$ for each prime number p.*

Proof: Since F is free, F is projective, so M is also projective [FrJ08, Prop. 22.4.7]. Thus, in order to prove that M is free of rank m, it suffices, by Proposition 10.1.14, to prove in each case that M is semi-free of rank m. By Proposition 10.1.13, F is semi-free. Therefore, (a) follows from Lemma 10.4.1, (b) follows from Lemma 10.4.2, (c) follows from Lemma 10.4.3, (d) follows from Theorem 10.5.3, (e), (f), (g), (h) follow from Theorem 10.5.4, and (i), (j) follow from Remark 10.5.5. □

Remark 10.5.7: C-semi-free groups. Let C be a **Melnikov formation**, that is C is a family of finite groups closed under taking quotients, normal subgroups, and extensions. For example the family of all finite groups, the family of all finite p-groups, and the family of all finite solvable groups are Melnikov formations. A C-**embedding problem** for a pro-C group F is an embedding problem $\mathcal{E} = (\varphi: F \to G, \alpha: H \to G)$ where H (hence also G) belongs to C.

We say that a profinite group F of rank m is C-**semi-free** if F is a pro-C group and each split C-embedding problem with a nontrivial kernel has m independent solutions. □

The results about subgroups of semi-free groups yield the corresponding results about subgroups of C-semi-free groups.

THEOREM 10.5.8: *Let C be a Melnikov formation of finite groups, F a C-semi-free group of infinite rank m, and M a closed subgroup of F. Then M is C-semi-free of rank m in each of the following cases:*
(a) *M is open in F.*
(b) *weight$(F/M) < m$.*
(c) *F/M is small.*
(d) *M is contained in an F-diamond.*
(e) *M is an open subgroup of a closed subgroup M_0 of F and M_0 contains a closed normal subgroup N of F not contained in M.*
(f) *M is a proper subgroup of finite index of a closed normal subgroup of F.*

(g) $M \lhd F$ and F/M is Abelian.

(h) $M \lhd F$ and F/M is pronilpotent of order divisible by at least two prime numbers.

(i) M is a sparse subgroup of F.

(j) $(F : M) = \prod p^{\alpha(p)}$, where $\alpha(p) < \infty$ for each prime number p.

Proof: Let $\mathcal{E} = (\varphi \colon M \to G, \alpha \colon H \to G)$ be a split $\mathcal{C}$-embedding problem with a nontrivial kernel. By [FrJ08, Prop. 17.4.8] there is a free profinite group $\hat{F}$ of rank m and an epimorphism $\psi \colon \hat{F} \to F$. For each closed subgroup L of M we let $\hat{L} = \psi^{-1}(L)$. In particular, $\hat{M}$ is a closed subgroup of $\hat{F}$ and $\hat{\mathcal{E}} = (\varphi \circ \psi|_{\hat{M}} \colon \hat{M} \to G, \alpha \colon H \to G)$ is a finite split embedding problem for $\hat{M}$. Moreover, $\mathrm{Ker}(\varphi \circ \psi|_{\hat{M}}) = \psi^{-1}(\mathrm{Ker}(\varphi))$.

We observe that in each of the cases (a)–(j), $\hat{M}$ satisfies the corresponding condition that M satisfies. Hence, by Theorem 10.5.6, the group $\hat{M}$ is semi-free of rank m. Hence, $\hat{\mathcal{E}}$ has m-independent solutions. The corresponding kernels $\hat{M}_i$ are normal subgroups of $\hat{M}$ such that $\hat{M}/\hat{M}_i$ are isomorphic to the $\mathcal{C}$-group H. It follows that those solutions define m-independent solutions to $\mathcal{E}$. Consequently, M is $\mathcal{C}$-semi-free of rank m. □

10.6 Quasi-Free Profinite Groups

If we drop the condition of the independence of the solutions from the definition of semi-free profinite group, we get the notion of "quasi-free profinite group" introduced in [HaS05]. In this section we introduce the latter notion and discuss its advantages and its limits.

Let G be a profinite group of infinite rank m. We say that G is **quasi-free** if every finite split embedding problem $\mathcal{E}$ for G with a nontrivial kernel has at least m solutions. In particular, every embedding problem of the form $(G \to 1, A \to 1)$, where A is a finite group is solvable, so A is a quotient of G. Unlike for semi-free profinite groups, we do not insist that the solutions will be independent. Still, by [FrJ08, Lemma 25.1.8], if G is projective and quasi-free of rank m, then G is free. By [RSZ07, Thm. 2.1], every open subgroup of a quasi-free profinite group is quasi-free. If G is quasi-free of rank m, then so is its commutator group [Hrb09, Thm. 2.4]. By [HaS05, Thm. 5.1], the absolute Galois group of the field $K((t_1, t_2))$ of formal power series in two variables over an arbitrary field K is quasi-free of rank equal to $\mathrm{card}(K)$. It follows that $\mathrm{Gal}(K((t_1, t_2))_{\mathrm{ab}})$ is quasi-free. If K is algebraically closed, then $\mathrm{Gal}(K((t_1, t_2))_{\mathrm{ab}})$ is projective [Hrb09, Thm. 4.4]. It follows that in this case $\mathrm{Gal}(K((t_1, t_2))_{\mathrm{ab}})$ is free of rank equal to $\mathrm{card}(K)$. Finally, Harbater proves that if K is ample, then $\mathrm{Gal}(K(x))$ is quasi-free [Hrb09, Thm. 3.4].

In view of Example 10.6.1 below, the latter result is weaker than Theorem 11.7.1 saying that $\mathrm{Gal}(K(x))$ is semi-free if K is ample. In addition, we have been able to prove the diamond theorem only for semi-free profinite groups but not for quasi-free profinite groups. Thus, using only the concept of quasi-free profinite groups, we would not be able to prove that the

K-radical extension of $K(x)$ (with K being PAC) that we have constructed in Theorem 11.7.6 has a free absolute Galois group. Finally, in contrast to semi-free groups (Theorem 10.5.8(b)), a closed subgroup N of a quasi-group G with weight$(G/N) < \mathrm{rank}(N)$ need not be quasi-free, as Lemma 10.6.4 below demonstrates.

Example 10.6.1: (Bary-Soroker, Haran, Harbater) Example of a quasi-free profinite group that is not semi-free. Let X be a set of uncountable cardinality m and let $C = \prod_p \mathbb{Z}/p\mathbb{Z}$ be the direct product of all cyclic groups of prime order. For each $x \in X$ let C_x be an isomorphic copy of C. We consider the free product $E = \Coprod_{x \in X} C_x$ in the sense of [BNW71]. Thus, each C_x is a closed subgroup of E and every family of homomorphisms $\psi_x \colon C_x \to \bar{C}$ into a finite group A, such that $\psi_x(C_x) = 1$ for all but finitely many $x \in X$, uniquely extends to a homomorphism $\psi \colon E \to \bar{C}$. Let $G = \Coprod_{x \in X} C_x * \hat{F}_\omega$.

CLAIM A: *G is quasi-free of rank m.* The rank of $\Coprod_{x \in X} C_x$ is m and the rank of $\hat{F}_\omega$ is $\aleph_0 < m$. Hence, rank$(G) = m$. Let

$$(1) \qquad\qquad\qquad (\varphi \colon G \to A, \ \alpha \colon B \to A)$$

be a finite split embedding problem with a nontrivial kernel and let $\alpha' \colon A \to B$ be its splitting. We need two auxiliary maps: First, there exists a nontrivial homomorphism $\pi \colon C \to \mathrm{Ker}(\alpha)$; namely, an epimorphism of C onto a subgroup of $\mathrm{Ker}(\alpha)$ of prime order. Secondly, there exists an epimorphism $\psi' \colon \hat{F}_\omega \to \alpha^{-1}(\varphi(\hat{F}_\omega))$ such that $\alpha \circ \psi'$ is the restriction of φ to $\hat{F}_\omega$ [FrJ08, Thm. 24.8.1]. In particular, $\psi'(\hat{F}_\omega)$ contains $\mathrm{Ker}(\alpha)$.

The definition of $\Coprod_{x \in X} C_x$ gives a subset Y of X such that $X \smallsetminus Y$ is finite and $\varphi(C_x) = 1$ for every $x \in Y$. For every $y \in Y$ we define a homomorphism $\psi_y \colon G \to B$ in the following manner:

$$\begin{aligned}
\psi_y|_{C_y} &= \pi \\
\psi_y|_{C_x} &= 1 \text{ if } x \in Y \text{ and } x \neq y \\
\psi_y|_{C_x} &= \alpha' \circ \varphi \text{ if } x \in X \smallsetminus Y \\
\psi_y|_{\hat{F}_\omega} &= \psi'
\end{aligned}$$

Then, $\alpha \circ \psi_y = \varphi$. Since $\psi_y(G) \geq \psi'(\hat{F}_\omega) \geq \mathrm{Ker}(\alpha)$, the map ψ_y is a solution of (1).

Since $\psi_{y_1} \neq \psi_{y_2}$ for distinct $y_1, y_2 \in Y$, (1) has at least $|Y| = m$ distinct solutions. Thus, G is quasi-free of rank m.

CLAIM B: *G is not semi-free.* Consider the finite split embedding problem

$$(2) \qquad\qquad\qquad (\varphi \colon G \to 1, \ \mathbb{Z}/4\mathbb{Z} \to 1)$$

with the nontrivial kernel $\mathbb{Z}/4\mathbb{Z}$. Let Ψ be an independent set of solutions of (2). The map $\mathbb{Z}/4\mathbb{Z} \to 1$ decomposes into $\alpha \colon \mathbb{Z}/4\mathbb{Z} \to \mathbb{Z}/2\mathbb{Z}$ and $\beta \colon \mathbb{Z}/2\mathbb{Z} \to 1$.

If $\psi_1, \psi_2 \in \Psi$ are independent, then $\alpha \circ \psi_1, \alpha \circ \psi_2$ are independent solutions of $(\varphi: G \to 1, \beta: \mathbb{Z}/2\mathbb{Z} \to 1)$ (Lemma 10.1.3(f)). In particular, $\alpha \circ \psi_1 \neq \alpha \circ \psi_2$. Thus, $\{\alpha \circ \psi \mid \psi \in \Psi\}$ has at least the cardinality of Ψ.

On the other hand, $\mathbb{Z}/4\mathbb{Z}$ is a 2-group and the 2-Sylow subgroup of C is of order 2. Hence, every $\psi \in \Psi$ maps each C_x into $\mathrm{Ker}(\alpha)$, the unique subgroup of $\mathbb{Z}/4\mathbb{Z}$ of order 2, so $\alpha \circ \psi$ is trivial on C_x. Therefore $\alpha \circ \psi$ is trivial on $\coprod_{x \in X} C_x$. It follows that $\alpha \circ \psi$ is determined by its restriction to $\hat{F}_\omega$. But there are only $\aleph_0$ homomorphisms $\hat{F}_\omega \to \mathbb{Z}/4\mathbb{Z}$. Therefore, $\mathrm{card}(\Psi) \leq \aleph_0$. $\square$

In order to prove the last piece of information about the group G of Example 10.6.1, we need a basic lemma about free products of two profinite groups.

LEMMA 10.6.2: *Let A and B be profinite groups, $A * B$ their free product, and $\pi: A * B \to B$ the homomorphism defined by $\pi(a) = 1$ for $a \in A$ and $\pi(b) = b$ for $b \in B$. Then, $A * B = B \ltimes \mathrm{Ker}(\pi)$ and $\mathrm{Ker}(\pi) = \langle A^b \mid b \in B \rangle$.*

Proof: Let $K = \langle A^b \mid b \in B \rangle$. Then K is a closed normal subgroup of $A * B$ and $K \leq \mathrm{Ker}(\pi)$.

If $a_1, \ldots, a_n \in K$ and $b_1, \ldots, b_n \in B$, then

$$a_1 b_1 a_2 b_2 \cdots a_n b_n = b_1 (a_1^{\hat{b}_1} a_2) b_2 \cdots a_n b_n$$

and $a_1^{b_1} a_2 \in K$. Induction on n gives a $k \in K$ such that

$$(a_1^{b_1} a_2) b_2 \cdots a_n b_n = b_2 \cdots b_n k.$$

Hence, $a_1 b_1 \cdots a_n b_n = bk$ with $b = b_1 \cdots b_n$.

Now let $g \in A * B$ and consider an open normal subgroup N of $A * B$. Since A and B generate $A * B$ there are $a_1, \ldots, a_n \in A$ and $b_1, \ldots, b_n \in B$ such that $g \equiv a_1 b_1 \cdots a_n b_n \bmod N$. By the preceding paragraph, $g \in BKN$. Intersecting on all possible N gives $g \in BK$ [FrJ08, Lemma 1.2.2(b)]. Thus, $A * B = BK$.

If $g \in \mathrm{Ker}(\pi)$, then writing $g = bk$ with $b \in B$ and $k \in K$ and applying π we get that $1 = b$, so $g = k \in K$. Therefore, $K = \mathrm{Ker}(\pi)$.

That $A * B = B \ltimes K$ follows now from the observation that $B \cap K = 1$. $\square$

Remark 10.6.3: It is further proved in [HJP09, Lemma 2.3] that $\mathrm{Ker}(\pi)$ in Lemma 10.6.2 is isomorphic to the free product $\coprod_{b \in B} A^b$ in the sense of Melnikov [Mel90]. However, we do not need here that extra information. $\square$

LEMMA 10.6.4: *Let $G = \coprod_{x \in X} C_x * \hat{F}_\omega$ be as in Example 10.6.1. Suppose $m > \aleph_0$. Then G has a closed normal subgroup K such that $\mathrm{weight}(G/K) < \mathrm{rank}(G)$ but K is not quasi-free.*

Proof: Let K be the kernel of the projection of G onto $\hat{F}_\omega$, mapping each element of $E = \coprod_{x \in X} C_x$ onto 1 and each element of $\hat{F}_\omega$ onto itself. Then

$G/K \cong \hat{F}_\omega$, so weight$(G/K) = $ rank$(\hat{F}_\omega) = \aleph_0 < m = $ rank(G). On the other hand, by Lemma 10.6.2, $K = \langle E^b \mid b \in \hat{F}_\omega \rangle$. Thus, K is generated by elements of prime order. Therefore, every finite quotient of K is generated by elements of prime order. In particular, $\mathbb{Z}/4\mathbb{Z}$ is not a quotient of K. Consequently, K is not quasi-free. □

Remark 10.6.5: Let K be an ample Hilbertian field. By Theorem 5.10.2(a), every finite split embedding problem for Gal(K) is solvable. Hence, if K is countable, or more generally, if rank$($Gal$(K)) = \aleph_0$, then Gal(K) is semi-free (Proposition 10.2.4). It follows from Theorem 10.5.8 that Gal(K') is semi-free in each of the following cases:

(3a) K' is a finite separable extension of K,

(3b) K' is a **small separable algebraic extension** of K (i.e. for each n there are only finitely many extensions of K in K' of degree at most n),

(3c) K' is contained in a K-diamond,

(3d) K' is a finite proper separable extension of a Galois extension of K, and

(3e) K' is an Abelian extension of K. □

Following Remark 10.6.5, it is tempting to conjecture that Gal(K) is semi-free if K is ample and Hilbertian. However, as Example 10.6.7 shows, this is not the case.

Example 10.6.6: A projective non-semi-free profinite group N for which each finite embedding problem is solvable.

Let m be an uncountable cardinal number and set $F = \hat{F}_m$. By [FrJ08, Prop. 25.7.7], F has a closed normal subgroup N that has m N-independent open normal subgroups M with $N/M \cong \mathbb{Z}/p\mathbb{Z}$ for each prime number p but only $\aleph_0$ N-independent open normal subgroups M with $N/M \cong S$ for each non-Abelian simple finite group S. By [FrJ08, Lemma 25.7.1], N is not free. As a closed subgroup of a free profinite group, N is projective [FrJ08, Cor. 22.4.6]. Hence, by Proposition 10.1.14, N is not semi-free. We prove that every finite embedding problem

(4) $$(\varphi: N \to A, \; \alpha: B \to A)$$

for N is solvable. By induction on the order of $C = \text{Ker}(\alpha)$, we may suppose that C is a minimal normal subgroup of B (see the proof of [FrJ08, Lemma 25.1.4]).

Let $N_1 = \text{Ker}(\varphi)$. By [FrJ08, Lemma 1.2.5], F has an open normal subgroup F_0 such that $N \cap F_0 \leq N_1$. Then $K = NF_0$ is an open normal subgroup of F and φ extends to a homomorphism $\kappa: K \to A$ by $\kappa(nf_0) = \varphi(n)$ for $f_0 \in F_0$ and $n \in N$.

By [FrJ08, Proposition 17.6.2], K is free of rank m. Hence, there exists an epimorphism $\theta: K \to B$ with $\alpha \circ \theta = \kappa$ [FrJ08, Lemma 25.1.2]. Set $K_1 = \text{Ker}(\kappa)$ and $K_2 = \text{Ker}(\theta)$. Then $N \cap K_1 = N_1$, $NK_1 = K$, $K/K_2 \cong B$, and $K_1/K_2 \cong C$. In particular, N_1 and K_2 are normal in K. Hence, $N_1 K_2/K_2$ is a normal subgroup of K/K_2 which is contained in K_1/K_2. The latter group is minimal normal in K/K_2 (because C is minimal normal in B), so either $N_1 K_2 = K_1$ or $N_1 \leq K_2$.

CASE 1: $N_1 K_2 = K_1$. Then $NK_2 = K$. Hence, $\theta(N) = \theta(K) = B$ and $\theta|_N \colon N \to B$ solves embedding problem (4).

CASE 2: $N_1 \leq K_2$. Then $L = NK_2$ is normal in K, $L/K_2 \cong N/N_1 \cong A$, and $L \cap K_1 = K_2$. Thus,

$$B \cong K/K_2 \cong L/K_2 \times K_1/K_2 \cong A \times C.$$

Since C is a minimal normal subgroup of B, it is isomorphic to a direct product $\prod_{i=1}^r S_i$ of isomorphic copies of a single finite simple group S [FrJ08, Remark 16.8.4]. By assumption, N has infinitely many N-independent open normal subgroups M with $N/M \cong S$. Hence, by Lemma 10.1.12, N has open normal subgroups $M_1, \ldots, M_r$ with $N/M_i \cong S$, $i = 1, \ldots, r$, and $M_1, \ldots, M_r, N_1$ are N-independent. Let $M = \bigcap_{i=1}^r M_i$ and $N_2 = M \cap N_1$. Then $N/M \cong S^r \cong C$, $N/N_2 \cong N/M \times N/N_1 \cong C \times A \cong B$ and the quotient map $\gamma \colon N \to N/N_2$ solves embedding problem (4). $\qquad\square$

Example 10.6.7: A Hilbertian ample field K with a non-semi-free absolute Galois group. Let N be the profinite group given by Example 10.6.6. In particular, N is projective. Hence, by Lubotzky-v.d.Dries, there exists a PAC field K with $\operatorname{Gal}(K) \cong N$ [FrJ08, Cor. 23.1.2]. In particular, K is ample (Example 5.6.1). Since every finite embedding problem for N is solvable, K is ω-free (Section 5.10). By Roquette, K is Hilbertian [FrJ08, Cor. 27.3.3]. Finally, by Example 10.6.6, $\operatorname{Gal}(K)$ is not semi-free. $\qquad\square$

Remark 10.6.8: Although the absolute Galois group of an arbitrary Hilbertian ample field F need not be semi-free, there are many cases where $\operatorname{Gal}(F)$ is semi-free and F is uncountable. See Theorem 12.4.1 and Example 12.4.4. $\square$

Remark 10.6.9: Non quasi-free fundamental groups. Let E be a function field of one variable over an algebraically closed field C of positive characteristic and S a finite nonempty set of prime divisors of E/C. Then $\operatorname{Gal}(E_S/E)$ is not quasi-free. Otherwise, since $\operatorname{Gal}(E_S/E)$ is projective (Theorem 9.5.7), [FrJ08, Lemma 25.1.8] will imply that $\operatorname{Gal}(E_S/E)$ is free. This will contradicts Proposition 9.9.4. $\qquad\square$

Notes

The notions of independent subgroups of a profinite group and of twist fiber products of several finite groups is used in [BHH10] in order to improve the criterion developed by Haran in his proof of the diamond theorem for profinite groups. While [FrJ08, Prop. 24.14.1] which reconstruct that proof gives a criterion for a closed subgroup M of a profinite group F to have all finite split embedding problems solvable once the same holds for F, Lemma

10.3.3 gives a criterion for M to have independent solutions of those problems once F has independent solutions.

We have placed the group G on the left side of the twisted wreath product and A on its right side in order to be consistent with the placement of the factors in the semidirect product $G \ltimes \operatorname{Ind}_{G_0}^{G}(A)$. Note however that this is inconsistent with the notation we use in [FrJ08], where the same group is denoted by $\operatorname{Awr}_{G_0} G$.

Most of Sections 10.1 – 10.5 is a workout of some parts of [BHH10].

Example 10.6.1 is a workout of [BHH10, Prop. 6.1].

The concept "sparse subgroup" of a profinite group F is introduced in [BSo06] in order to prove the diamond theorem for free profinite groups of finite rank.

Chapter 11.
Function Fields of One Variable over PAC Fields

We prove that if K is an ample field of cardinality m and E is a function field of one variable over K, then $\mathrm{Gal}(E)$ is semi-free of rank m (Theorem 11.7.1). It follows from Theorem 10.5.4 that if F is a finite extension of E, or an Abelian extension of E, or a proper finite extension of a Galois extension of E, or F is "contained in a diamond" over E, then $\mathrm{Gal}(F)$ is semi-free.

We apply the latter results to the case where K is PAC and $E = K(x)$, where x is an indeterminate. We construct a K-radical extension F of E in a diamond over E and conclude that F is Hilbertian and $\mathrm{Gal}(F)$ is semi-free and projective (Theorem 11.7.6), so $\mathrm{Gal}(F)$ is free. In particular, if K contains all roots of unity of order not divisible by $\mathrm{char}(K)$, then $\mathrm{Gal}(E)_{\mathrm{ab}}$ is free of rank equal to $\mathrm{card}(K)$ (Theorem 11.7.6).

11.1 Henselian Fields

We give a sufficient condition for the absolute Galois group of a Henselian field (M, v) to be projective. Our proof is valuation theoretic and starts almost from the basic definitions. In particular, we do not use the connection between projectivity and the vanishing of the Brauer groups.

Let p be a prime number and A an Abelian group. We say that A is p'-**divisible**, if for each $a \in A$ and every positive integer n with $p \nmid n$ there exists $b \in A$ such that $a = nb$. Note that if $p = 0$, then "p'-divisible" is the same as "divisible".

LEMMA 11.1.1: *Let p be 0 or a prime number, B a torsion free Abelian group, and A a p'-divisible subgroup of finite index. Then B is also p'-divisible.*

Proof: First suppose that $p = 0$ and let $m = (B : A)$. Then, for each $b \in B$ and a positive integer n there exists $a \in A$ such that $mb = mna$. Since B is torsion free, $b = na$. Thus, B is divisible.

Now suppose p is a prime number, let $mp^k = (B : A)$, with $p \nmid m$ and $k \geq 0$, and consider $b \in B$. Then $mp^k b \in A$. Hence, for each positive integer n with $p \nmid n$ there exists $a \in A$ with $mp^k b = mna$. Thus, $p^k b = na$. Since $p \nmid n$, there exist $x, y \in \mathbb{Z}$ such that $xp^k + yn = 1$. It follows from $xp^k b = xna$ that $b = n(xa + yb)$, as claimed. $\square$

COROLLARY 11.1.2: *Let L/K be an algebraic extension, v a valuation of L, and $p = 0$ or p is a prime number. Suppose that $v(K^\times)$ is p'-divisible. Then $v(L^\times)$ is p'-divisible.*

Proof: Let $x \in L^\times$ and n a positive integer with $p \nmid n$. Then $v(K(x)^\times)$ is a torsion free Abelian group and $v(K^\times)$ is a subgroup of index at most

M. Jarden, *Algebraic Patching*, Springer Monographs in Mathematics,
DOI 10.1007/978-3-642-15128-6_11, © Springer-Verlag Berlin Heidelberg 2011

$[L : K]$. Since $v(K^\times)$ is p'-divisible, Lemma 11.1.1 gives $y \in K(x)^\times$ such that $v(x) = nv(y)$. It follows that $v(L^\times)$ is p'-divisible.   □

Given a Henselian valued field (M, v), we use v also for its unique extension to M_s. We use a bar to denote the residue with respect to v of objects associated with M, let O_M be the valuation ring of M, and let $\Gamma_M = v(M^\times)$ be the value group of M.

PROPOSITION 11.1.3: *Let (M, v) be a Henselian valued field. Suppose $p = \mathrm{char}(M) = \mathrm{char}(\bar{M})$, $\mathrm{Gal}(\bar{M})$ is projective, and Γ_M is p'-divisible. Then $\mathrm{Gal}(M)$ is projective.*

Proof: We denote the **inertia field** of M by M_u. It is determined by its absolute Galois group: $\mathrm{Gal}(M_u) = \{\sigma \in \mathrm{Gal}(M)\,|\, v(\sigma x - x) > 0$ for all $x \in M_s$ with $v(x) \geq 0\}$. The map $\sigma \mapsto \bar{\sigma}$ of $\mathrm{Gal}(M)$ into $\mathrm{Gal}(\bar{M})$ such that $\bar{\sigma}\bar{x} = \overline{\sigma x}$ for each $x \in O_{M_s}$ is a well defined epimorphism [Efr06, Thm. 16.1.1] whose kernel is $\mathrm{Gal}(M_u)$. It therefore defines an isomorphism

(1) $$\mathrm{Gal}(M_u/M) \cong \mathrm{Gal}(\bar{M}).$$

CLAIM A: *$\bar{M}_u$ is separably closed.* Let $g \in \bar{M}_u[X]$ be a monic irreducible separable polynomial of degree $n \geq 1$. Then there exists a monic polynomial $f \in O_{M_u}[X]$ of degree n such that $\bar{f} = g$. We observe that f is also irreducible and separable. Moreover, if $f(X) = \prod_{i=1}^n (X - x_i)$ with $x_1, \ldots, x_n \in M_s$, then $g(X) = \prod_{i=1}^n (X - \bar{x}_i)$. Given $1 \leq i, j \leq n$ there exists $\sigma \in \mathrm{Gal}(M_u)$ such that $\sigma x_i = x_j$. By definition, $\bar{x}_j = \overline{\sigma x_i} = \bar{\sigma}\bar{x}_i = \bar{x}_i$. Since g is separable, $i = j$, so $n = 1$. We conclude that $\bar{M}_u$ is separably closed.

CLAIM B: *Each l-Sylow group of $\mathrm{Gal}(M_u)$ with $l \neq p$ is trivial.* Indeed, let L be the fixed field of an l-Sylow group of $\mathrm{Gal}(M_u)$ in M_s. If $l = 2$, then $\zeta_l = -1 \in L$. If $l \neq 2$, then $[L(\zeta_l) : L]|l - 1$ and $[L(\zeta_l) : L]$ is a power of l, so $\zeta_l \in L$.

Assume that $\mathrm{Gal}(L) \neq 1$. By the theory of finite l-groups, L has a cyclic extension L' of degree l. By the preceding paragraph and Kummer theory, there exists $a \in L$ such that $L' = L(\sqrt[l]{a})$. By Corollary 11.1.2, there exists $b \in L^\times$ such that $lv(b) = v(a)$. Then $c = \frac{a}{b^l}$ satisfies $v(c) = 0$. By Claim A, $\bar{L}$ is separably closed. Therefore, $\bar{c}$ has an lth root in $\bar{L}$. By Hensel's lemma, c has an lth root in L. It follows that a has an lth-root in L. This contradiction implies that $L = M_s$, as claimed.

Having proved Claim B, we consider again a prime number $l \neq p$ and let G_l be an l-Sylow subgroup of $\mathrm{Gal}(M)$. By the claim, $G_l \cap \mathrm{Gal}(M_u) = 1$, hence the map res: $\mathrm{Gal}(M) \to \mathrm{Gal}(M_u/M)$ maps G_l isomorphically onto an l-Sylow subgroup of $\mathrm{Gal}(M_u/M)$. By (1), G_l is isomorphic to an l-Sylow subgroup of $\mathrm{Gal}(\bar{M})$. Since the latter group is projective, so is G_l, i.e. $\mathrm{cd}_l(\mathrm{Gal}(M)) \leq 1$ [Ser79, p. 58, Cor. 2].

Finally, if $p \neq 0$, then $\mathrm{cd}_p(\mathrm{Gal}(M)) \leq 1$ [Ser79, p. 75, Prop. 3], because then $\mathrm{char}(M) = p$. It follows that $\mathrm{cd}(\mathrm{Gal}(M)) \leq 1$ [Ser79, p. 58, Cor. 2].   □

11.2 Brauer Groups of Henselian Fields

We establish a short exact sequence for the Brauer group of a finite unramified extension of a Henselian field. That sequence will be used in the proof of Lemma 11.5.1.

Again, when (M, v) is a Henselian field, we denote its valuation ring by O_M, the maximal ideal of O_M by $\mathfrak{m}_M$, the group of units of M by U_M, the value group of (M, v) by Γ_M, and use a bar to denote reduction modulo $\mathfrak{m}_M$

PROPOSITION 11.2.1: *Let (M, v) be a Henselian valued field and (N, v) a finite Galois extension with a trivial inertia group. Set $G = \mathrm{Gal}(N/M)$. Then the G-module $1 + \mathfrak{m}_N$ is G-cohomologically trivial, that is $H^i(G, 1 + \mathfrak{m}_N) = 0$ for all positive integers i.*

Proof: By Subsection 9.3.13, it suffices to prove the following equalities:

(1)
$$(1 + \mathfrak{m}_N)^G = \mathrm{norm}_{N/M}(1 + \mathfrak{m}_N)$$
$$Z^1(G, 1 + \mathfrak{m}_N) = B^1(G, 1 + \mathfrak{m}_N).$$

Since the right hand sides of (1) are contained in the left hand sides, it suffices to prove only the other inclusions. This is done in two parts.

PART A: *Proof that $(1 + \mathfrak{m}_N)^G \le \mathrm{norm}_{N/M}(1 + \mathfrak{m}_N)$.* Note that $(1 + \mathfrak{m}_N)^G = 1 + \mathfrak{m}_M$. Thus, we have to prove that $1 + \mathfrak{m}_M \le \mathrm{norm}_{N/M}(1 + \mathfrak{m}_N)$.

Since $G_0(N/M) = 1$, (1) of Section 11.1 implies that the map $\sigma \mapsto \bar{\sigma}$ is an isomorphism $\mathrm{Gal}(N/M) \cong \mathrm{Gal}(\bar{N}/\bar{M})$. By the normal basis theorem there exists $x \in O_N$ such that $\{\bar{\sigma}\bar{x} \mid \sigma \in G\}$ is a basis of $\bar{N}/\bar{M}$ [Lan93, p. 312 for the case where M is infinite and Jac64, p. 61 for M finite]. Then the elements σx, $\sigma \in G$, are linearly independent over M, so they form a basis of N/M. If $\mathrm{trace}_{N/M}(x) = 0$, then $\mathrm{trace}_{N/M}(\sigma x) = 0$ for each $\sigma \in G$, so $\mathrm{trace}_{N/M}(y) = 0$ for all $y \in N$. This contradiction to the fact that $\mathrm{trace}_{N/M} \colon N \to M$ is a nonzero M-linear function [Lan93, p. 286, Thm. 5.2] proves that $a = \mathrm{trace}_{N/M}(x) \ne 0$. Dividing x by a, we may assume that $\mathrm{trace}_{N/M}(x) = 1$.

Now let $n = [N : M] = |G|$ and consider $y \in \mathfrak{m}_M$ and the polynomial

$$f(Z) = -y + Z + a_2 Z^2 + \cdots + a_{n-1} Z^{n-1} + \mathrm{norm}_{N/M}(x) Z^n$$

with $a_k = \sum_\sigma x^{\sigma_1} \cdots x^{\sigma_k}$, where σ ranges over all injections from $\{1, \ldots, k\}$ into G. In particular, $f \in O_M[Z]$. For each $z \in O_M$ we have

$$\mathrm{norm}_{N/M}(1 + xz) = \prod_{\sigma \in G} (1 + x^\sigma z)$$

$$= 1 + \mathrm{trace}_{N/M}(x)z + a_2 z^2 + \cdots + a_{n-1} z^{n-1} + \mathrm{norm}_{N/M}(x) z^n$$

$$= 1 + z + a_2 z^2 + \cdots + a_{n-1} z^{n-1} + \mathrm{norm}_{N/M}(x) z^n,$$

so $f(z) = \mathrm{norm}_{N/M}(1 + xz) - 1 - y$.

Since $y \in \mathfrak{m}_M$, we have

$$f(y) = a_2 y^2 + \cdots + a_{n-1} y^{n-1} + \mathrm{norm}_{N/M}(x) y^n \equiv 0 \bmod \mathfrak{m}_M^2$$

and

$$f'(y) = 1 + 2a_2 y^2 + \cdots + (n-1)a_{n-1}y^{n-2} + n \cdot \mathrm{norm}_{N/M}(x) y^{n-1} \equiv 1 \bmod \mathfrak{m}_M^2.$$

The Henselianity of (M, v) gives a $z \in \mathfrak{m}_M$ with $f(z) = 0$, that is

$$\mathrm{norm}_{N/M}(1 + xz) = 1 + y,$$

as desired.

PART B: $Z^1(G, 1 + \mathfrak{m}_N) \leq B^1(G, 1 + \mathfrak{m}_N)$. Consider a 1-cocycle

$$a \in Z^1(G, 1 + \mathfrak{m}_N).$$

Then $a \in Z^1(G, N^\times)$. Since $H^1(G, N^\times) = 1$ (Hilbert's theorem 90, Subsection 9.3.17), there exists $b \in N^\times$ such that $a_\sigma = (\sigma - 1)b$ for each $\sigma \in G$. Since $v(N^\times) = v(M^\times)$, there exists $b' \in M^\times$ with $v(b') = v(b)$. Then $c = \frac{b}{b'}$ satisfies $v(c) = 0$ and $a_\sigma = (\sigma - 1)c$ for each $\sigma \in G$. Since $a_\sigma \in 1 + \mathfrak{m}_N$, we have $1 = (\bar{\sigma} - 1)\bar{c}$, hence $\bar{\sigma}\bar{c} = \bar{c}$ for all $\sigma \in G$. Therefore, $\bar{c} \in \bar{M}$, so there exists $c' \in O_M$ with $\bar{c'} = \bar{c}$. The element $d = \frac{c}{c'}$ is in $1 + \mathfrak{m}_N$ and satisfies $a_\sigma = (\sigma - 1)d$ for all $\sigma \in G$. This means that $a \in B^1(G, 1 + \mathfrak{m}_N)$, as contended. □

Proposition 11.2.1 has a series of consequences expressed in the following lemmas.

LEMMA 11.2.2: *Let M, N, and G be as in Proposition 11.2.1. Then, for each positive integer i there is a natural isomorphism, $H^i(G, U_N) \cong H^i(G, \bar{N}^\times)$.*

Proof: The short exact sequence $1 \to 1 + \mathfrak{m}_N \to U_N \to \bar{N}^\times \to 1$ of G-modules, in which $U_N \to \bar{N}^\times$ is the reduction map, induces a natural long exact sequence

$$H^i(G, 1 + \mathfrak{m}_N) \to H^i(G, U_N) \to H^i(G, \bar{N}^\times) \to H^{i+1}(G, 1 + \mathfrak{m}_N)$$

(Subsection 9.3.4). The first and the fourth terms of that sequence are trivial by Proposition 11.2.1. Hence the second and the third terms of that sequence are naturally isomorphic. □

LEMMA 11.2.3: *Let M, N, v, and G be as in Proposition 11.2.1. Then for each positive integer i there is a natural short exact sequence*

$$1 \to H^i(G, \bar{N}_v^\times) \to H^i(G, N^\times) \xrightarrow{v} H^i(G, \Gamma_M) \to 0.$$

In particular, for $i = 2$ the following short sequence is exact:

$$0 \to \mathrm{Br}(\bar{N}_v/\bar{M}_v) \to \mathrm{Br}(N/M) \to H^2(G, \Gamma_M) \to 0$$

Proof: The short exact sequence $1 \to U_N \to N^\times \xrightarrow{v} \Gamma_N \to 0$ gives rise to a long exact sequence

$$(2) \qquad \cdots \xrightarrow{\delta} H^i(G, U_N) \to H^i(G, N^\times) \xrightarrow{v} H^i(G, \Gamma_N) \xrightarrow{\delta} \cdots.$$

By Lemma 11.2.2, we may replace $H^i(G, U_N)$ by $H^i(G, \bar{N}^\times)$. Since N/M is unramified, $\Gamma_N = \Gamma_M$. Hence, (2) simplifies to a long exact sequence

$$(3) \qquad \cdots \xrightarrow{\delta} H^i(G, \bar{N}^\times) \to H^i(G, N^\times) \xrightarrow{v} H^i(G, \Gamma_M) \xrightarrow{\delta} \cdots$$

of cohomology groups. We have to prove that each of the homomorphisms δ is the zero map. This is equivalent to proving that the map v in (3) is surjective for each $i \geq 0$.

To this end we consider a finitely generated subgroup A of Γ_M. Since Γ_M is torsion free, A is free. Lifting free generators of A to elements of $M^\times$ gives generators of a subgroup B of $M^\times$ that v maps isomorphically onto A. Since G acts trivially both on M and on Γ_M, $v|_B$ is a G-isomorphism.

$$
\begin{array}{ccc}
N^\times & \xrightarrow{v} & \Gamma_N \\
\uparrow & & \| \\
M^\times & \xrightarrow{v} & \Gamma_M \\
\uparrow & & \uparrow \\
B & \xrightarrow{v} & A
\end{array}
$$

Ignoring the second row and taking cohomology gives a commutative diagram

$$
\begin{array}{ccc}
H^i(G, N^\times) & \xrightarrow{v} & H^i(G, \Gamma_M) \\
\uparrow & & \uparrow \\
H^i(G, B) & \xrightarrow{v} & H^i(G, A)
\end{array}
$$

in which the lower arrow v is an isomorphism. In particular, each element of $H^i(G, A)$ lies in the image of v. Since $H^i(G, \Gamma_M)$ is the inductive limit of all of the groups $H^i(G, A)$ (Subsection 9.3.10), the upper arrow of the preceding diagram is surjective. $\qquad\square$

11.3 Local-Global Theorems for Brauer Groups

We establish a commutative diagram for the Brauer group of a generalized function field of one variable over a field K relating it to the product of the Brauer groups of the Henselizations.

Remark 11.3.1: Let K be a perfect field and F a generalized function field of one variable over K, that is a regular extension of K of transcendence degree 1. We denote the set of all equivalence classes of valuations of F that are trivial on K by $\mathbb{P}(F/K)$. We choose a representative $v_{\mathfrak{p}}$ in each $\mathfrak{p} \in \mathbb{P}(F/K)$ and a Henselian closure $F_{\mathfrak{p}}$ of F at $v_{\mathfrak{p}}$. Then the residue fields $\bar{F}_{\mathfrak{p}}$ of both F and $F_{\mathfrak{p}}$ are the same and so are the value groups $\Gamma_{\mathfrak{p}}$. We extend the residue map of $F_{\mathfrak{p}}$ to a place $x \mapsto \bar{x}$ of F_s onto $\tilde{K} \cup \{\infty\}$ that fixes the elements of $\tilde{K}$. Then the map $\sigma \mapsto \bar{\sigma}$ defined by $\bar{\sigma}\bar{x} = \overline{\sigma x}$ is an epimorphism of $\mathrm{Gal}(F_{\mathfrak{p}})$ onto $\mathrm{Gal}(\bar{F}_{\mathfrak{p}})$. In particular, $\bar{\sigma}x = \sigma x$ for each $\sigma \in \mathrm{Gal}(F_{\mathfrak{p}})$ and every $x \in \tilde{K}$, that is the map $\sigma \to \bar{\sigma}$ is the restriction map. It follows that $F_{\mathfrak{p}} \cap \tilde{K} = \bar{F}_{\mathfrak{p}}$. Moreover, if $\sigma \in \mathrm{Gal}(F_{\mathfrak{p}})$, then $\bar{\sigma}\bar{x} = \bar{x}$ for all $x \in F_s$ with $\bar{x} \in \tilde{K}$ if and only if $\sigma \in \mathrm{Gal}(F_{\mathfrak{p}}\tilde{K})$. Thus, $\mathrm{Gal}(F_{\mathfrak{p}}\tilde{K})$ is the inertia group of the extension of $\mathfrak{p}$ to F_s and the restriction map $\mathrm{Gal}(F_{\mathfrak{p}}\tilde{K}/F_{\mathfrak{p}}) \to \mathrm{Gal}(\bar{F}_{\mathfrak{p}})$ is an isomorphism. $\square$

LEMMA 11.3.2: *Let F be a generalized function field of one variable over a field K and let p be a prime number. Suppose for each function field E of one variable over K in F the map*

$$(1) \qquad\qquad \mathrm{res:} \ \mathrm{Br}(E)_{p^\infty} \to \prod_{\mathfrak{p} \in \mathbb{P}(E/K)} \mathrm{Br}(E_{\mathfrak{p}})_{p^\infty}$$

is injective and its image lies in $\bigoplus_{\mathfrak{p} \in \mathbb{P}(E/K)} \mathrm{Br}(E_{\mathfrak{p}})_{p^\infty}$. Then the map

$$(2) \qquad\qquad \mathrm{res:} \ \mathrm{Br}(F)_{p^\infty} \to \prod_{\mathfrak{p} \in \mathbb{P}(F/K)} \mathrm{Br}(F_{\mathfrak{p}})_{p^\infty}$$

is injective.

Proof: Given an algebraic extension of fields $E \subseteq E'$, we denote the restriction map $\mathrm{Br}(E)_{p^\infty} \to \mathrm{Br}(E')_{p^\infty}$ by $\mathrm{res}^E_{E'}$. Now we consider a function field E of one variable over K in F, let $\mathfrak{p} \in \mathbb{P}(E/K)$, and let $x \in \mathrm{Br}(E_{\mathfrak{p}})_{p^\infty}$. Suppose $\mathrm{res}^{E_{\mathfrak{p}}}_{F_{\mathfrak{q}}}(x) = 0$ for each $\mathfrak{q} \in \mathbb{P}(F/K)$ over $\mathfrak{p}$. Let $\mathcal{E}$ be the set of all finite extensions of E in F. We prove there exists $E' \in \mathcal{E}$ such that $\mathrm{res}^{E_{\mathfrak{p}}}_{E'_{\mathfrak{q}}}(x) = 0$ for each $\mathfrak{q} \in \mathbb{P}(E'/K)$ lying over $\mathfrak{p}$.

To this end we recall that for each $E' \in \mathcal{E}$ the set of prime divisors of E'/K that lie over $\mathfrak{p}$ bijectively corresponds to the set of all $E_{\mathfrak{p}}$-isomorphisms of $E'E_{\mathfrak{p}}$ into E_s. If σ' is such an isomorphism and $\mathfrak{q}'$ is the corresponding prime divisor of E'/K, we choose $\sigma'(E'E_{\mathfrak{p}})$ as the Henselian closure $E'_{\mathfrak{q}'}$ of

E' at $\mathfrak{q}'$. This choice ensures that if E'' is a finite extension of E' in F and $\mathfrak{q}''$ is a prime divisor of E''/K that lies over $\mathfrak{q}'$, then $E'_{\mathfrak{q}'} \subseteq E''_{\mathfrak{q}''}$.

Now assume E has no extension E' as in the first paragraph of the proof. Then for each $E' \in \mathcal{E}$ the finite set $Q(E')$ of all prime divisors $\mathfrak{q} \in \mathbb{P}(E'/K)$ lying over $\mathfrak{p}$ such that $\mathrm{res}_{E'_{\mathfrak{q}}}^{E_{\mathfrak{p}}}(x) \neq 0$ is nonempty. If E'' is a finite extension of E' in F, then restriction of divisors maps $Q(E'')$ into $Q(E')$. Since the inverse limit of nonempty finite sets is nonempty [FrJ08, Cor. 1.1.4], there exists a set $\mathfrak{Q} = \{\mathfrak{q}_{E'} \in Q(E') \mid E' \in \mathcal{E}\}$ such that $\mathfrak{q}_{E'}$ is the restriction of $\mathfrak{q}_{E''}$ for all $E', E'' \in \mathcal{E}$ with $E' \subseteq E''$. The set $\mathfrak{Q}$ determines an element $\mathfrak{q}$ of $\mathbb{P}(F/K)$ such that $\mathrm{res}_{F_{\mathfrak{q}}}^{E_{\mathfrak{p}}}(x) \neq 0$, in contrast to the assumption made in the first paragraph of the proof.

CLAIM: *The map (2) is injective.* Otherwise, there exists $z \in \mathrm{Br}(F)_{p^\infty}$ such that $z \neq 0$ and $\mathrm{res}_{F_{\mathfrak{q}}}^{F}(z) = 0$ for every $\mathfrak{q} \in \mathbb{P}(F/K)$. Since F is the union of function fields E of one variable over K and $\mathrm{Br}(F)_{p^\infty}$ is the direct limit of the groups $\mathrm{Br}(E)_{p^\infty}$ (Subsections 9.3.10 and 9.3.18), there exist such a field E and an element $x \in \mathrm{Br}(E)_{p^\infty}$ with $x \neq 0$ and $\mathrm{res}_F^E(x) = z$. By our assumption on the image of the map (1), $\mathrm{res}_{E_{\mathfrak{p}}}^{E}(x) = 0$ for all but finitely many $\mathfrak{p} \in \mathbb{P}(E/K)$. We denote the exceptional set by P. For each $\mathfrak{p} \in P$ let $x_{\mathfrak{p}} = \mathrm{res}_{E_{\mathfrak{p}}}^{E}(x)$. Then $\mathrm{res}_{F_{\mathfrak{q}}}^{E_{\mathfrak{p}}}(x_{\mathfrak{p}}) = \mathrm{res}_{F_{\mathfrak{q}}}^{F}(z) = 0$ for each $\mathfrak{q} \in \mathbb{P}(F/K)$ lying over $\mathfrak{p}$. By what we have proved above, E has a finite extension $E(\mathfrak{p})$ in F such that $\mathrm{res}_{E(\mathfrak{p})_{\mathfrak{q}}}^{E_{\mathfrak{p}}}(x_{\mathfrak{p}}) = 0$ for each $\mathfrak{q} \in \mathbb{P}(E(\mathfrak{p})/K)$ lying over $\mathfrak{p}$. Let $E' = \prod_{\mathfrak{p} \in P} E(\mathfrak{p})$. Then E' is a finite extension of E in F and $\mathrm{res}_{E'_{\mathfrak{q}}}^{E}(x) = 0$ for each $\mathfrak{p} \in P$ and every $\mathfrak{q} \in \mathbb{P}(E'/K)$ lying over $\mathfrak{p}$. It follows from the definition of P that $\mathrm{res}_{E'_{\mathfrak{q}}}^{E}(x) = 0$ for each $\mathfrak{p} \in \mathbb{P}(E/K)$ and every $\mathfrak{q} \in \mathbb{P}(E'/K)$ lying over $\mathfrak{p}$. Finally, let $y = \mathrm{res}_{E'}^{E}(x)$. Then $\mathrm{res}_{F}^{E'}(y) = z \neq 0$, so $y \neq 0$. On the other hand, $\mathrm{res}_{E'_{\mathfrak{q}}}^{E'}(y) = 0$ for all $\mathfrak{q} \in \mathbb{P}(E'/K)$. This contradicts the injectivity of the map (1). $\square$

LEMMA 11.3.3: *In the notation of Remark 11.3.1 and with $\mathbb{P} = \mathbb{P}(F/K)$ there is a natural commutative diagram*
(3)

$$
\begin{array}{ccc}
\mathrm{Br}(F) \xrightarrow{\;\beta\;} H^2(\mathrm{Gal}(K), (F\tilde{K})^\times) \xrightarrow{\;\gamma\;} H^2(\mathrm{Gal}(K), (F\tilde{K})^\times/\tilde{K}^\times) \\
\downarrow{\scriptstyle\mathrm{res}} \qquad\qquad \downarrow{\scriptstyle\mathrm{res}} \qquad\qquad\qquad\qquad \downarrow \\
\prod_{\mathfrak{p}\in\mathbb{P}} \mathrm{Br}(F_{\mathfrak{p}}) \xrightarrow{\;\beta'\;} \prod_{\mathfrak{p}\in\mathbb{P}} H^2(\mathrm{Gal}(\bar{F}_{\mathfrak{p}}), (F_{\mathfrak{p}}\tilde{K})^\times) \xrightarrow{\;\gamma'\;} \prod_{\mathfrak{p}\in\mathbb{P}} H^2(\mathrm{Gal}(\bar{F}_{\mathfrak{p}}), \Gamma_{\mathfrak{p}})
\end{array}
$$

where β and β' are isomorphisms.

Proof: The inflation-restriction sequence for Brauer groups (Subsection 9.3.18) applied to $\mathrm{Gal}(F)$ and $\mathrm{Gal}(F\tilde{K})$ is
(4)
$$1 \to H^2(\mathrm{Gal}(F\tilde{K}/F), (F\tilde{K})^\times) \xrightarrow{\;\inf\;} H^2(\mathrm{Gal}(F), F_s^\times) \xrightarrow{\;\mathrm{res}\;} H^2(\mathrm{Gal}(F\tilde{K}), F_s^\times).$$

Since F/K is regular, the map res: $\mathrm{Gal}(F\tilde{K}/F) \to \mathrm{Gal}(K)$ is an isomorphism. By Proposition 9.4.6(b), $\mathrm{cd}(\mathrm{Gal}(F\tilde{K})) \leq 1$, so $H^2(\mathrm{Gal}(F\tilde{K}), F_s^\times) = 1$ (Subsection 9.3.18). Thus, inf in (4) is an isomorphism. We denote its inverse by β to get the left upper map in Diagram (3). The homomorphism γ in (3) is induced by the quotient map $(F\tilde{K})^\times \to (F\tilde{K})^\times/\tilde{K}^\times$.

For each $\mathfrak{p} \in \mathbb{P}$ we replace F and K in the preceding argument by $F_\mathfrak{p}$ and $\bar{F}_\mathfrak{p}$, respectively, and use that $F_\mathfrak{p}/\bar{F}_\mathfrak{p}$ is a regular extension (Remark 11.3.1) to produce an isomorphism $\beta_\mathfrak{p}: \mathrm{Br}(F_\mathfrak{p}) \to H^2(\mathrm{Gal}(\bar{F}_\mathfrak{p}), (F_\mathfrak{p}\tilde{K})^\times)$ that commutes with the restriction map. Then we define β' as the product of all the $\beta_\mathfrak{p}$'s.

Similarly, for each $\mathfrak{p} \in \mathbb{P}$, the quotient map $(F_\mathfrak{p}\tilde{K})^\times \to (F_\mathfrak{p}\tilde{K})^\times/\tilde{K}^\times$ yields a homomorphism

$$\gamma_\mathfrak{p}: H^2(\mathrm{Gal}(\bar{F}_\mathfrak{p}), (F_\mathfrak{p}\tilde{K})^\times) \to H^2(\mathrm{Gal}(\bar{F}_\mathfrak{p}), (F_\mathfrak{p}\tilde{K})^\times/\tilde{K}^\times).$$

The valuation $v_\mathfrak{p}$ extended to $F_\mathfrak{p}\tilde{K}$ maps $(F_\mathfrak{p}\tilde{K})^\times$ onto the valuation group $\Gamma_\mathfrak{p}$ and vanish on $\tilde{K}^\times$. So it defines a homomorphism

$$\gamma_\mathfrak{p}': H^2(\mathrm{Gal}(\bar{F}_\mathfrak{p}), (F_\mathfrak{p}\tilde{K})^\times/\tilde{K}^\times) \to H^2(\mathrm{Gal}(\bar{F}_\mathfrak{p}), \Gamma_\mathfrak{p}).$$

We let $\gamma' = \prod_{\mathfrak{p} \in \mathbb{P}} \gamma_\mathfrak{p}' \circ \gamma_\mathfrak{p}$. Finally, noting that $F\tilde{K} = F_\mathfrak{p}\tilde{K}$, the third vertical arrow in (3) is just $\prod_{\mathfrak{p} \in \mathbb{P}} \gamma_\mathfrak{p}' \circ$ res. $\square$

11.4 Picard Groups

Let F be a function field of one variable over a perfect field K. Thus, F/K is a finitely generated regular extension of transcendence degree 1. Let $\mathbb{P} = \mathbb{P}(F/K)$ be the set of prime divisors of F/K (Remark 11.3.1). Using the notation of Remark 5.8.1, we recall that each $\mathfrak{a} \in \mathrm{Div}(F/K)$ has a unique representations as $\mathfrak{a} = \sum_{\mathfrak{p} \in \mathbb{P}} v_\mathfrak{p}(\mathfrak{a})\mathfrak{p}$, with integers $v_\mathfrak{p}(\mathfrak{a})$, all but finitely many are 0. In particular, $\mathrm{div}(f) = \sum_{\mathfrak{p} \in \mathbb{P}} v_\mathfrak{p}(f)\mathfrak{p}$ for each $f \in F^\times$. Thus, the map $\mathfrak{a} \mapsto (v_\mathfrak{p}(\mathfrak{a}))_{\mathfrak{p} \in \mathbb{P}}$ is a natural isomorphism,

$$(1) \qquad\qquad \mathrm{Div}(F/K) \cong \bigoplus_{\mathfrak{p} \in \mathbb{P}} \Gamma_\mathfrak{p}$$

The map div: $F^\times \to \mathrm{Div}(F/K)$ is a homomorphism with $\mathrm{Ker}(\mathrm{div}) = K^\times$ [Deu73, p. 25]. The **Picard group** of F/K is the cokernel of div, also called the **group of divisor classes** of F/K, that is, $\mathrm{Pic}(F/K) = \mathrm{Div}(F/K)/\mathrm{div}(F^\times)$. Since $\mathrm{div}(F^\times) \cong F^\times/K^\times$, we get the following natural short exact sequence:

$$(2) \qquad\qquad 1 \to F^\times/K^\times \xrightarrow{\mathrm{div}} \mathrm{Div}(F/K) \to \mathrm{Pic}(F/K) \to 0.$$

Recall that the group of divisors of degree 0, $\mathrm{Div}_0(F/K)$, contains $\mathrm{div}(F^\times)$ (Remark 5.8.1(a)). Hence, $\mathrm{Pic}_0(F/K) = \mathrm{Div}_0(F/K)/\mathrm{div}(F^\times)$ is a subgroup of $\mathrm{Pic}(F/K)$ and (2) yields the following natural short exact sequence:

$$(3) \qquad 1 \to F^\times / K^\times \xrightarrow{\mathrm{div}} \mathrm{Div}_0(F/K) \to \mathrm{Pic}_0(F/K) \to 0.$$

Analogous convention and rules hold for the function field $F\tilde{K}/\tilde{K}$. Here we write $\tilde{\mathbb{P}} = \mathbb{P}(F\tilde{K}/\tilde{K})$.

LEMMA 11.4.1: *There is a natural isomorphism*

$$\mathrm{Div}(F\tilde{K}/\tilde{K}) \cong \bigoplus_{\mathfrak{p} \in \mathbb{P}} \mathrm{Ind}_{\mathrm{Gal}(\bar{F}_{\mathfrak{p}})}^{\mathrm{Gal}(K)}(\Gamma_{\mathfrak{p}})$$

of $\mathrm{Gal}(K)$-*modules.*

Proof: Consider a prime divisor $\mathfrak{p} \in \mathbb{P}$ and a prime divisor $\mathfrak{P} \in \tilde{\mathbb{P}}$ lying over $\mathfrak{p}$. We identify $\mathrm{Gal}(F\tilde{K}/\tilde{K})$ with $\mathrm{Gal}(K)$ via restriction. For each $\sigma \in \mathrm{Gal}(K)$ the prime divisor $\sigma\mathfrak{P}$ is the equivalence class of the valuation $v_{\sigma\mathfrak{P}}$ of $F\tilde{K}$ defined by $v_{\sigma\mathfrak{P}}(x) = v_{\mathfrak{P}}(\sigma^{-1}x)$. When σ ranges over $\mathrm{Gal}(K)$, the divisor $\sigma\mathfrak{P}$ ranges over all extensions of $\mathfrak{p}$ to $F\tilde{K}$. By Remark 11.3.1, the stabilizer of $\mathfrak{P}$ under this action is $\mathrm{Gal}(\bar{F}_{\mathfrak{p}})$. Hence, $\bigoplus_{\mathfrak{Q}|\mathfrak{p}} \Gamma_{\mathfrak{Q}} = \bigoplus_{\sigma \in S} \Gamma_{\sigma\mathfrak{P}}$, where S is a subset of $\mathrm{Gal}(K)$ satisfying $\mathrm{Gal}(K) = \bigcup_{\sigma \in S} \mathrm{Gal}(\bar{F}_{\mathfrak{p}})\sigma$. Note that for each $\mathfrak{Q} \in \tilde{\mathbb{P}}$ lying over $\mathfrak{p}$ the value group $\Gamma_{\mathfrak{Q}}$ is $\mathbb{Z}$, so we may identify it with $\Gamma_{\mathfrak{p}}$. It follows from Subsection 9.3.12 that $\bigoplus_{\mathfrak{Q}|\mathfrak{p}} \Gamma_{\mathfrak{Q}} = \bigoplus_{\sigma \in S} \Gamma_{\sigma\mathfrak{P}} = \bigoplus_{\sigma \in S} \Gamma_{\mathfrak{p}} = \mathrm{Ind}_{\mathrm{Gal}(\bar{F}_{\mathfrak{p}})}^{\mathrm{Gal}(K)}(\Gamma_{\mathfrak{p}})$. Consequently, $\mathrm{Div}(F\tilde{K}/\tilde{K}) \cong \bigoplus_{\mathfrak{p} \in \mathbb{P}} \bigoplus_{\mathfrak{Q}|\mathfrak{p}} \Gamma_{\mathfrak{Q}} = \bigoplus_{\mathfrak{p} \in \mathbb{P}} \mathrm{Ind}_{\mathrm{Gal}(\bar{F}_{\mathfrak{p}})}^{\mathrm{Gal}(K)}(\Gamma_{\mathfrak{p}})$, as claimed. $\qquad\square$

LEMMA 11.4.2: *Let G be a profinite group acting trivially on a discrete torsion free Abelian group A. Then $H^1(G, A) = \mathrm{Hom}(G, A) = 0$.*

Proof: The left equality follows from the definition of H^1 (Subsection 9.3.2). Each element of $\mathrm{Hom}(G, A)$ is a continuous homomorphism $f\colon G \to A$. Its image is a compact subgroup, so must be finite. Since A is torsion free, $f(G) = 0$. Therefore, $\mathrm{Hom}(G, A) = 0$. $\qquad\square$

LEMMA 11.4.3: *Let F be a function field of one variable over a perfect field K. Then there is a natural exact sequence*

$$0 \to H^1(\mathrm{Gal}(K), \mathrm{Pic}(F\tilde{K}/\tilde{K})) \to H^2(\mathrm{Gal}(K), (F\tilde{K})^\times/\tilde{K}^\times)$$

$$(4) \qquad \to \bigoplus_{\mathfrak{p} \in \mathbb{P}(F/K)} H^2(\mathrm{Gal}(\bar{F}_{\mathfrak{p}}), \Gamma_{\mathfrak{p}}) \to H^2(\mathrm{Gal}(K), \mathrm{Pic}(F\tilde{K}/\tilde{K}))$$

$$\to H^3(\mathrm{Gal}(K), (F\tilde{K})^\times/\tilde{K}^\times).$$

Proof: As above we set $\mathbb{P} = \mathbb{P}(F/K)$ and start from the short exact sequence for $F\tilde{K}/\tilde{K}$ analogous to (2):

$$1 \to (F\tilde{K})^\times/\tilde{K}^\times \xrightarrow{\mathrm{div}} \mathrm{Div}(F\tilde{K}/\tilde{K}) \to \mathrm{Pic}(F\tilde{K}/\tilde{K}) \to 0.$$

It induces a long exact sequence:

$$H^1(\mathrm{Gal}(K), \mathrm{Div}(F\tilde{K}/\tilde{K})) \to H^1(\mathrm{Gal}(K), \mathrm{Pic}(F\tilde{K}/\tilde{K}))$$

(5) $$\to H^2(\mathrm{Gal}(K), (F\tilde{K})^\times/\tilde{K}^\times) \to H^2(\mathrm{Gal}(K), \mathrm{Div}(F\tilde{K}/\tilde{K}))$$

$$\to H^2(\mathrm{Gal}(K), \mathrm{Pic}(F\tilde{K}/\tilde{K})) \to H^3(\mathrm{Gal}(K), (F\tilde{K})^\times/\tilde{K}^\times).$$

By Lemma 11.4.1 and by Shapiro's Lemma (Subsection 9.3.12), we have for $i = 1, 2$ natural isomorphism

$$H^i(\mathrm{Gal}(K), \mathrm{Div}(F\tilde{K}/\tilde{K})) \cong \bigoplus_{\mathfrak{p} \in \mathbb{P}} H^i(\mathrm{Gal}(K), \mathrm{Ind}_{\mathrm{Gal}(\bar{F}_\mathfrak{p})}^{\mathrm{Gal}(K)}(\Gamma_\mathfrak{p}))$$

$$\cong \bigoplus_{\mathfrak{p} \in \mathbb{P}} H^i(\mathrm{Gal}(\bar{F}_\mathfrak{p}), \Gamma_\mathfrak{p}),$$

where the action of $\mathrm{Gal}(\bar{F}_\mathfrak{p})$ on $\Gamma_\mathfrak{p}$ is trivial. Since $\Gamma_\mathfrak{p}$ is a torsion free discrete Abelian group, $H^1(\mathrm{Gal}(K), \mathrm{div}(F\tilde{K}/\tilde{K})) = 0$ (Lemma 11.4.2). Collecting this information into (5) gives the exact sequence (4).   □

LEMMA 11.4.4: *Let F be a function field of one variable over a perfect field K and let p be a prime number. Then:*
(a) *The natural map*

$$H^1(\mathrm{Gal}(K), \mathrm{Pic}_0(F\tilde{K}/\tilde{K})) \to H^1(\mathrm{Gal}(K), \mathrm{Pic}(F\tilde{K}/\tilde{K}))$$

is surjective.
(b) *If F/K has a prime divisor of degree 1, then*

$$H^i(\mathrm{Gal}(K), \mathrm{Pic}_0(F\tilde{K}/\tilde{K}))_{p^\infty} = 0$$

for each $i > \mathrm{cd}_p(\mathrm{Gal}(K))$ and
(c) *there is a natural isomorphism*

$$H^i(\mathrm{Gal}(K), \mathrm{Pic}(F\tilde{K}/\tilde{K}))_{p^\infty} \cong H^{i-1}(\mathrm{Gal}(K), \mathbb{Q}/\mathbb{Z})_{p^\infty}$$

for each $i > \max(1, \mathrm{cd}_p(\mathrm{Gal}(K)))$.

Proof of (a): The definition of the Picard groups gives rise to a short exact sequence

(6) $$0 \to \mathrm{Pic}_0(F\tilde{K}/\tilde{K}) \to \mathrm{Pic}(F\tilde{K}/\tilde{K}) \xrightarrow{\deg} \mathbb{Z} \to 0$$

of $\mathrm{Gal}(K)$-modules. We consider a segment of the corresponding long exact sequence of cohomology groups:

$$H^1(\mathrm{Gal}(K), \mathrm{Pic}_0(F\tilde{K}/\tilde{K})) \to H^1(\mathrm{Gal}(K), \mathrm{Pic}(F\tilde{K}/\tilde{K})) \to H^1(\mathrm{Gal}(K), \mathbb{Z}).$$

Since $\mathrm{Gal}(K)$ acts trivially on $\mathbb{Z}$, Lemma 11.4.2 implies that $H^1(\mathrm{Gal}(K),\mathbb{Z}) = 0$, hence (a) is true.

Proof of (b): Let J be the Jacobian variety of F/K. By Subsection 6.3.1, $J(\tilde{K})$ is a divisible Abelian group. Hence, multiplication by p^n gives a short exact sequence:

$$0 \to J(\tilde{K})_{p^n} \to J(\tilde{K}) \xrightarrow{p^n} J(\tilde{K}) \to 0,$$

which in turn gives for each positive integer i a long exact sequence

(7) $\qquad H^i(\mathrm{Gal}(K), J(\tilde{K})_{p^n}) \to H^i(\mathrm{Gal}(K), J(\tilde{K}))$
$$\xrightarrow{p^n} H^i(\mathrm{Gal}(K), J(\tilde{K}))$$
$$\to H^{i+1}(\mathrm{Gal}(K), J(\tilde{K})_{p^n}).$$

If $i > \mathrm{cd}_p(\mathrm{Gal}(K))$, then both the first and the last groups in (7) are zero. Therefore multiplication with p^n is an automorphism of $H^i(\mathrm{Gal}(K), J(\tilde{K}))$. In particular, $H^i(\mathrm{Gal}(K), J(\tilde{K}))_{p^\infty} = 0$. Finally, by Subsection 6.3.2,

$$\mathrm{Pic}_0(F\tilde{K}/\tilde{K}) \cong J(\tilde{K})$$

as $\mathrm{Gal}(K)$-modules. Consequently, $H^i(\mathrm{Gal}(K), \mathrm{Pic}_0(F\tilde{K}/\tilde{K}))_{p^\infty} = 0$, as claimed.

Proof of (c): For each $i \geq 0$ the short exact sequence (6) induces an exact sequence

$$H^i(\mathrm{Gal}(K), \mathrm{Pic}_0(F\tilde{K}/\tilde{K})) \to H^i(\mathrm{Gal}(K), \mathrm{Pic}(F\tilde{K}/\tilde{K}))$$
(8) $$\to H^i(\mathrm{Gal}(K), \mathbb{Z})$$
$$\to H^{i+1}(\mathrm{Gal}(K), \mathrm{Pic}_0(F\tilde{K}/\tilde{K})).$$

By Subsection 9.3.10, the p-primary part

$$H^i(\mathrm{Gal}(K), \mathrm{Pic}_0(F\tilde{K}/\tilde{K}))_{p^\infty} \to H^i(\mathrm{Gal}(K), \mathrm{Pic}(F\tilde{K}/\tilde{K}))_{p^\infty}$$
$$\to H^i(\mathrm{Gal}(K), \mathbb{Z})_{p^\infty}$$
$$\to H^{i+1}(\mathrm{Gal}(K), \mathrm{Pic}_0(F\tilde{K}/\tilde{K}))_{p^\infty}$$

of (8) is also exact. By (b), the first and the last groups in the latter sequence are zero if $i > \mathrm{cd}_p(\mathrm{Gal}(K))$. In addition, by Lemma 9.3.11, there is a natural isomorphism $H^i(\mathrm{Gal}(K), \mathbb{Z}) \cong H^{i-1}(\mathrm{Gal}(K), \mathbb{Q}/\mathbb{Z})$ if $i \geq 2$. Hence, there is a natural isomorphism as in (c) if $i > \max(1, \mathrm{cd}_p(\mathrm{Gal}(K)))$. $\square$

11.5 Fields of Cohomological Dimension at most 1

We analyze the exact sequence of Lemma 11.4.3 in the case where $\mathrm{cd}(\mathrm{Gal}(K)) \leq 1$ and prove a local-global principle for Brauer groups of generalized function fields of one variable over perfect PAC fields.

LEMMA 11.5.1: *Let F be a generalized function field over a perfect field K with $\mathrm{cd}(\mathrm{Gal}(K)) \leq 1$. Then:*
(a) *The natural homomorphism*

$$\gamma\colon H^2(\mathrm{Gal}(K), (F\tilde{K})^\times) \to H^2(\mathrm{Gal}(K), (F\tilde{K})^\times/\tilde{K}^\times)$$

induced by the quotient map $(F\tilde{K})^\times \to (F\tilde{K})^\times/\tilde{K}^\times$ is an isomorphism.
(b) $H^i(\mathrm{Gal}(K), (F\tilde{K})^\times/\tilde{K}^\times) = 0$ *for $i \geq 3$.*
(c) *For each $\mathfrak{p} \in \mathbb{P}(F/K)$, the valuation map*

$$H^2(\mathrm{Gal}(\bar{F}_\mathfrak{p}), (F_\mathfrak{p}\tilde{K})^\times) \to H^2(\mathrm{Gal}(\bar{F}_\mathfrak{p}), \Gamma_\mathfrak{p})$$

is an isomorphism.

Proof: The short exact sequence

$$1 \to \tilde{K}^\times \to (F\tilde{K})^\times \to (F\tilde{K})^\times/\tilde{K}^\times \to 1$$

of $\mathrm{Gal}(K)$-modules gives rise to an exact sequence

$$\begin{aligned}(1) \quad &H^2(\mathrm{Gal}(K), \tilde{K}^\times) \to H^2(\mathrm{Gal}(K), (F\tilde{K})^\times) \\ &\to H^2(\mathrm{Gal}(K), (F\tilde{K})^\times/\tilde{K}^\times) \to H^3(\mathrm{Gal}(K), \tilde{K}^\times)\end{aligned}$$

of cohomology groups. Since $\mathrm{cd}(\mathrm{Gal}(K)) \leq 1$, we have $H^2(\mathrm{Gal}(K), \tilde{K}^\times) \cong \mathrm{Br}(K) = 0$ (Subsection 9.3.18) and $H^3(\mathrm{Gal}(K), \tilde{K}^\times) = 0$ for each $i \geq 3$ (Subsection 9.3.15). Thus, (a) follows from (1). Moreover, (b) holds.

Finally, let $\mathfrak{p} \in \mathbb{P}(F/K)$ and apply Lemma 11.2.3 for $F_\mathfrak{p}$, $F_\mathfrak{p}\tilde{K}$, and $v_\mathfrak{p}$ rather than to M, N, and v. Recall that we have identified $\mathrm{Gal}(F_\mathfrak{p}\tilde{K}/F_\mathfrak{p})$ with $\mathrm{Gal}(\bar{F}_\mathfrak{p})$ (Remark 11.3.1). Hence, that lemma gives a short exact sequence

$$0 \to \mathrm{Br}(\bar{F}_\mathfrak{p}) \to H^2(\mathrm{Gal}(\bar{F}_\mathfrak{p}), (F_\mathfrak{p}\tilde{K})^\times) \to H^2(\mathrm{Gal}(\bar{F}_\mathfrak{p}), \Gamma_\mathfrak{p}) \to 0.$$

Now we use that $\mathrm{Gal}(\bar{F}_\mathfrak{p})$ as a closed subgroup of $\mathrm{Gal}(K)$ has cohomological dimension at most 1 to deduce that $\mathrm{Br}(\bar{F}_\mathfrak{p}) = 0$ and conclude the proof of (c). $\square$

LEMMA 11.5.2: *Let F be a generalized function field over a perfect field K with $\mathrm{cd}(K) \leq 1$. Then there is a natural commutative square*

$$\begin{array}{ccc} \mathrm{Br}(F) & \longrightarrow & H^2(\mathrm{Gal}(K), (F\tilde{K})^\times/\tilde{K}^\times) \\ \downarrow{\scriptstyle \mathrm{res}} & & \downarrow \\ \prod_{\mathfrak{p}\in\mathbb{P}(F/K)} \mathrm{Br}(F_\mathfrak{p}) & \longrightarrow & \prod_{\mathfrak{p}\in\mathbb{P}(F/K)} H^2(\mathrm{Gal}(\bar{F}_\mathfrak{p}), \Gamma_\mathfrak{p}) \end{array}$$

where the horizontal arrows are isomorphisms.

Proof: By Lemma 11.5.1, the maps γ and γ' of Lemma 11.3.3 are isomorphisms. Hence, so are the maps $\gamma \circ \beta$ and $\gamma' \circ \beta'$ of the latter lemma. Therefore, the diagram of Lemma 11.3.3 shrinks to the diagram of our lemma.   □

LEMMA 11.5.3: *Let F be a function field of one variable over a perfect field K. Suppose F/K has a prime divisor of degree 1 and $\operatorname{cd}(\operatorname{Gal}(K)) \leq 1$. Then there exists a natural exact sequence*

$$0 \to H^1(\operatorname{Gal}(K), \operatorname{Pic}(F\tilde{K}/\tilde{K})) \to \operatorname{Br}(F)$$

$$\xrightarrow{\text{res}} \bigoplus_{\mathfrak{p} \in \mathbb{P}(F/K)} \operatorname{Br}(F_{\mathfrak{p}})$$

$$\to \operatorname{Hom}(\operatorname{Gal}(K), \mathbb{Q}/\mathbb{Z}) \to 0$$

Proof: We apply Lemma 11.5.2 to replace

$$H^2(\operatorname{Gal}(K), (F\tilde{K})^\times / \tilde{K}^\times) \quad \text{and} \quad \bigoplus_{\mathfrak{p} \in \mathbb{P}(F/K)} H^2(\operatorname{Gal}(\bar{F}_{\mathfrak{p}}), \Gamma_{\mathfrak{p}})$$

in the exact sequence of Lemma 11.4.3 by $\operatorname{Br}(F)$ and $\bigoplus_{\mathfrak{p} \in \mathbb{P}(F/K)} \operatorname{Br}(F_{\mathfrak{p}})$, respectively. Since $\operatorname{cd}(\operatorname{Gal}(K)) \leq 1$ and each cohomology group of positive degree is the sum of its primary parts, Lemma 11.4.4(c) implies that

$$H^2(\operatorname{Gal}(K), \operatorname{Pic}(F\tilde{K}/\tilde{K})) \cong H^1(\operatorname{Gal}(K), \mathbb{Q}/\mathbb{Z}) = \operatorname{Hom}(\operatorname{Gal}(K), \mathbb{Q}/\mathbb{Z}).$$

By Lemma 11.5.1(b), $H^3(\operatorname{Gal}(K), (F\tilde{K})^\times / \tilde{K}^\times) = 0$. Consequently, the exact sequence of Lemma 11.4.3 becomes the sequence of our lemma.   □

LEMMA 11.5.4: *Let F be a function field of one variable over a perfect PAC field K. Then there is a natural exact sequence*

$$0 \to \operatorname{Br}(F) \xrightarrow{\text{res}} \bigoplus_{\mathfrak{p} \in \mathbb{P}(F/K)} \operatorname{Br}(F_{\mathfrak{p}}) \to \operatorname{Hom}(\operatorname{Gal}(K), \mathbb{Q}/\mathbb{Z}) \to 0$$

Proof: Let J be the Jacobian variety of F/K. Since K is PAC,

$$H^1(\operatorname{Gal}(K), J(\tilde{K})) = 0.$$

(Subsection 6.3.3). By Subsection 6.3.2, $\operatorname{Pic}_0(F\tilde{K}/\tilde{K}) \cong J(\tilde{K})$ as $\operatorname{Gal}(K)$-modules. Hence,

$$H^1(\operatorname{Gal}(K), \operatorname{Pic}_0(F\tilde{K}/\tilde{K})) = 0.$$

Therefore, by Lemma 11.4.4(a), $H^1(\operatorname{Gal}(K), \operatorname{Pic}(F\tilde{K}/\tilde{K})) = 0$. Consequently, the exact sequence of Lemma 11.5.3 shortens to the exact sequence of the present lemma.   □

Using lemma 11.3.2, we extract the following result for generalized function fields from Lemma 11.5.4:

PROPOSITION 11.5.5 (Efrat): *Let F be a generalized function field of one variable over a perfect PAC field K. Then the restriction map*

$$\text{res: Br}(F) \rightarrow \prod_{\mathfrak{p} \in \mathbb{P}(F/K)} \text{Br}(F_{\mathfrak{p}})$$

is injective

COROLLARY 11.5.6: *Let F be a generalized function field of one variable over a perfect PAC field K. Suppose $\text{Gal}(F_{\mathfrak{p}})$ is projective for each $\mathfrak{p} \in \mathbb{P}(F/K)$. Then $\text{Gal}(F)$ is projective.*

Proof: For each $\mathfrak{p} \in \mathbb{P}(F/K)$ the group $\text{Gal}(F_{\mathfrak{p}})$ is projective, hence $\text{cd}(\text{Gal}(F_{\mathfrak{p}})) \leq 1$ (Subsection 9.3.16). Since K is perfect, $\text{Br}(F_{\mathfrak{p}}) = 0$ (Subsection 9.3.18). Therefore, by Proposition 11.5.5, $\text{Br}(F) = 0$.

The same conclusion holds for every finite separable extension F' of F, because the algebraic closure of K in F' is perfect and PAC [FrJ08, Corollary 11.2.5] and closed subgroups of projective groups are projective [FrJ08, Prop. 22.4.7]. By Subsection 9.3.18, $\text{Gal}(F)$ is projective. □

11.6 Radical Extensions

We call an algebraic extension F/E of fields of characteristic p **radical** if for each $a \in E$ and every positive integer n with $p \nmid n$ there exists $x_{a,n} \in F$ such that $x_{a,n}^n = a$ and $F = E(x_{a,n})_{a \in E, \, p \nmid n}$. The following conjecture is a variant of a conjecture of Bogomolov-Positselski [BoP05, Conjecture 1.1]:

CONJECTURE 11.6.1: *Let E be an extension of a field K with trans.deg$(E/K) = 1$ and F an algebraic extension of E. Suppose F contains a radical algebraic extension of E. Then $\text{Gal}(F)$ is projective.*

We prove Conjecture 11.6.1 in the special case where K is PAC. It turns out that in this case it suffices to adjoin much less radicals to E than demanded by the definition of the radical extension.

Definition 11.6.2: K-radical extensions. Let E/K be a function field of one variable and F an algebraic extension of E. In the notation of Remark 11.3.1 we say that F/E is a **K-radical extension** if for each $\mathfrak{p} \in \mathbb{P}(E/K)$ and for each positive integer n with $\text{char}(K) \nmid n$ there exists an element $x_{\mathfrak{p},n} \in F$ such that $x_{\mathfrak{p},n}^n \in E$, $v_{\mathfrak{p}}(x_{\mathfrak{p},n}^n) = 1$, and $F = K(x_{\mathfrak{p},n})_{\mathfrak{p} \in \mathbb{P}(E/K), \text{char}(K) \nmid n}$.

In particular, if F/E is a radical extension, then F/E is also a K-radical extension. □

Definition 11.6.3: Let K be a field of characteristic p and F an extension of K of transcendence degree 1. We say that F has **p'-divisible K-functional valuation groups** if the value group of F at each valuation trivial on K is p'-divisible.

Note that in that case each algebraic extension F' of F also has p'-divisible K-functional valuation groups (Remark 11.1.2). □

LEMMA 11.6.4: *Let p be either 0 or a prime number and let Γ be an additive subgroup of $\mathbb{Q}$. Suppose $\frac{1}{n} \in \Gamma$ for each positive integer n with $p \nmid n$. Then Γ is p'-divisible.*

Proof: We consider $\gamma \in \Gamma$. If $p = 0$, we write $\gamma = \frac{a}{b}$, with $a \in \mathbb{Z}$ and $b \in \mathbb{N}$. Given $n \in \mathbb{N}$, we have $\frac{\gamma}{n} = a \cdot \frac{1}{nb} \in \Gamma$.

If $p > 0$, we write $\gamma = \frac{a}{bp^k}$, where $a \in \mathbb{Z}$, $b \in \mathbb{N}$, $k \in \mathbb{Z}$, and $p \nmid a, b$. Let $n \in \mathbb{N}$ with $p \nmid n$. If $k < 0$, then $\frac{\gamma}{n} = ap^{-k} \cdot \frac{1}{nb} \in \Gamma$. If $k > 0$, we may choose $x, y \in \mathbb{Z}$ such that $xp^k + ynb = 1$. Then $\frac{\gamma}{n} = \frac{a}{nbp^k} = \frac{axp^k + aynb}{nbp^k} = ax \cdot \frac{1}{nb} + by \cdot \frac{a}{bp^k} \in \Gamma$, as claimed. $\quad\square$

LEMMA 11.6.5: *Let E/K be a function field of one variable of characteristic p and F a K-radical extension of E. Then F has a p'-divisible K-functional valuation groups.*

Proof: Let $\mathfrak{p} \in \mathbb{P}(E/K)$ and consider a valuation w of F extending $v_\mathfrak{p}$. Thus, $w(y) = v_\mathfrak{p}(y)$ for each $y \in E$. Since F is an algebraic extension of E the value group Γ of w is contained in $\mathbb{Q}$. On the other hand, for each $\mathfrak{p} \in \mathbb{P}(E/K)$ and each n not divisible by p there is $x_{\mathfrak{p},n} \in F$ such that $\frac{1}{n} = \frac{1}{n} v_\mathfrak{p}(x_{\mathfrak{p},n}^n) = w(x_{\mathfrak{p},n}) \in \Gamma$. By Lemma 11.6.4, Γ is p'-divisible. $\quad\square$

PROPOSITION 11.6.6: *Let K be a PAC field of characteristic p, F an extension of K of transcendence degree 1 with p'-divisible K-functional valuation groups. Then $\mathrm{Gal}(F)$ is projective.*

Proof: By assumption, the value group of each valuation of F/K is p'-divisible. Hence, so is the value group of each valuation of every algebraic extension F' of F, therefore also of each Henselian closure of F'.

By Ax-Roquette, each algebraic extension of a PAC field is again PAC [FrJ08, Cor. 11.2.5]. Hence, we may first replace K by K_{ins} and F by FK_{ins} to assume that K is perfect. Then, we may replace K by $F \cap \tilde{K}$ to assume that F is a generalized function field of one variable over K.

Now we consider a prime divisor $\mathfrak{p}$ of F/K and its Henselization $F_\mathfrak{p}$. The residue field $\bar{F}_\mathfrak{p}$ is an algebraic extension of K, so $\bar{F}_\mathfrak{p}$ is PAC. Hence, by [FrJ08, Thm. 11.6.2], $\mathrm{Gal}(\bar{F}_\mathfrak{p})$ is projective. It follows from Proposition 11.1.3 that $\mathrm{Gal}(F_\mathfrak{p})$ is projective. By Corollary 11.5.6, $\mathrm{Gal}(F)$ is projective. $\quad\square$

COROLLARY 11.6.7: *Let K be a PAC field, E a function field of one variable over K, and F an algebraic extension of a K-radical extension of E. Then $\mathrm{Gal}(F)$ is projective.*

Proof: Let $p = \mathrm{char}(K)$. By Lemma 11.6.5 and Definition 11.6.3, F has a p'-divisible K-functional valuation groups. Hence, by Proposition 11.6.6, $\mathrm{Gal}(F)$ is projective. $\quad\square$

11.7 Semi-Free Absolute Galois Groups

The chapter culminates with its main new result. We construct for each PAC field K of cardinality m an algebraic extension F of $K(x)$ in $K_{\text{cycl}}(x)_{\text{ab}}$ such that F is Hilbertian and $\text{Gal}(F) \cong \hat{F}_m$. If K contains all roots of unity, then $\text{Gal}(K(x)_{\text{ab}}) \cong \hat{F}_m$. The latter result, can be considered as an analog of a well known conjecture of Shafarevich saying that $\text{Gal}(\mathbb{Q}_{\text{ab}}) \cong \hat{F}_\omega$.

First of all we apply the main result of Chapter 8 to the absolute Galois group of a function field of one variable over an ample field.

THEOREM 11.7.1: *Let E be a function field of one variable over an ample field K of cardinality m. Then $\text{Gal}(E)$ is semi-free of rank m.*

Proof: We choose a separating transcendence element x for E/K. Since K is ample, m is infinite and $m = \text{card}(K(x)) = \text{card}(E)$. Hence, $\text{rank}(\text{Gal}(K(x))) \leq m$. By Proposition 8.6.3, each finite split embedding problem for $\text{Gal}(K(x))$ with a nontrivial kernel has m linearly disjoint solutions. Thus, $\text{Gal}(K(x))$ is semi-free of rank m (Remark 10.1.6). Since $\text{Gal}(E)$ is an open subgroup of $\text{Gal}(K(x))$, $\text{Gal}(E)$ is semi-free of rank m (Lemma 10.4.1). $\qquad\square$

The combination of Theorems 10.5.8 and 11.7.1 gives the following result:

THEOREM 11.7.2: *Let K be an ample field with $\text{rank}(\text{Gal}(K)) = m$, E a function field of one variable over K, and F a separable algebraic extension of E. Then $\text{Gal}(F)$ is a semi-free profinite group of rank m in each of the following cases:*
(a) $[F : E] < \infty$.
(b) $\text{weight}(F/E) < m$.
(c) F/E is small.
(d) F is contained in an E-diamond.
(e) F is a proper finite extension of an extension E_0 of E and E_0 is contained in a Galois extension N of E that does not contain F.
(f) F is a proper finite extension of a Galois extension of E.
(g) F/E is Abelian.

The next construction will allow us to move from a function field F of one variable to infinite extensions of F that are not too large.

LEMMA 11.7.3: *Let E be a function field of one variable over a field K, F a finite extension of E, and $\mathfrak{p}$ a prime divisor of E/K tamely and totally ramified in F. Then F is a regular extension of K.*

Proof: The extension F/E is separable, because $\mathfrak{p}$ is tamely and totally ramified in F. Since E/K is separable, also F/K is separable.

It remains to prove that K is algebraically closed in F. Thus, it suffices to prove that $F \cap EL = E$ for each finite extension L of K. Indeed, let L_0 be the maximal separable extension of K in L. Then $\mathfrak{p}$ is unramified in EL_0. Hence, $F \cap EL_0 = E$ and each extension $\mathfrak{p}'$ of $\mathfrak{p}$ to EL_0 is tamely and totally ramified in FL_0. Since EL/EL_0 is purely inseparable, $\mathfrak{p}'$ is either unramified or wildly ramified in EL. Therefore, $FL_0 \cap EL = EL_0$. Consequently, $F \cap EL = E$. $\qquad\square$

Given a field E and a prime number p, we write $E_{\mathrm{ab}}^{(p')}$ for the maximal Abelian extension of E of degree prime to p.

Construction 11.7.4: Special K-radical extensions. Let K be a field of characteristic p and infinite cardinality m. Let x be a variable and set $E = K(x)$. We denote the set of all monic irreducible polynomials of $K[x]$ by $\mathcal{F}$. Let $\mathcal{F} = \bigcup_{i=1}^{r} \mathcal{F}_i$ be a partition of $\mathcal{F}$ such that $\mathrm{card}(\mathcal{F}_i) = \mathcal{F} = m$ for $i = 1, \ldots, r$. For each i we choose a wellordering $\mathcal{F}_i = (f_{i,\alpha})_{\alpha < m}$. Then, for each $\alpha < m$ and every positive integer n with $p \nmid n$ we choose a root $(f_{1,\alpha} \cdots f_{r,\alpha})^{1/n}$ in E_s such that if $n = dd'$, then $\big((f_{1,\alpha} \cdots f_{r,\alpha})^{1/n}\big)^d = (f_{1,\alpha} \cdots f_{r,\alpha})^{1/d'}$. Then we consider the separable algebraic field extension

$$F_0 = E\big((f_{1,\alpha} \cdots f_{r,\alpha})^{1/n}\big)_{\alpha < m,\, p \nmid n}$$

of E and call F_0 a **special K-radical extension** of E. Note that $F_0 K_{\mathrm{cycl}}$ is an Abelian extension of $K_{\mathrm{cycl}}(x)$ of degree not divisible by p. Hence, $F_0 \subseteq K_{\mathrm{cycl}}(x)_{\mathrm{ab}}^{(p')}$.

In the special case where $r = 1$, the presentation of F_0 is simplified to $F_0 = E(f^{1/n})_{f \in \mathcal{F},\, p \nmid n}$. $\qquad\square$

LEMMA 11.7.5: *Let K, x, E, and F_0 be as in Construction 11.7.4. Then:*
(a) *F_0/E is a K-radical extension (Definition 11.6.2).*
(b) *F_0/K is regular, thus F_0/K is a generalized function field of one variable.*
(c) *Every extension F of F_0 in $K_{\mathrm{cycl}}(x)_{\mathrm{ab}}^{(p')}$ is contained in an E-diamond, hence F is Hilbertian.*
(d) *If K contains no primitive root of order l for some prime number $l \neq \mathrm{char}(K)$, then F_0/E is not Galois.*

Proof of (a): For each prime divisor $\mathfrak{p} \neq \mathfrak{p}_{x,\infty}$ of $K(x)/K$ there exist (unique) $1 \leq j \leq r$ and $\alpha < m$ such that $v_{\mathfrak{p}}(f_{j,\alpha}) = 1$. Since the $\mathcal{F}_i$'s are disjoint, $v_{\mathfrak{p}}(f_{i,\alpha}) = 0$ if $i \neq j$. For $p \nmid n$, let $x_{\mathfrak{p},n} = (f_{1,\alpha} \cdots f_{r,\alpha})^{1/n}$. Then $x_{\mathfrak{p},n} \in F_0$, $x_{\mathfrak{p},n}^n \in E$, and $v_{\mathfrak{p}}(x_{\mathfrak{p},n}^n) = 1$. Next, for $\mathfrak{p} = \mathfrak{p}_{x,\infty}$ we set $\mathfrak{p}' = \mathfrak{p}_{x,0}$ and $x_{\mathfrak{p},n} = x_{\mathfrak{p}',n}^{-1}$. Then $x_{\mathfrak{p},n} \in F_0$, $x_{\mathfrak{p},n}^n \in E$, and $v_{\mathfrak{p}}(x_{\mathfrak{p},n}^n) = 1$. Finally, by construction, F_0 is the field obtained from E by adjoining all $x_{\mathfrak{p},n}$ where $\mathfrak{p} \in \mathbb{P}(E/K)$ and $p \nmid n$. Thus, F_0 is a K-radical extension of E.

Proof of (b): Every finite extension E' of E in F_0 is contained in a field

$$E_r = E(f_1^{1/n_1}, \ldots, f_r^{1/n_r}),$$

where $f_1, \ldots, f_r$ are distinct elements of $\mathcal{F}$ and $n_1, \ldots, n_r$ are positive integers not divisible by p. Inductively assume $E_{r-1} = E(f_1^{1/n_1}, \ldots, f_{r-1}^{1/n_{r-1}})$ is a regular extension of K. For $i = 1, \ldots, r$, let v_i be the valuation of E/K satisfying $v_i(f_i) = 1$. Then $v_i(f_j) = 0$ for $i \neq j$. By [FrJ08, Example 2.3.8], v_r is unramified in E_{r-1}. Let w be an extension of v_r to E_{r-1}. Then $w(f_r) = 1$, so again by [FrJ08, Example 2.3.8], w tamely and totally ramifies in E_r. By Lemma 11.7.3, E_r/K is regular. Consequently, F_0 is a regular extension of K.

Proof of (c): Let N_1 be the field obtained from E by adjoining all roots of unity ζ_n and all roots $x^{1/n}$ with $p \nmid n$. Let N_2 be the field obtained from E by adjoining all ζ_n and all roots $f^{1/n}$ with $f \in \mathcal{F} \smallsetminus \{x\}$ and $p \nmid n$. Then both N_1 and N_2 are Galois extensions of E and $N_1 N_2 = K_{\mathrm{cycl}}(x)_{\mathrm{ab}}^{(p')}$, so $F \subseteq N_1 N_2$. Moreover, $\mathfrak{p}_{x,1}$ is ramified in F_0 but unramified in N_1, so $F \not\subseteq N_1$. Similarly, $\mathfrak{p}_{x,0}$ is ramified in F_0 but unramified in N_2, so $F_0 \not\subseteq N_2$, hence $F \not\subseteq N_2$. Thus, F is contained in a diamond over E. By [FrJ08, Thm. 13.4.2], E is Hilbertian. Hence, by Haran's diamond theorem [FrJ08, Thm. 13.8.3], F is Hilbertian.

Proof of (d): Now we assume that $\zeta_l \notin K$ for some prime number $l \neq p$. Then, by (b), $\zeta_l \notin F_0$. Let $f = f_{1,0} \cdots f_{r,0}$. Then $f^{1/l} \in F_0$. If F_0/E is Galois, then also $\zeta_l f^{1/l} \in F_0$, hence $\zeta_l \in F_0$. It follows from this contradiction that F_0 is not a Galois extension of E. $\qquad\square$

THEOREM 11.7.6: *Let K be a PAC field of characteristic p and cardinality m and let F_0 be a special K-radical extension of $E = K(x)$ (Construction 11.7.4). Then:*

(a) *Every extension F of F_0 in $K_{\mathrm{cycl}}(x)_{\mathrm{ab}}^{p'}$ is Hilbertian and $\mathrm{Gal}(F) \cong \hat{F}_m$.*
(b) *If K contains no primitive root of order l for some prime number $l \neq p$, then F_0/E is not Galois.*
(c) *If K contains all roots of unity, then E_{ab} is a Hilbertian field with $\mathrm{Gal}(E_{\mathrm{ab}}) \cong \hat{F}_m$.*

Proof: By Lemma 11.7.5(a), F_0 is indeed a K-radical extension of E. Let F be as in (a). By Lemma 11.7.5(c), F is contained in an E-diamond, in particular F is Hilbertian. By Theorem 11.7.1, $\mathrm{Gal}(E)$ is semi-free of rank m. Hence, by Theorem 11.7.2(d), $\mathrm{Gal}(F)$ is semi-free of rank m. By Corollary 11.6.7, $\mathrm{Gal}(F)$ is projective. Hence, by Proposition 10.1.14, $\mathrm{Gal}(F)$ is free of rank m as claimed in (a).

Statement (b) is a special case of Lemma 11.7.5(d). To prove (c) note that since F_0 is generated by radicals of elements of E and all roots of unity of order prime to p are contained in E, we have $F_0 \subseteq E_{\mathrm{ab}}$. In particular, E_{ab} is an Abelian extension of F. Since F is Hilbertian, so is E_{ab} [FrJ08, Thm. 16.11.3]. Since $\mathrm{Gal}(F)$ is isomorphic to $\hat{F}_m$, so is $\mathrm{Gal}(E_{\mathrm{ab}})$ [FrJ08, Cor. 25.4.8]. $\qquad\square$

Remark 11.7.7: Note that Theorem 11.7.6(c) follows already from the results of David Harbater quoted in the second paragraph of Section 10.6. Indeed according to those results, if K is a PAC field that contains all roots of unity, then $\mathrm{Gal}(K(x))$ is quasi-free of rank m, hense so is $\mathrm{Gal}(K(x)_{\mathrm{ab}})$. In addition, by Corollary 11.6.7, $\mathrm{Gal}(K(x)_{\mathrm{ab}})$ is projective. Hence, by Proposition 9.4.7, $\mathrm{Gal}(K(x)_{\mathrm{ab}}) \cong \hat{F}_m$.

The condition that K contains all roots of unity of order not divisible by $\mathrm{char}(K)$ is necessary for Theorem 11.7.6(c) to hold. In fact given an odd prime number l, we have examples of Hilbertian PAC fields K that contain all roots of unity of order not divisible by n with $\zeta_l \notin K$ such that $\mathrm{Gal}(K(x)_{\mathrm{ab}})$ is not projective. In particular $\mathrm{Gal}(K(x)_{\mathrm{ab}})$ is not free. We will publish those examples elsewhere. $\qquad\square$

Example 11.7.8: Starting from a PAC field K of cardinality m, Theorem 11.7.6 gives an extension F of $K(x)$ in $K(x)_{\mathrm{ab}}$ such that $\mathrm{Gal}(F) \cong \hat{F}_m$ and F is Hilbertian. It is however not clear to us whether F is ample. We suspect it is not.

However, [GeJ01, Thm. 2.6] gives an example of a Hilbertian field F with $\mathrm{Gal}(F) \cong \hat{F}_\omega$ (in particular, $\mathrm{Gal}(F)$ is projective), but F is nonample. $\square$

Notes

Proposition 11.1.3 about the projectivity of $\mathrm{Gal}(M)$ for a Henselian field M under appropriate assumptions on the residue field and the value group reproduces [JaP09, Lemma 1.3].

The results about the cohomology of local Galois groups appearing in Section 11.2 are taken from [Pop88, §2].

Sections 11.3, 11.4, and 11.5 are a work out of part of Efrat's work [Efr01]. The main result of [Efr01] we use is Proposition 11.5.5.

Lemma 11.3.2 is a special case of a more general lemma on a local-global principle for the Brauer group of a field that is a directed union of fields satisfying a local-global principle for their Brauer groups (see [Pop88, Lemma 4.4], or [Efr01, Lemma 3.3]).

Proposition 11.6.6 is [JaP09, Lemma 1.4].

One of the main results of the chapter is Theorem 11.7.1. It also appears as [BHH10, Thm. 7.2]. The proof of the latter theorem is an adjustment of the proof of [HaS05, Thm. 4.3] about quasi-freeness.

We note that [BHH10, Section 8] gives an account of Construction 11.7.4 and of Theorem 11.7.6 with a reference to our book. That work also refers to Theorem 11.7.1 (see [BHH08, comment following the proof of Thm. 7.2]).

Chapter 12.
Complete Noetherian Domains

Following [Pop10], we generalize and strengthen Theorem 5.11.3 and prove
that the absolute Galois group of a Noetherian domain which is complete with
respect to a prime ideal of height at least 2 is semi-free (Theorem 12.4.3).

12.1 Perseverance of Inertia Groups

Let K be a Hilbertian field, x an indeterminate, and F a finite Galois extension of $K(x)$. Then K has a separable Hilbert subset H such that if $a \in H$,
then the specialization $x \to a$ extends to a place of F/K that induces an
isomorphism of $\mathrm{Gal}(F/K(x))$ onto the residue Galois group $\mathrm{Gal}(\bar{F}/K)$. We
prove in this section that H can be chosen such that, under certain conditions, the inertia groups of $\mathrm{Gal}(F/K(x))$ are mapped isomorphically onto
inertia groups of $\mathrm{Gal}(\bar{F}/K)$.

Let $(F/w)/(E, v)$ be a finite Galois extension of valued fields. We recall
the notation $O_w = \{x \in F \mid w(x) \geq 0\}$ for the valuation ring of w, $e_{w/v} =
(w(F^\times) : w(E^\times))$ for the ramification index of w/v (so, $w(a) = e_{w/v}v(a)$ for
each $a \in E$),

$$D_{w/v} = \{\sigma \in \mathrm{Gal}(F/E) \mid \sigma(O_w) = O_w\}$$

for their decomposition group, and

$$I_{w/v} = \{\sigma \in \mathrm{Gal}(F/E) \mid w(\sigma x - x) > 0 \text{ for each } x \in O_w\}$$

for their inertia group.

LEMMA 12.1.1: Let $(F, w)/(E, v)$ be a finite Galois extension of valued fields.
Suppose $e_{w/v} = |D_{w/v}|$ and let z be an element of F satisfying
(1) $w(z) = 1$ and $w(\sigma(z - 1)) = 1$ for each $\sigma \in \mathrm{Gal}(F/E) \smallsetminus D_{w/v}$.

Then z is integral over O_v and the coefficients of the polynomial

(2) $$X^n + a_{n-1}X^{n-1} + \cdots + a_0 = \prod_{\sigma \in \mathrm{Gal}(F/E)} (X - \sigma z)$$

satisfy
(a1) $v(a_k) \geq 0$ for each $0 \leq k \leq n - 1$,
(a2) $v(a_0) = 1$,
(a3) $v(a_k) \geq 1$ for each $0 \leq k \leq e_{w/v} - 1$, and
(a4) $v(a_{e_{w/v}}) = 0$.

Proof: Set $G = \mathrm{Gal}(F/E)$, $D = D_{w/v}$, $e = e_{w/v}$, and note that $n = |G|$. By
Lemma (after going to the Henselian closure), $D = \{\sigma \in G \mid w \circ \sigma = w\}$.

M. Jarden, *Algebraic Patching*, Springer Monographs in Mathematics,
DOI 10.1007/978-3-642-15128-6_12, © Springer-Verlag Berlin Heidelberg 2011

By (1), $w(\sigma z) = 1$ for each $\sigma \in D$. Moreover, $w(\sigma z - 1) = 1$, so $w(\sigma z) = 0$, for each $\sigma \in G \smallsetminus D$. Hence, $v(a_k) \geq 0$ for each k, in particular, z is integral over O_v.

It follows from the equality $a_0 = (-1)^n \prod_{\sigma \in D} \sigma z \prod_{\sigma \in G \smallsetminus D} \sigma z$ that

$$ev(a_0) = w(a_0) = \sum_{\sigma \in D} w(\sigma z) + \sum_{\sigma \in G \smallsetminus D} w(\sigma z) = e,$$

so $v(a_0) = 1$.

In order to prove (a3) and (a4) we denote for each $0 \leq k \leq n - 1$ the collection of all subsets of G of $n - k$ elements by $\mathcal{S}_{n-k}$. Then

$$a_k = \sum_{S \in \mathcal{S}_{n-k}} (-1)^{n-k} \prod_{\sigma \in S} \sigma z.$$

If $0 \leq k < e$ and $S \in \mathcal{S}_{n-k}$, then $|S| = n - k > n - e = |G \smallsetminus D|$, so $S \cap D \neq \emptyset$. Hence, $w(\prod_{\sigma \in S} \sigma z) \geq 1$, so $w(a_k) \geq 1$.

Finally we write

$$(3) \qquad a_e = (-1)^{n-e} \Big(\prod_{\sigma \in G \smallsetminus D} \sigma z + \sum_{\substack{S \in \mathcal{S}_{n-e} \\ S \neq G \smallsetminus D}} \prod_{\sigma \in S} \sigma z \Big).$$

The w-value of the first term in the right hand side of (3) is 0. If $S \in \mathcal{S}_{n-e}$ and $S \neq G \smallsetminus D$, then $S \cap D \neq \emptyset$. Hence, $w(\prod_{\sigma \in S} \sigma z) \geq 1$. Therefore, the w-value of the second term in the right hand side of (3) is at least 1. We conclude that $w(a_e) = 0$, as asserted by (a4). $\qquad \square$

Setup 12.1.2: Let K be a Hilbertian field, x an indeterminate, and F a finite Galois extension of $E = K(x)$ with Galois group G. We fix an element $b_0 \in K$, set $\mathfrak{p} = \mathfrak{p}_{x,b_0}$ to be the corresponding prime divisor of E/K, let $v_{\mathfrak{p}} = v_{x,b_0}$ be the normalized discrete valuation of E/K associated with $\mathfrak{p}$ (thus, $v_{\mathfrak{p}}(x - b_0) = 1$), and consider a normalized valuation $w_{\mathfrak{p}}$ of F lying over $v_{\mathfrak{p}}$. We assume that the residue field of F under $w_{\mathfrak{p}}$ is also K. Thus, the inertia group $I = I_{w_{\mathfrak{p}}/v_{\mathfrak{p}}}$ coincides with the decomposition group $D = D_{w_{\mathfrak{p}}/v_{\mathfrak{p}}}$ of $w_{\mathfrak{p}}/v_{\mathfrak{p}}$.

For each $b \in K$ we may extend the specialization $x \to b$ to a K-place φ of F. We let $\bar{F}$ be the residue field of φ. For each $y \in F$ we set $\bar{y} = \varphi(y)$. There is a separable Hilbert subset H_0 of K such that if $b \in H_0$, then $\bar{F}/K$ is a Galois extension and φ induces an isomorphism $\varphi_*: \mathrm{Gal}(F/E) \to \mathrm{Gal}(\bar{F}/K)$ mapping each $\sigma \in \mathrm{Gal}(F/E)$ onto an element $\bar{\sigma} \in \mathrm{Gal}(\bar{F}/K)$ such that $\bar{\sigma}\bar{y} = \overline{\sigma y}$ for each $y \in F$ with $\bar{y} \in \bar{F}$ [FrJ08, Lemma 16.1.1]. Then φ_* maps I isomorphically onto a subgroup $\bar{I}$ of $\mathrm{Gal}(\bar{F}/K)$. $\qquad \square$

LEMMA 12.1.3: *Under Setup 12.1.2, F/E has a primitive element z that satisfies (1) and is integral over $K[x]$.*

Proof: Since $w_{\mathfrak{p}} \circ \sigma \neq w_{\mathfrak{p}}$ for each $\sigma \in G \smallsetminus D$, the strong approximation theorem [FrJ08, Prop. 3.3.1] gives $z_0 \in F$ satisfying $w_{\mathfrak{p}}(\sigma z_0) = 1$ for each $\sigma \in D$ and $w_{\mathfrak{p}}(\sigma(z_0 - 1)) = 1$ for each $\sigma \in G \smallsetminus D$, and $w_{\mathfrak{q}}(z) \geq 0$ for each prime divisor of F/K that lies neither over $\mathfrak{p}$ nor over $\mathfrak{p}_{x,\infty}$. In particular, z_0 is integral over $K[x]$.

Now we choose a primitive element z_1 for F/E, integral over $K[x]$. Then $z_0 + cz_1$ is a primitive element for each $c \in K[x]$ satisfying $\sigma z_0 + c\sigma z_1 \neq \tau z_0 + c\tau z_1$ for all distinct $\sigma, \tau \in G$. In addition, $z_0 + cz_1$ is integral over $K[x]$. Since $\sigma z_1 \neq \tau z_1$ if σ, τ are distinct, there exist such c. Moreover, we may choose c such that, in addition, $w_{\mathfrak{p}}(c) + w_{\mathfrak{p}}(\sigma z_1) \geq 2$ for each $\sigma \in G$. Then $z = z_0 + cz_1$ is a primitive element for F/E that satisfies (1).   □

LEMMA 12.1.4: *Under Setup 12.1.2, K has a separable Hilbert set H contained in H_0 and a finite subset A, such that if v is a discrete valuation of K satisfying $v(a) = 0$ for each $a \in A$ and if $b \in H$ satisfies $v(b - b_0) = 1$, then $\bar{I}$ is an inertia group of an extension w of v to $\bar{F}$.*

Proof: Suppose we have proved the lemma for a K-place φ of F with $\varphi(x) = b$ and φ' is another K-place of F satisfying $\varphi'(x) = b$. Then φ' is conjugate to φ over E. Hence, φ'_* maps I onto a subgroup $\bar{I}'$ of $\mathrm{Gal}(\bar{F}/K)$ which is conjugate to $\bar{I}$. Thus, $\bar{I}$ is the inertia group of another extension w' of v to $\bar{F}$. Therefore, in order to prove the lemma, we may choose φ at will.

PART A: *A good primitive element for F/E.* We set $G = \mathrm{Gal}(F/E)$, $n = [F : E] = |G|$, and $e = e_{w_{\mathfrak{p}}/v_{\mathfrak{p}}}$. Since $\bar{F}_{w_{\mathfrak{p}}} = \bar{E}_{v_{\mathfrak{p}}}$, we have $|D| = |I| = e$. By Lemma 12.1.3, there exists a primitive element z for F/E, integral over $K[x]$, such that
(4) $w_{\mathfrak{p}}(\sigma z) = 1$ for each $\sigma \in I$ and $w_{\mathfrak{p}}(\sigma z - 1) = 1$ for each $\sigma \in G \smallsetminus I$.

It follows that
(5) $w_{\mathfrak{p}}(\sigma z - 1) = 0$ for each $\sigma \in I$ and $w_{\mathfrak{p}}(\sigma z) = 0$ for each $\sigma \in G \smallsetminus I$.

Since z is integral over $K[x]$,

$$(6) \qquad f = \mathrm{irr}(z, E) = Z^n + a_{n-1}(x)Z^{n-1} + \cdots + a_0(x) \in K[x, Z].$$

By Lemma 12.1.1, $v_{\mathfrak{p}}(a_0(x)) = 1$. Since $v_{\mathfrak{p}}(x - b_0) = 1$, we have:
(7) $a_0(x) = (x - b_0)g_0(x)$ with $g_0 \in K[x]$ satisfying $g_0(b_0) \neq 0$.

Let R be the integral closure of $K[x]$ in F. Then R is the intersection of all valuation rings of F/K that contain x [FrJ08, Prop. 2.4.1]. Thus, if $y \in F$ and every pole of y is a pole of x, then $y \in R$. Similarly, if every pole of y is a pole of x or a zero of $g_0(x)$, then y belongs to the integral closure $R' = R[g_0(x)^{-1}]$ of $K[x, g_0(x)^{-1}]$ in F.

In particular, since $\sigma z \in R$ for each $\sigma \in G$, each pole $\mathfrak{q}$ of $\frac{\sigma z}{z}$ or of $\frac{\sigma z - 1}{z}$ which is not a pole of x is a zero of z. Hence, $\mathfrak{q}$ is a zero of $a_0(x)$ (by (6)), so

$\mathfrak{q}$ is a zero of $x - b_0$ or of $g_0(x)$ (by (7)). In the first case, $\mathfrak{q}$ lies over $\mathfrak{p}$, hence $w_{\mathfrak{q}} = w_{\mathfrak{p}} \circ \tau$ for some $\tau \in G$.

First suppose $\sigma \in I$. If $\tau \in I$, then $\tau\sigma \in I$, so by (4), $w_{\mathfrak{q}}\left(\frac{\sigma z}{z}\right) = w_{\mathfrak{p}}(\tau\sigma z) - w_{\mathfrak{p}}(\tau z) = 1 - 1 = 0$. If $\tau \in G \smallsetminus I$, then $\tau\sigma \in G \smallsetminus I$, so by (5), $w_{\mathfrak{q}}\left(\frac{\sigma z}{z}\right) = w_{\mathfrak{p}}(\tau\sigma z) - w_{\mathfrak{p}}(\tau z) = 0 - 0 = 0$. Hence, $\frac{\sigma z}{z} \in R'$ and $\frac{z}{\sigma z} \in R'$. Thus, in this case, $\frac{\sigma z}{z}$ is a unit of R'.

Now suppose $\sigma \in G \smallsetminus I$. If $\tau \in I$, then $\tau\sigma \in G \smallsetminus I$. Hence, by (4) and (5), $w_{\mathfrak{q}}\left(\frac{\sigma z - 1}{z}\right) = w_{\mathfrak{p}}(\tau\sigma z - 1) - w_{\mathfrak{p}}(\tau z) = 1 - 1 = 0$. If $\tau \in G \smallsetminus I$, then $w_{\mathfrak{q}}\left(\frac{\sigma z - 1}{z}\right) = w_{\mathfrak{p}}(\tau\sigma z - 1) - w_{\mathfrak{p}}(\tau z) \geq 0 - 0 = 0$. Thus, $\frac{\sigma z - 1}{z} \in R'$.

It follows that for each $\sigma \in I$ there exist elements $u_\sigma, u'_\sigma \in R$ and nonnegative integers $k(\sigma), k'(\sigma)$, and for each $\sigma \in G \smallsetminus I$ there exist an element $u_\sigma \in R$ and a nonnegative integer $k(\sigma)$ such that

(8a) $\sigma z = \frac{u_\sigma}{g_0(x)^{k(\sigma)}} z$ if $\sigma \in I$,

(8b) $\sigma z = 1 + \frac{u_\sigma}{g_0(x)^{k(\sigma)}} z$ if $\sigma \in G \smallsetminus I$, and

(8c) $u_\sigma u'_\sigma = g_0(x)^{k'(\sigma)}$ if $\sigma \in I$.

PART B: *A Hilbert set.* For each $\sigma \in I$ we set $f_\sigma = \mathrm{irr}(u_\sigma, E)$ and $f'_\sigma = \mathrm{irr}(u'_\sigma, E)$. For each $\sigma \in G \smallsetminus I$ we let $f_\sigma = \mathrm{irr}(u_\sigma, E)$. Since $u_\sigma, u'_\sigma \in R$, we may consider f_σ and f'_σ as elements of the ring $K[x, Z]$ that satisfy

(9) $f_\sigma(x, u_\sigma) = 0$ for each $\sigma \in G$ and $f'_\sigma(x, u'_\sigma) = 0$ for each $\sigma \in I$.

We choose a finite subset A of K and let $O = \mathbb{Z}[A]$ if $\mathrm{char}(K) = 0$ and $O = \mathbb{F}_p[A]$ if $\mathrm{char}(K) = p > 0$ such that

(10a) $g_0(x) \in O[x]$, $f, f_\sigma \in O[x, Z]$ for each $\sigma \in G$, and $f'_\sigma \in O[x, Z]$ for each $\sigma \in I$, and

(10b) $b_0, g_0(b_0) \in A$.

Let H_0 be the separable Hilbert subset of K given in Setup 12.1.2 and consider the separable Hilbert set $H = H_0 \smallsetminus \{a \in K \mid g_0(a) \neq 0\}$. For each $b \in H$ let φ and φ_* be as in Setup 12.1.2. Since $\varphi(x) = b$ and $g_0(b) \neq 0$, φ is finite on $K[x, g_0(x)^{-1}]$, hence φ is also finite on R'. We apply φ and φ_* on the equalities appearing in Conditions (8) and (10a) to get

(11a) $\bar{\sigma}\bar{z} = \frac{\bar{u}_\sigma}{g_0(b)^{k(\sigma)}} \bar{z}$ if $\sigma \in I$,

(11b) $\bar{\sigma}\bar{z} = 1 + \frac{\bar{u}_\sigma}{g_0(b)^{k(\sigma)}} \bar{z}$ if $\sigma \in G \smallsetminus I$,

(11c) $\bar{u}_\sigma \bar{u}'_\sigma = g_0(b)^{k'(\sigma)}$ if $\sigma \in I$,

(11d) $g_0(b) \in O[b]$, $g_0(b) \neq 0$, and $f(b, Z), f_\sigma(b, Z) \in O[b, Z]$ for all $\sigma \in G$, and $f'_\sigma(b, Z) \in O[b, Z]$ for all $\sigma \in I$.

PART C: *A valuation of K.* We consider a discrete valuation v of K that satisfies $v(a) = 0$ for each $a \in A$. Then $O \subseteq O_v$ and, by (10),

(12a) $g_0(x) \in O_v[x]$, $f, f_\sigma \in O_v[x, Z]$ for each $\sigma \in G$, and $f'_\sigma \in O_v[x, Z]$ for each $\sigma \in I$.

(12b) $v(b_0) = 0$ and $v(g_0(b_0)) = 0$.

Let w be a normalized valuation of $\bar{F}$ over v and set $\bar{e} = e_{w/v}$. Suppose $b \in H$ satisfies $v(b - b_0) = 1$. By (12),

(13) $v(b) = 0$ and $v(g_0(b)) = 0$.

By (7), $a_0(b) = (b - b_0)g_0(b)$. Hence, by (13),
(14) $v(a_0(b)) = 1$.
Since $f(b, Z) = Z^n + a_{n-1}(b)Z^{n-1} + \cdots + a_0(b) \in O_v[Z]$ (by (12a)), all of
the roots of $f(b, Z)$ belong to O_w and their product is $a_0(b)$. By (14), one of
them, which we denote by $\bar{z}$, satisfies
(15) $w(\bar{z}) > 0$.

We choose a K-place φ as in Setup 12.1.2 such that $\varphi(x) = b$ and
$\varphi(z) = \bar{z}$.

Let $\sigma \in I$. Since $f_\sigma(b, \bar{u}_\sigma) = 0$ and $f'_\sigma(b, \overline{u'_\sigma}) = 0$ (by (9)) and since
$f_\sigma(b, Z)$ and $f'_\sigma(b, Z)$ are monic polynomials with coefficients in O_v (by (12a)),
we have $\bar{u}_\sigma, \overline{u'_\sigma} \in O_w$. Hence, by (11c) and (13), $w(\bar{u}_\sigma) = 0$ if $\sigma \in I$.
Therefore, by (11a) and (13),
(16a) $w(\bar{\sigma}\bar{z}) = w(\bar{z})$ if $\sigma \in I$.

If $\sigma \in G \smallsetminus I$, then by (12a) and (9), $f_\sigma(b, Z) \in O_v[Z]$ and $f_\sigma(b, \bar{u}_\sigma) = 0$,
so $w(\bar{u}_\sigma) \geq 0$. Therefore, by (11b), (13), and (15),
(16b) $w(\bar{\sigma}\bar{z} - 1) \geq w(\bar{z}) > 0$, so $w(\bar{\sigma}\bar{z}) = 0$ if $\sigma \in G \smallsetminus I$.

By (15) and (16b), we have for $\sigma \in G \smallsetminus I$ that $w(\bar{\sigma}\bar{z}) \neq w(\bar{z})$, hence
$w \circ \bar{\sigma} \neq w$. Therefore, $\bar{\sigma} \notin D_{w/v}$, hence $\bar{\sigma} \notin I_{w/v}$. This implies that $\bar{G} \smallsetminus \bar{I} \subseteq$
$\bar{G} \smallsetminus I_{w/v}$, hence $I_{w/v} \subseteq \bar{I}$. On the other hand, $a_0(b) = \prod_{\sigma \in G} \bar{\sigma}\bar{z}$. Hence,
by (14), (16a), and (16b), $\bar{e} = \bar{e} \cdot v(a_0(b)) = w(a_0(b)) = |I|w(\bar{z})$. Therefore,
$|\bar{I}| = |I| \leq \bar{e} \leq |I_{w/v}|$, so $\bar{I} = I_{w/v}$, as claimed.   □

LEMMA 12.1.5: *Let $(K, v) \subseteq (K', v')$ and $(L, w) \subseteq (L', w')$ be extensions of
discrete valued fields such that $(K, v) \subseteq (L, w)$. Suppose L/K is a Galois
extension, K'/K is a purely inseparable extension, and $L' = LK'$. Then
res: $\mathrm{Gal}(L'/K') \to \mathrm{Gal}(L/K)$ maps $I_{w'/v'}$ bijectively onto $I_{w/v}$.*

Proof: By assumption, the map res is an isomorphism. Let $\sigma' \in \mathrm{Gal}(L'/K')$
and set $\sigma = \sigma'|_L$. If $\sigma' \in I_{w'/v'}$ and $x \in O_w$, then $x \in O_{w'}$, so $w(\sigma x - x) =$
$e_{w'/w}^{-1}w'(\sigma'x - x) > 0$. Hence, $\sigma \in I_{w/v}$.

Conversely, we may assume that $p = \mathrm{char}(K) > 0$. If $\sigma \in I_{w/v}$ and
$y \in O_{w'}$, then there exists a power q of p such that $y^q \in L$. Hence, $y^q \in O_w$
and $qw'(\sigma'y - y) = e_{w'/w}w(\sigma y^q - y^q) > 0$. Therefore, $w'(\sigma'y - y) > 0$. It
follows that $\sigma' \in I_{w'/v'}$. Consequently, res: $I_{w'/v'} \to I_{w/v}$ is a bijective map.
□

12.2 Krull Fields

One of the consequences of the Chebotarev density theorem for a finite Galois
extension K'/K of global fields is that there are infinitely many primes of K
that totally split in K'. We prove an analog of that statement that applies
in particular to fields of formal power series in several variables.

Remark 12.2.1: Splitting of prime ideals. Let S be an integrally closed domain with quotient field K. Let K' be a Galois extension of K of degree n

and let S' be the integral closure of S in K'. We introduce some notation that concern total splitting of prime ideals of S in S'.

(a) Let $z \in S'$ be a primitive element of K'/K and

$$p_z(X) = X^n + a_{n-1}X^{n-1} + \cdots + a_0 = \mathrm{irr}(z, K) \in S[X].$$

Consider a prime ideal $\mathfrak{p}$ of S. If $\mathrm{discr}(p_z) \not\equiv 0 \mod \mathfrak{p}$, then $S_{\mathfrak{p}}[z]$ is the integral closure of $S_{\mathfrak{p}}$ in K' [FrJ08, Lemma 6.1.2]. Hence, if $\mathfrak{p}'$ is a prime ideal of S' lying over $\mathfrak{p}$ and we use a bar to denote reduction modulo $\mathfrak{p}'$, we have $\overline{K'}_{\mathfrak{p}'} = S'_{\mathfrak{p}'}/\mathfrak{p}'S'_{\mathfrak{p}'} = \mathrm{Quot}(\overline{S_{\mathfrak{p}}[z]}) = \bar{K}_{\mathfrak{p}}(\bar{z})$. Moreover, the distinct roots $z_1, \ldots, z_n$ of p_z in K' belong to S', their reductions $\bar{z}_1, \ldots, \bar{z}_n$ are distinct and belong to $\bar{K}_{\mathfrak{p}}[\bar{z}]$. In addition, $\overline{K'}_{\mathfrak{p}}/\bar{K}_{\mathfrak{p}}$ is a Galois extension whose Galois group is isomorphic to the decomposition group $D_{\mathfrak{p}'/\mathfrak{p}}$.

(b) In particular, if $\bar{z} \in \bar{K}_{\mathfrak{p}}$, then $\overline{K'}_{\mathfrak{p}'} = \bar{K}_{\mathfrak{p}}$. Hence, $\bar{z}_1, \ldots, \bar{z}_n \in \bar{K}_{\mathfrak{p}}$ and $D_{\mathfrak{p}'/\mathfrak{p}} = 1$. Since the prime ideals of S' that lie over $\mathfrak{p}$ are conjugate and their number is $(\mathrm{Gal}(K'/K) : D_{\mathfrak{p}'/\mathfrak{p}})$, this number is n. In other words, $\mathfrak{p}$ totally splits in S'. In this case we also say that $\mathfrak{p}$ **totally splits** in K'.

(c) One way to achieve total splitting of $\mathfrak{p}$ is to consider the homogeneous polynomial

$$p_z^*(T, U) = T^n + a_{n-1}T^{n-1}U + \cdots + a_1TU^{n-1} + a_0U^n$$

associated with p_z. For each $r, s \in S$ with $s \neq 0$ we have $s^n p_z\left(\frac{r}{s}\right) = p_z^*(r, s)$. It follows from the factorization $p_z(X) = \prod_{i=1}^n (X - z_i)$ that $p_z^*(r, s) = \prod_{i=1}^n (r - z_is)$.

If $s, \mathrm{discr}(p_z) \notin \mathfrak{p}$ and $p_z^*(r, s) \in \mathfrak{p}$, then p_z has a root modulo $\mathfrak{p}$ and $\mathfrak{p}$ splits completely in K' (by (b)).

(d) Let O_v be the valuation ring of a discrete valuation v of K and let w be a valuation of K' lying over v. If v splits completely in K', then $\overline{K'}_w = \bar{K}_v$ and v is unramified in K'. It follows from the equality $\hat{K}_v = \widehat{K'}_w$ for the completions that K is w-dense in K'. More directly, we consider $x \in K' \smallsetminus K$ and inductively assume that there exists $a \in K$, such that $w(x - a) \geq m$. Since v is unramified in K' there exists $b \in K^\times$ such that $w(b) = w(x - a)$. Then $w(b^{-1}(x - a)) = 0$, so from $\bar{K}_v = \overline{K'}_w$ there exists $c \in K$ with $w(b^{-1}(x - a) - c) \geq 1$. Hence, $w(x - (a + bc)) \geq m + 1$. $\square$

We prove that the quotient field of a complete Krull domain with respect to a maximal ideal of height at least 2 is a Krull field in the following sense:

Definition 12.2.2: Let K be a field and let $\mathcal{V}$ be a set of discrete valuations of K. We say that $(K, \mathcal{V})$ is a **Krull field** (or also that K is a **Krull field with respect to $\mathcal{V}$**) if

(1a) for each $a \in K^\times$ the set $\mathcal{V}_a = \{v \in \mathcal{V} \mid v(a) \neq 0\}$ is finite, and

(1b) for each finite Galois extension K' of K the set $\mathrm{Spl}_{\mathcal{V}}(K'/K)$ of all $v \in \mathcal{V}$
    that totally split in K' has the same cardinality as of K.

We say that K is a **Krull field** if K is a Krull field with respect to some set of discrete valuations. $\square$

Remark 12.2.3: Finite ramification. Let $(K, \mathcal{V})$ be a Krull field and K' a finite Galois extension of K. Choose a primitive element z for K'/K and let $\mathrm{irr}(z, K) = \sum_{i \in I} a_i X^i$, where I is a finite set of nonnegative integers and $a_i \in K^\times$ for each $i \in I$. Note that $d = \mathrm{discr}(\mathrm{irr}(z, K)) \in K^\times$. By (1a), $\mathcal{V}' = \mathcal{V}_d \cap \bigcup_{i \in I} \mathcal{V}_{a_i}$ is a finite set. Since each $v \in \mathcal{V}$ is a discrete valuation of K, each $v \in \mathcal{V} \smallsetminus \mathcal{V}_d$ is unramified in K' [Lan70, p. 62]. Hence, the set $\mathrm{Ram}_\mathcal{V}(K'/K)$ of all $v \in \mathcal{V}$ that ramify in K' is finite.   □

The following lemma says that finite extensions of Krull fields are again Krull fields.

LEMMA 12.2.4: *Let K be a Krull field with respect to a set $\mathcal{V}$ of discrete valuations and let K' be a finite extension of K. Then K' is a Krull field with respect to the set $\mathcal{V}'$ of all extensions of the valuations in $\mathcal{V}$ to K'.*

Proof: Note that each $v \in \mathcal{V}$ has finitely many extensions to K' and each of them is discrete. Every $x \in (K')^\times$ satisfies an equation $x^n + a_{n-1} x^{n-1} + \cdots + a_0 = 0$ with $a_0, \ldots, a_{n-1} \in K$ and $a_0 \neq 0$. Also, there are $b_0, \ldots, b_{n-1} \in K$ with $b_0 \neq 0$ such that $(x^{-1})^n + b_{n-1}(x^{-1})^{n-1} + \cdots + b_0 = 0$. Let W be the set of all $w \in \mathcal{V}'$ such that $w(a_i) = 0$ when $a_i \neq 0$ and $w(b_j) = 0$ when $b_j \neq 0$. Each $w \in W$ satisfies $w(x) \geq 0$ and $w(x^{-1}) \geq 0$, so $w(x) = 0$. Thus, $\mathcal{V}' \smallsetminus W$ is a finite set that contains $\mathcal{V}'_x$. Therefore, $\mathcal{V}'_x$ is finite.

In order to prove (1b) for K' it suffices to consider only the cases where K'/K is separable or K'/K is purely inseparable.

CASE A: *K'/K is separable.* Let L' be a finite Galois extension of K' and choose a finite Galois extension N of K that contains L'. Then, $\mathrm{Spl}_{\mathcal{V}'}(L'/K')$ is contained in the set of all extensions of the valuations of $\mathrm{Spl}_\mathcal{V}(N/K)$ to N. Since the latter set has cardinality $\mathrm{card}(K) = \mathrm{card}(K')$, so does the former.

CASE B: *K'/K is purely inseparable.* Let L' be a finite Galois extension of K'. Then L' is a normal extension of K. Hence, K has a Galois extension L such that $LK' = L'$ and $L \cap K' = K$. Thus, $n = [L : K] = [L' : K']$.

If $v \in \mathcal{V}$ splits into distinct valuations $w_1, \ldots, w_n$ of L, then their respective extensions $w'_1, \ldots, w'_n$ to L' are distinct and lie over the unique extension v' of v to K'. It follows from $\mathrm{card}(\mathrm{Spl}_\mathcal{V}(L/K)) = \mathrm{card}(K)$ that $\mathrm{card}(\mathrm{Spl}_{\mathcal{V}'}(L'/K')) = \mathrm{card}(K')$.

Consequently, in all cases $(K', \mathcal{V}')$ is a Krull field.   □

LEMMA 12.2.5: *Let R be an integral domain with a prime ideal $\mathfrak{m}$ of height at least 2. Let S be the integral closure of R in K and choose a prime ideal $\mathfrak{n}$ of S lying over $\mathfrak{m}$ of height at least 2. Denote the set of all nonzero minimal prime ideals of S contained in $\mathfrak{n}$ by $\mathcal{P}$. Suppose S is a Krull domain. Then:*
(a) *For each $\mathfrak{p} \in \mathcal{P}$, the local ring $S_\mathfrak{p}$ of S at $\mathfrak{p}$ is a discrete valuation ring of K, each nonzero $a \in S$ is contained in only finitely many $\mathfrak{p} \in \mathcal{P}$, and $S_\mathfrak{n} = \bigcap_{\mathfrak{p} \in \mathcal{P}} S_\mathfrak{p}$.*
(b) *Each $a \in \mathfrak{n}$ is contained in some $\mathfrak{p} \in \mathcal{P}$.*

(c) *For each nonzero $r_0 \in \mathfrak{n}$ there exists $r_1 \in \mathfrak{n}$ such that r_0 and r_1 are not contained in a common $\mathfrak{p} \in \mathcal{P}$.*

(d) *$\mathcal{P}$ is an infinite set and $\mathrm{card}(\mathcal{P}) \leq \mathrm{card}(S)$.*

Proof: By assumption, $\mathfrak{m}$ properly contains a nonzero prime ideal $\mathfrak{p}$. Since S is integral over R, the going up theorem gives prime ideals $\mathfrak{q}$ and $\mathfrak{n}$ of S lying over $\mathfrak{p}$ and $\mathfrak{m}$, respectively, [Mats94, p. 68, Thm. 9.4(i)] such that $\mathfrak{q} \subseteq \mathfrak{n}$. It follows that $\mathfrak{q} \subset \mathfrak{n}$, so $\mathrm{height}(\mathfrak{n}) \geq 2$. This justifies the choice of $\mathfrak{n}$ as in the lemma.

Proof of (a): Since S is a Krull domain, $S_{\mathfrak{p}}$ is a discrete valuation ring of K for every $\mathfrak{p} \in \mathcal{P}$ and each nonzero $a \in S$ lies in only finitely many $\mathfrak{p} \in \mathcal{P}$ (Definition 5.11.1). Moreover, $S_{\mathfrak{n}}$ is also a Krull domain (Remark 5.11.2(b)) and $\{\mathfrak{p} S_{\mathfrak{n}} \mid \mathfrak{p} \in \mathcal{P}\}$ is the set of all nonzero minimal prime ideals of $S_{\mathfrak{n}}$. Now note that $S_{\mathfrak{p}} = (S_{\mathfrak{n}})_{\mathfrak{p} S_{\mathfrak{n}}}$. Hence, by Definition 5.11.1(1c), $S_{\mathfrak{n}} = \bigcap_{\mathfrak{p} \in \mathcal{P}} S_{\mathfrak{p}}$.

Proof of (b): Consider $a \in \mathfrak{n}$. If $a \notin \mathfrak{p}$ for all $\mathfrak{p} \in \mathcal{P}$, then by (a), $a^{-1} \in \bigcap_{\mathfrak{p} \in \mathcal{P}} S_{\mathfrak{p}} = S_{\mathfrak{n}}$. Therefore, $1 = aa^{-1} \in \mathfrak{n} S_{\mathfrak{n}}$, which is a contradiction.

Proof of (c): By (a), only finitely many prime ideals $\mathfrak{p} \in \mathcal{P}$ contain r_0. List them as $\mathfrak{p}_1, \ldots, \mathfrak{p}_m$. Since $\mathrm{height}(\mathfrak{n}) \geq 2$, each $\mathfrak{p} \in \mathcal{P}$ is properly contained in $\mathfrak{n}$. Hence, there exists $r_1 \in \mathfrak{n} \smallsetminus \bigcup_{i=1}^{m} \mathfrak{p}_i$ [AtM69, Prop. 1.11(i)]. It follows that no $\mathfrak{p} \in \mathcal{P}$ contains both r_0 and r_1.

Proof of (d): Suppose we have already proved the existence of m distinct prime ideals $\mathfrak{p}_1, \ldots, \mathfrak{p}_m \in \mathcal{P}$. In each $\mathfrak{p}_i$ we choose a nonzero element r_i. Then, $r_0 = r_1 \cdots r_m \in \mathfrak{p}_1 \cap \cdots \cap \mathfrak{p}_m$. By (c), there exists $r \in \mathfrak{n}$ which is contained in no $\mathfrak{p}$ belonging to $\mathcal{P}$ that contains r_0. By (b), there exists $\mathfrak{p}_{m+1} \in \mathcal{P}$ that contains r. In particular, $\mathfrak{p}_{m+1} \neq \mathfrak{p}_i$ for $i = 1, \ldots, m$. Consequently, $\mathcal{P}$ is infinite.

In order to bound the cardinality of $\mathcal{P}$ we note that for each nonzero $a \in S$ the set $\mathcal{P}(a) = \{\mathfrak{p} \in \mathcal{P} \mid a \in \mathfrak{p}\}$ is finite (by (a)) and $\mathcal{P} = \bigcup_{\substack{a \in S \\ a \neq 0}} \mathcal{P}(a)$. Hence, $\mathrm{card}(\mathcal{P}) \leq \aleph_0 \cdot \mathrm{card}(S) = \mathrm{card}(S)$. $\qquad\square$

Setup 12.2.6: We consider an integral domain R with quotient field K, a prime ideal $\mathfrak{m}$ of R, a nonzero element $t \in \mathfrak{m}$, and a system of representatives R_0 for $R/\mathfrak{m}$. We assume that R is Hausdorff and complete with respect to the Rt-adic topology. In particular, $\bigcap_{n=0}^{\infty} Rt^n = 0$. Thus, each series $\sum_{n=0}^{\infty} c_n t^n$ with $c_n \in R$ for each n converges to a unique element of R in the Rt-adic topology. Let R_1 be the set of the sums of all those series for which $c_n \in R_0$ for all n and $c_0 \notin \mathfrak{m}$. By assumption, $\mathfrak{m} \neq 0$, so R is infinite, hence R_1 is infinite.

Let S be the integral closure of R in K. We assume that $\mathrm{height}(\mathfrak{m}) \geq 2$ and use Lemma 12.2.5 to choose a prime ideal $\mathfrak{n}$ of S over $\mathfrak{m}$ such that $\mathrm{height}(\mathfrak{n}) \geq 2$.

Now we assume that S is a Krull domain. Let $\mathcal{P}$ be the set of all nonzero minimal prime ideals of S in $\mathfrak{n}$, For each nonzero $x \in S$, we write

$\mathcal{P}(x) = \{\mathfrak{p} \in \mathcal{P} \mid x \in \mathfrak{p}\}$. By Lemma 12.2.5(a),(b), $\mathcal{P}(x)$ is a finite nonempty set if $x \in \mathfrak{n}$. If $x \notin \mathfrak{n}$, then $\mathcal{P}(x) = \emptyset$.

For each $\mathfrak{p} \in \mathcal{P}$ we let $v_\mathfrak{p}$ be the normalized discrete valuation of K with valuation ring $S_\mathfrak{p}$ (Lemma 12.2.5(a)). Then let $\mathcal{V} = \{v_\mathfrak{p} \mid \mathfrak{p} \in \mathcal{P}\}$.

Finally, for a finite Galois extension K' of K, let S' be the integral closure of R in K'. Then $S \subseteq S'$, S' is integral over S, and S' is integrally closed, hence S' is also the integral closure of S in K'. Let $\mathrm{Spl}_\mathcal{P}(K'/K)$ be the set of all $\mathfrak{p} \in \mathcal{P}$ that totally split in S'. Further, let $\mathrm{Spl}_\mathcal{V}(K'/K)$ be the set of all $v_\mathfrak{p} \in \mathcal{V}$ with $\mathfrak{p} \in \mathcal{P}$ that totally split in K'. □

LEMMA 12.2.7: *The map $\mathfrak{p} \mapsto v_\mathfrak{p}$ is a bijection of $\mathcal{P}$ onto $\mathcal{V}$. Moeover, for each finite Galois extension K' of K that map maps $\mathrm{Spl}_\mathcal{P}(K'/K)$ onto $\mathrm{Spl}_\mathcal{V}(K'/K)$.*

Proof: By definition, the map $\mathfrak{p} \mapsto v_\mathfrak{p}$ maps $\mathcal{P}$ onto $\mathcal{V}$. Since $S_\mathfrak{p} = O_{v_\mathfrak{p}}$, we have $\mathfrak{p} = \{x \in S \mid v_\mathfrak{p}(x) > 0\}$, so the map is also injective.

Now we consider a finite Galois extension K' of K. Let S' be the inegral closure of S in K' and let $\mathfrak{p} \in \mathcal{P}$. By Remark 5.11.2(d), S' is a Krull domain. Since there are no inclusions among prime ideals of S' that lie over the same prime ideal of S [Mats94, p. 66, Thm. 9.3(ii)], each prime ideal of S' that lies over $\mathfrak{p}$ is nonzero and minimal. If S' has $n = [K' : K]$ prime ideals lying over $\mathfrak{p}$, then the local rings of S' at those ideals are distinct discrete valuation rings whose intersections with K are $S_\mathfrak{p} = O_{v_\mathfrak{p}}$.

Conversely, suppose K' has n discrete valuation rings $O_{w_1}, \ldots, O_{w_n}$ lying over $O_{v_\mathfrak{p}}$. Then each i contains the integral closure of $O_{v_\mathfrak{p}}$ in K', hence $S' \subseteq O_{w_i}$. The set $\mathfrak{p}_i = \{x \in S' \mid w_i(x) > 0\}$ is a prime ideal of S' that lie over $\mathfrak{p}$. Moreover, since $\mathfrak{p}$ is a nonzero minimal prime ideal of S, the same argument as in the preceding paragraph implies that the $\mathfrak{p}_i$'s are nonzero minimal prime ideals of S'. Since S' is Krull, $S'_{\mathfrak{p}_i}$ is a discrete valuation ring contained in O_{w_i}. Hence, $S'_{\mathfrak{p}_i} = O_{w_i}$. It follows that $\mathfrak{p}_1, \ldots, \mathfrak{p}_n$ are distinct, so $\mathfrak{p} \in \mathrm{Spl}_\mathcal{P}(K'/K)$. □

LEMMA 12.2.8: *Let A be an infinite set and let $\{A_i \mid i \in I\}$ be a family of finite subsets of A with $A = \bigcup_{i \in I} A_i$. Then $\mathrm{card}(A) \leq \mathrm{card}(I)$.*

Proof: Since A is infinite, the cardinality of $\mathcal{B} = \{A_i \mid i \in I\}$ is infinite. Since the map $i \mapsto A_i$ maps I onto $\mathcal{B}$ and $A = \bigcup_{B \in \mathcal{B}} B$, we have $\mathrm{card}(A) \leq \mathrm{card}(\mathcal{B})\aleph_0 = \mathrm{card}(\mathcal{B}) \leq \mathrm{card}(I)$, as claimed. □

LEMMA 12.2.9: *The following statements on R_1 are true under Setup 12.2.6:*
(a) $R_1 \cap \mathfrak{m} = \emptyset$, in particular $\mathcal{P}(u) = \emptyset$ for each $u \in R_1$.
(b) If $u, v \in R_1$ and $u \neq v$, then $\mathcal{P}(u - v) = \mathcal{P}(t)$.
(c) $\mathrm{card}(R/\mathfrak{m})^{\aleph_0} = \mathrm{card}(R_1) \leq \mathrm{card}(\mathrm{Spl}_\mathcal{P}(K'/K))$ for each finite Galois extension K' of K.

Proof of (a): We consider $u \in R_1$ and write $u = \sum_{k=0}^\infty u_k t^k$ as a converging series in the Rt-adic topology, where $u_k \in R_0$ for each $k \geq 0$ and $u_0 \notin \mathfrak{m}$. Then $u - u_0 \in Rt \subseteq \mathfrak{m}$, so $u \notin \mathfrak{m}$, as (a) claims.

Proof of (b): We write $v = \sum_{k=0}^{\infty} v_k t^k$ as a converging series in the Rt-adic topology with $v_k \in R_0$ for each $k \geq 0$ and $v_0 \notin \mathfrak{m}$. Suppose $u \neq v$. Then

$$(2) \qquad u - v = \sum_{k=m}^{\infty} (u_k - v_k)t^k = t^m \left((u_m - v_m) + \sum_{k=m+1}^{\infty} (u_k - v_k)t^{k-m} \right)$$

for some $m \geq 0$ with $u_m - v_m \neq 0$. Since R_0 is a system of representatives of $R/\mathfrak{m}$, the latter condition means that $u_m \not\equiv v_m \mod \mathfrak{m}$. It follows that the element in the parentheses in the right hand side of (2) does not belong to $\mathfrak{m}$, hence nor to $\mathfrak{n}$. In particular, that element does not belong to any $\mathfrak{p} \in \mathcal{P}$. Therefore, $\mathcal{P}(u - v) = \mathcal{P}(t^m) = \mathcal{P}(t)$, as claimed in (b).

Proof of (c): Let K' be a finite Galois extension of K of degree n. One notes, as in the proof of (b), that since R_0 is a set of representatives of $R/\mathfrak{m}$, the map $(c_0, c_1, c_2, \ldots) \mapsto \sum_{n=0}^{\infty} c_n t^n$ from $(R_0 \smallsetminus \mathfrak{m}) \times R_0^{\mathbb{N}}$ onto R_1 is bijective. Hence, $\mathrm{card}(R/\mathfrak{m})^{\aleph_0} = \mathrm{card}(R_0)^{\aleph_0} = \mathrm{card}(R_1)$. This proves the equality of (c).

In order to prove the right inequality of (c) we use the notation of Remark 12.2.1 and set $a = a_0 \cdot \mathrm{discr}(p_z)$. Then observe that $at \in \mathfrak{n}$ and use Lemma 12.2.5(d) to choose an $r \in \mathfrak{n}$ such that $\mathcal{P}(r) \cap \mathcal{P}(at) = \emptyset$.

For each $\mathfrak{p} \in \mathcal{P}$, we consider the set $R(\mathfrak{p}) = \{u \in R_1 \mid p_z^*(ru, at) \in \mathfrak{p}\}$. Thus, if $u \in R(\mathfrak{p})$, then in the notation of Remark 12.2.1(c),

$$(3) \qquad r^n u^n + a_{n-1} r^{n-1} u^{n-1} at + \cdots + a_1 r u a^{n-1} t^{n-1} + a_0 a^n t^n \equiv 0 \mod \mathfrak{p}.$$

CLAIM A: *If $\mathfrak{p} \in \mathcal{P}$ and $u \in R(\mathfrak{p})$, then $ruat \notin \mathfrak{p}$.* Otherwise, $ru \in \mathfrak{p}$ or $at \in \mathfrak{p}$. If $at \in \mathfrak{p}$, then by (3), $ru \in \mathfrak{p}$. Since $\mathcal{P}(r) \cap \mathcal{P}(at) = \emptyset$, this implies that $u \in \mathfrak{p}$, so $u \in \mathfrak{n} \cap R = \mathfrak{m}$. Since $u \in R_1$, this contradicts (a). We conclude that $at \notin \mathfrak{p}$.

If $ru \in \mathfrak{p}$, (3) implies that $a_0(at)^n \in \mathfrak{p}$. It follows from the preceding paragraph that $a_0 \in \mathfrak{p}$. Hence, $at = a_0 \cdot \mathrm{discr}(p_z)t \in \mathfrak{p}$. This contradicts the preceding paragraph and concludes the proof of Claim A.

CLAIM B: *If $\mathfrak{p} \in \mathcal{P}$ and $R(\mathfrak{p}) \neq \emptyset$, then $\mathfrak{p} \in \mathrm{Spl}_{\mathcal{P}}(K'/K)$.* Indeed, let $u \in R(\mathfrak{p})$. Since $a = a_0 \cdot \mathrm{discr}(p_z)$, Claim A implies that $\mathrm{discr}(p_z) \notin \mathfrak{p}$. Hence, by Remark 12.2.1(c), applied to ru, at rather than r, s, we have $\mathfrak{p} \in \mathrm{Spl}_{\mathcal{P}}(K'/K)$.

CLAIM C: *For each $\mathfrak{p} \in \mathcal{P}$, the set $R(\mathfrak{p})$ contains at most n elements.*

Let S' be the integral closure of S in K'. Assume $R(\mathfrak{p})$ contains at least $n + 1$ elements. For each $u \in R(\mathfrak{p})$ we have $p_z^*(ru, at) \equiv 0 \mod \mathfrak{p}$. By Remark 12.2.1(a), there exist $z_1, \ldots, z_n \in S'$ such that $p_z^*(ru, at) = \prod_{i=1}^n (ru - atz_i)$. Hence, for each prime ideal $\mathfrak{p}'$ of S' lying over $\mathfrak{p}$ we have $\prod_{i=1}^n (ru - atz_i) \equiv 0 \mod \mathfrak{p}'$. Therefore, there exists $1 \leq i(u) \leq n$ such that $ru - atz_{i(u)} \equiv 0 \mod \mathfrak{p}'$. Since $R(\mathfrak{p})$ contains at least $n + 1$ elements, there exist distinct $u, v \in R(\mathfrak{p})$ such that $i(u) = i(v)$. Thus, $rv - atz_{i(u)} \equiv 0 \mod \mathfrak{p}'$. Therefore, $ru - rv \equiv 0 \mod \mathfrak{p}'$, so $r(u - v) \equiv 0 \mod \mathfrak{p}$. By Claim A, $r \notin \mathfrak{p}$, hence $u - v \in \mathfrak{p}$. It follows from (b) that $t \in \mathfrak{p}$, in contrast to Claim A. We conclude that Claim C is true.

CONCLUSION OF PROOF OF (c): Since $r, t \in \mathfrak{n}$, each $u \in R_1$ satisfies $p_z^*(ru, at) \in \mathfrak{n}$. Hence, by Lemma 12.2.5(b), we may choose $\mathfrak{p}_u \in \mathcal{P}$ that contains $p_z^*(ru, at)$. Thus, $u \in R(\mathfrak{p}_u)$. Therefore, setting $\mathcal{P}_1 = \{\mathfrak{p}_u \mid u \in R_1\}$, we have $R_1 = \bigcup_{\mathfrak{p} \in \mathcal{P}_1} R(\mathfrak{p})$.

By Claim B, $\mathcal{P}_1 \subseteq \mathrm{Spl}_\mathcal{P}(K'/K)$. Since $R(\mathfrak{p})$ is finite for each $\mathfrak{p} \in \mathcal{P}_1$ (Claim C) and R_1 is infinite (Setup 12.2.6), it follows from Lemma 12.2.8 that $\mathrm{card}(R_1) \leq \mathrm{card}(\mathcal{P}_1)$. Hence, $\mathrm{card}(R_1) \leq \mathrm{card}(\mathrm{Spl}_\mathcal{P}(K'/K))$. $\square$

PROPOSITION 12.2.10 (Pop): *Let R be an integral domain, $K = \mathrm{Quot}(R)$, $\mathfrak{m}$ a prime ideal of R of height at least 2, and t a nonzero element of $\mathfrak{m}$. Let S be the integral closure of R in K and choose a prime ideal $\mathfrak{n}$ of S of height at least 2 lying over $\mathfrak{m}$. Let $\mathcal{P}$ be the set of all nonzero minimal prime ideals of S in $\mathfrak{n}$. Suppose S is a Krull domain. For each $\mathfrak{p} \in \mathcal{P}$ let $v_\mathfrak{p}$ be the unique discrete valuation of K whose valuation ring is $S_\mathfrak{p}$. Set $\mathcal{V} = \{v_\mathfrak{p} \mid \mathfrak{p} \in \mathcal{P}\}$. Suppose $\mathfrak{m}$ is generated by $\aleph_0$ elements, $\bigcap_{k=1}^\infty \mathfrak{m}^k = 0$, and R is complete with respect to the Rt-adic topology for some nonzero $t \in \mathfrak{m}$. Then $(K, \mathcal{V})$ is a Krull field.*

Proof: By Lemma 12.2.5(a), each nonzero $a \in S$ belongs to only finitely many $\mathfrak{p} \in \mathcal{P}$. Since $K = \mathrm{Quot}(S)$, every $x \in K^\times$ satisfies $v_\mathfrak{p}(x) \neq 0$ for only finitely many $\mathfrak{p} \in \mathcal{P}$.

Next let K' be a finite Galois extension of K and let S' be the integral closure of S in K'. Then the discriminant d of S'/S is a nonzero element of S [Ser79, p. 53, Cor. 1] and $\mathfrak{p} \in \mathcal{P}$ ramifies in S' if and only if $d \in \mathfrak{p}$. By the first paragraph, there are only finitely many such $\mathfrak{p}$'s. It follows that $\mathrm{Ram}_\mathcal{P}(K'/K)$ is finite. Hence, by Lemma 12.2.7, $\mathrm{Ram}_\mathcal{V}(K'/K)$ is finite.

Finally we choose a system of representatives R_0 for $R/\mathfrak{m}$ and, as in Setup 12.2.6, let R_1 be the set of all series $\sum_{k=0}^\infty c_k t^k$ with $c_0 \in R_0 \smallsetminus \mathfrak{m}$ and $c_k \in R_0$, for all $k \geq 1$. Then, by Lemma 12.2.9(c), $\mathrm{card}(R/\mathfrak{m})^{\aleph_0} = \mathrm{card}(R_1) \leq \mathrm{card}(\mathrm{Spl}_\mathcal{P}(K'/K))$. By Lemma 12.2.5(d), $\mathrm{card}(\mathcal{P}) \leq \mathrm{card}(S) = \mathrm{card}(K)$. Hence, $\mathrm{card}(\mathrm{Spl}_\mathcal{P}(K'/K)) \leq \mathrm{card}(\mathcal{P}) \leq \mathrm{card}(K)$.

By assumption $\mathfrak{m} = \sum_{i \in I} Rx_i$, where $\mathrm{card}(I) = \aleph_0$. For each $e \geq 0$ let M_e be the set of all monomials in the x_i's of degree e. Then $\mathrm{card}(M_e) \leq \mathrm{card}(I) = \aleph_0$. Since R_0 represents R modulo $\mathfrak{m}$, induction on n proves that for each $x \in R$ and every positive integer n there exists a unique presentation $x \equiv \sum_{e=0}^n \sum_{m \in M_d} c_{x,e,m} m \mod \mathfrak{m}^{n+1}$ with $c_{x,e,m} \in R_0$ for all e, m. Since $\bigcap_{n=1}^\infty \mathfrak{m}^n = 0$, the map $x \to (c_{x,e,m})_{e,m}$ is injective. Thus, $\mathrm{card}(K) = \mathrm{card}(R) \leq \mathrm{card}(R_0)^{\aleph_0} = \mathrm{card}(R_1)$.

It follows that $\mathrm{card}(\mathrm{Spl}_\mathcal{P}(K'/K)) = \mathrm{card}(K)$. Hence, by Lemma 12.2.7, $\mathrm{card}_\mathcal{V}(K'/K) = \mathrm{card}(K)$. Consequently, K is a Krull field with respect to $\mathcal{V}$. $\square$

Remark 12.2.11: *Function fields of one variable.* Let E be a function field of one variable over a Hilbertian field K, let F_0 be a finite separable extension of E, and let F be a finite Galois extension of E that contains F_0. Consider also $u_1, \ldots, u_m \in F^\times$. F. K. Schmidt proved in [Sch34] that F has a K-place

φ unramified over E such that $\varphi(u_1), \ldots, \varphi(u_m) \in K_s^\times$ and $\mathrm{Gal}(F/F_0)$ is the decomposition group of φ over E. If K is countable, this implies that E is a Krull field with respect to the set $\mathcal{V}$ of all valuations of E/K.

Next recall that each rational function field $K_0(t)$ is Hilbertian. Moreover, if K_0 is infinite, then each separable Hilbert subset of $K_0(t)$ contains a set of the form $\{a + bt \mid (a, b) \in U(K_0)\}$ for some nonempty Zariski-open subset U of $\mathbb{A}^2$ [FrJ08, Prop. 13.2.1]. Using transfinite induction, it is possible then to prove that every function field of one variable over an arbitrary field K is Krull with respect to the set $\mathcal{V}$ of all its valuations over K.

We will not use this result, so we do not go into the details of the proof.

$\square$

12.3 Density of Hilbert Sets

Let $v_1, \ldots, v_m, <_1, \ldots, <_n$ be an independent set of valuations and orderings of a Hilbertian field K. We prove that the diagonal map $\mathbf{x} \mapsto (\mathbf{x}, \ldots, \mathbf{x})$ maps each separable Hilbert subset H of K^r into a dense subset of $(K^r)^{m+n}$. Here, the ith copy of K^d is equipped with the v_i-topology, $i = 1, \ldots, m$, while the jth copy of K^d is equipped with the $<_j$-topology, $j = 1, \ldots, n$. Each ordering $<$ of a field K yields a generalized absolute value $|\ |$ on K defined by $|x| = \max(x, -x)$. The above valuations and orderings are **independent** if the topologies of K they induce are distinct.

The proof depends on the weak approximation theorem for valuations and orderings. It has two versions. The first one holds for arbitrary valuations and orderings while the approximated element is common to all coordinates.

PROPOSITION 12.3.1 ([Jar94, Prop. 17.1]): *Let $v_1, \ldots, v_m, <_1, \ldots, <_n$ be valuations and orderings of a field K. Consider $a, a_1', \ldots, a_m', b_1', \ldots, b_n' \in K$ such that $a_i' \neq 0$ and $b_j' >_j 0$ for all i and j. Then*
(a) *there exists $x \in K$, $x \neq a$, such that $v_i(x - a) \geq v_i(a_i')$, $i = 1, \ldots, m$, and $|x - a| <_j b_j'$, $j = 1, \ldots, n$, and*
(b) *there exists $y \in K^\times$ such that $v_i(y) < v_i(a_i')$, $i = 1, \ldots, m$, and $y >_j b_j'$, $j = 1, \ldots, n$.*

The second version gives approximations to various elements, for that it holds only if the valuations and orderings are independent.

PROPOSITION 12.3.2 (Weak approximation theorem for independent valuations and orderings [Jar94, Prop. 17.4]): *Let $v_1, \ldots, v_m, <_1, \ldots, <_n$ be an independent set of valuations and orderings of a field K. Let*

$$a_1, \ldots, a_m, a_1', \ldots, a_m', b_1, \ldots, b_n, b_1', \ldots, b_n' \in K$$

such that $a_i' \neq 0$ and $b_j' >_j 0$ for all i and j. Then there exists $x \in K$ such that

$$v_i(x - a_i) \geq v(a_i'), \quad i = 1, \ldots, m, \quad \text{and} \quad |x - b_j| \leq_j b_j', \quad j = 1, \ldots, n.$$

Next we prove several results on irreducible polynomials.

LEMMA 12.3.3: *Let K be a Hilbertian field and $f \in K[T_1, \ldots, T_r, X]$ an absolutely irreducible polynomial separable in X. Then, there exist $\mathbf{a}_1, \mathbf{a}_2, \mathbf{a}_3, \ldots \in K^r$ and $c_1, c_2, c_3, \ldots \in K_s$ such that $f(\mathbf{a}_i, c_i) = 0$ and $K(c_1), K(c_2), K(c_3), \ldots$ is a linearly disjoint sequence of separable extensions of K of degree $n = \deg_X f$.*

Proof: Assume inductively that $\mathbf{a}_i, c_i$ have been found for $i = 1, \ldots, m$. Then $L = K(c_1, \ldots, c_m)$ is a finite separable extension of K. Since $f(\mathbf{T}, X)$ is irreducible over L, [FrJ08, Cor. 12.2.3] gives an $\mathbf{a}_{m+1} \in K^r$ such that $f(\mathbf{a}_{m+1}, X)$ is irreducible over L of degree n. Choose a root c_{m+1} of $f(\mathbf{a}_{m+1}, X) = 0$. Then $K(c_{m+1})$ is a separable extension of K of degree n which is linearly disjoint from L over K. Consequently, $K(c_1), \ldots, K(c_{m+1})$ are linearly disjoint over K. $\square$

LEMMA 12.3.4: *Let $F_1, F_2, F_3, \ldots$ be a linearly disjoint sequence of extensions of a field K. Let L be a finite separable extension of K and let $f \in K[X]$ be an irreducible separable polynomial. Then there exists k such that*
(a) *f is irreducible over F_i, $i = k, k+1, k+2, \ldots$, and*
(b) *the sequence $LF_k, LF_{k+1}, LF_{k+2}, \ldots$ is linearly disjoint over L.*

Proof of (a): Let N be the splitting field of f over K. Then N has only finitely many subfields that contain K. If $N \cap F_i$ were a proper extension of K for infinitely many i's, then there would exist $i < j$ such that $N \cap F_i = N \cap F_j$ is a proper extension of K. This would however contradict $F_i \cap F_j = K$. Hence, there exists k such that for each $i \geq k$, $N \cap F_i = K$, hence by [FrJ08, Lemma 2.5.3], f_i is irreducible over F_i.

Proof of (b): Replace L, if necessary, by its Galois closure over K to assume that L is Galois over K. Assume that the sequence $L, F_k, F_{k+1}, F_{k+2}, \ldots$ is linearly disjoint over K for no k. Then, for each k there exists an integer $f(k) \geq k$ such that $L \cap (F_k \cdots F_{f(k)})$ is a proper extension of K. Again, since L has only finitely many subfields that contain K, there exists a proper extension K' of K such that $L \cap (F_k \cdots F_{f(k)}) = K'$ for infinitely many k's. Fix one of those k's and take $l > f(k)$ such that $L \cap (F_l \cdots F_{f(l)}) = K'$. Then, $K' \subseteq (F_k \cdots F_{f(k)}) \cap (F_l \cdots F_{f(l)})$. This contradiction to the linear disjointness of $F_1, F_2, F_3, \ldots$ over K proves the existence of k such that $L, F_k, F_{k+1}, F_{k+2}, \ldots$ is linearly disjoint over K. Consequently, $LF_k, LF_{k+1}, LF_{k+2}, \ldots$ are linearly disjoint over L. $\square$

LEMMA 12.3.5: *Let $f \in K[X_1, \ldots, X_r, Y]$ be an irreducible polynomial over K and let n be a positive integer which is not a multiple of $\operatorname{char}(K)$. For $i = 1, 2, 3, \ldots$ and $j = 1, \ldots, r$ let $c_{ij} \in K$ and choose an nth root $\sqrt[n]{c_{ij}}$. Suppose*

$$\{K(\sqrt[n]{c_{ij}}) \mid i = 1, 2, 3, \ldots; \ j = 1, \ldots, r\}$$

is a linearly disjoint set of extensions of K of degree n. Then for all but *finitely many i's, the polynomial*

$$(1) \qquad f_i(X_1, \ldots, X_r, Y) = f\left(\frac{1}{X_1^{n-1} - c_{i1}X_1^{-1}}, \ldots, \frac{1}{X_r^{n-1} - c_{ir}X_r^{-1}}, Y\right)$$

is irreducible in the ring $K(X_1, \ldots, X_r)[Y]$.

Proof: We choose r algebraically independent elements $t_1, \ldots, t_r$ over K. For each i and j we choose $x_{ij} \in K(t_j)_s$ such that

$$(2) \qquad x_{ij}^n - t_j^{-1}x_{ij} - c_{ij} = 0.$$

Then

$$(3) \qquad t_j^{-1} = x_{ij}^{n-1} - c_{ij}x_{ij}^{-1}, \qquad j = 1, \ldots, r.$$

Hence, with $\mathbf{x}_i = (x_{i1}, \ldots, x_{ir})$, $K(\mathbf{x}_i)$ is an algebraic extension of $K(\mathbf{t})$, of degree at most n^r. Since $t_1, \ldots, t_r$ are algebraically independent over K, so are $x_{i1}, \ldots, x_{ir}$.

CLAIM: *For each m, $[K(\mathbf{x}_1, \ldots, \mathbf{x}_m) : K(\mathbf{t})] = n^{rm}$.* For all i, j the polynomial $X^n - c_{ij}$ is irreducible over K, so $K(\sqrt[n]{c_{ij}}) \cong K(\zeta_n \sqrt[n]{c_{ij}})$, where ζ_n is a root of 1 of order n. Therefore, by (2), the specialization $t_j^{-1} \to 0$, $j = 1, \ldots, r$, extends to a homomorphism $\varphi_1: K[\mathbf{x}_i]_{i=1,\ldots,m} \to K[\zeta_{ij} \sqrt[n]{c_{ij}}]_{i,j}$, where $\zeta_{ij}^n = 1$ for all i, j. Since the fields $K(\sqrt[n]{c_{ij}})$ are linearly disjoint over K of degree n, the K-isomorphisms $\varphi_2': K[\sqrt[n]{c_{ij}}] \to K[\zeta_{ij} \sqrt[n]{c_{ij}}]$ combine to a K-isomorphism $\varphi_2': K[\sqrt[n]{c_{ij}}]_{i,j} \to K[\zeta_{ij} \sqrt[n]{c_{ij}}]_{i,j}$. Let $\varphi = (\varphi_2')^{-1} \circ \varphi_1$. Then

$$\varphi: K[\mathbf{x}_1, \ldots, \mathbf{x}_m] \to K[\sqrt[n]{c_{ij}} \mid i = 1, \ldots, m; \ j = 1, \ldots, r]$$

is a K-homomorphism. By assumption, the latter ring (which is actually a field) has, as a vector space over K, dimension n^{rm}. Hence, the dimension of $K(\mathbf{x}_1, \ldots, \mathbf{x}_m)$ as a $K(\mathbf{t})$-vector space is at least n^{rm}. Since $[K(\mathbf{x}_1, \ldots, \mathbf{x}_m) : K(\mathbf{t})] \leq n^{rm}$ (by (2)), equality must hold.

By the claim, $K(\mathbf{x}_1), K(\mathbf{x}_2), K(\mathbf{x}_3), \ldots$ is a linearly disjoint sequence of extensions of $K(\mathbf{t})$ of degree n^r. Since $f(t_1, \ldots, t_r, Y)$ is irreducible over $K(\mathbf{t})$, Lemma 12.3.4(a) provides a k such that for each $i \geq k$ we have, by (3),

$$f(t_1, \ldots, t_r, Y) = f\left(\frac{1}{x_{i1}^{n-1} - c_{i1}x_{i1}^{-1}}, \ldots, \frac{1}{x_{ir}^{n-1} - c_{ir}x_{ir}^{-1}}, Y\right)$$

is irreducible over $K(\mathbf{x}_i)$. Since $x_{i1}, \ldots, x_{ir}$ are algebraically independent over K, this means that the polynomial (1) is irreducible over $K(\mathbf{X})$. $\square$

LEMMA 12.3.6: *Let K be a Hilbertian field with valuations and orderings $v_1, \ldots, v_m, <_{m+1}, \ldots, <_n$. Then, for each irreducible polynomial $f \in K[X_1, \ldots, X_r, Y]$ separable in Y there exists $\mathbf{a} \in K^r$ such that $f(\mathbf{a}, Y)$ is irreducible over K,*

(4) $\qquad v_i(\mathbf{a}) \geq 0, \ i = 1, \ldots, m, \text{ and } |\mathbf{a}| \leq_j 1, \ j = m+1, \ldots, n.$

Proof: Let $p = 2$ if $\mathrm{char}(K) \neq 2$ and $p = 3$ if $\mathrm{char}(K) = 2$. Since the polynomial $Y^p - X$ is absolutely irreducible and separable in Y and K is Hilbertian, Lemma 12.3.3 gives a set $\{c_{kl} \in K^\times \mid k = 0, 1, 2, \ldots; \ l = 1, \ldots, r\}$ such that $\{K(\sqrt[p]{c_{kl}}) \mid k = 0, 1, 2, \ldots; \ l = 1, \ldots, r\}$ is a linearly disjoint set of extensions of K of degree p.

For each nonnegative integer k there are unique $\varepsilon_{kil} \in \{0, 1\}$ such that

(5) $$k \equiv \sum_{l=1}^{r} \sum_{i=1}^{n} \varepsilon_{kil} 2^{i-1+(l-1)n} \mod 2^{rn}.$$

Each c_{kl} can be multiplied by a p-power b_{kl}^p of $K^\times$ without changing $K(\sqrt[p]{c_{kl}})$. Thus, using Proposition 12.3.1, we may choose the c_{kl}'s such that for $1 \leq i \leq m$ and $m+1 \leq j \leq n$

(6a) $\qquad \varepsilon_{kil} = 0$ implies $v_i(c_{kl}) > 0$, and $\varepsilon_{kjl} = 0$ implies $|c_{kl}| \leq_j 1$

(6b) $\qquad \varepsilon_{kil} = 1$ implies $v_i(c_{kl}) \leq 0$, and $\varepsilon_{kjl} = 1$ implies $|c_{kl}| \geq_j 6$.

By Lemma 12.3.5, all but finitely many of the polynomials

$$f_k(X_1, \ldots, X_r, Y) = f\left(\frac{1}{X_1^{p-1} - c_{k1}X_1^{-1}}, \ldots, \frac{1}{X_r^{p-1} - c_{kr}X_r^{-1}}, Y\right)$$

are irreducible in $K(\mathbf{X})[Y]$. We omit the first $e \cdot 2^{rn}$ of them for e large enough, if necessary, to assume that each f_k is irreducible. Then, we use the Hilbertianity of K to choose $\mathbf{b} \in K^r$ such that $f_k(\mathbf{b}, Y)$ is irreducible in $K[Y]$ for each $0 \leq k < 2^{rn}$.

For all $1 \leq i \leq m$, $m+1 \leq j \leq n$, and $1 \leq l \leq r$ we define ε_{il} and ε_{jl} in the following way:

(7a) $\qquad \varepsilon_{il} = 0$ if $v_i(b_l) \leq 0$; $\quad \varepsilon_{jl} = 0$ if $|b_l| \geq_j 2$;

(7b) $\qquad \varepsilon_{il} = 1$ if $v_i(b_l) > 0$; $\quad \varepsilon_{jl} = 1$ if $|b_l| <_j 2$.

Then we set

$$k = \sum_{l=1}^{r} \left[\sum_{i=1}^{m} \varepsilon_{il} 2^{i-1+(l-1)n} + \sum_{j=m+1}^{n} \varepsilon_{jl} 2^{j-1+(l-1)n} \right]$$

and

$$a_l = \frac{1}{b_l^{p-1} - c_{kl} b_l^{-1}}, \qquad l = 1, \ldots, r.$$

Thus, ε_{il} is equal to the coefficient ε_{kil} appearing in (5), $i = 1, \ldots, n$. It follows that $f(\mathbf{a}, Y) = f_k(\mathbf{b}, Y)$ is irreducible in $K[Y]$. We prove that $\mathbf{a}$ satisfies (4).

If $1 \le i \le m$ and $v_i(b_l) \le 0$, then $\varepsilon_{il} = 0$ (by (7a)). Hence, by (6a), $v_i(c_{kl}) > 0$. Thus, $v_i(c_{kl} b_l^{-1}) > v_i(b_l^{p-1})$. Therefore, $v_i(b_l^{p-1} - c_{kl} b_l^{-1}) = v_i(b_l^{p-1}) \le 0$ and $v_i(a_l) \ge 0$.

If $v_i(b_l) > 0$, then $\varepsilon_{il} = 1$ (by (7b)). Hence, by (6b), $v_i(c_{kl}) \le 0$. Thus, $v_i(b_l^{p-1}) > v_i(c_{kl}) - v_i(b_l)$. Therefore, $v_i(b_l^{p-1} - c_{kl} b_l^{-1}) = v_i(c_{kl}) - v_i(b_l) < 0$ and $v_i(a_l) > 0$.

If $m + 1 \le j \le n$ and $|b_l| \ge_j 2$, then $\varepsilon_{jl} = 0$ (by (7a)). Also, $p = 2$, because otherwise $\mathrm{char}(K) = 2$ and K has no orderings. Hence, by (6a), $|c_{kl}| \le_j 1$. Thus, $|b_l - c_{kl} b_l^{-1}| \ge_j |b_l| - |c_{kl}||b_l^{-1}| >_j 1$, so $|a_l| <_j 1$.

Finally, if $|b_l| <_j 2$, then $\varepsilon_{jl} = 1$ (by (7b)). Hence, by (6b), $|c_{kl}| \ge_j 6$. Thus, $|b_l - c_{kl} b_l^{-1}| \ge_j |c_{kl}||b_l^{-1}| - |b_l| >_j 1$ and $|a_l| <_j 1$. Therefore, (4) is satisfied in each case. $\qquad\square$

PROPOSITION 12.3.7 (Approximation of zero theorem for separable Hilbert sets): *Let K be a Hilbertian field equipped with valuations and orderings $v_1, \ldots, v_m, <_1, \ldots, <_n$. Denote the valuation ring of v_i by Γ_i. Let H be a separable Hilbert subset of K^r. Then, for all $\mathbf{a} \in K^r$, $b_1, \ldots, b_m \in K^\times$, and all $c_1, \ldots, c_n \in K$ with $c_j >_j 0$ for $j = 1, \ldots, n$, there exists $\mathbf{x} \in H$ such that*

(8) $\quad v_i(\mathbf{x} - \mathbf{a}) \ge v_i(b_i),\ i = 1, \ldots, m, \qquad$ and $\qquad |\mathbf{x} - \mathbf{a}| <_j c_j,\ j = 1, \ldots, n.$

Proof: By [FrJ08, Lemma 12.1.6], there exists an irreducible polynomial g in $K[X_1, \ldots, X_r, Y]$ separable in Y such that $H_K(g) \subseteq H$. By Proposition 12.3.1, there exists $d \in K^\times$ such that

$$v_i(d) \ge v_i(b_i),\ i = 1, \ldots, m, \text{ and } |d| <_j c_j,\ j = 1, \ldots, n.$$

Applying Lemma 12.3.6 to the polynomial $f(\mathbf{T}, Y) = g(\mathbf{a} + d\mathbf{T}, Y)$, we find $\mathbf{t} \in K^r$ such that $g(\mathbf{a} + d\mathbf{t}, Y)$ is irreducible over K,

$$v_i(\mathbf{t}) \ge 0,\ i = 1, \ldots, m, \text{ and } |\mathbf{t}| \le_j 1,\ j = 1, \ldots, n.$$

Then $\mathbf{x} = \mathbf{a} + d\mathbf{t}$ belongs to H and satisfies (8). $\qquad\square$

A combination of the weak approximation theorem for independent valuations and orderings (Proposition 12.3.2) and Proposition 12.3.7 gives the density theorem for separable Hilbert sets.

PROPOSITION 12.3.8 (Density theorem for separable Hilberts sets): *Let K be a Hilbertian field equipped with independent valuations and orderings $v_1, \ldots, v_m, <_1, \ldots, <_n$. Let H be a separable Hilbert subset of K^r. Then, for all $\mathbf{a}_1, \ldots, \mathbf{a}_m, \mathbf{b}_1, \ldots, \mathbf{b}_n \in K^r$, $a_i' \in K^\times$, $i = 1, \ldots, m$, and $b_j' \in K$ with $b_j' >_j 0$, $j = 1, \ldots, n$, there exists $\mathbf{x} \in H$ such that*

$$v_i(\mathbf{x} - \mathbf{a}_i) \ge v_i(a_i'),\ i = 1, \ldots, m, \qquad \text{and} \qquad |\mathbf{x} - \mathbf{b}_j| <_j b_j',\ j = 1, \ldots, n.$$

12.4 Krull Hilbertian Ample Fields

We prove that the absolute Galois group of a Hilbertian ample Krull field is semi-free and deduce that the absolute Galois group of the quotient field of each Noetherian domain which is complete with respect to a prime ideal of dimension at least 2 is semi-free.

THEOREM 12.4.1 (Pop): *Let K be a Hilbertian ample field of cardinality m. Suppose K is Krull with respect to a set $\mathcal{V}$ of discrete valuations. Then $\mathrm{Gal}(K)$ is semi-free of rank m.*

Proof: By Theorem 5.10.2(a), every finite split embedding for $\mathrm{Gal}(K)$ is solvable. Hence, by Proposition 10.2.4, every finite split embedding problem for $\mathrm{Gal}(K)$ has infinitely many linearly disjoint solutions. Thus, $\mathrm{rank}(\mathrm{Gal}(K)) \geq \aleph_0$. Hence, by [FrJ08, Prop. 17.1.2], $\mathrm{rank}(\mathrm{Gal}(K))$ is the cardinality of the set of all finite extensions of K, so $\mathrm{rank}(\mathrm{Gal}(K)) \leq m$. Therefore, if K is countable, then $\mathrm{Gal}(K)$ is semi-free of countable rank.

In order to complete the proof of the theorem, we may assume that $m > \aleph_0$ and we have to prove that each finite split embedding problem

$$\mathcal{E}: \quad G \xrightarrow{\ \alpha\ } \mathrm{Gal}(L/K)$$

has m linearly disjoint solutions.

PART A: *Regular solution.* Since K is ample, $\mathcal{E}$ has a rational solution (Theorem 5.9.2). In other words, given a variable x, the field $E = K(x)$ has a Galois extension F that contains L, there exists an isomorphism $\gamma\colon \mathrm{Gal}(F/E) \to G$, such that $\alpha \circ \gamma = \mathrm{res}_{F/L}$, and F has an L-rational place. In particular, F is a regular extension of L [FrJ08, Lemma 2.6.9]. Let $P_1, \ldots, P_r$ be the prime divisors of E/K that ramify in F. For each i between 1 and r we choose a prime divisor Q_i of F/L that lies over P_i. Then we let L' be the normal closure of $L\bar{F}_{Q_1} \cdots \bar{F}_{Q_r}/K$.

PART B: *Assuming L'/K is Galois.* Let K' be the maximal purely inseparable extension of K in L'. Then, res: $\mathrm{Gal}(LK'/K') \to \mathrm{Gal}(L/K)$ and res: $\mathrm{Gal}(K') \to \mathrm{Gal}(K)$ are isomorphisms. So, $\mathcal{E}$ yields an embedding problem $G \to \mathrm{Gal}(LK'/K')$. If the latter has card$(K)$ linearly disjoint solutions, so does $\mathcal{E}$. Next note that K' is Hilbertian [FrJ08, Prop. 12.3.3], ample (Lemma 5.5.1(b)), and Krull with respect to the set of all unique extensions of the valuations belonging to $\mathcal{V}$ to K' (Lemma 12.2.4). Therefore, replacing K by K' and F by FK', if necessary, we may assume that L' is a Galois extension of K.

PART C: *Transfinite induction.* Let $\lambda < \mathrm{card}(K)$ be a cardinal number and suppose we have already constructed a transfinite sequence $(M_\kappa)_{\kappa < \lambda}$ of linearly disjoint solutions for $\mathcal{E}$. Then $M = \prod_{\kappa < \lambda} M_\kappa$ is a Galois extension of K that contains L. We have to construct an additional solution M_λ for $\mathcal{E}$ such that $M_\lambda \cap M = L$. To this end we let $\mathrm{Ram}_\mathcal{V}(M_\kappa/K)$ and $\mathrm{Ram}_\mathcal{V}(M/K)$

be the sets of all $v \in V$ that ramify in M_κ and M, respectively, and note that $\mathrm{Ram}_V(M/K) = \bigcup_{\kappa<\lambda} \mathrm{Ram}_V(M_\kappa/K)$. By Remark 12.2.3, each of the sets $\mathrm{Ram}_V(M_\kappa/K)$ is finite. Hence,

(1)
$$\mathrm{card}(\mathrm{Ram}_V(M/K)) = \mathrm{card}\Big(\bigcup_{\kappa<\lambda} \mathrm{Ram}_V(M_\kappa/K)\Big)$$
$$\leq \sum_{\kappa<\lambda} \mathrm{card}(\mathrm{Ram}_V(M_\kappa/K)) \leq \aleph_0 \cdot \mathrm{card}(\lambda) < \mathrm{card}(K).$$

PART D: *Fiber product.* We consider the fiber product $G' = G \times_{\mathrm{Gal}(L/K)} \mathrm{Gal}(L'/K)$ and the function field $F' = FL'$ of one variable over L'. They give rise to a commutative diagram of groups

(2)

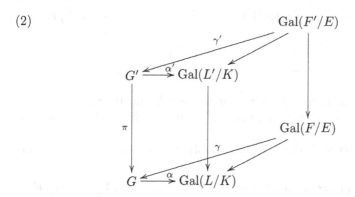

in which γ' is an isomorphism, α' and π are epimorphism, and the four remaining maps are restrictions.

We let $E' = L'(x)$ and choose prime divisors $Q'_1, \ldots, Q'_r$ of F'/L' that lie respectively over $Q_1, \ldots, Q_r$. For each $1 \leq i \leq r$ let P'_i be the prime divisor of E'/L' that lies under Q'_i. Since L' is a separable extension of K, $\overline{F}'_{Q'_i} = \overline{F}_{Q_i}L' = L' = \overline{E}'_{P'_i}$. Let b_i be the residue of x under P'_i and note that $b_i \in L'$. Then, the assumptions of Setup 12.1.2 are satisfied with respect to L', E', P'_i, b_i replacing $K, E, \mathfrak{p}, b_0$. By Lemma 12.1.4, there exists for each $1 \leq i \leq r$ a separable Hilbert subset H_i of L' and a finite subset A_i of L', such that if v' is a discrete valuation of L' satisfying $v'(a) = 0$ for each $a \in A_i$, if $b \in H_i$ satisfies $v'(b - b_i) = 1$, if φ is an L'-place of F' that satisfies $\varphi(x) = b$ and we denote reduction of objects of F'/L' under φ by a bar, then φ induces an isomorphism $\varphi'_*\colon \mathrm{Gal}(F'/E') \to \mathrm{Gal}(\overline{F}'/L')$ that commutes with reduction of elements and maps each inertia group of P'_i in $\mathrm{Gal}(F'/L'(x))$ onto an inertia group of v' in $\mathrm{Gal}(\overline{F}'/L')$.

PART E: *Descent to K.* We may replace each element of A_i by the set of all nonzero coefficients of its irreducible polynomial over K, if necessary, to assume that $A_i \subseteq K$. By [FrJ08, Prop. 12.3.3], we may assume that H_i is a Hilbert subset of K, $[\overline{F}' : K] = [F' : E]$, $\overline{F}$ is linearly disjoint from L' over

L, and $\bar{F}L' = \overline{F'}$. Note that under this assumption, each b in H_i is also in K, so under the additional assumption that $v'(b - b_i) = 1$, the isomorphism φ'_* extends to an isomorphism $\varphi'_*\colon \operatorname{Gal}(F'/E) \to \operatorname{Gal}(\overline{F'}/K)$ compatible with φ, and φ'_* induces an isomorphism $\varphi_*\colon \operatorname{Gal}(F/E) \to \operatorname{Gal}(\bar{F}/K)$ such that the following diagram of groups is commutative:

(3)

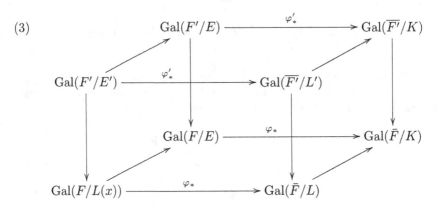

In this diagram, the horizontal arrows are isomorphisms, the vertical arrows are restriction maps and they are surjective, and the diagonal maps are inclusions.

Combining (2) with the back side of (3), we find that $\bar{F}$ is a solution field of $\mathcal{E}$.

PART F: *Valuations that totally split in L'.* We set $A = \bigcup_{i=1}^r A_i$ and $H = \bigcap_{i=1}^r H_i$. By assumption $(K, \mathcal{V})$ is a Krull field, so $\operatorname{card}(\operatorname{Spl}_{\mathcal{V}}(L'/K)) = \operatorname{card}(K)$. Hence, we may use (1) to choose distinct

$$v_1, \ldots, v_r \in \operatorname{Spl}_{\mathcal{V}}(L'/K) \smallsetminus \operatorname{Ram}_{\mathcal{V}}(M/K)$$

such that $v_i(a) = 0$ for each i and for all $a \in A$. Let $v'_1, \ldots, v'_r$ be normalized discrete valuations of L' lying over $v_1, \ldots, v_r$. Then v'_i/v_i is unramified and $v'_i(a) = 0$ for each i and all $a \in A$.

It follows from Part D that if $b \in H$ satisfies $v'_i(b - b_i) = 1$ for $i = 1, \ldots, r$, φ is an L'-place of F' that satisfies $\varphi(x) = b$, and we use a bar to denote reduction of objects of F'/L' with respect to φ, then $\bar{F}$ is a solution field of $\mathcal{E}$, linearly disjoint from L' over L, φ induces the commutative diagram (3), φ'_* maps each inertia group over P'_i in $\operatorname{Gal}(F'/E')$ onto an inertia group of v'_i in $\operatorname{Gal}(\overline{F'}/L')$, $i = 1, \ldots, r$. Since E'/E and L'/K are unramified extensions, each inertia group of P'_i in $\operatorname{Gal}(F'/E')$ is also an inertia group of P'_i in $\operatorname{Gal}(F'/E)$ and each inertia group of v'_i in $\operatorname{Gal}(\overline{F'}/L')$ is an inertia group of v'_i in $\operatorname{Gal}(\overline{F'}/K)$. Finally, since both restriction maps in the back rectangle of (3) map inertia groups onto the corresponding inertia groups [Ser79, p. 22, Prop. 22], φ_* maps each inertia group of P_i in $\operatorname{Gal}(F/E)$ isomorphically onto an inertia group of v_i in $\operatorname{Gal}(\bar{F}/K)$.

PART G: *Density of Hilbertian sets.* Since $v_1, \ldots, v_r$ totally splits in L', the field K is v_i'-dense in L' (Remark 12.2.1(d)). Hence, by Proposition 12.3.8, there exists $b \in H$ such that $v_i'(b - b_i) = 1$ for $i = 1, \ldots, r$. Let φ and $\bar{F}$ be as in Part F and set $M_\lambda = \bar{F}$. Then M_λ is a solution field of embedding problem $\mathcal{E}$. Since $L(x)/K(x)$ is a Galois extension, P_i is unramified in $L(x)$, so all of its inertia groups in $\mathrm{Gal}(F/E)$ are contained in $\mathrm{Gal}(F/L(x))$. Since F/L is regular, $\mathrm{Gal}(F/L(x))$ is generated by all of the inertia groups of $P_1, \ldots, P_r$ in $\mathrm{Gal}(F/K(x))$ [FrJ08, Remark 3.6.2(e)]. Therefore, $\mathrm{Gal}(M_\lambda/L)$ is generated by all of the inertia groups of $v_1, \ldots, v_r$. Since $v_1, \ldots, v_r$ are unramified in M, we get $M_\lambda \cap M = L$, as desired. $\qquad\square$

We need the following result from commutative algebra [ZaS75, p. 275, Thm. 14]:

LEMMA 12.4.2: *Let A be a Noetherian domain and $\mathfrak{b} \subseteq \mathfrak{a}$ proper ideals of A. Suppose A is complete with respect to $\mathfrak{a}$. Then A is also complete with respect to $\mathfrak{b}$.*

Proof: We start the proof with two claims:

CLAIM A: $\mathfrak{a}$ *is contained in every maximal ideal of A.* Otherwise there exists a maximal ideal $\mathfrak{m}$ that does not contain $\mathfrak{a}$. Thus, $\mathfrak{a} + \mathfrak{m} = A$. In other words, there exist $a \in \mathfrak{a}$ and $m \in \mathfrak{m}$ with $a + m = 1$. Hence, $m = 1 - a$ is not not a unit of A. However, since A is complete with respect to $\mathfrak{a}$, the series $\sum_{n=0}^{\infty} a^n$ converges in A and $(1 - a) \sum_{n=0}^{\infty} a^n = 1$ (Here we use Krull's lemma saying that $\bigcap_{n=0}^{\infty} \mathfrak{a}^n = 0$ [AtM69, Cor. 10.18]). This contradiction proves our claim.

CLAIM B: $\bigcap_{n=1}^{\infty} (\mathfrak{a}^n + \mathfrak{b}^m) = \mathfrak{b}^m$ *for each $m \geq 1$.* Indeed, the A-modul $M = A/\mathfrak{b}^m$ is generated by one element, namely $1 + \mathfrak{b}^m$. Let $N = \bigcap_{n=0}^{\infty} \mathfrak{a}^n M$. Then $\mathfrak{a}N = N$. It follows from Claim A and from a consequence of the lemma of Artin-Rees [AtM69, Cor. 10.19] that $N = 0$. Therefore, $\bigcap_{n=1}^{\infty} (\mathfrak{a}^n + \mathfrak{b}^m) = \mathfrak{b}^m$.

In order the conclude the proof of the lemma, we consider a Cauchy sequence $x_1, x_2, x_3, \ldots$ in A in the $\mathfrak{b}$-adic topology. Then

$$(4) \qquad x_j - x_i \in \mathfrak{b}^{m(i)}$$

for each $j \geq i \geq 1$ and $m(i)$ tends to infinity as $i \to \infty$. Since $\mathfrak{b}^{m(i)} \subseteq \mathfrak{a}^{m(i)}$, the sequence $x_1, x_2, x_3, \ldots$ is also Cauchy in the $\mathfrak{a}$-adic topology. By assumption, the sequence converges $\mathfrak{a}$-adically to an element $x \in A$. In other words,

$$(5) \qquad x_j - x \in \mathfrak{a}^{n(j)}$$

and $n(j)$ tends to infinity as $j \to \infty$. It follows from (4) and (5) and from Claim B that

$$x_i - x \in \bigcap_{j \geq i} (\mathfrak{b}^{m(i)} + \mathfrak{a}^{n(j)}) = \mathfrak{b}^{m(i)}.$$

Consequently, $x_i \to x$ in the $\mathfrak{b}$-topology, as desired. $\qquad\square$

THEOREM 12.4.3 (Pop): *Let R be a Noetherian domain of dimension at least 2. Then $K = \text{Quot}(R)$ is Hilbertian. If in addition, R is complete with respect to a nonzero ideal $\mathfrak{a}$ and $\mathfrak{a}$ is contained in a prime ideal $\mathfrak{m}$ of height at least 2, then K is ample and Krull. Moreover, $\text{Gal}(K)$ is semi-free of rank $\text{card}(K)$.*

Proof: By Mori-Nagata, the integral closure R' of R in K is a Krull domain (Remark 5.11.2). By the going up theorem [Mats94, p. 68, Thm. 9.4(i)] $\dim(R') = \dim(R) \geq 2$. Hence, by Weissauer, K is Hilbertian [FrJ08, Thm. 15.4.6].

Now we assume that R is complete with respect to a nonzero ideal $\mathfrak{a}$ of R. Then, by Proposition 5.7.7, K is ample. In addition, let t be a nonzero element of $\mathfrak{a}$. Then, by Lemma 12.4.2, R is complete with respect to the ideal Rt.

Finally, we assume that, in addition, $\mathfrak{a}$ is contained in a prime ideal $\mathfrak{m}$ of height at least 2. Since R is Noetherian, $\mathfrak{m}$ is finitely generated and $\bigcap_{k=0}^{\infty} \mathfrak{m}^k = 0$. By Proposition 12.2.10, K is a Krull field. Consequently, by Theorem 12.4.1, $\text{Gal}(K)$ is semi-free of rank $\text{card}(R) = \text{card}(K)$. $\quad\square$

As an application of Theorem 12.4.3, we strengthen Theorem 5.11.3.

THEOREM 12.4.4: *Each of the following fields K is Hilbertian, ample, and Krull. Thus, $\text{Gal}(K)$ is semi-free of rank $\text{card}(K)$.*
(a) $K = K_0((X_1, \ldots, X_n))$, *where K_0 is an arbitrary field and $n \geq 2$.*
(b) $K = \text{Quot}(R_0[[X_1, \ldots, X_n]])$, *where R_0 is a Noetherian domain which is not a field and $n \geq 1$.*

Proof: Note that (a) is a special case of (b), since $K_0[[X_1, \ldots, X_n]] = R_0[[X_2, \ldots, X_n]]$, where $R_0 = K_0[[X_1]]$ is a complete discrete valuation ring, hence also a Noetherian domain which is not a field. So it suffices to prove (b).

To this end we set $R = R_0[[X_1, \ldots, X_n]]$ and $\mathfrak{a} = \sum_{i=1}^{n} R_0 X_i$. Then we choose a nonzero prime ideal $\mathfrak{p}$ of R_0 and let $\mathfrak{m} = R\mathfrak{p} + \mathfrak{a}$. We observe that $R/\mathfrak{a} \cong R_0$ is an integral domain, so $\mathfrak{a}$ is a nonzero prime ideal. Also, $R/(R\mathfrak{p} + \mathfrak{a}) \cong R_0/\mathfrak{p}$ is a nonzero integral domain, so $\mathfrak{m}$ is a prime ideal of height at least 2.

Since R_0 is a Noetherian ring, so is R [AtM69, p. 113, 10.27]. Finally, $\mathfrak{a} \subseteq \mathfrak{m}$ and R is complete with respect to $\mathfrak{a}$ (Example 5.7.5). It follows from Theorem 12.4.3 that K is Hilbertian, ample, Krull, and $\text{Gal}(K)$ is semi-free of rank $\text{card}(K)$. $\quad\square$

Another interesting special case of Theorem 12.4.3 occurs when R is local.

THEOREM 12.4.5 (Paran): *Let R be a complete local Noetherian domain of dimension at least 2. Then $K = \text{Quot}(R)$ is Hilbertian, ample, and Krull. Moreover, $\text{Gal}(K)$ is semi-free of rank $\text{card}(K)$.*

Proof: Let $\mathfrak{m}$ be the unique maximal ideal of R and let $\mathfrak{a} = \mathfrak{m}$. Then height$(\mathfrak{m}) \geq 2$ and R is complete with respect to $\mathfrak{a}$. Thus, our theorem is a special case of Theorem 12.4.3. □

In the second paragraph of Section 10.6 we point out that Harbater uses quasi-freeness in order to prove that $\mathrm{Gal}(C((x,y))_{\mathrm{ab}})$ is a free profinite group if C is separably closed. Here, we reprove that result using semi-freeness. Both proofs rely on a deep result that $\mathrm{Gal}(C((x,y))_{\mathrm{ab}})$ is projective. Unfortunately, the proof of that result lies beyond the scope of our book.

THEOREM 12.4.6: *Let C be a separably closed field of characteristic p, $K = C((x,y))$ the field of formal power series in x, y over C. Denote the maximal Abelian extension of K (resp. of degree not divisible by p) by K_{ab} (resp. K'_{ab}). Then $\mathrm{Gal}(K_{\mathrm{ab}})$ and $\mathrm{Gal}(K'_{\mathrm{ab}})$ are free of rank card(K).*

Proof: Set $m = \mathrm{card}(K)$. By Example 12.4.4, $\mathrm{Gal}(K)$ is semi-free of rank m. Hence, by Theorem 10.5.4(c), both $\mathrm{Gal}(K_{\mathrm{ab}})$ and $\mathrm{Gal}(K'_{\mathrm{ab}})$ are semi-free of rank m. By a theorem of Harbater [Hrb09, Thm. 4.4] that generalizes a theorem of Colliot-Thélène, Ojanguren, and Parimala [COP02, Thm. 2.2], both groups are projective. By Proposition 9.4.7, both $\mathrm{Gal}(K_{\mathrm{ab}})$ and $\mathrm{Gal}(K'_{\mathrm{ab}})$ are free. □

Notes

Most of Chapter 12 except of Section 12.3 is a workout of [Pop10]. The basic result of the Chapter, Theorem 12.4.1, is [Pop10, Thm. 1.1]. It says that if K is a Hilbertian ample Krull field of cardinality m, then $\mathrm{Gal}(K)$ is semi-free of rank m. Thus, each finite split embedding problem $\mathcal{E}$ over K with a nontrivial kernel has m linearly disjoint solutions.

The conditions on K in Theorem 12.4.1 of being ample and Krull hold under standard assumption of commutative algebra. This leads to the main result of the chapter: Let R be a Noetherian ring, $\mathfrak{m}$ a prime ideal of height at least 2, and $\mathfrak{a}$ a nonzero ideal in $\mathfrak{m}$ such that R is complete with respect to $\mathfrak{a}$. Then $K = \mathrm{Quot}(R)$ is Hilbertian, ample, and Krull, so $\mathrm{Gal}(K)$ is semi-free of rank card(K) (Theorem 12.4.3).

The most striking examples of Theorem 12.4.3 occur when $R = K_0[[X_1, \ldots, X_n]]$ with K_0 an arbitrary field and $n \geq 2$, or $R = \mathbb{Z}[[X_1, \ldots, X_n]]$ (Example 12.4.4). In each case we conclude that $\mathrm{Gal}(K)$ is semi-free. There are two independent approaches to the proof of this case.

Paran first realizes cyclic extensions with control on ramification over $K(x)$. Then he uses his variant of algebraic patching to patch the latter cyclic extensions to m linearly disjoint solutions of a given finite split embedding problem $\mathcal{E}$ over $K(x)$, again with control on ramification and without using the ampleness of K. Finally he uses the Hilbertianity of K to specialize the solutions of $\mathcal{E}$ he constructed over $K(x)$ to distinct solutions of $\mathcal{E}$ over K [Par10].

Pop on the other hand uses the ampleness of K in [Pop10] to solve $\mathcal{E}$ over $K(x)$. Then he uses the Hilbertianity of K to specialize the solution of $\mathcal{E}$ over $K(x)$ to m linearly disjoint solutions over K (Theorem 12.4.1).

Pop's proof essentially shows that K is a **fully Hilbertian field**, that is a field satisfying a strong form of Hilbertianity that allows specializations of each finite separable extension F of $K(x)$ into card(K) extensions of K that are linearly disjoint over the algebraic closure of K in F. This notion is introduced and studied in a general context in [BSP09]. Among others, [BSP09] gives an independent proof of the above mentioned theorem about rings of formal power series.

Using Cohen's structure theorem and Lemma 12.2.4, Example 12.4.4 implies that the absolute Galois group of the quotient field of a complete local Noetherian domain of dimension at least 2 is also semi-free with rank card(K) (Theorem 12.4.5), as noticed in [Par10].

Theorem 12.4.3 generalizes Paran's result, mainly by dropping the condition on R to be local.

Proposition 12.2.10 captures the special case of [Pop10, Thm. 3.4] we need in order to prove Theorem 12.4.3.

Section 12.3 is a rewrite of [Jar94, Sec. 19]. The main result of the latter section is the density theorem for Hilbert sets with respect to independent valuations and orderings. That result generalizes Lemma 3.4 of [Gey78] about the density of Hilbert sets with respect to distinct absolute values. Note however that the only application of the density theorem appears in Part G of Proposition 12.4.1 and it applies only the special case proven by Geyer.

Theorem 12.4.6 is due to Harbater [Hrb09, Thm. 4.6]. Theorem 1.3 of [Pop10] generalizes that result.

Open Problems

1. Let K be a field such that the order of $\mathrm{Gal}(K)$ is divisible by only finitely many prime numbers. Is K ample (Problem 5.8.4)?

2. Let K be a field and x a variable. Which of the following statements is true:
(a) Every constant finite split embedding problem over $K(x)$ has a regular solution.
(b) If K is Hilbertian, then every finite split embedding problem over K is solvable.
(c) If every finite split embedding problem over K is solvable, then K is ample.
(Remark 5.10.4)

3. Is the field $\mathbb{Q}_{\mathrm{ab}}$ ample (Example 5.10.5)?

4. Is the field $\mathbb{Q}_{\mathrm{solv}}$ ample (Example 5.10.6)?

5. Let K be a field such that $\mathrm{Gal}(K)$ is finitely generated. Prove that
(a) If K is infinite, then K is ample (Koenigsmann [JuK09, discussion proceeding Question 8]), and
(b) if K is not an algebraic extension of a finite field and A is a nonzero Abelian variety defined over K, then $\mathrm{rr}(A(K)) = \infty$ (Conjecture 6.5.8).

6. The maximal pro-p extension of a number field is nonample (Conjecture 6.5.10).

7. Let $F/\mathbb{Q}$ be a function field of one variable with a prime divisor of degree 1. Is $\mathrm{gon}(\bar{F}_p/\mathbb{F}_p) = \mathrm{gon}(F/\mathbb{Q})$ for almost all prime numbers p, where $\bar{F}_p$ is the residue field of F under a good reduction that extends the p-adic reduction of $\mathbb{Q}$ in the sense of Section 8.1? (Problem 6.7.6).

8. Give an algebraic proof to Proposition 9.1.1(a): Let F be a finite Galois extension of $\mathbb{C}(x)$. Let $\mathfrak{p}_1, \ldots, \mathfrak{p}_r$ be the prime divisors of $\mathbb{C}(x)/\mathbb{C}$ that are ramified in F. Then there exist generators $\sigma_1, \ldots, \sigma_r$ of $\mathrm{Gal}(F/\mathbb{C}(x))$ satisfying $\sigma_1 \cdots \sigma_r = 1$ such that σ_i generates an inertia group over $\mathfrak{p}_i$, $i = 1, \ldots, r$ (Section 9.1).

9. Give an algebraic proof to Proposition 9.1.1(b): Let G be a finite group generated by $\sigma_1, \ldots, \sigma_r$ with $\sigma_1 \cdots \sigma_r = 1$. Then $\mathbb{C}(x)$ has a finite Galois extension F ramified at most over $\{\mathfrak{p}_1, \ldots, \mathfrak{p}_r\}$ such that σ_i generates an inertia group over $\mathfrak{p}_i$, $i = 1, \ldots, r$ (Section 9.1).

10. (Bogomolov-Positselski) Let E be an extension of a field K such that $\mathrm{trans.deg}(E/K) = 1$ and F an algebraic extension of E. Suppose F contains a radical algebraic extension of E. Then $\mathrm{Gal}(F)$ is projective (Conjecture 11.6.1).

M. Jarden, *Algebraic Patching*, Springer Monographs in Mathematics, DOI 10.1007/978-3-642-15128-6, © Springer-Verlag Berlin Heidelberg 2011

References

[Abh57]   S. S. Abhyankar, *Coverings of algebraic curves*, American Journal of Mathematics **79** (1957), 825–856.

[Art67]   E. Artin, *Algebraic Numbers and Algebraic Functions*, Gordon and Breach, New York, 1967.

[AtM69]   M. F. Atiyah and I. G. Mackdonald, *Introduction to Commutative Algebra*, Addison–Wesley, Reading, 1969.

[BSo06]   L. Bary-Soroker, *Diamond theorem for a finitely generated free profinite group*, Mathematische Annalen **336** (2006), 949–961.

[BHH08]   L. Bary-Soroker, Dan Haran, and David Harbater, *Permanence criteria for semi-free profinite groups*, arXiv:0810.0845, v2, 1 April 2010.

[BHH10]   L. Bary-Soroker, Dan Haran, and David Harbater, *Permanence criteria for semi-free profinite groups*, Mathematische Annalen, DOI 10.1007/s00208-010-0484-8, 9 February 2010.

[BSP09]   L. Bary-Soroker and E. Paran, *Fully Hilbertian fields*, arXiv:0907.0343

[Bel80]   G. V. Belyi, *On extensions of the maximal cyclotomic field having a given classical Galois group*, Journal für die reine und angewandte Mathematik **341**, (1980), 147-158.

[BNW71]   E. Binz, J. Neukirch, and G. H. Wenzel, *A subgroup theorem for free products of profinite groups*, Journal of Algebra **19** (1971), 104–109.

[Bla99]   E. V. Black, *Deformations of dihedral 2-group extensions of fields*, Transactions of the AMS **351** (1999), 3229–3241.

[Bou58]   N. Bourbaki, *Algeébre, chapitr 8* Springer, Hermann, 1958.

[Bou89]   N. Bourbaki, *Commutative Algebra, Chapters 1–7*, Springer, Berlin, 1989.

[Car40]   H. Cartan, *Sur les matrices holomorphes de n variables complexes*, Journal de Mathématique Pures et Appliquées **19** (1940), 1–26.

[CaF67]   J. W. S. Cassels and A. Fröhlich, *Algebraic Number Theory*, Academic Press, London, 1967.

[Che51]   C. Chevalley, *Introduction to the Theory of Algebraic Functions of One Variable*, Mathematical Surveys VI, AMS, Providence, 1951.

[CoT00]   J.-L. Colliot-Thélène, *Rational connectedness and Galois covers of the projective line*, Annals of Mathematics **151** (2000), 359-373.

[COP02]   J.-L. Colliot-Thélène, M. Ojanguren, and R. Parimala, *Quadratic forms over fraction fields of two-dimensional Henselian rings and Brauer groups of related schemes*, In: "Proceedings of the International Colloquium on Algebra, Arithmetic and Geometry", Tata Institute of Fundamental Research, Studies in Mathematics **16**, 185–217, Narosa Publ. Co., 2002.

[CoS86]   G. Cornell and J. H. Silverman, *Arithmetic Geometry*, Springer-Verlag, New York, 1986.

M. Jarden, *Algebraic Patching*, Springer Monographs in Mathematics,
DOI 10.1007/978-3-642-15128-6, © Springer-Verlag Berlin Heidelberg 2011

[DeD97]   P. Dèbes and B. Deschamps, *The regular inverse Galois problem over large fields*, Geometric Galois actions **2**, London Mathematical Society Lecture Notes Series, **243**, Cambridge University Press, 1997, 119–138.

[Deu42]   M. Deuring, *Reduktion algebraischer Funtionenkörper nach Primdivisoren des Konstantenkörper*, Mathematische Zeitschrift **47** (1942), 643–654.

[Deu73]   M. Deuring, *Lectures on the Theory of Algebraic Functions of One Variable*, Lecture Notes in Mathematics **314**, Springer, Berlin, 1973.

[Dou64]   A. Douady, *Détermination d'un groupe de Galois*, Comptes Rendus de l'Académie des Sciences **258** (1964), 5305–5308.

[Dou79]   A. Douady and R. Douady, *Algébre et thèorie Galoissiennes II*, Paris: Cedic/Fernand Nathan 1979.

[Efr01]   I. Efrat, *A Hasse principle for function fields over PAC fields*, Israel Journal of Mathematics **122** (2001), 43–60.

[Efr06]   I. Efrat, *Valuations, Orderings, and Milnor K-Theory*, Mathematical surveys and monogrphs **124**, American Mathematical Society, Providence, 2006.

[End72]   O. Endler, *Valuation Theory*, Springer, Berlin, 1972.

[Fal83]   G. Faltings, *Endlichkeitssätze für abelsche Varietäten über Zahlkörpern*, Inventiones mathmaticae **73** (1983), 349–366.

[Fal91]   G. Faltings, *Diophantine approximation on abelian varieties*, Annals of Mathematics **133** (1991), 549–576.

[Fal94]   G. Faltings, *The general case of S. Lang's conjecture*, Perspectives in Mathematics, **15** (1994), 575–603.

[Feh10]   A. Fehm, *Subfields of ample fields, Rational maps and definability.* Journal of Algebra **323** (2010) 1738–1744.

[Feh11]   A. Fehm, *Embeddings of function fields into ample fields*, Manuscript, Tel Aviv 2009.

[FeP10]   A. Fehm and S. Petersen, *On the rank of Abelian varieties over ample fields*, International Journal of Number Theory **6** (2010), 1–8.

[FrP81]   J. Fresnel, M. v.d. Put, *Géométrie Analytique Rigide et Applications*, Progress in Mathematics **18**, Birkhäuser, Boston, 1981.

[FrP04]   J. Fresnel, M. v.d. Put, *Rigid Analytic Geometry and its Applications*, Progress in Mathematics **218**, Birkhäuser, Boston, 2004.

[Fre94]   G. Frey, *Curves with infinitely many points of fixed degree*, Israel Journal of Mathematics **85** (1994), 79–83.

[FyJ74]   G. Frey and M. Jarden, *Approximation theory and the rank of abelian varieties over large algebraic fields*, Proceedings of the London Mathematical Society **28** (1974), 112–128.

[FrJ86]   M. D. Fried and M. Jarden, *Field Arithmetic*, Ergebnisse der Mathematik (3) **11**, Springer-Verlag, Heidelberg, 1986.

[FrJ08]   M. D. Fried and M. Jarden, *Field Arithmetic, Third Edition, revised by Moshe Jarden*, Ergebnisse der Mathematik (3) **11**, Springer, Heidelberg, 2008.

[FrV91]   M. D. Fried and H. Völklein, *The inverse Galois problem and rational points on moduli spaces*, Mathematische Annalen **290** (1991), 771–800.

[FrV92]   M. D. Fried and H. Völklein, *The embedding problem over a Hilbertian PAC-field*, Annals of Mathematics **135** (1992), 469–481.

[Ful89]   W. Fulton, *Algebraic Curves*, Addison Weseley, Redwood City, 1989.

[Gey78]   W.-D. Geyer, *Galois groups of intersections of local fields*, Israel Journal of Mathematics **30** (1978), 382–396.

[GeJ78]   W.-D. Geyer and M. Jarden, *Torsion points of elliptic curves over large algebraic extensions of finitely generated fields*, Israel Journal of Mathematics **31** (1978), 157–197.

[GeJ89]   W.-D. Geyer and M. Jarden, *On stable fields in positive characteristic*, Geometriae Dedicata **29** (1989), 335–375.

[GeJ98]   W.-D. Geyer and M. Jarden, *Bounded realization of l-groups over global fields*, Nagoya Mathematical Journal **150** (1998), 13–62.

[GeJ01]   W.-D. Geyer and M. Jarden, *Non-PAC fields whose Henselian closures are separably closed*, Mathematical Research Letters **8** (2001), 509–519.

[GeJ02]   W.-D. Geyer and M. Jarden, *PSC Galois extensions of Hilbertian fields*, Mathematische Nachrichten **236** (2002), 119–160.

[GeJ06]   W.-D. Geyer and M. Jarden, *The rank of Abelian varieties over large algebraic fields*, Archiv der Mathematik **86**, (2006), 211–216.

[GPR95]   B. Green, F. Pop, and P. Roquette, *On Rumely's local-global principle*, Jahresbericht der Deutschen Mathematiker-Vereinigung **97** (1995), 43–74.

[GrR]   B. Green and P. Roquette, *Good Reduction*, a manuscript in preperation.

[Gre01]   R. Greenberg, *Introduction to Iwasawa theory for elliptic curves*, in Arithmetic Algebraic Geometry (B. Conrad and K. Rubin editors), IAS / Park City Mathematics Series **9** (1999), 407– 467,

[Gro61]   A. Grothendieck, *Éléments de Géométrie Algébrique III*, Publications Mathématiques, IHES **11** (1961),

[Gro04]   A. Grothendieck, *Revêtements étale et groupe fondamental (SGA 1)*, Séminaire de Géométrie algébrique du Bois Marie 1960-61, Augmenté de deus exposé de Mme M. Raynaud, Édition recomposée et annotée du volume 224 des Lecutre Notes in Mathematics publié en 1971 par Springer-Verlag, http://arxiv.org/PS_cache/math/pdf/0206/0206203v2.pdf

[Har99a] D. Haran, *Hilbertian fields under separable algebraic extensions*, Inventiones Mathematicae **137** (1999), 113–126.

[Har99b] D. Haran, *Free subgroups of free profinite groups*, Journal of Group Theory **2** (1999), 307-317.

[Har05]   D. Haran, *Regular split embedding problems over complete valued fields without restrictions on branch points*, manuscript, Tel Aviv, 2005.

[HaJ98a]   D. Haran and M. Jarden, *Regular split embedding problems over complete valued fields,* Forum Mathematicum **10** (1998), 329–351.

[HaJ98b]   D. Haran and M. Jarden, *Regular split embedding problems over function fields of one variable over ample fields,* Journal of Algebra **208** (1998), 147-164.

[HaJ00a]   D. Haran and M. Jarden, *The absolute Galois group of C(x),* Pacific Journal of Mathematics **196** (2000), 445–459.

[HaJ00b]   D. Haran and M. Jarden, *Regular lifting of covers over ample fields,* manuscript, Tel Aviv, 2000.

[HJP09]   D. Haran, M. Jarden, and F. Pop, *The absolute Galois group of subfields of the fields of totally S-adic number,* manuscript, Tel Aviv, 2008.

[HaV96]   D. Haran and H. Völklein, *Galois groups over complete valued fields,* Israel Journal of Mathematics, **93** (1996), 9–27.

[Hrb87]   D. Harbater, *Galois coverings of the arithmetic line,* in Number Theory — New York 1984–85, ed. by D. V. and G. V. Chudnovsky, Lecture Notes in Mathematics **1240**, Springer, Berlin, 1987, pp. 165–195.

[Hrb94a]   D. Harbater, *Abhyankar's conjecture on Galois groups over curves,* Inventiones Mathematicae **117** (1994), 1–25.

[Hrb94b]   D. Harbater, *Galois groups with prescribed ramifcation,* Contemporary Mathematics **174** (1994), 35–60.

[Hrb95]   D. Harbater, *Fundamental groups and embedding problems in characteristic p,* Contemporary Mathematics **186** (1995), 353-369.

[Hrb03]   D. Harbater, *Patching and Galois theory,* In "Galois Groups and Fundamental Groups" (L. Schneps, ed.), MSRI Publications series **41**, Cambridge University Press, 2003, pp. 313–424.

[Hrb09]   D. Harbater, *On function fields with free absolute Galois groups,* Journal für die reine und andgewandte Mathematik **632** (2009), 85–103.

[HaH10]   D. Harbater and J. Hartmann, *Patching over Fields,* Israel Journal of Mathematics **176** (2010), 61–108.

[HHK09]   D. Harbater, J. Hartmann, and D. Krashen, *Applications of patching to quadratic forms and central simple algebras,* Inventiones Mathematicae **178** (2009), 231-263.

[HaS05]   D. Harbater and K. Stevenson, *Local Galois theory in dimension two,* Advances in Mathematics **198** (2005), 623–653.

[Hrt77]   R. Hartshorne, *Algebraic Geometry,* Graduate Texts in Mathematics **52**, Springer, New York, 1977.

[Has80]   H. Hasse, *Number Theory,* Grundlehren der mathematischen Wissenschaften **229**, Springer-Verlag, Berlin, 1980.

[Hec23]   E. Hecke, *Vorlesungen über die Theorie der algebraischen Zahlen,* Leipzig 1923.

[Hil1892] D. Hilbert, *Über die Irreduzibilität ganzer rationaler Funktionen mit ganzzahligen Koeffizienten,* Journal für die reine und angewandte Mathematik **110** (1892), 104–129.

[Hin88]   M. Hindry, *Autour d'une conjecture de Serge Lang*, Inventiones Mathematicae **94** (1988), 575–603.

[ImB06]   B.-H. Im, Mordell-Weil groups and the rank of elliptic curves over large fields, Canadian Journal of Mathematics (2006)

[ImL08]   B.-H. Im and M. Larsen, *Abelian varieties over cyclic fields*, American Journal of Mathematics **130** (2008), 1195-1210.

[Iwa53]   K. Iwasawa, *On On solvable extensions of algebraic number fields*, Annals of Mathematics **58** (1953), 548–572.

[Jac64]   N. Jacobson, Lectures in Abstract Algebra III — Theory of Fields and Galois Theory, D. Van Nostrand Company, Princeton, 1964.

[Jac96]   N. Jacobson, *Finite Dimensional Division Algebras over Fields*, Springer-Verlag, Berlin, 1996.

[Jar72]   M. Jarden, *Elementary statements over large algebraic fields*, Transactions of AMS **164** (1972), 67–91.

[Jar75]   M. Jarden, *Roots of unity over large algebraic fields*, Mathematische Annalen **213** (1975), 109–127.

[Jar79]   M. Jarden, *Torsion in linear groups over large algebraic fields*, Archiv der Mathematik **32** (1979), 445–451.

[Jar82]   M. Jarden, *The elementary theory of large e-fold ordered fields*, Acta mathematica **149** (1982), 239–260.

[Jar91]   M. Jarden, *Algebraic realization of p-adically projective groups*, Compositio Mathematica **79** (1991), 21–62.

[Jar94]   M. Jarden, *Intersection of local algebraic extensions of a Hilbertian field* (A. Barlotti et al., eds), NATO ASI Series C **333**, 343–405, Kluwer, Dordrecht, 1991.

[Jar97]   M. Jarden, *Large normal extensions of Hilbertian fields*, Mathematische Zeitschrift **224** (1997), 555-565.

[Jar99]   M. Jarden, *The projectivity of the fundamental group of an affine line*, The Turkish Journal of Mathematics **23** (1999), 531–547.

[Jar03]   M. Jarden, *Ample fields*, Archiv der Mathematik **80** (2003), 475–477.

[JaP09]   M. Jarden and F. Pop, *Function fields of one Variable over PAC Fields*, Documenta Mathematica **14** (2009) 517–523.

[JaR94]   M. Jarden and A. Razon, *Pseudo algebraically closed fields over rings*, Israel Journal of Mathematics **86** (1994), 25–59.

[JaR98]   M. Jarden and A. Razon, *Rumely's local global principle for algebraic PSC fields over rings*, Transactions of AMS **350** (1998), 55-85.

[JaRo80]   M. Jarden and P. Roquette, *The Nullstellensatz over p-adically closed fields*, Journal of the Mathematical Society of Japan **32** (1980), 425–460.

[JuK09]   M. Junker and J. Koenigsmann, *Schlanke Körper (Slim Fields)*, Journal of Symbolic Logic **75** (2010), 481–500.

[Kle76]   S. L. Kleiman, *r-Special subschemes and an argument of Severi's*, Advances in Mathematics **22** (1976), 1–31.

[KlL72]   S. L. Kleiman and D. Laksov, *On the existence of special divisors*, American Journal of Mathematics **94** (1972), 431–436.

[Koc70]   H. Koch, *Galoissche Theorie der p-Erweiterungen*, Mathematische Monographien **10**, Berlin 1970.

[Kol07]   J. Kollár, *A conjecture of Ax and degenerations of Fano varieties*, Israel Journal of Mathematics **162** (2007), 235–251.

[Koe02]   J. Koenigsmann, *Defining transcendentals in function fields*, The Journal of Symbolic Logic **67** (2002), 947–956.

[KuR96]   F. V. Kuhlmann and P. Roquette, Convergent power series, http://www.rzuser.uni-heidelberg.de/ ci3/KONVPOTREIHEN.pdf

[KPR75]   H. Kurke, G. Pfister, and M. Roczen, *Henselsche Ringe*, VEB Deutscher Verlag der Wissenschaften, Berlin, 1975.

[Laf63]   J.-P. Lafon, *Anneaux Henséliens*, Bulletin de la Societé Mathematique Francaise **91** (1963), 77–107.

[Lan58]   S. Lang, *Introduction to Algebraic Geometry*, Interscience Publishers, New York, 1958.

[Lan59]   S. Lang, *Abelian Varieties*, Interscience Tracts in Pure and Applied Mathematics **7**, Interscience Publishers, New York, 1959.

[Lan70]   S. Lang, *Algebraic Number Theory*, Addison-Wesley, Reading, 1970.

[Lan93]   S. Lang, *Algebra, Third Edition*, Eddison-Wesley, Reading, 1993.

[LaT58]   S. Lang and J. Tate, *Principal homogeneous spaces over abelian varieties*, American Journal of Mathematics **80** (1958), 659–684.

[Lar03]   M. Larsen, *Rank of elleiptic curves over almost separably closed fields*, Bulletin of the London Mathematical Society **35** (2003), 817–820.

[Liu95]   Q. Liu, *Tout groupe fini est un groupe de Galois sur $\mathbb{Q}_p(T)$*, Proceedings of the 1993 Seattle AMS Summer Conference, "Recent Development in the Inverse Galois Problem." Contemporary Mathematics **186** (1995), 261–265.

[Lor08]   F. Lorenz, *Algebra II, Fields with Structure, Algebras and Advanced Topices*, Springer 2008.

[Lut37]   E. Lutz, *Sur l'équation $y^2 = x^3 - Ax - B$ dan les corps p-adiques*, Journal für die reine und angewandte Mathematik **177** (1937), 237–247.

[MaM99]   G. Malle and B. H. Matzat, *Inverse Galois Theory*, Springer Monographs in Mathematics, Springer-Verlag, Berlin 1999.

[Mats94]   H. Matsumura, *Commutative ring theory*, Cambridge studies in advanced mathematics **8**, Cambridge University Press, Cambridge, 1994.

[Mat55]   A. Mattuck, *Abelian varieties over a p-adic ground field*, Annals of Mathematics **62** (1955), 92–119.

[Matz87]   B. H. Matzat, *Konstruktive Galoistheorie*, Lecture Notes in Mathematics **1284**, Springer, Berlin, 1987.

[Maz00]   B. Mazur, *Abelian varieties and the Mordell-Lang conjecture*, Model Theory, Algebra, and Geometry, MSRI Publications **39** (2000), 199–227.

[MaR03]   B. Mazur and K. Rubin, *Studying the growth of Mordell-Weil*, Documenta Mathematica, Extra Volume Kato (2003), 585–607.

[Mel90]   O. V. Melnikov, *Subgroups and homology of free products of profinite groups* Mathematical USSR Izvestija **34** (1990), 97–119.

[MoB01]   L. Moret-Bailly, *Constructions de revêtements de courbe pointées*, Journal of Algebra **240** (2001), 505–534.

[Mum74]   D. Mumford, *Abelian Varieties*, Oxford University Press, London, 1974.

[Mum88]   D. Mumford, *The Red Book of Varieties and Schemes*, Lecture Notes in Mathematics **1358**, Springer, Berlin, 1988.

[Nag62]   M. Nagata, *Local Rings*, Interscience Publishers, New York, 1962.

[Nag51]   T. Nagell, *Introduction to Number Theory*, Almqvist & Wiksell, Stockholm; John Wiley & Sons, New York, 1951.

[Neu99]   J. Neukrich, *Algebraic Number Theory*, Grundlehren der mathematischen Wissenschaften **322**, Springer, 1999.

[Noe27]   E. Noether, *Der Diskriminantensatz für die Ordnungen eines algebraischen Zahl- oder Funktionenkörpers,* Journal für die reine und angewandte Mathematik **157** (1927), 82–104.

[Ohm83]   J. Ohm, *The ruled residue theorem for simple transcendental extensions of valued fields*, Proceedings of the American Mathematical Society **89** (1983), 16–18.

[Par08]   E. Paran, *Algebraic patching over complete domains*, Israel Journal of Mathematics **166** (2008), 185–219.

[Par09]   E. Paran, *Split embedding problems over complete domains*, Annals of Mathematics (170) (2009), 899–914.

[Par10]   E. Paran, *Galois theory over complete local domains*, Mathematische Annalen

[PaS07]   R. Parimala and V. Suresh, *The u-invariant of the function fields of p-adic curves*, Annals of Mathematics, Preprint arXiv:0708.3128 (2007).

[Pet06]   S. Petersen, *On a question of Frey and Jarden about the rank of Abelian varieties*, Journal of Number Theory **120** (2006), 287–302.

[Poo07]   B. Poonen, *Gonality of modular cureves in characteristic p*, Mathematical Research Letters **14** (2007), 691–701.

[Pop88]   F. Pop, *Galoissche Kennzeichnung p-adisch abgeschlossener Körper*, Journal für die reine und angewandte Mathematik **392** (1988), 145–175.

[Pop93]   F. Pop, *The geometric case of a conjecture of Shafarevich,* — $G_{\tilde{\kappa}(t)}$ is profinite free —, preprint, Heidelberg, 1993.

[Pop94]   F. Pop, $\frac{1}{2}$ *Riemann Existence Theorem with Galois Action*, in: Algebra and Number Theory, pp. 1–26 (ed. G. Frey – J. Ritter), de Gruyter Proceedings in Mathemathics, Berlin 1994.

[Pop95]   F. Pop, *Étale Galois covers of affine smooth curves. The geometric case of a conjecure of Shafarevich. On Abhyankar's conjecture.* Inventiones mathematicae **120** (1995), 555–578.

[Pop96]   F. Pop, *Embedding problems over large fields*, Annals of Mathematics **144** (1996), 1–34.

[Pop10]   F. Pop, *Henselian implies large*, Annals of Mathematics **172** (2010), 101–113.

[Pos05] L. Positselski, *Koszul property and Bogomolov's conjecture*, International Mathematics Research Notices **31** (2005), 1901–1936.

[Ray70] M. Raynaud, *Anneaux Locaux Henséliens*, Lecture Notes in Mathematics **169**, Springer, Berlin, 1970.

[Ray83] M. Raynaud, *Around the Mordell conjecture for function fields and a conjecture of Serge Lange*, Algebraic geometry (Tokyo/Kyota, 1982) Lecture Notes in Mathematics **1016**, Springer, Berlin, 1983, pp. 1–19.

[Ray94] M. Raynaud, *Revêtements de la droite affine en caractéristique $p > 0$ et conjecture d'Abhyankar*, Inventiones Mathematicae **116** (1994), 425–462.

[Rbn64] P. Ribenboim, *Théorie des valuations*, Les Presses de l'Université de Montréal, Montréal, 1964.

[Rib70] L. Ribes, *Introduction to Profinite Groups and Galois Cohomology*, Queen's papers in Pure and Applied Mathematics **24**, Queen's University, Kingston, 1970.

[RSZ07] L. Ribes, K. Stevenson, and P. Zalesski, *On Quasifree profinite groups*, Proceedings of the American Mathematical Society **135** (2007), 2669–2676.

[Sam66] P. Samuel, *Lectures on Old and New Results on Algebraic Curves*, Tata Institute of Fundamental Research, Bombay, 1966.

[Sch34] F. K. Schmidt, *Über die Kennzeichnung algebraischer Funtionenkörper durch ihren Regularitätsbereich*, Journal für die reine und angewandte Mathematik **171** (1934), 162–169.

[Ser56] J.-P. Serre, *Géométrie algébrique et géométrie analytique*, Annales de l'Institut Fourier **6** (1956), 1–42.

[Ser79] J.-P. Serre, *Local Fields*, Graduate Text in Mathematics **67** Springer, New York, 1979.

[Ser88] J.-P. Serre, *Algebraic Groups and Class Fields*, Graduate Texts in Mathematics **118** (1988), Springer-Verlag.

[Ser90] J.-P. Serre, *Construction de revêtement étales de la droite affine en caractéristique p*, Comptes Rendus de l'Académie des Sciences **311** (1990), 341–346.

[Ser92] J.-P. Serre, *Topics in Galois Theory*, Jones and Barlett, Boston 1992.

[Sha77] I. R. Shafarevich, *Basic Algebraic Geometry*, Grundlehren der mathematischen Wissenschaften **213**, Springer Berlin, 1977.

[Sha89] I.R. Shafarevich, *On the construction of fields with a given Galois group of order l^a*, Collected Mathematical Papers 107–142, Springer, Berlin, 1989.

[Shi98] G. Shimura, *Abelian Varieties with Complex Multiplication and Modular Functions*, Princeto University Press, Princeton, 1998.

[Sil86] J. H. Silverman, *The Arithmetic of Elliptic Curves*, Graduate texts in Mathematics **106**, Springer, New York, 1986.

[Son94] J. Sonn, *Brauer groups, embedding problems, and nilpotent groups as Galois groups*, Israel Journal of Mathematics **85** (1994), 391–405.

[Sti93]   H. Stichtenoth, *Algebraic Function Fields and Codes*, Springer-Verlag, Berlin, 1993.

[Tse33]   C. Tsen, *Divisionsalgebren über Funktionenkörpern*, Nachrichten von der Gesellschaft der Wissenschaften zu Göttingen, Mathematisch-Physikalische Klasse (1933), 335–339.

[Voj91]   P. Vojta, *Siegel's theorem in the compact case*, Annals of Mathematics **133** (1991), 509–548.

[Voe96]   H. Völklein, *Groups as Galois groups – an Introduction*, Cambridge Studies in Advanced Mathematics **53**, Cambridge University Press 1996.

[ZaS75]   O. Zariski and P. Samuel, *Commutative Algebra II*, Springer, New York, 1975.

Glossary of Notation

See also Notation and Convention on page xxiii.

P_i' $(= \bigcap_{j \neq i} P_j)$ 1

$\Gamma \ltimes G$ (semi-direct product) 5

$\mathbb{R}^+$ (the additive group of the real numbers) 10

$A\{x\}$ (ring of convergent power series at 1) 14

pseudo.deg(f) (pseudo degree) 15

$K((x))_0$ (field of convergent power series) 21

$w_i = \frac{r}{x - c_i} \in K(x)$ 31

$K\{w_i \mid i \in I\}$ (ring of convergent power series in several dependent variables) 33

P_J (quotient field of $K\{w_j \mid j \in J\}$) 37

P_i (the field $P_{I \smallsetminus \{i\}}$) 37

$e_{\mathfrak{P}/\mathfrak{p}}$ (ramification index) 43

Branch$(F/E, x)$ (or Branch(F/E)) 44

Ram(F/E) (set of prime divisors of E/K that ramify in F) 44

$\varphi_{x,a} \colon K(x) \to \tilde{K} \cup \{\infty\}$ (K-place of $K(x)$ with $\varphi_{x,a}(x) = a$) 44

$v_{x,a}$ (normalized valuation corresponding to $\varphi_{x,a}$) 44

$\mathfrak{p}_{x,a}$ (prime divisor corresponding to $v_{x,a}$) 44

$\mathfrak{p}_{K,x,a}$ (prime divisor corresponding to $v_{x,a}$) 44

ζ_n (root of unity of order n) 45

card(K) (cardinality of K) 54

$V_{\mathrm{simp}}(L)$ (L-rational simple points of a variety V) 62

$\mathcal{L}(\mathrm{ring}, K)$ (first order language of rings with constants for the elements of K) 65

$K_s(\boldsymbol{\sigma})$ (fixed field in K_s of $\sigma_1, \ldots, \sigma_e$) 74

$K_s[\boldsymbol{\sigma}]$ (the maximal Galois extension of K in $K_s(\boldsymbol{\sigma})$) 74

K_{symm} (compositum of all symmetric extensions of K) 74

$K_{\mathrm{tot},S}[\boldsymbol{\sigma}] = K_{\mathrm{tot},S} \cap K_s[\boldsymbol{\sigma}]$ 75

P$\mathcal{K}$C, (pseudo $\mathcal{K}$ closed) 75

PSC (pseudo S closed) 75

$K_{\mathrm{tot},S}(\boldsymbol{\sigma}) = K_{\mathrm{tot},S} \cap K_s(\boldsymbol{\sigma})$ 76

deg$(\mathfrak{p})$ (degree of the prime divisor $\mathfrak{p}$) 81

deg$(\mathfrak{a})$ (degree of the divisor $\mathfrak{a}$) 81

div(f) (divisor of the function f) 81

div$_0(f)$ (divisor of zeros of f) 82

div$_\infty(f)$ (divisor of poles of f) 82

$v_{\mathfrak{p}}(\mathfrak{a})$ ($\mathfrak{p}$-th value of $\mathfrak{a}$) 82

$\mathfrak{a} \leq \mathfrak{b}$ (inequality between divisors) 82

$\mathcal{L}(\mathfrak{a})$ (vector space attached to $\mathfrak{a}$) 82

dim$(\mathfrak{a})$ (dimension of $\mathcal{L}(\mathfrak{a})$) 82

Div(F/K) (group of divisors of F/K) 82

M. Jarden, *Algebraic Patching*, Springer Monographs in Mathematics,
DOI 10.1007/978-3-642-15128-6, © Springer-Verlag Berlin Heidelberg 2011

$\mathbb{P}(F/K)$ (set of prime divisors of F/K) 82

$\mathrm{Div}_0(F/K)$ (group of divisors of degree 0 of F/K) 82

$\mathbb{Q}_{\mathrm{ab}}$ (maximal Abelian extension of $\mathbb{Q}$) 90

$\mathbb{Q}_{p,\mathrm{ur}}$ (maximal unramified extension of $\mathbb{Q}_p$) 90

$\mathbb{Q}_{\mathrm{solv}}$ (maximal prosolvable extension of $\mathbb{Q}$) 90

K_{cycl} (extension of K by all roots of unity) 91

$A \subset B$ (A is a proper subset of B) 93

$K \models \varphi(x)$ (the formula $\varphi(x)$ holds in K) 100

$\Gamma^{\mathrm{div}}(L)$ (divisible hull) 106

$\mathrm{rr}(\Gamma)$ (rational rank) 109

$\mathrm{Div}(C)$ (group of divisors of C) 123

$C^{(d)}$ (symmetric product) 123

$K^{(d)}$ (set of elements of degree at most d) 123

$[x, y] = x^{-1}y^{-1}xy$ (commutator in a group) 165

E_S (maximal extension of E ramified at most over S) 165, 169

$\mathcal{F}(E, S, G)$ (set of Galois extensions of E with Galois group G, ramification over S) 166

$\mathrm{Zero}(h)$ (set of zeros of a polynomial h) 166

$E_{S,\mathrm{tr}}$ (maximal extension of E tamely ramified at most over S) 172

$H^q(G, A)$ (q-th cohomology group) 176

$\varinjlim A_i$ (direct limit) 178

A_n (n-torsion of an Abelian group) 179

A_{p^∞} (p-primary part of A) 179

$\mathrm{Ind}_H^G(A)$ (induced module) 180

$\mathrm{Br}(K)$ (Brauer group) 182

$\mathrm{Br}(L/K)$ (relative Brauer group) 182

C_i (field) 183

C_0 (field) 183

C_1 (field) 183

C_2 (field) 183

μ_p (group of roots of unity or order dividing p) 185

$\hat{F}_m$ (free profinite group of rank m) 186

$\wp(x)$ (Artin-Schreier operator) 190

E_{ur} (maximal unramified extension of E) 193

$E_{\mathrm{ab}}^{(p')}$ (maximal Abelian extension of E of degree not divisible by p) 249

$\mathrm{Ram}_\mathcal{V}(K'/K)$ (the set of all valuations $v \in \mathcal{V}$ that ramify in K') 258

Index

M. Jarden, *Algebraic Patching*, Springer Monographs in Mathematics,
DOI 10.1007/978-3-642-15128-6, © Springer-Verlag Berlin Heidelberg 2011